Lecture Notes on Multidisciplinary Industrial Engineering

Series Editor

J. Paulo Davim, *Department of Mechanical Engineering, University of Aveiro, Aveiro, Portugal*

"Lecture Notes on Multidisciplinary Industrial Engineering" publishes special volumes of conferences, workshops and symposia in interdisciplinary topics of interest. Disciplines such as materials science, nanosciences, sustainability science, management sciences, computational sciences, mechanical engineering, industrial engineering, manufacturing, mechatronics, electrical engineering, environmental and civil engineering, chemical engineering, systems engineering and biomedical engineering are covered. Selected and peer-reviewed papers from events in these fields can be considered for publication in this series.

Genti Guxho · Tatjana Kosova Spahiu ·
Valma Prifti · Ardit Gjeta · Eralda Xhafka ·
Anis Sulejmani
Editors

Proceedings of the Joint International Conference: 10th Textile Conference and 4th Conference on Engineering and Entrepreneurship

Springer

Editors
Genti Guxho
Polytechnic University of Tirana
Tirana, Albania

Tatjana Kosova Spahiu
Polytechnic University of Tirana
Tirana, Albania

Valma Prifti
Polytechnic University of Tirana
Tirana, Albania

Ardit Gjeta
Polytechnic University of Tirana
Tirana, Albania

Eralda Xhafka
Polytechnic University of Tirana
Tirana, Albania

Anis Sulejmani
Polytechnic University of Tirana
Tirana, Albania

ISSN 2522-5022 ISSN 2522-5030 (electronic)
Lecture Notes on Multidisciplinary Industrial Engineering
ISBN 978-3-031-48932-7 ISBN 978-3-031-48933-4 (eBook)
https://doi.org/10.1007/978-3-031-48933-4

This Springer imprint is published by the registered company Springer Nature Switzerland AG
The registered company address is: Gewerbestrasse 11, 6330 Cham, Switzerland

Paper in this product is recyclable.

Preface

The Proceedings of the "Joint International Conference: 10th Textile Conference and 4th Conference on Engineering and Entrepreneurship" book contains keynotes and papers presented at the Joint International Conference: 10th Textile Conference and 4th Conference on Engineering and Entrepreneurship (10th ITC and 4th ICEE 2023), organized by the Faculty of Mechanical Engineering at the Polytechnic University of Tirana, Albania (PUT), from 19 to October 20, 2023.

This joint event aimed to focus on important topics and issues of European and global science in the fields of textile materials, technologies, engineering, management and production, entrepreneurship, fashion, etc. Therefore, the conference papers presented give an essential reading for all those working on topics as: business or e-commerce, constructions and transport, energetics, management and production, material science, mechatronics, and textiles (garments, footwear, and leather).

It delves into various areas of energetics, exploring aspects such as power production, solar power, wind turbines, advanced energetics technologies, energy resource efficiency, emissions, climate change, clean and renewable energies, as well as economic development, and environmental protections. The constructions and transport section features discussions on numerical methods for data manipulation, construction science and technology, transport systems, modeling of transport systems, intelligent transport, and traffic management and safety. The materials segment addresses materials science and application, biopolymers and biotechnology, metallic and composite materials, metallurgical engineering, recycling, manufacturing, and processing of various materials such as paper, plastics, rubber, glass, ceramics, and more. Management and production topics include technology management, logistic and supply chain management, total quality management, knowledge and innovation management, financial management, marketing research and strategy, industrial marketing, operational research, project management, as well as information technology in enterprises, e-activities, and e-commerce. The book also features an extensive section dedicated to textiles, covering textile processing and testing, technological advances in the textile industry, ecology and environment in textile production, fiber physics and textile mechanics, finishing, dyeing, and treatment techniques, modeling and simulation, smart and interactive textiles, technical and protective textiles, textile design, fashion, and garment manufacturing, innovations in textile education, as well as leather and footwear technologies.

This joint international conference was also a great contribution to the internationalization of the Faculty of Mechanical Engineering, Polytechnic University of Tirana.

We are deeply grateful to everyone as authors, members of scientific committee from different European countries who with their contributions made it possible to have a great Joint International Conference and a pleasant time filled with discussions for future collaborations.

Prof. Dr. Genti Guxho
Chairman of this Joint Conference
Dean of the Faculty of Mechanical Engineering
Polytechnic University of Tirana
Tirana, Albania

Conference Committees

Conference Chair

Genti Guxho

Organizing Committee

Odhisea Koça
Altin Dorri
Ermira Shehi
Majlinda Alcani
Ilo Bodi
Tatjana Spahiu
Valma Prifti
Eralda Xhafka
Ardit Gjeta
Anis Sulejmani

Scientific Committee

Ada Trauman	Institute of Engineering and Circular Economy, Estonia
Ágota Drégelyi-Kiss	Óbuda University, Hungary
Alma Afezolli	Polytechnic University of Tirana, Albania
Amine Hadj Taieb	Superior Institute of Arts and Design, Sfax, Tunisia
Andonaq Londo	Polytechnic University of Tirana, Albania
Andrea Ehrmann	Mathematics, Bielefeld University of Applied Sciences and Arts, Germany
Anita Tarbuk	University of Zagreb, Croatia
Antonio Jimeno-Morenilla	University of Alicante, Spain
Artan Sinoimeri	Université de Haute-Alsace, France
Arzu Marmarali	Ege University, Turkey
Adnan Mazari	Technical University of Liberec, Czech Republic
Astrit Bardhi	Polytechnic University of Tirana, Albania
Aura Mihai	"Gheorghe Asachi" Technical University of Iasi, Rumania

Acknowledgments and Sponsors

The editors and the conference chair wish to acknowledge for the support and sponsorship given in the organization of the 10th ITC and 4th ICEE 2023—Joint International Conference—10th Textile Conference and 4th Conference on Engineering and Entrepreneurship.

AUTEX Association of Universities for Textiles

NETFAS.eu

BALKAN SOCIETY of TEXTILE ENGINEERS

PEA

DBS group INDUSTRIES

Invited Lectures

Prof. Dr. Luigi Maria Galantucci Dr. h.c, Politecnico di Bari, Italy, "Doctor Honoris Causa" from Polytechnic University of Tirana

Keynotes

Resource efficiency and clean production assessment for the Albanian textile manufacturing sector
Ermira Shehi, Elmira Dumishllari and Tatiana Chernyavskaya

Systems engineering for aerospace: new challenges in modern on-board systems
Mateo D. L. Dalla Vedova, Politecnico di Torino, Italy

Contents

Energetics

Business or E-Commerce

Albania's Unsuccessful Attempts to Rapidly Develop the Industrial Sector (1920–1990)

Genti Guxho(✉)

Textile Engineering and Fashion Department, Polytechnic University of Tirana, UPT Tirana, Tirana, Albania
gguxho@fim.edu.al

Abstract. During the period 1920–1990, Albania experienced two political regimes that differed greatly in ideological and socio-economic attitudes, as well as in the form of government and geopolitical situation. In the first period (1920–1939) Albania was under an autocratic regime that nevertheless allowed relative freedom of speech and a free market, the other was a communist dictatorship (1944–1991) that implemented a centralized and planned economy, and brutally suppressed basic human rights. Surprisingly, both regimes implemented the same plan to rapidly modernize and industrialize the most backward country in Europe. It may seem utopian, driven by miscalculations or ideology, for a tiny and backward country like Albania to embark on a rapid industrialization program. However, the existence of abundant mineral resources (chrome, copper, iron, oil, and natural gas) and a potentially large amount of hydroelectric power generation capacity were factors that could justify the choices made by both political regimes. In the first case, Albania's vulnerability to foreign financing made its economy increasingly dependent on Fascist Italy and eventually led to the country's occupation. While in the case of the communist regime, the dogmatic policy led the country's economy to autarky and collapse. Today, Albania has another chance to gradually develop a normal and balanced economy with the perspective of EU integration.

Keywords: Albania · Economy · Political regime · Natural resources · Rapid industrialization

1 Introduction

Formally, the history of Albania as an independent state begins in December 1912 when the Great Powers agreed to recognize an autonomous Albania. Subsequently at the Conference of Ambassadors in London were defined its borders with the agreement of July 30, 1913. However, due to the beginning of the Great War (1914–1918), and the chaos that followed, Albania started functioning as a state only from 17 December 1920 when its independence reconfirmed by League of Nations. After "*a troubled and short experiment with democracy*" [1], as Edwin Jacque calls the period between 1920 and 1924, in 1925 in Albania was established a dictatorial regime like the dictatorships in other newly created states in Europe.

G. Guxho et al. (Eds.): ITC-ICEE 2023, LNMIE, pp. 3–11, 2024.
https://doi.org/10.1007/978-3-031-48933-4_1

The newly created state came from a 500-year domination of the Ottoman Empire, which had left Albania in the same socio-economic conditions and with the same road infrastructure as it had found in the 15th century. Albania was the most backward country in Europe economically and culturally, and the Albanian nation lacked a true national consciousness. According to the economic expert Professor Albert Calmès, who compiled a report for the League of Nations about economic situation in Albania in September 1922: "*Politically, Albania is a State, but economically it is not...*" [2] In fact, at the beginning of the 1920s Albania was not only the most backward country in Europe economically, but it had no economy at all, by the standards of developed western countries.

Albania in 1925 have neither a paper currency in circulation, nor a bank of issue, meanwhile the ways of communication and trade between urban centers was almost non-existent. For example, it happened that the southern regions were exporting olive oil to Italy while, say, Shkodra in the north of Albania was importing it. Also, Albania suffered from a passive foreign trade balance.

According to the report of Prof. *Albert Calmès* the communication network, the development of agriculture, the establishment of a national bank, and the development of a mining industry required an urgent intervention through a loan from the League of Nations or from the European Banks. This report estimated that the value of the financial support required for this was around 120 million gold francs [2] (Fig. 1).

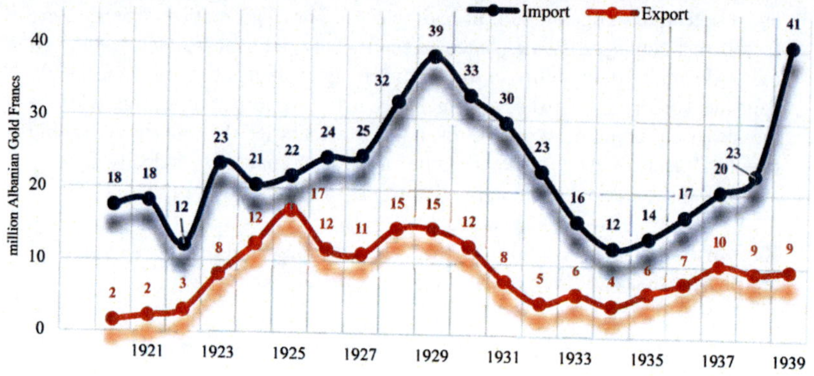

Fig. 1. Exports and imports of Albania in the period 1920–1939 in Albanian Gold Francs [3, 4]

2 The Effort to Modernize Albania (1925–1939)

The authoritarian regime of Ahmet bey Zogu, first as the president of the republic (1925–1928) and then as the king of the Albanians [*sic*] followed an economic policy that generally was in accordance with *Calmès*' blueprint (Fig. 2).

Based on the agreement between the Albanian government and an Italian financial group in 1925, Banka e Shqipnis (Bank of Albania) was founded as a National Bank of

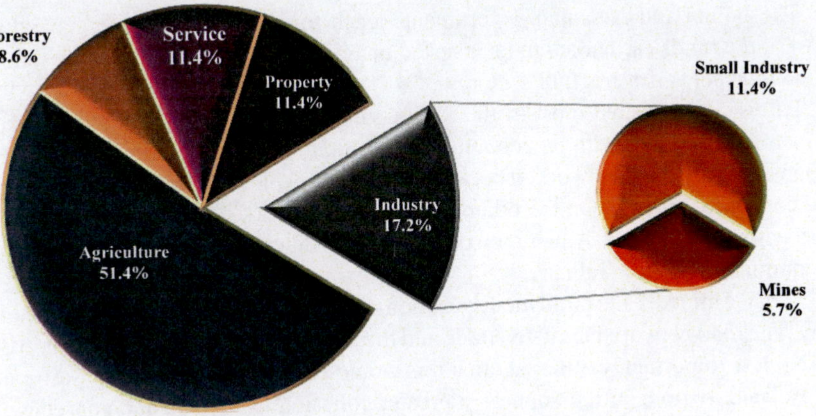

Fig. 2. Share of each sector in national income (1938) [5, 6]

Issue. At the same time, a financial institution was created for the necessary investments in agriculture, road and railway network and other public works—Società di Sviluppo Economico della Albania SVEA (Society for the Economic Development of Albania). For this purpose, SVEA offered the Albanian government a loan of 62.7 million gold francs with 7.5% interest for 40 years.

Table 1. Share by sector (in %) in investments from the loan of SVEA [6, 7]

Destination	Plan (% of the total)	Fact (% of the total)
Transport infrastructure	*42.4*	*67.8*
Roads	18.7	31.0
Bridges	5.7	23.0
Ports	15.2	13.8
Railroads	2.9	0.0
Agriculture	*49.8*	*5.4*
Development	15.90	3.0
Reclamation and Embankment, draining canals	32.30	2.4
Irrigation works	1.60	0.0
Public administrative buildings	*7.8*	*17.4*
Other	*0.0*	*9.4*
Total	100.0	100.0

The share of each economic sector in the planned and realized investments from the loan granted by SVEA is given in Table 1. There is a big difference between the original planning in 1926 and the way the investment was actually distributed.

The Italian and Albanian governments, each for their own reasons, have diverted most of the funds earmarked to be invested in agriculture (almost 50% of the total loan) to the transport infrastructure sector.

Fascist Italy wanted most of the SVEA funds to be invested in roads and seaports infrastructure in line with its economic and military interests in Albania. Through the expansion of the road network, it became possible to exploit the resources of oil, chrome, and copper in Albania. On the other hand, a road network directed to the southern and northeastern borders of Albania made the later a suitable base for the fascist plans for the domination of the Balkans.

Albania aimed at the rapid modernization of the society and development of the economy. The road network facilitated trade and the penetration of foreign capital in Albania. [8] No less important was the fact that this would ensure the connection of different, so far isolated, regions of the country, a greater cohesion of society, and consequently to the increase of national consciousness in the new Albanian state.

The neglect of agriculture and the desire to invest in infrastructure for the development of the mining industry, make it clear that modernization and industrialization were becoming the main objective of the government during the period of the Albanian Kingdom (1928–1939) (Fig. 3).

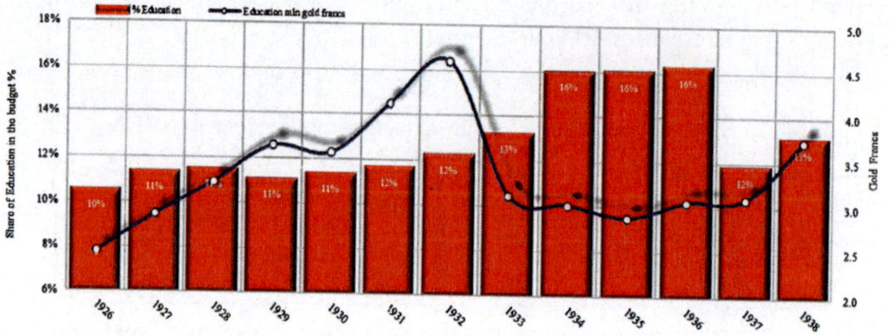

Fig. 3. Albania's education budget as % to the total budget and in Gold Francs (1926–1938) [9, 10]

A developed economy needs qualified people. In the 1920s, Albania was classified as the most backward country in Europe in the education sector (it counted about 90% of its population illiterate). The efforts of the Albanian government were aiming at establishing a complete network of primary schools and vocational secondary education. The share of the budget devoted to education increased year after year to reach 16% in the mid-1930s [10]. Within a few years the number of primary schools tripled, and a national, secular education system was established.

In the establishment of secondary schools, greater attention was given to vocational industrial schools. In the end of interwar period the number of them was 5 against 2 agricultural secondary schools. Government university scholarships (about 60 per year) enabled the graduation of young people in engineering and agriculture [11], in the universities of Italy, France and Austria.

As a result of government investments, in 1939, Albania had 2224 km of national roads (of which 350 were completely new) and 767 secondary local roads. For Albania this was extraordinary if we consider that in 1923 there were only 695 km of roads and in 1913–140 km. In addition to them, about 4000 small bridges with a length of 1–10 m and 100 bridges with a length of over 30 m were built [12]. Thanks to domestic and foreign capital investment, Albania had 244 industrial enterprises in 1939 compared to 165 in 1927, while the contribution of industry to national income doubled at the same period of time [13].

The desire to rapidly modernize the country through Italian loans, the penetration mostly of Italian capital in the economy, as well as the signing of military and political agreements with Fascist Italy, turned Albania into a semi-protectorate of Italy. Eventually Albania in 1939 became one of the first victims of fascist aggression.

3 Albania as a Communist Country (1944–1991)

At the end of the Second World War, in Albania the National Liberation Movement, led by the communists, succeeded in liberating the country and emerged victorious in the struggle for power. Immediately after the war, the first measures that were taken in the economic recovery phase in accordance with Stalinist manual were:

- Nationalization of all companies with foreign capital without compensation.
- Nationalization of all industrial enterprises, and banks.
- Establishing a state monopoly on foreign trade.

The second stage was the organization of a centralized and planned economy. From 1951, the Albanian government operated in the field of economy and investments based on a 5-year planning.

The economy of the period 1945–1990 in Albania was characterized by centralized planning and forced industrialization according to the Soviet model. Albanian communists in their economic decisions were guided by "*ideological considerations*" [14], adding to them later the autarky-oriented policy, which followed the Maoist principle of self-reliance or so called "*relying on one's own forces*" [15].

It may seem absurd, driven by communist ideology, for a small country like Albania to start an industrialization program, even with the ambitious objective of developing heavy industry [16]. However, if some determining factors of such an economic orientation are considered, then it seems that the choice made was logical. Albania was rich in natural resources (chrome, copper, iron, oil and natural gas, bauxite, phosphate, nickel, coal, etc.) and had potentially large capacities to produce hydroelectric power.

Albania's economy experienced extraordinary progress in the period 1948–1961, especially in the industrial sector with financial and technical assistance from Eastern European countries and the Soviet Union (Table 2).

The annual average rate of growth of gross industrial production in Albania in the period 1948–1950 was 10.8% and in the First Five Year Plan—22.4% [18]. This rapid development came first from the low level of industrialization of Albania after the second world war and secondly from the financial and technical support of the Soviet Union and other socialist countries. Albania benefited from this in aid as a Soviet

Union' satellite country. Also, Albania was founding member of military organization of socialist countries of Central and Eastern Europe countries (Warsaw Treaty 1955) and member of Council of Mutual Economic Assistance (1949)—an economic coordination organization of Soviet Union and other socialist countries.

Table 2. Annual average growth rate of gross industrial production (1951–1959) [17]

No	Country (soviet bloc)	Annual average growth rates (%)	No	Country (free world)	Annual average growth Rates (%)
1	P.R. of China	28.4	11	West Germany,	9.4
2	North Korea	22.7	12	Greece	8.8
3	*Albania*	*20.6*	13	Italy	8.3
4	Bulgaria	14.4	14	France	6.3
5	Poland	13.1	15	Netherlands	5.3
6	Rumania	12.8	16	India	5.0
7	USSR	11.9	17	USA	3.9
8	East Germany	11.6	18	Belgium	2.8
9	Czechoslovakia	10.8	19	Great Britain	2.7
10	Hungary	10.2	20	Sweden	2.5

Albania in the 1950s invested in the industrial sector an average of 50% of the investment fund, while investments for agriculture constituted only 14–15% of this fund. Due to the neglect of agriculture, agricultural and livestock production grew at lower rates than population growth, and Albania exported ever larger quantities of food commodities. Also, Albania has continued to have a negative balance in foreign trade.

Thus, loans were used not only for industrial development but also for the purchase of foodstuffs. Thus, loans were used not only for industrial development but also for the purchase of foodstuffs. Over 95% of Albanian turnover in foreign trade was covered by communist countries where the dominant position was occupied by Soviet Union, which covered 45–50% of imports and around 50% of Albania's exports [20].

As can be seen from Fig. 4 during the 1950s, on average, only 30% of imports were covered by exports. However, the strategy of the Albanian government was to develop the industry in order to export processed goods and not raw materials. On the other hand, according to the Albanian government, the development of the industry would bring the mechanization of agriculture, i.e., the increase of the general agricultural product. The obsession of the Albanian communists for the development of heavy industry as a means to ensure the economic independence of Albania and consequently its political independence was also one of the causes of the conflict with the Soviet leadership.

The Albanian-Soviet conflict ended in the total severance of relations with the USSR at the end of 1961. But unlike the cases of Hungary in 1956 and Czechoslovakia in 1968, the soviets and Warshaw Treaty Organization did not intervene militarily. This is due to Albania's isolated geographical position and its alignment with Mao Zedong's China.

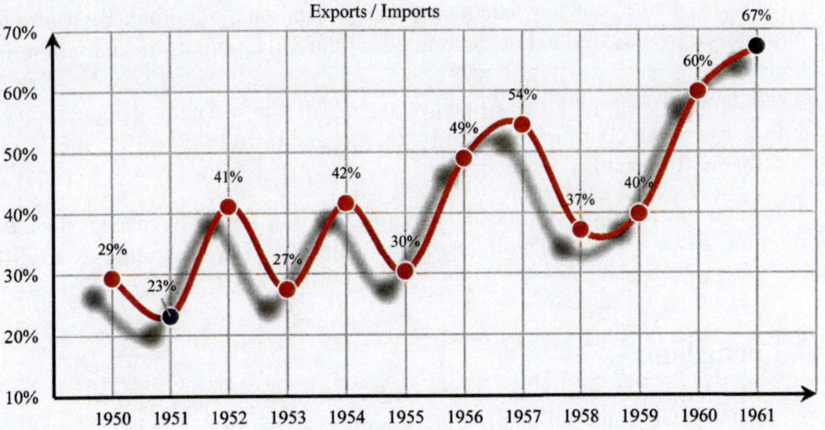

Fig. 4. Export to import ratio in % (1950–1961) [19]

The latter has been in conflict with the Soviet Union since 1959. From this moment until 1978 the period of financing the Albanian economy by the People's Republic of China began.

Here we have the case when the geopolitics of the cold war combined with ideological militancy, which was supported by propaganda with dogmatic and nationalist slogans, surprisingly worked in favor, at least in the short term, of the goals of the Albanian leaders, and they took advantage of this, as much as they could [21].

4 The Similarities of the Two Periods and the Causes of the Failure

In both political periods of Albania, before and after the Second World War, one can distinguish similarities in the economic policies of the respective ruling elites:

1. Political-economic decisions were primarily made under the pressure of fear of losing political independence.
2. The political leaders in both cases tried to forcefully modernize the society, bring Albania out of the backwardness, and rapidly develop the industry.
3. Whether they were motivated by noble intentions or were motivated simply by opportunistic objectives, these efforts led to almost the same end. In both cases, the disintegration of the Albanian society, which even then had weak ties that held it together, was at risk.
4. Similarly, the regimes used the geopolitical situation to survive and at the same time to provide the necessary financial assistance to Albania's non-self-sufficient economy. Alongside the focus on industrialization, the emancipation of society and establishing a national education, both regimes had in common large military expenditures.

Rapid industrialization failed in both cases, among others, for the following reasons:

- Albania could not afford an rapid development of industry at the same time when it had an 80–90% illiterate population as well as lack of skilled workers and managers. Also, a massive education of the population depended on the staff of available teachers.

- Albania, as a rural country, had its economy based on agriculture. By trying to give importance to industry and neglecting agriculture, the regimes risked failure in both sectors.
- A rapid growth of population. Albania had the fastest rate of natural population growth among European countries. In a period of 30years, from 1945 to 1975 the population of Albania was doubled.

Of course, we mentioned some of the factors that may have influenced in the failure of the attempts to make Albania an industrialized country, in two seemingly very different times.

5 Conclusions

Being economically and politically independent was the principle that always guided the political class in Albania in choosing alliances with foreign powers and that in the first case (1920–1938) failed and ended with the invasion of the country by foreigners and in the second case (1945–1991) led the country's economy to autarky and collapse.

Not only that, but both regimes fell victim to their ambition beyond the limits of the country's real possibilities, to emancipate the Albanian society, to overcome its centuries-old backwardness and to economically modernize and industrialize Albania rapidly.

This is evidence of the existence of a historical and long-term exigency among Albanians to build their national identity, which has been endangered throughout their history. Therefore, paradoxically Albania simultaneously tended towards isolation, which was driven by the language, mountainous relief, and historical experiences of the Albanians. The political objective that Albania has chosen today, to be part of the European community, can once and for all heal these contradictions in the identity psyche of Albanians.

References

1. Jacques, E.: The Albanians. An ethnic history from prehistoric times to the present, p. 384. McFarland & Company, Inc. Publishers, USA (1995)
2. Calmès, A.: The economic and financial situation in Albania. In: League of Nations, C. 706. M, 417, Geneva (1922)
3. Caselli, G.P., Thoma, G.: Lo storia economica albanese 1912–1939 e lo stabilirsi dell'egemonia italiana. In: Rivista di Storia Economica, vol. XIX, p. 1 (2001)
4. Baldacci, A.: Studi Speciali Albanesi **II**, 586 (1933)
5. Demaria, G.: Il problema della politica economica albanese e lo sviluppo del reddito nazionale. In: Giornale degli economisti e annali di economia, July–October 1940, pp. 399–402
6. Roselli, A.: Italy and Albania: financial relations in the fascist period. I. B. Taurus London-New York (2006)
7. Groß, H.: Wirtschaftsstruktur und Wirtschaftsbeziehungen Albaniens. In: Weltwirtschaftliches Archiv, 38 Bd (1933 II), p. 505 (1933)
8. Czekalski, T.: The shining bacon of socialism in Europe. The Albanian state and society in the period of communist dictatorship 1944–1992, p. 66. Jagiellonian University Press, Kraków Poland (2013)

9. Kaser, M.C., Radice, E.A.: The economic history of eastern Europe 1919–1975, vol II. Clarendon Press-Oxford (1984)
10. Fishta, I., Toci, V.: Gjendja ekonomike e Shqipërisë në vitet 1912–1944, prapambetja e saj, shkaqet dhe pasojat. Tiranë: Shtëpia Botuese "8 Nëntori", pp. 73, 110 and 156 (1983)
11. Misja, V., Teta, A., Kallulli, A.: Higher education in Albania, monographs on higher education UNESCO, CEPES Bucharest, p. 45 (1986)
12. Annuario del Regno di Albania Anno 1940-XVIII, p. 68. Casa Editrice Ravagnani Milano, Italy
13. Demaria, G.: Il problema della politica economica albanese e lo sviluppo del reddito nazionale. In: Giornale degli economisti e annalisti di economia, pp. 399–402 (1940)
14. East European economies post-Helsinki. In: A compendium of papers submitted to the Joint Economic Committee Congress of the United States, p. 581. U.S. Government Printing Office, Washington (1977)
15. Kopsidis, M., Ivanov, M.: Modern Industry in Southeast Europe 1945–2007. From rapid industrialization to deindustrialization, p. 37. Regensburg, Germany (2013)
16. O'Donnell, J.S.: A coming age, Albania under Enver Hoxha, p. 152. East European Monographs, Boulder, New York, USA
17. Oleinik, I.P., et al.: Economic development of the socialist countries (Statistical Data). In: Institute of Economics of the Soviet Academy of Sciences. Published by S. C. Easton, 16 King Street, London, p. 7 (1961)
18. Spiridonov, M.F.: Stroitel'stvo sotsializma v yevropeyskikh stranakh narodnoy demokratii. Izdatel'stvo VPSH pri TSK KPSS, Moskva, p. 40 (1959)
19. An Economic Survey of Albania 1950–1964, pp. 3–4. Joseph Crosfield & Sons Limited, Warrington (1967)
20. Freedman, R.O.: Economic Warfare in the Communist Bloc; a Study of Soviet Economic Pressure Against Yugoslavia, Albania, and Communist China, p. 70. Praeger Publishers, New York (1970)
21. Logoreci, A.: Albania and China: the incongruous alliance. In: Current History, vol. 52, no. 308, pp. 231, 245 (1967)

Implementation of Industry 4.0 Challenges in Developing Countries: A Case Study of Albania

Irma Shyle[1(✉)] and Erisjena Rruplli[2]

[1] Polytechnic University of Tirana, Bulevardi Dëshmorët E Kombit Nr. 4, 1000 Tiranë, Albania
ishyle@fim.edu.al

[2] Bentley University, 175 Forest St, Waltham, MA 02452, USA

Abstract. The Fourth Industrial Revolution, also known as "Industry 4.0", presents a critical opportunity for fostering social and technological innovation. Embracing advanced cybernetic systems enables companies to enhance efficiency and productivity significantly. Digital trust plays a crucial role in promoting open innovation in this era of rapid transformation. This research paper focuses on the challenges faced by developing countries, using Albania as a case study, in adopting Industry 4.0 on a large scale. We analyze the factors hindering comprehensive implementation and propose strategies to overcome these challenges. As the world becomes more globalized and digitized, organizations must swiftly plan their digital transformations to remain competitive, emphasizing increased flexibility. The study employs survey data from various companies to assess current Industry 4.0 implementation levels and identify obstacles to further development. Key terms related to the industrial revolution are introduced, followed by an analysis of digitalization's primary aspects, including cloud systems, IoT, and cybernetics. The study highlights the significance of digital transformation, focusing on providing closer and more secure services aligned with consumers' needs. Emphasizing social and organizational dimensions, the research underscores the concept of digitalization and organizational changes supported by previous studies. The paper discusses opportunities and challenges digitalization poses to societies and organizations globally. Ultimately, this research aims to illuminate barriers to Industry 4.0 adoption in developing nations and propose practical strategies to facilitate successful integration into this transformative era.

Keywords: Industry 4.0 · Technological innovation · Developing countries · Albania · Implementation challenges

1 Introduction

The story of energy consumption began with the advent of the Industrial Revolution in the 1770s, completely transforming multiple facets of human life, spanning from homes and factories to offices. As technology continued to advance, the rise of Industry 4.0 marked a revolutionary transformation, driven by the fusion of digital technologies,

G. Guxho et al. (Eds.): ITC-ICEE 2023, LNMIE, pp. 12–23, 2024.
https://doi.org/10.1007/978-3-031-48933-4_2

artificial intelligence (AI), the Internet of Things (IoT), and advanced automation. This era held the promise of global economic growth and significantly enhanced industrial efficiency. One of the pivotal technologies shaping Industry 4.0 is the Internet of Things (IoT), enabling efficient management of homes, businesses, and factories. IoT's ability to facilitate machine interactions and informed decision-making is instrumental in achieving optimal energy usage. Another promising innovation is 3D printing, which revolutionizes industries through cost-effective prototyping and streamlined manufacturing processes. Utilizing 3D printing for low-cost prototypes enables companies to save time and resources during product development.

The integration of 5G technology is set to revolutionize data transfer speeds, communication, and data storage capabilities. This faster internet connection will enhance operations across various sectors, including smart manufacturing and predictive maintenance. Additionally, advancements in cloud technology are tailoring solutions to specific organizational needs, further facilitating the widespread adoption of Industry 4.0 practices. Adaptive artificial intelligence (AI) empowers businesses to adapt swiftly, learn from past experiences, and make informed decisions, particularly valuable in navigating disruptions and uncertainties. Smart manufacturing is gaining prominence, emphasizing reduced operational costs and improved visibility. With smart manufacturing systems optimizing workforce utilization and enhancing automation, clear-cut procedures and minimal production delays become achievable.

Predictive maintenance, utilizing data for proactive maintenance planning, mitigates downtime costs and unexpected interruptions in manufacturing processes. This proactive approach allows for timely replacement of defective parts, resulting in cost and time savings. Incorporating Augmented Reality (AR) and Virtual Reality (VR) technologies in Industry 4.0 offers practical benefits, enabling practitioners to closely interact with products and work practices. Companies can utilize AR and VR for remote assistance, walkthroughs, troubleshooting, and repairs, enhancing customer experiences and streamlining production processes. Next-generation Manufacturing ERP systems automate repetitive tasks, provide real-time information, and enhance overall operational efficiency, contributing to better value and increased competitiveness in the industry 4.0 era.

In conclusion, realizing optimal energy usage within the framework of Industry 4.0 requires harnessing cutting-edge technologies and embracing innovative strategies [1]. However, the benefits of Industry 4.0 are not evenly distributed across nations due to socio-economic disparities and technological readiness. Developing countries, including Albania, face unique challenges in adopting Industry 4.0 practices. This research paper aims to shed light on the multifaceted aspects of Industry 4.0 in developing countries, with a special focus on Albania as a case study. Through an in-depth exploration of Albania's efforts to navigate Industry 4.0, we aim to identify specific challenges and propose strategies and recommendations to guide successful integration of Industry 4.0 technologies. Ultimately, this research endeavors to provide valuable insights into the transformative potential of Industry 4.0 in the digital era.

2 Literature Review

While developed countries are successfully transitioning from Industry 4.0 to Industry 5.0, many developing countries still face difficulties in implementing Industry 4.0 technologies [2]. A survey from United Nations Industrial Development Organization [3] conducted in Ghana, Thailand, and Vietnam reveals that less than 4% of firms in these countries have adopted transformative technologies associated with Industry 4.0. This limited diffusion of technology poses challenges to their manufacturing sectors, hindering their ability to harness the full potential of digitalization, artificial intelligence, and other advancements. Industry 4.0, also known as the Fourth Industrial Revolution (4IR), represents the ongoing digital transformation of the manufacturing sector [4]. It is characterized by the convergence of digital, biological, and physical innovations, including artificial intelligence, advanced analytics, robotics, and 3D printing [5]. This era of connectivity and advanced manufacturing technology offers significant potential for operational improvements and the future of production. However, developing countries face challenges in fully integrating Industry 4.0 into their manufacturing processes, with issues related to access, adoption, and skills development [6].

One of the critical factors hindering the implementation of Industry 4.0 in developing countries is the need for workforce engagement and upskilling [7]. To thrive in the Fourth Industrial Revolution, companies must ensure that their employees are equipped with the necessary skills for this era of digitization and automation [8]. This involves upskilling, where employees learn new skills to adapt to evolving job requirements [9], as well as reskilling [10], which re-trains workers for different positions within their organizations [11]. Developing countries have the opportunity to embark on their Industry 4.0 journey on a smaller scale and scale up quickly, but workforce engagement and training are essential for successful transformation. To address these challenges, companies should adopt an agile approach that incorporates quick iterations, fast learning, and continuous improvement to drive innovation. Additionally, establishing a transformation office can support the launch and scale-up of Industry 4.0 initiatives. UNCTD' study [12] discusses the concept of Industry 4.0, or the Fourth Industrial Revolution, and its potential implications for developing countries. While Industry 4.0 offers transformative opportunities for increased productivity, energy efficiency, and sustainability in the manufacturing sector, the author highlights concern about its potential to exacerbate inequalities between developed and developing nations. Access and adoption of these technologies remain uneven, emphasizing the need for empowering developing countries to harness the potential of Industry 4.0 for inclusive development. The study emphasizes the importance of strengthening manufacturing sectors, promoting technology transfer, and adopting a participatory approach to technological adoption. International collaboration is deemed crucial in bridging the gap between nations and ensuring that Industry 4.0 contributes to a more equitable and sustainable global development landscape.

Mourtzis et al. [12, 13] introduces the concept of Industry 5.0, the upcoming phase of manufacturing that seeks to integrate human workers and machines to create a more efficient and productive process. Industry 5.0 is characterized by collaboration, flexibility, personalization, and sustainability, harnessing artificial intelligence, machine learning, and the Internet of Things (IoT) to revolutionize manufacturing [14, 15].

Industry 4.0 and Industry 5.0 represent distinct technological waves with the potential to reshape the manufacturing landscape globally. While some countries have made strides in transitioning to Industry 5.0, many developing countries face hurdles in fully embracing Industry 4.0 technologies [2]. Overcoming these challenges through workforce engagement, upskilling, and international collaboration can empower developing nations to harness the benefits of these transformative revolutions for inclusive and sustainable growth [16]. Challenges for implementing Industry 4.0 independent of company size [17].

3 Methodology

World economies are ranked based on their abilities and innovation results through the global index [18]. This index (GII) is published by the World Intellectual Property Organization (WIPO), a specialized agency of the USA. The measurement of this index is based on several predetermined criteria such as human resources, institutions, research capital, credit, investments, infrastructure, connections, knowledge creation, dissemination, and utilization of results. The main purpose of the GII is to provide rankings of innovation and detailed analysis of 130 economies. Recently, the Global Competitiveness Index is being considered as a key reference tool regarding innovation and a crucial instrument for economies that prioritize it on their agenda [18].

Category	Score (out of 100)	Ranking (out of 141 countries)
Financial system	53	102
Innovation capability	30	110
Macroeconomic stability	70	104
Infrastructure	58	81
Transparency	36	85

- The Innovation Capability of Albania received a low score of 30 out of 100, ranking 110th out of 141 countries, indicating a *need for improvement in fostering innovation*, research, and intellectual property protection.
- The Transparency score is also relatively low at 36 out of 100, with a ranking of 85th out of 141 countries, suggesting a *need for enhancing transparency and accountability* in government and business practices.

Data According to INSTAT (2022) [19].

Year	Percentage (%)	Indicator
2022	99.0	Businesses using computers
2022	31.3	Workers using computers for work
2022	78.9	Workers in information and communication activities using computers

(continued)

(*continued*)

Year	Percentage (%)	Indicator
2022	26.3	Businesses employing ICT experts
2022	99.0	Businesses with fixed internet connection
2022	63.0	Businesses with download speeds > 30 Mbit/s
2022	22.1	Businesses with download speeds > 100 Mbit/s
2022	5.5	Businesses engaged in autonomous jobs
2022	42.0	Businesses with access to email and business applications
2022	38.6	Businesses conducting virtual meetings
2022	13.8	Businesses selling goods/services online
2022	66.2	Enterprises implementing data security measures

Based on the data provided by INSTAT (2022) [19], it can be inferred that Albania still faces some gaps in terms of Industry 4.0 readiness and adoption. Let's interpret the gaps in the context of Industry 4.0 for Albania.

- **Low Percentage of Businesses Using Computers**: While 99% of businesses in Albania use computers, it is essential to assess the quality of computer usage. Industry 4.0 relies heavily on digital technologies and data-driven decision-making. Businesses need to not only use computers but also embrace advanced digital tools and technologies to gain a competitive edge.
- **Limited Workers Using Computers for Work**: The percentage of workers using computers for work is 31.3%. This suggests that a significant portion of the workforce may not be adequately equipped with digital skills required for Industry 4.0. Upskilling the workforce in digital literacy and relevant technologies is crucial to enhance productivity and efficiency.
- **Lack of Workers in Information and Communication Activities**: The percentage of workers in information and communication activities using computers is 78.9%. This indicates a potential shortage of skilled workers in crucial sectors like IT and communication, which are essential for driving digital transformation.
- **Low Percentage of Businesses with Fixed Internet Connection**: While 99% of businesses have fixed internet connections, it is essential to ensure that these connections are robust and high-speed to support the increased data transfer and communication requirements of Industry 4.0 technologies.
- **Limited Businesses with High-Speed Download**: Only 63% of businesses have download speeds greater than 30 Mbit/s, and only 22.1% have speeds greater than 100 Mbit/s. Faster internet speeds are necessary for efficient data processing and real-time communication, which are critical in the Industry 4.0 environment.
- **Low Adoption of Autonomous Jobs**: Only 5.5% of businesses are engaged in autonomous jobs, indicating limited investment in automation and advanced technologies. Embracing automation can increase productivity, reduce costs, and improve competitiveness.
- **Limited Businesses Conducting Virtual Meetings**: The percentage of businesses conducting virtual meetings is 38.6%. Expanding the use of virtual meetings can

improve collaboration and reduce the need for physical presence, supporting flexible and agile work practices.

- **Low Percentage of Businesses Selling Goods/Services Online**: Only 13.8% of businesses sell goods/services online. Embracing e-commerce and digital marketplaces can open new revenue streams and expand market reach.
- **Data Security Measures Implementatio**n: While 66.2% of enterprises implement data security measures, there is a need to ensure that businesses have robust cybersecurity protocols in place to safeguard sensitive data and protect against cyber threats, which are prevalent in the digital era.

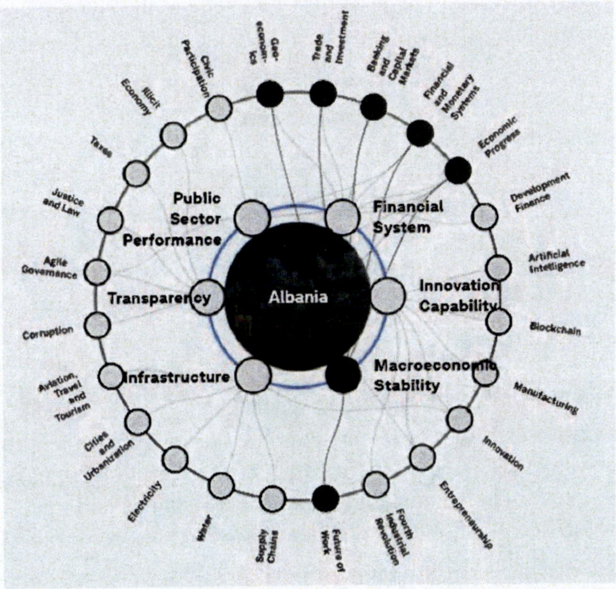

Fig. 1. Global competitiveness index—Albania. *Source* World Economic Forum [18]

Overall, the data indicates that Albania has made some progress in adopting Industry 4.0 technologies but still faces gaps in workforce skills, infrastructure, and digital adoption. To accelerate Industry 4.0 readiness, there is a need for targeted investment in digital infrastructure, upskilling the workforce, and encouraging businesses to embrace advanced digital technologies and data-driven practices.

The quantitative method will be used to analyze the gathered data, focusing on numerical and statistical techniques to quantify and measure various aspects of the study. Descriptive statistics will be employed to summarize the main characteristics of the data, such as mean, median, standard deviation, and frequency distributions. These statistics will provide a comprehensive overview of the data and highlight central tendencies and variabilities within different variables. Inferential statistics will also be applied to draw conclusions and make inferences about the larger population from which the data was collected. Quantitative analysis will be complemented by data visualization methods, including charts and graphs, to present the findings in a visually engaging and understandable manner. Overall, the quantitative approach will provide rigorous and empirical insights into the impact of Industry 4.0 implementation in the company, enhancing the validity and reliability of the study's outcomes.

4 Data Analysis

The range of values for the first variable is from 1.00 to 8.00, indicating that there are eight distinct categories or types of economic activities among the enterprises in the sample.

The mean value of 5.5844 suggests that, on average, the enterprises in the sample are engaged in economic activities categorized around level 5.

Descriptive Statistics

	N	Minimum	Maximum	Mean	Std. Deviation
Type of economic activity of the enterprise.	154	1.00	8.00	5.5844	2.07286
How many employees are there in the company where you work?	154	1.00	4.00	2.9935	1.04473
Valid N (listwise)	154				

- The standard deviation of 2.07286 indicates that the types of economic activities vary among the enterprises, with some being closer to the mean and others more dispersed.
- The range of values for the second variable is from 1.00 to 4.00, indicating that the number of employees in the companies varies between one and four.
- The mean value of 2.9935 indicates that, on average, the companies in the sample have approximately three employees.
- The standard deviation of 1.04473 suggests that there is some variation in the number of employees among the companies, with some having more or fewer employees than the average.

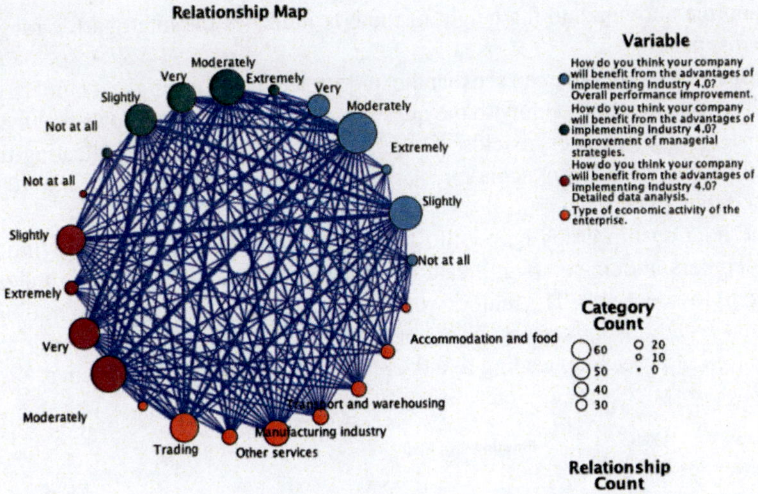

From this concept map, it becomes evident that there is a notable lack of awareness regarding the advantages of adopting Industry 4.0 technologies. The majority of companies tend to hold a moderately optimistic view, anticipating slight to moderate performance enhancements through the adoption of Industry 4.0. This sentiment is similarly reflected in the expectation of strategic improvements. Companies primarily operating in the trading and manufacturing sectors place significance on the integration of Industry 4.0 for streamlined data accessibility and analysis. The prevalent "low trust" perception concerning the benefits could potentially signal a deficiency in comprehensive information dissemination.

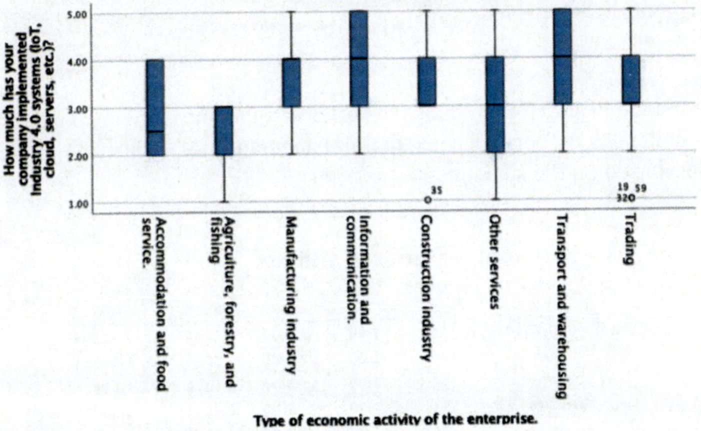

The box plot results show the distribution of responses for the question "How much has your company implemented Industry 4.0 systems (IoT, cloud, servers, etc.)?" across different types of economic activities of the enterprises. Each box represents a category of economic activity, and the height of the box indicates the interquartile range (IQR) of the responses.

From the box plot, we can see that the trading sector has the highest number of cases, with 40 companies responding to the question. The box plot also shows that the median response for each category is close to the "4" on the Likert scale, indicating that most companies across different economic activities have implemented Industry 4.0 systems to a moderate extent.

The relationship graph below illustrates that larger companies employing over 50 staff members anticipate a significant and even substantial impact of digitalization on the unemployment rate. The majority of these enterprises are engaged in trading, transport, and warehousing sectors. This correlation arises primarily due to the extensive automation of processes, leading to a decrease in the workforce's influence.

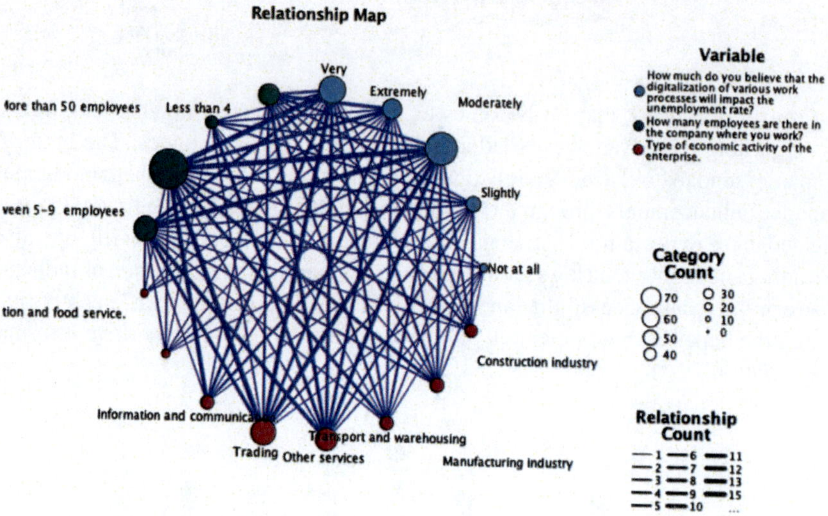

The provided "Descriptive Statistics" offer insights into how respondents perceive different challenges associated with the implementation of Industry 4.0. Here are the key findings based on the statistics.

Descriptive Statistics

	N	Minimum	Maximum	Mean	Std. Deviation
Cybersecurity.	154	1.00	5.00	2.5000	1.11584
Piracy.	154	1.00	5.00	2.5065	1.09214
Risk assessment.	154	1.00	5.00	2.5325	1.12121
Legal challenges.	154	1.00	5.00	2.4545	1.13236
Lack of specialists.	154	1.00	5.00	2.6234	1.19955
Valid N (listwise)	154				

Cybersecurity

- Respondents' average perception of cybersecurity challenges is relatively moderate, with a mean score of approximately 2.5000 on a scale from 1 to 5.
- The responses have a relatively low standard deviation (1.11584), indicating that there is not a wide variation in how respondents view cybersecurity challenges.

Piracy

- Similar to cybersecurity, the perception of piracy challenges is moderately scored, with a mean of about 2.5065.
- The standard deviation (1.09214) suggests that the responses are relatively consistent and not widely dispersed.

Risk Assessment

- Respondents perceive risk assessment challenges slightly higher than the previous two categories, with a mean of around 2.5325.
- The standard deviation (1.12121) indicates a moderate level of variability in respondents' views on risk assessment challenges.

Legal Challenges

- The perception of legal challenges appears to be slightly lower, with a mean score of about 2.4545.
- The standard deviation (1.13236) suggests a moderate degree of variability in responses regarding legal challenges.

Lack of Specialists

- Respondents express relatively higher concern about the lack of specialists, as indicated by the mean score of approximately 2.6234.
- The standard deviation (1.19955) suggests a noticeable degree of variability in respondents' perceptions of this challenge.

In summary, the findings suggest that respondents perceive the challenges of "Lack of Specialists" and "Risk Assessment" with slightly higher concern compared to the challenges of "Cybersecurity," "Piracy," and "Legal Challenges." However, it's important to note that the variations in perceptions across these challenges are not extremely wide, indicating a certain level of consistency in responses. Overall, these statistics provide a snapshot of how respondents assess and prioritize different challenges related to Industry 4.0 implementation.

Suggestions

To mitigate the negative findings and improve Albania's performance in the areas of Innovation Capability and Transparency, several strategies and actions can be considered.

- **Investment in Research and Development (R&D)**: Increasing funding and support for R&D initiatives can foster innovation and create an environment that encourages the development of new technologies, products, and services.

- **Strengthen Intellectual Property Protection**: Implementing and enforcing robust intellectual property laws and regulations can incentivize businesses and individuals to invest in innovation and protect their creations.
- **Promote Entrepreneurship and Startups**: Creating a supportive ecosystem for startups and entrepreneurs can drive innovation and new business ideas, encouraging a culture of innovation.
- **Enhance Education and Skills Development**: Investing in education and skills training programs can improve the workforce's capabilities, enabling them to participate more effectively in research and innovation activities.
- **Streamline Regulatory Processes**: Simplifying and digitizing bureaucratic processes can improve transparency and reduce opportunities for corruption, making it easier for businesses to operate ethically and transparently.
- **Encourage Open Data and Information Sharing**: Promoting open data initiatives and information sharing can increase transparency in government operations and decision-making processes.
- **Promote Accountability and Anti-Corruption Measures**: Implementing strong anti-corruption measures and enforcing accountability can enhance transparency in public institutions and foster a culture of integrity.
- **Engage Stakeholders in Policymaking**: Involving various stakeholders, including businesses, civil society, and academia, in the policymaking process can lead to more comprehensive and effective strategies to address the challenges.
- **International Collaboration and Best Practice Sharing**: Learning from the experiences of other countries and collaborating with international organizations can provide valuable insights and best practices for enhancing innovation and transparency.

References

1. Bajic, B., Rikalovic, A., Suzic, N., Piuri, V.: Industry 4.0 implementation challenges and opportunities: a managerial perspective. IEEE Syst. J. **15**(1), 546–559 (2021). https://doi.org/10.1109/JSYST.2020.3023041
2. Zizic, M.C., Mladineo, M., Gjeldum, N., Celent, L.: From industry 4.0 towards industry 5.0: a review and analysis of paradigm shift for the people, organization and technology. Energies **15**(14), 5221 (2022)
3. United Nations Industrial Development Organization, Industrial Development Report. Industrializing in the digital age. Vienna (2020)
4. Castano, F., Cruz, Y.J., Villalonga, A., Haber, R.E.: Data-driven insights on time-to-failure of electromechanical manufacturing devices: a procedure and case study. IEEE Trans. Ind. Inform. **19**, 7190–7200 (2022)
5. Vrontis, D., Christofi, M., Pereira, V., Tarba, S., Makrides, A., Trichina, E.: Artificial intelligence, robotics, advanced technologies and human resource management: a systematic review. Int. J. Hum. Resour. Manag. **33**, 1237–1266 (2021)
6. Saniuk, S., Grabowska, S., Straka, M.: Identification of social and economic expectations: contextual reasons for the transformation process of industry 4.0 into the industry 5.0 concept. Sustainability **14**, 1391 (2022)

7. Sayem, A., Biswas, P.K., Khan, M.M.A., Romoli, L., Dalle Mura, M.: Critical barriers to industry 4.0 adoption in manufacturing organizations and their mitigation strategies. J. Manuf. Mater. Proc. **6**(6), 136 (2022). https://doi.org/10.3390/jmmp6060136
8. Bakhtari, A.R., Kumar, V., Waris, M.M., Sanin, C., Szczerbicki, E.: Industry 4.0 implementation challenges in manufacturing industries: an interpretive structural modelling approach. Procedia Comput. Sci. **176**, 2384–2393 (2020)
9. Stentoft, J., Adsbøll Wickstrøm, K., Philipsen, K., Haug, A.: Drivers and barriers for industry 4.0 readiness and practice: empirical evidence from small and medium-sized manufacturers. Prod. Plan. Control **32**, 811–828 (2021)
10. Obiso, J.J.A., Himang, C.M., Ocampo, L.A., Bongo, M.F., Caballes, S.A.A., Abellana, D.P.M., Deocaris, C.C., Padua, R., Ancheta, R.A.: Management of Industry 4.0–reviewing intrinsic and extrinsic adoption drivers and barriers. Int. J. Technol. Manag. **81**, 210–257 (2019)
11. Kumar, P., Bhamu, J., Sangwan, K.S.: Analysis of barriers to industry 4.0 adoption in manufacturing organizations: an ISM approach. Procedia CIRP **98**, 85–90 (2021)
12. United Nations Conference on Trade and Development (UNCTD), Industry 4.0 for Inclusive Development, Geneva, 2022, ISBN: 978-92-1-113036-2, e-ISBN: 978-92-1-001444-1
13. Mourtzis, D., Angelopoulos, J., Panopoulos, N.: A literature review of the challenges and opportunities of the transition from industry 4.0 to society 5.0. Energies **15**(17), 6276 (2022)
14. Aheleroff, S., Huang, H., Xu, X., Zhong, R.Y.: Toward sustainability and resilience with industry 4.0 and industry 5.0. Front. Manuf. Technol. **2**, 951643 (2022)
15. Bi, Z., Da Xu, L., Wang, C.: Internet of things for enterprise systems of modern manufacturing. IEEE Trans. Industr. Inf. **10**(2), 1537–1546 (2014)
16. Kolade, O., Owoseni, A.: Employment 5.0: the work of the future and the future of work. Technol. Soc. 102086 (2022)
17. Müller, J.M., Kiel, D., Voigt, K.-I.: What drives the implementation of industry 4.0? The role of opportunities and challenges in the context of sustainability. Sustainability **10**(1), 247 (2018)
18. https://www.weforum.org/reports/annual-report-2021-2022/
19. INSTAT: Përdorimi i Teknologjisë së Informacionit dhe Komunikimit në ndërmarrje. ghttps://www.instat.gov.al/ (2022)

Using "Machine Learning" Techniques in Increasing the Efficiency of Sales Forecasting in Albania

Valma Prifti(✉), Dea Sinoimeri, Armira Lazaj, Betina Dini, and Kevin Luniku

Polytechnic University of Tirana, 1010 Tirana, Albania
vprifti@fim.edu.al

Abstract. This paper investigates the utilization of a Machine Learning (ML) approach with the objective of selecting an appropriate model for sales forecasting. Within this study, three ML algorithms are examined: Simple Linear Regression, Gradient Boosting Regression, and Random Forest Regression. A comparative analysis of these algorithms is conducted using two performance metrics: Accuracy Score and Max Error. The significance of sales forecasting cannot be overstated, as it plays a critical role across various industries. Therefore, the application of ML technology is essential to mitigate potential financial losses resulting from inaccurate demand assessments. A retail company based in Albania, which provided historical data as input for the model, is utilized as a case study. The Random Forest Model demonstrates exceptional performance, characterized by minimal deviations between predicted and actual values. The findings of this research endeavor present a pioneering initiative that holds significant potential for enhancing the forecasting of future sales and delivering substantial benefits to firms operating in the Albanian market.

Keywords: Technology · Machine learning · Algorithms · Sales forecast

1 Introduction

Data Mining is described as a process of extracting valuable information from a large collection of raw data, using statistics, artificial intelligence, machine learning, and pattern recognition methods. Machine Learning (ML) is a field of study that enables machines to learn without being explicitly programmed. ML is defined as computer programs that learn from experience. There are three categories of machine learning algorithms. Figure 1 provides an overview of Machine Learning and the techniques applied to solve various tasks. In the fast-paced world of sales forecasting, accurate predictions are crucial for entrepreneurs to optimize their operations, improve inventory management, and maximize profits. Traditional sales forecasting methods often struggle to capture the complex dynamics of consumer behavior and the intricate relationships between various market factors. However, with the advent of Machine Learning (ML) techniques, there is a significant opportunity to use advanced algorithms and engineering principles to revolutionize the efficiency of sales forecasting processes. The essence of

G. Guxho et al. (Eds.): ITC-ICEE 2023, LNMIE, pp. 24–34, 2024.
https://doi.org/10.1007/978-3-031-48933-4_3

this paper lies in the application aspects of ML techniques for sales forecasting. By employing principles such as data processing, feature engineering, model selection, and optimization, engineers can develop powerful and efficient ML models that surpass traditional forecasting approaches. This focused approach aims to address the challenges entrepreneurs face in handling large volumes of data, managing complex models [1], and ensuring scalability and reliability in real-world sales forecasting scenarios. By analyzing historical sales data, customer demographics, and relevant market variables, we aim to evaluate current forecasting practices and identify areas where ML algorithms can significantly enhance accuracy and efficiency in a real retail company.

The findings of this research will provide valuable insights for engineers, data scientists, and entrepreneurs seeking to improve their sales forecasting capabilities. By embracing machine learning techniques with a focus on engineering principles, organizations can utilize advanced analytics to gain a competitive advantage, improve resource allocation, and foster informed decision-making processes.

2 Materials and Methods

In Supervised Learning, the algorithm utilizes labeled data and attempts to find the label that corresponds to these data based on given features. If the label is continuous, it involves linear regression. In the case of categorical labels, classification algorithms are used. The Unsupervised Learning technique is employed when only feature data is available [3]. The most common application is clustering, where data with only features are divided into groups based on similarities. The Reinforcement Learning technique is used to solve much more complex problems, such as teaching a computer how to play music or drive a car. It involves three components: the Agent, which makes decisions; the Environment, with which the agent interacts, and the Action, which is taken by the agent.

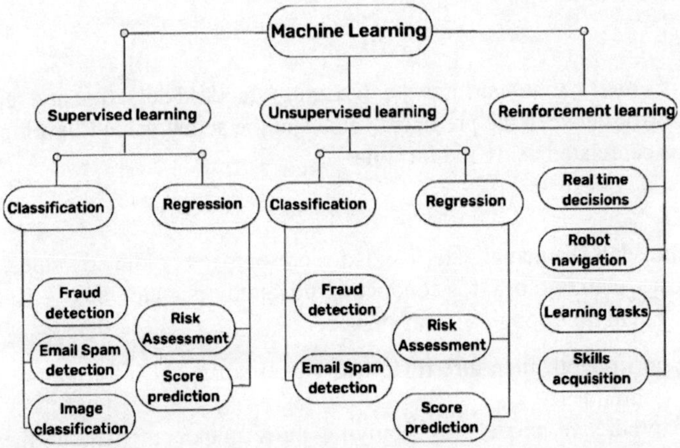

Fig. 1. Tree representation of machine learning approach, techniques, and tasks for which they are used.

2.1 Selection of Machine Learning Algorithms

Forecasting involves discovering possible future events, usually based on past data. In this paper, the algorithms used belong to the category of supervised learning, such as Simple Linear Regression, Random Forest Regression, and Gradient Boosting Regression. These algorithms [5], can facilitate finding better results compared to traditional analytical time series techniques.

- Simple Linear Regression is useful for determining the relationship between two continuous variables. This type of regression requires a non-deterministic statistical relationship.
- Gradient Boosting Regression is based on the premise that, when combined with previous techniques, iterative refinement minimizes the maximum prediction error.
- Random Forest Regression is a type of ensemble method that allows predictions by integrating decisions from a series of simple models.

2.2 Metrics for Evaluating Model Effectiveness

To assess the effectiveness of the models in a more objective manner, several metrics are used. The main objective of this study was to compare the performance of Machine Learning techniques by applying performance metrics such as accuracy score and maximum error.

Accuracy Score

This metric is known as the ratio of correct predictions to the total number of predictions (data points) (Developers 2020a), and is calculated using the formula:

$$Accuracy\ Score = \frac{TN + TP}{TN + TP + FN + FP}$$

where TN is true negative, TP is true positive, and (TN + TP + FN + FP) is the total number of predictions.

Max Error

Max Error is a metric that measures the maximum standard deviation and represents the worst- case error between the predicted value and the actual value (Developers 2020b). Max Error is calculated using the formula:

$$Max\ Error(y, x) = \max(|y_i - x_i|)$$

where y_i represents the actual values and xi represents the predicted values.

The chosen methodology for conducting this study is a case study. To achieve the final goal, we will go through several steps:

1. Conduct an in-depth literature review to gather sufficient information on machine learning techniques.
2. Evaluate predictive models by applying performance metrics to measure their efficiency.
3. Identify the suitable algorithm for sales forecasting in the selected company.
4. Gather data from the company for the application of the ML predictive model.
5. Visualize the predicted results of the model on future sales of the company.

3 Results and Discussions

The Machine Learning predictive models considered for comparison were three: Simple Linear Regression, Gradient Boosting Regression, and Random Forest Regression. The most suitable model for achieving the objective of this research demonstrates the highest value of the selected metrics. The data collected for this research work is confidential, and to maintain the company's privacy, the five considered articles are labeled with Arabic numerals. They are presented in Table 1, where for each article, the price in Euros, the quantity sold in each year, and the revenue generated because of sales are recorded.

Table 1. Data collected from the wholesale company for the years 2015–2018.

Year	Product	Price (Euro)	Sale (Pcs)	Income (Euro)
2015	1	15	6210	93150
	2	25	5422	135550
	3	10	6005	60050
	4	37	7115	263255
	5	99	7250	717750
Total			*32002*	*1269755*
2016	1	15	6821	102315
	2	25	7340	183500
	3	10	7410	74100
	4	37	7500	277500
	5	99	7415	734085
Total			*36486*	*1371500*
2017	1	15	7601	114015
	2	25	7589	189725
	3	10	7840	78400
	4	37	7795	288415
	5	99	8005	792495
Total			*38830*	*1463050*
2018	1	15	9550	143250
	2	25	9883	247075
	3	10	1058	10580
	4	37	11000	407000
	5	99	11010	1089990
Total			*42501*	*1897895*

The first step of the analysis is to study the dataset, which contains information on sales from the wholesale company. The graph presented in the figure illustrates the behavior of the company's current sales (Fig. 2).

On the other hand, the revenue follows a proportional trend with the sales. As shown in Fig. 3, most of the graph is dominated by the year 2018, represented in blue, followed by 2017 in yellow, 2016 in orange, and finally 2015, depicted in green.

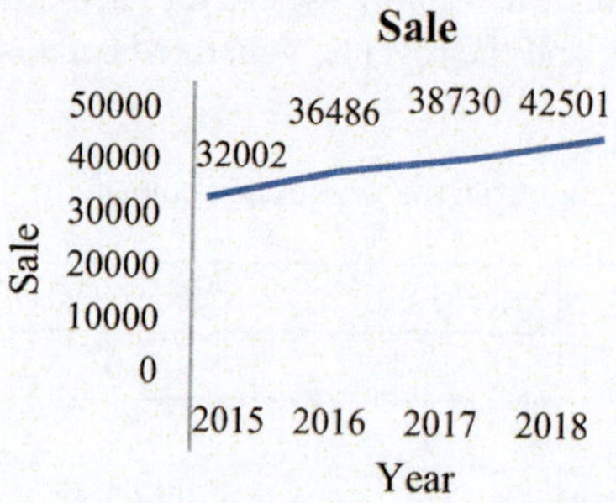

Fig. 2. The sales trend of the company from 2015 to 2018.

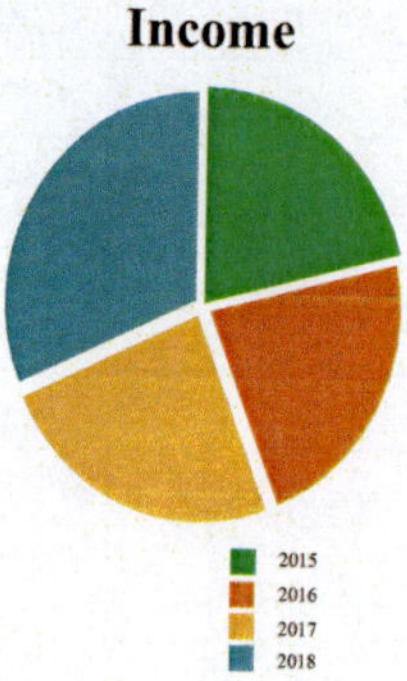

Fig. 3. Distribution of company revenue by year.

The final step is the use of the model for predicting the revenue from wholesale sales. The chosen [14], Machine Learning algorithm, which was deemed most suitable for the task, was Random Forest. Based on the provided historical data, the model successfully made sales predictions for the next two years, specifically 2019 and 2020.

After applying the metrics to each of the three Machine Learning algorithms, the following results were obtained.

3.1 Simple Linear Regression

The graph presented in Fig. 4 shows the accuracy score results obtained from the Simple Linear Regressor. It can be observed that the maximum accuracy score is 84.099%, the average accuracy score is 81.2%, and the minimum accuracy score is 73.95%.

The graph presented in Fig. 5 displays the Maximum Error (ME) obtained from the Simple Linear Regressor, where the maximum value is 0.5118%, the average value is 0.4917%, and the minimum value is 0.4731%.

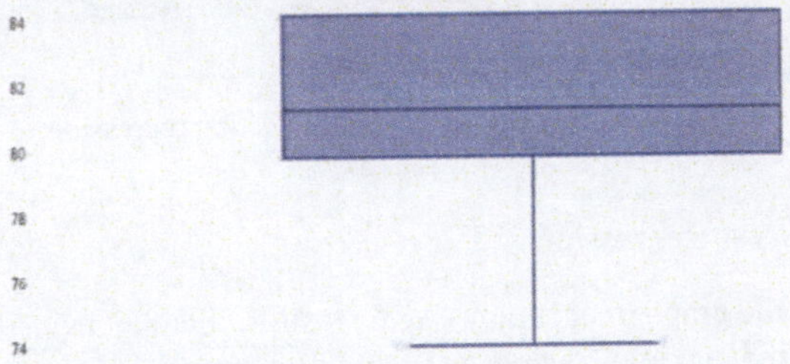

Fig. 4. ACC graph for simple linear regression.

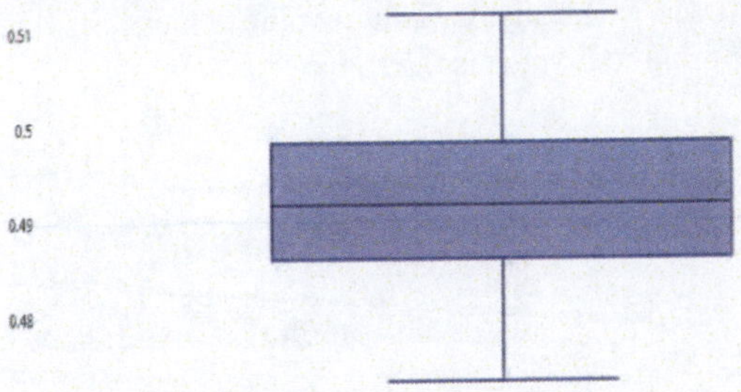

Fig. 5. ME graph for simple linear regression.

3.2 Gradient Boosting Regression

The graph presented in Fig. 6 shows the accuracy (ACC) results obtained from the Gradient Boosting Regressor, where the maximum accuracy value is 91.2%, the average accuracy is 86.27%, and the minimum accuracy is 78.3%.

In the graph of Fig. 7, the maximum error (ME) values are provided, where the highest value is 0.464, the average ME value is 0.441, and the achieved minimum value is 0.425.

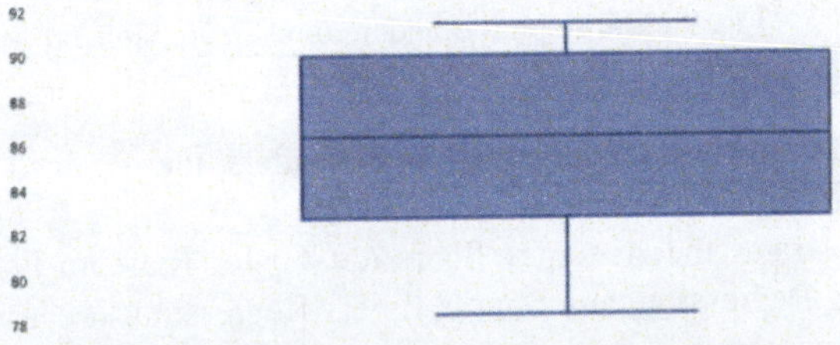

Fig. 6. ACC graph for gradient boosting regression.

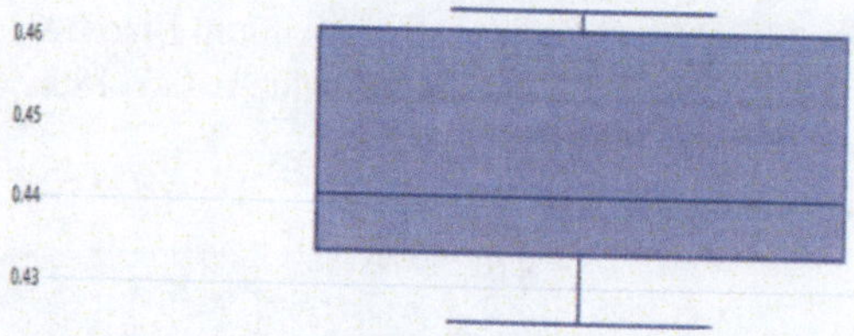

Fig. 7. ME graph for gradient boosting regression.

3.3 Random Forest Regression

Figure 8 presents the graph of accuracy (ACC) results obtained from the Random Forest Regressor, where the maximum achieved accuracy value is 91.35%, the average accuracy is 87.72%, and the minimum accuracy is 78.31%.

In the graph presented in Fig. 9, the Maximum Error (ME) obtained from the Random Forest Regressor is shown, where the maximum value achieved is 0.6568, the average ME is 0.6135, and the minimum value is 0.5964.

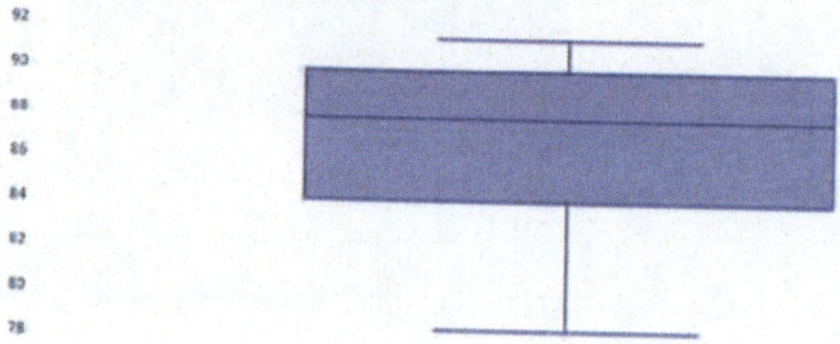

Fig. 8. ACC graph for random forest regression.

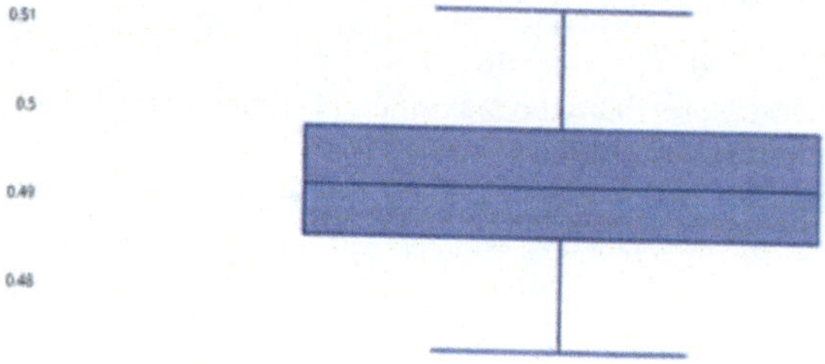

Fig. 9. ME graph for random forest regression.

To gather and simplify the understanding of readers, the obtained results from the comparison of algorithms are presented in Table 2.

Figure 10 illustrates the Accuracy Score (ACC) for Random Forest Regression, Gradient Boosting Regression, and Simple Linear Regression. It can be observed that Random Forest Regression shows the highest accuracy score. Figure 11 presents the Max Error (ME) value for each selected Machine Learning algorithm. Random Forest Regression exhibits the lowest error value.

Table 2. Results of accuracy score and max error metrics for corresponding machine learning algorithms

Algorithms	Accuracy scor (%)e	Max error
Random forest regression	87.72	0.441
Gradient boosting regression	86.27	0.491
Simple linear regression	81.20	0.491

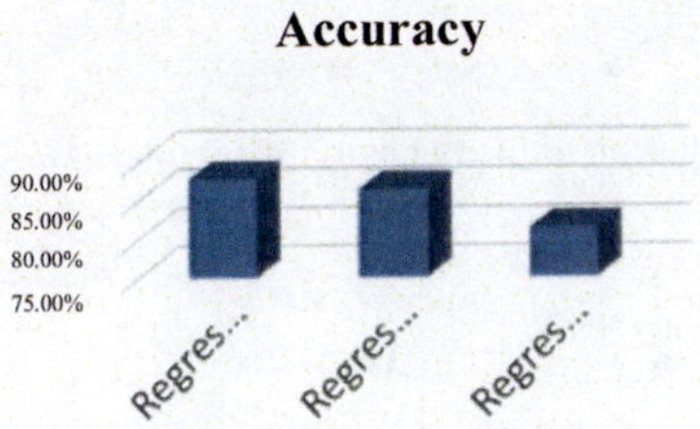

Fig. 10. Accuracy values

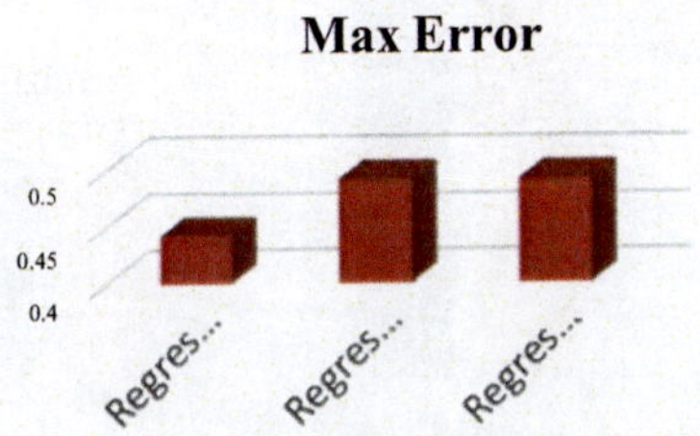

Fig. 11. Max serror values

Table 3. Sales and revenue values

Year	Actual sales	Predicted sales	Actual revenue	Predicted revenue
2019	42993	43070	2000102	2003684
2020	42840	42904	1901152	1903992

The sales and revenue graphs are constructed separately because their units of measurement and value ranges are completely different from each other (Table 3).

To visualize the sales and revenue values of the company presented in the table and to understand the proximity of the actual values provided by the company to those predicted by the Random Forest Model, the following two graphs have been constructed. Figure 12 illustrates the behavior of sales, where the columns in blue represent the actual sales, while the columns in red represent the sales predicted by the Model (Fig. 13).

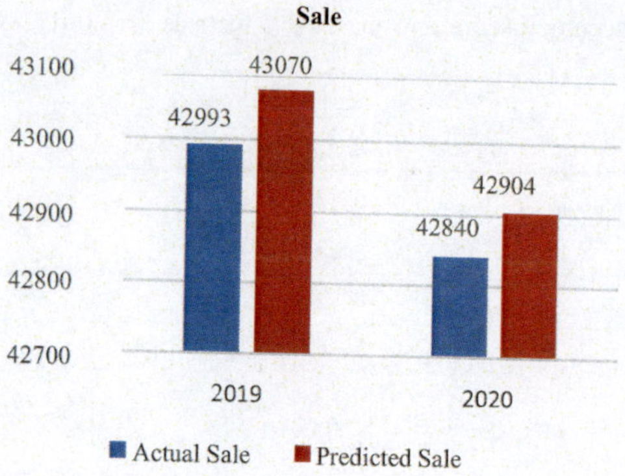

Fig. 12. Graphical representation of the behavior of actual and predicted sales for the years 2019 and 2020.

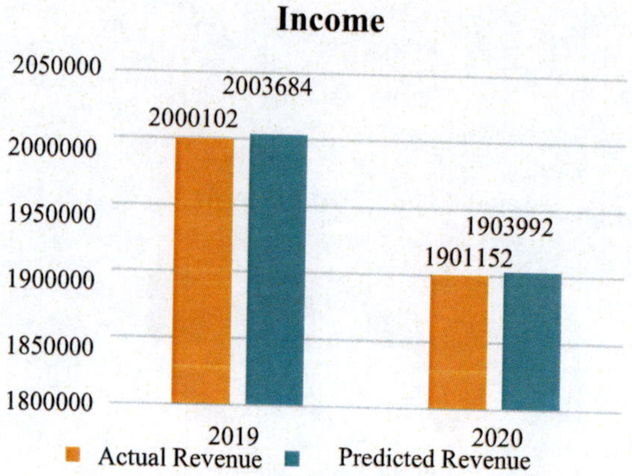

Fig. 13. Graph of the trend of actual and predicted revenues for the years 2019 and 2020.

4 Conclusion

The comparative analysis of these Machine Learning models: Simple Linear Regression Gradient Boosting Regression, and Random Forest Regression; yielded important results for sales forecasting. In terms of accuracy scores, the Simple Linear Regression model achieved a maximum score of 84.099%, with an average of 81.2% and a minimum of 73.95%. The Gradient Boosting Regression model performed even better, with a maximum accuracy score of 91.2%, an average of 86.27%, and a minimum of 78.3%. However, the Random Forest Regression model outperformed both, obtaining the highest maximum accuracy score of 91.35%, an average of 87.72%, and a minimum of 78.31%. When considering maximum error (ME), the Simple Linear Regression model exhibited a maximum error of 0.5118%, an average of 0.4917%, and a minimum of 0.4731%. The Gradient Boosting Regression model showcased a higher level of accuracy, with a maximum error of 0.464, an average of 0.441, and a minimum of 0.425. The Random Forest Regression model, however, demonstrated the lowest maximum error of 0.6568, an average of 0.6135, and a minimum of 0.5964. Based on these findings, it can be concluded that the Random Forest Regression model offers the most accurate and precise sales forecasting capabilities among the three models evaluated. With a maximum accuracy score of 91.35% and the lowest maximum error, this model proves to be the most suitable choice for accurately predicting sales in the wholesale industry.

5 Acronyms

ML Machine Learning
ACC Accuracy Score
ME Maximum Error
TN True Negative Numbers
TP True Positive Numbers
FN False Negative numbers
FP False Positive numbers

References

1. Bangdiwala, S.I.: Regression: simple linear. Int. J. Inj. Contr. Saf. Promot. **25**(1), 113–115 (2018)
2. Hanssens, D.M.: Order forecasts, retail sales, and the marketing mix for consumer durables. J. Forecast. **17**(3–4), 327–346 (1998)
3. Prifti, V., Dhoska, K.: Information systems in project management and their role in decision making. Int. J. Tech. Phys. Prob. Eng. **14**(4), 189–194 (2022)
4. Kahn, K., Adams, M.: Sales forecasting as a knowledge management process. J. Bus. Forecast. Meth. Syst. **19**(4), 19 (2001)
5. Harrison, P.J.: Short-term sales forecasting. J. Royal Stat. Soc. Series C Appl. Stat. **14**(2/3), 102–139 (1965)
6. Prifti, V., Aranitasi, M.: E-commerce business model in KLER enterprise for shirt manufacturing. Int. J. Innov. Technol. Interdiscipl. Sci. **5**(1), 858–864 (2022)
7. Prifti, V., Sinoimeri, D., Lazaj, A., Keci, J.: Impact of the information systems and technology on enterprises. J. Integr. Eng. Appl. Sci. **1**(1), 23–31 (2023)
8. Mentzer, J.T., Cox J.R., J.E.: Familiarity, application, and performance of sales forecasting techniques. J. Forecast. **3**(1), 27–36 (1984)
9. Ray, S.A.: Quick review of machine learning algorithms. In: International Conference on Machine Learning, Big Data, Cloud and Parallel Computing (COMITCon), pp. 35–39 (2019)
10. Prifti, V., Dervishi, I., Dhoska, K., Markja , I., Pramono, A.: Minimization of transport costs in an industrial company through linear programming. IOP Conf. Ser.: Mater. Sci. Eng. **909**, 012040 (2020)
11. Morgan, M.S., Chintagunta, P.K.: Forecasting restaurant sales using self-selectivity models. J. Retail. Consum. Serv. **4**(2), 117–128 (1997)
12. Pavlyshenko, B.M.: Machine-learning models for sales time series forecasting. In: Data, vol. 4 (2019)
13. Skorikov, M., Momen, S.: Machine learning approach to predicting the acceptance of academic papers. In: International Proceedings of Conference on Industry 4.0, Artificial Intelligence, and Communications Technology, pp.113–117 (2020)

14. Prifti, V., Markja, I., Dhoska, K., Pramono, A.: Management of information systems, implementation, and their importance in Albanian enterprises. IOP Conf. Ser.: Mater. Sci. Eng. **909**, 012047 (2020)
15. Ma, X.U., Tian, Y., Luo, C.H.U., Zhang, Y.: Predicting future visitors using big data. In: International Proceedings of Conference on Machine Learning and Cybernetics, pp. 269–274. IEEE (2018)

Information Technology in Supply Chain Management. Case Study

Dea Sinoimeri(✉), Jonida Teta, Valma Prifti, and Armira Lazaj

Polytechnic University of Tirana, 1010 Tirana, Albania
dsinoimeri@fim.edu.al

Abstract. Information technology advancements have made communicating information easier for supply chain participants. The performance of the supply chain may be improved by raising the quality of the information. This article aimed to identify information technology tools/instruments companies can use to manage effective supply chains. The data collected for this article was obtained from 100 companies located in Tiranë and Durrës, in Albania, during the mid-year of 2022. The relationship between CEOs/managers, information sharing, information quality, and information technology tools/instruments was examined in this article. The results showed that information technology tools/instruments significantly affect information quality and information sharing for managing the internal work of companies. The results gained from this article benefit managers, information technology specialists, and CEOs of companies. Relationships between supply chain partners and companies with the help of information technology in information sharing can help companies operate better and gain a competitive edge.

Keywords: Information technology · Supply chain management · Survey · Case study · Data collection

1 Introduction

Nowadays, companies participate in supply chain management competitions instead of individual market management competitions. The last decades have been a period of rapid changes for organizations, especially for businesses, forcing them to be much more innovative, flexible, responsive, and focus on the actual demands and needs of consumers to survive and compete in the intensely competitive global environment [1]. Supply chain management is a science that discusses suppliers and customers from upstream to downstream to achieve lower costs and superior customer value [2]. Supply chain management is an essential company process planned through systemic coordination, such as procurement, purchasing, conversion, and logistics [3].

Coordination between all participants in supply chain management depends primarily on information sharing. The basis of integration in the supply chain is information sharing; therefore, decisions about what information to share and how to share it are directly related to conclusions concerning the level of integration.

G. Guxho et al. (Eds.): ITC-ICEE 2023, LNMIE, pp. 35–44, 2024.
https://doi.org/10.1007/978-3-031-48933-4_4

Knowledge transfer in supply chain management has significantly transformed in recent years. Companies now know how crucial it is for their supply chain partners to use information technology [4]. Sharing information with all supply chain members can increase efficiency, effectiveness, and ability to respond more rapidly to shifting client demands [5].

As a result of globalization, companies are now starting to change their supply chain relationships, including the ways of communication, the use of technology, the increase of the diversity of business operations, and process optimization.

A company's supply chain management must be responsive to respond as rapidly as possible to different customer demands and short product life cycles. Supply chain management has evolved and been more widely used across companies, attributable to the constant development and application of information technology. The supply chain now mainly consists of intelligence, networking, and digitization.

De Barros [6] emphasized the benefits generated by companies for using information technology tools/instruments, such as cost reduction, time of delays reduction, inventory control, quality and efficiency improvement, innovation and distinction of product, reliability, and accuracy of the information, etc. Suppliers, manufacturers, distributors, retailers, and customers can reduce delays, documentation, and other unnecessary processes using information technology tools/instruments.

The article aims to identify information technology tools/instruments companies can use to manage their supply chain more effectively. This article has the following structure. The second section demonstrates a brief review of the literature. The third section presents the methodology used in this article. The fourth section provides the results of the research in this article. The fifth and final section concludes with the conclusions.

2 Literature Review

With the advancement of both hardware and software in the 1980s, companies started utilizing computer applications in their support functions.

The conventional approaches to managing the supply chain have significantly modified during the past few decades.

Li [7] classifies the information chain systems into decision-making systems, executive information systems, information management systems, communication systems, and transaction processing systems.

Carr and Kaynak [8] evaluated the ways of communication buyers use to interact with their suppliers and divided them into conventional and cutting-edge techniques. Conventional techniques include the use of telephone, e-mail, and face-to-face communication, while cutting-edge techniques include the use of Electronic Data Interchange (EDI), Computer-to-computer links, and Enterprise Resource Planning (ERP).

The most recent technologies utilized in logistics and supply chain management have been divided into three categories [9]: Automatic Identification Technology, Communication Technology, and Information Technology.

2.1 Supply Chain Management with Simple Information Technology

Electronic Data Interchange, EDI—permits data transfer between computers owned by different companies in a uniform electronic format, such as invoices and purchase orders.

Electronic Funds Transfer, EFT—refers to the automatic transmission of invoice messages and payments via electronic technologies.

Enterprise Resource Planning, ERP—combines every aspect of company operation, such as marketing, production, sale, finance, accounting, distribution, warehousing, planning, etc.

Scanner—is used to identify and monitor commodities at every procedure stage.

Barcode—is how a number/code is presented so the computer can read it.

Inventory Management Systems, IMS—determines what inventory levels should be maintained, when they should be placed, and how big orders should be based on the rules and procedures that monitor inventory levels.

Warehouse Management Systems, WMS—includes receiving products, allocating storage locations, filling empty spaces, producing lists, order taking, and order picking.

Transport Management Systems, TMS—increases the order and delivery visibility.

2.2 Supply Chain Management with advanced Information Technology

Radio Frequency Identification, RFID—is a technology used in supply chain management to identify products uniquely, communicate intelligently, and share real-time information.

Decision Making Systems, DMS—is a computer program or knowledge-based system that facilitates automated, manual, or combined decision-making at a company's managerial, planning, and operational levels.

Electronic Supply Chains, ESC—refers to the supply chain between companies maintained electronically based on EDI or the Internet.

Customer Relationship Management Systems, CRM—refers to companies' procedures, plans, and tools to control, analyze, and manage customer interactions and lifecycle data to enhance customer relationships, maintain clients, and increase revenue.

Advanced Planning and Scheduling, APS—is a group of programs used to control the three parts of the supply chain: planning, scheduling, and execution.

Technology and Web Services—allows companies to work with partners for lower costs and more customer satisfaction.

Human Resources Management System, HRMS—helps with payroll management, performance evaluation, benefits administration, employee schedules, recruiting, collecting, storing, and accessing employee information, etc.

Distribution Requirement Planning, DRP—connects warehouse operations and transportation needs.

Computerized Maintenance Management System, CMMS—analyzes maintenance and information about equipment to optimize management and support for strategic, tactical, and operational decisions.

E-commerce—describes a company's capacity to purchase and resell products online.

E-business—defines a collection of company's practices enabled by Internet technologies focused on networking consumers, suppliers, and manufacturers to improve supply chain performance continuously.

3 Methodology

The article was created using data analysis collected through a survey given to companies operating in Albania (respectively in Tiranë and Durrës) that are crucial links in supply chain management. A concrete overview of the present condition of supply chain management in Albania is revealed by the questionnaire given to the CEO and managers (in two ways: electronically through email and face-to-face meetings) of a sample size of 100 companies that operate in a variety of sectors, including agriculture, construction, industry, trade, transport/warehousing, accommodation/food service, information/communication, etc. The sample size is a critical factor that can affect research results. A large sample size brings more confidence in the generalizability of statistics [10].

3.1 Establishment of the Survey

The survey given to companies includes 20 questions in total. The questions throughout the survey are formulated in a clear, concise, and understandable way for all company managers/CEO so that the result obtained from it is reliable.

In the first part, the survey contains questions about the profile of the company:

- company name,
- location,
- website/e-mail,
- the position of the respondent,
- type of sector,
- number of employees,
- position of the company in the supply chain.

In the second part, the survey contains questions about the employers:

- in managing the internal company work,
- in sharing information with supply chain partners (the level of information sharing and the level of information quality),
- in creating relations with supply chain partners (trust, commitment, and vision),
- in showing the benefits/barriers of supply chain management from information technology.

Indicators obtained from the survey are measured on a 5-Level Likert scale, which means:

1—not at all, 2—a little, 3—enough, 4—a lot, and 5—completely.

3.2 Methodology Framework

More details on the methodological steps in this article are presented in Fig. 1.

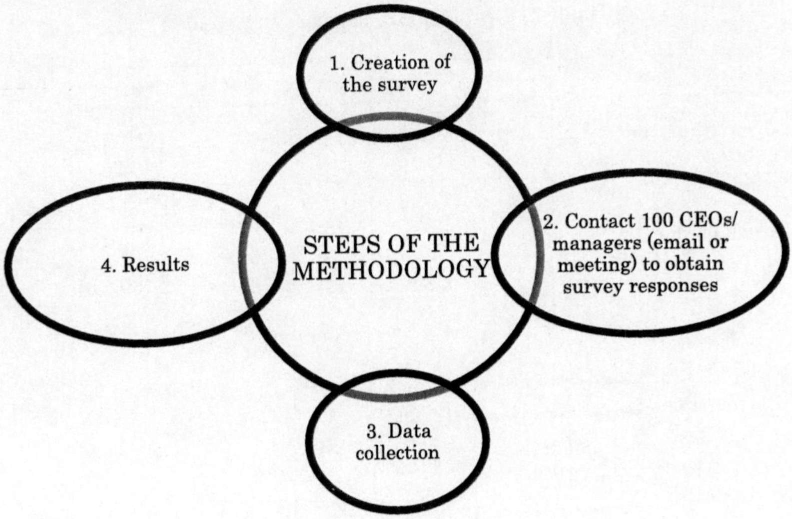

Fig. 1. Methodology framework.

Table 1. Distribution of companies by city.

City	Frequency	%
Tiranë	72	72
Durrës	28	28
Total	*100*	*100*

4 Results

The distribution of the companies taken into consideration in this article can be seen from the collection and evaluation of data from the survey reflected in Table 1.

Tirana has the most significant number of companies, with 72 companies or 72% of the total number of companies, followed by Durrës with 28 companies or 28% of the total number of companies.

The operation sector of the companies taken into consideration in this article can be seen from the collection and evaluation of data from the survey reflected in Table 2.

52% of companies operate in the service sector, 26% in the trade sector, 14% in the production sector, and only 8% in the service & production sector.

The technologies used by companies taken into consideration in this article for sharing information with supply chain partners can be seen from the collection and evaluation of data from the survey reflected in Fig. 2.

93% of respondents use information technology to share information with supply chain partners. The results show that most employers for sharing information with supply chain partners use internet services such as e-mails (46%), social networks (42%), and

Table 2. Distribution of companies by sector.

Sector	Frequency	%
Service	52	52
Production	14	14
Trade	26	26
Service and production	8	8
Total	*100*	*100*

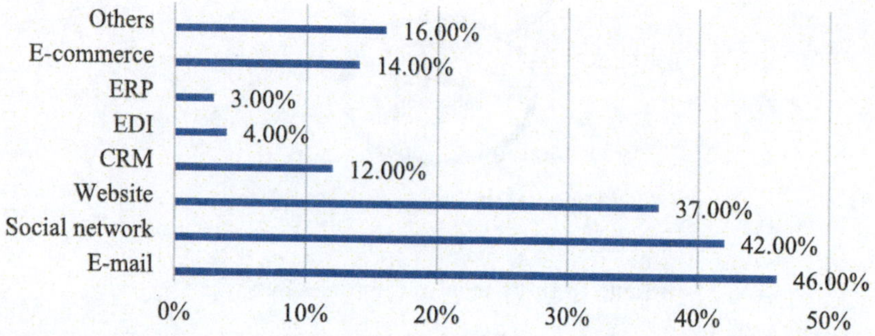

Fig. 2. Distribution of technologies for sharing information with supply chain partners.

websites (37%). (14%) make purchases online, (12%) use CRM, (4%) use EDI, (3%) use ERP, and (16%) use other information technology tools/instruments.

The technologies used by companies taken into consideration in this article for managing the internal work of companies can be seen from the collection and evaluation of data from the survey reflected in Fig. 3.

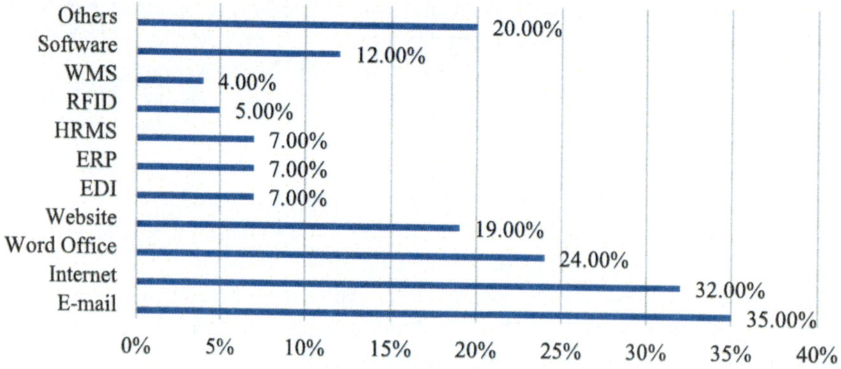

Fig. 3. Distribution of technologies for managing the internal work of companies.

The most used technologies for managing internal work for companies are e-mail (35%), internet (32%), word office (24%), other information technology tools/instruments (20%), website (19%), and software (12%). The least used technologies for managing internal work for companies are ERP (7%), EDI (7%), HRMS (7%), RFID (5%), and WMS (4%).

Table 3 shows the benefits of the companies taken into consideration using information technology.

Table 3. Benefits of indicators using information technology for companies.

Indicators	Not at all (%)	A little (%)	Enough (%)	A lot (%)	Completely (%)
Increase the sales	2	13	30	**35**	20
Increase coordination with customers	4	14	30	**33**	19
Increase coordination with suppliers	13	20	**36**	16	15
Increase coordination inside the company	**27**	20	22	15	16
Increase the quality of information	4	13	32	**37**	14
Increase the quantity of information	7	16	**32**	23	22
Increase the forecasts	14	28	**30**	16	12

The use of information technology has brought the following changes to the companies taken into consideration:

- 35% have benefited a lot in increasing their sales,
- 33% have benefited a lot in increasing the coordination with their customers,
- 36% have benefited enough in increasing the coordination with their suppliers,
- 27% have not benefited at all in increasing the coordination inside the company,
- 37% have benefited a lot in increasing the quality of information,
- 32% have benefited enough in increasing the quantity of information,
- 30% have benefited enough in increasing their forecasts.

Table 4 shows the relationship of companies taken into consideration with supply chain partners.

The use of information technology has brought the following changes to the relationship of supply chain partners with companies taken into consideration:

Table 4. Relationship of companies with supply chain partners using information technology.

Indicators	Not at all (%)	A little (%)	Enough (%)	A lot (%)	Completely (%)
Partners have shown in the past that companies can have faith	6	14	***52***	18	10
Partners have been honest in their dealing with companies	4	19	22	***28***	27
Partners respect the confidentially of the information they receive from companies	9	8	22	***49***	12
Partners respect the agreement very well	9	6	25	***42***	18
Companies and partners always try to keep each other's promises	5	10	19	***41***	25
Companies and partners have a similar understanding of supply chain vision and objectives	10	18	24	***30***	18
Companies and partners must review their relationship to improve it continuously	13	19	***29***	19	20

- 52% of supply chain partners have shown in the past that companies can have faith,
- 28% of supply chain partners have been honest in their dealing with companies,
- 49% of supply chain partners respect the confidentially of the information they receive from companies,
- 42% of supply chain partners respect the agreement very well,
- 41% of supply chain partners and companies always try to keep each other's promises,
- 30% of supply chain partners and companies have a similar understanding of supply chain vision and objectives,

- 29% of supply chain partners and companies must review their relationship to improve it continuously.

Most of the companies considered have understood the role of information technology in the success of their companies (68%). In the next 2–4 years, the companies intend to implement:

- Website (27%),
- Purchase system and online booking (8%),
- Customer management system (8%),
- Other technologies not on the survey list (mobile application, advanced computer networking technologies, etc.) (13%).

5 Conclusions

Companies nowadays modify their interactions due to globalization by utilizing information technology in supply chain management. Technology and information utilization are required to manage intricate supply chains efficiently. To improve product/service levels, decrease inventory levels, and reduce supply chain costs, information technologies can help in coordinating all supply chain processes. This article uses data collection analysis to determine which information technology companies use to manage their supply chains. The results of the data collection analysis revealed that email is the most popular information technology tool for managing tasks with partners in the supply chain and tasks within companies. The benefits of using information technology are significant. The participants benefited from better quality information, increased coordination with customers, and increased sales.

References

1. Coyle, J.J., Langley, C.J., Novack, R.A., Gibson, B.: Supply chain management: a logistics perspective. In: Nelson Education (2016)
2. Boateng, A.: Supply chain management and lean concept in construction: a case of Ghanaian building construction industry. Organ. Technol. Manag. Constr. **11**, 2034–2043 (2019)
3. Chakraborty, S., Gonzalez, J.: An integrated lean supply chain framework for US hospitals. Oper. Supply Chain Manag. Int. J. **11**, 98–109 (2018)
4. Santin, K., Sousa, L.S.G., Cardoso, B.M., Antoniolli, P.D., Argoud, A.R.T.T.: MRP implementation on supply management process: a Brazilian furniture industry case study. Euro. J. Bus. Soc. Sci. **4**, 58–173 (2015)
5. Li, S., Lin, B.: Accessing information sharing and information quality in supply chain management. Decis. Support Syst. **42**, 1641–1656 (2006)
6. De Barros, A.P., Ishikiriyama, C.S., Peres, R.C., Gomes, C.F.S.: Processes and benefits of the application of information technology in supply chain management: an analysis of the literature. Procedia Comput. Sci. **55**, 698–705 (2015)
7. Li, D.: e-supply chain management. WIT Trans. State Art Sci. Eng. **16** (2005)
8. Carr, A.S., Kaynak, H.: Communication methods, information sharing, supplier development and performance: an empirical study of their relationships. Int. J. Oper. Prod. Manag. **27**, 346–370 (2007)

9. Bhandari, R.: Impact of technology on logistics and supply chain management. IOSR J. Bus. Manag. **1**, 19–24 (2014)
10. Grinnell., Unrau, Y.A.: Social work research and evaluation: foundations of evidence-based practice. Oxford University Press (2010)

Applying the Ishikawa Diagram and Pareto Chart for Defect Reduction in the Manufacturing Industry. A Case Study from a Textile-Producing Company

Eralda Xhafka[1](✉), Elidon Avrami[2], Gabriele Canton[2], and Artemisa Mazreku[2]

[1] Department of Production and Management, Mechanical Engineering Faculty, Polytechnic University of Tirana, Tirana, Albania
exhafka@fim.edu.al

[2] DBS Group, Tirana, Albania

Abstract. The most difficult aspect of production is consistently generating high-quality products. Improving product quality is critical to satisfy consumers, minimize production costs, and conserve energy. Improving product quality requires intentional and ongoing attention and devotion from every individual inside the production organization. Producing a high-quality product is essential for survival in this globally competitive sector. In this paper, the application of two TQM techniques in a garment factory will show an important role in preserving garment quality. Stitching is a critical process in the garment and textile industries. As a result, any defects must be avoided when stitching. This research looks at many faults or defects that can occur when sewing and potential tactics for re-solving these concerns. According to the Pareto Chart, three problems accounted for more than 80% of all errors in the DBS garment factory's sewing area for an item. Following identifying major defects, a brainstorming tool was utilized to generate possible explanations, and probable root causes were identified using online inspections and root cause analysis. The Taguchi approach is extremely effective in defect reduction. Using these strategies regularly will help the company reduce defect rates while enhancing productivity. While this study focuses just on sewing, an integrated evaluation of other production processes like knitting, dyeing, cutting, and finishing may be more beneficial in reducing errors. Many Albanian textile firms should be better aware of Six Sigma concepts, and this study will trigger a good surge throughout the garment sector, making them more competitive.

Keywords: Cause and effect diagram · Pareto chart · Efficiency · Textile industry

1 First Section

1.1 Introduction

Many Albanian firms have started to implement Total Quality Management Methodology in recent years, and there have been numerous successful examples. To remain competitive in the global market, a company must have a quality system that consistently

G. Guxho et al. (Eds.): ITC-ICEE 2023, LNMIE, pp. 45–55, 2024.
https://doi.org/10.1007/978-3-031-48933-4_5

improves quality and productivity. Many businesses in Albania have recently committed to using best operating practices and quality tools in order to decrease waste, improve quality, and boost operational standards. The textile and garment industry represents about 39.9% of total exports (2018) and more than 40% of the labor force in the processing industry [1]. Global clothing marketplaces operate with competitive pricing and the lowest lead time feasible.

The performance of the textile business is determined by how efficiently are implemented the methods to prevent defect reduction. The fundamental advantage of various TQM methodology is that it eliminates subjectivity in decision-making by establishing a system in which everyone in the business gathers, analyzes, and presents data consistently. The immediate benefit of the Pareto chart and cause and effect diagram for manufacturing organizations is a reduction in the number of faults as a result of enhanced production processes. The high-quality level is a measure of the process defect rate for these organizations and can thus be utilized to evaluate the quality of the manufacturing process. Automatic production control is an important step to ensure quality and the main reason is to prevent the delivery of defective products to the customer. In the textile industry, since all operations are handmade, high productivity and quality can be achieved only by intensive quality inspection before and between the manufacturing stages. The high production speed and the large flexibility required by customers induce automated defect detection for the quality assurance system. In the traditional textile industry, this task is carried out by human visual inspection [2]. Due to the small scale of defects, and the enormous scale of inspected surfaces, manual inspection is typically a tough process. In the textile industry, inspection occurs at the conclusion of the production process. Large batches of cloth rolls are manually examined, and actions are carried out outside of the manufacturing system. Using computer vision automation directly in the manufacturing phase will increase the online responsiveness of the production workforce and minimize the number of defects.

This article discusses numerous critical components of quality control in the textile sector using the Ishikawa diagram and the Pareto chart. The general objective is to investigate the defects in the sewing section and propose a solution to minimize the defects. Some other specific objectives of this work are:

- To identify and measure the frequency of occurrence of critical defects.
- To assess the contributing factors for each defect type.
- To maintain improvements using appropriate control charts.

1.2 Literature Review

The Pareto chart is useful for non-numeric data, 'type', or classification, and is useful to prioritize where action and process changes. It uses bar diagrams to sort problems based on frequency, severity, nature, or source and displays the size of problems that are vital. It ranks the data from highest frequency to lowest frequency of occurrences. The principle was developed by Vilfredo Pareto, an Italian economist, and sociologist who conducted a study in Europe in the early 1900s on wealth and poverty. It is a universal principle of the 'vital few and trivial many'. According to this principle, the 80/20 rule has been formulated with the following meaning. For many phenomena, 80% of the consequences originate from 20% of the causes.

Cause-and-effect diagrams show the causes of a specific event. Since its inception, the cause-and-effect diagram has gained enormous popularity for identifying the root causes of a variety of problems. A cause-and-effect diagram, often called a 'fishbone' diagram, can help in brainstorming to identify possible causes of a problem and in sorting ideas into useful categories. It is a more structured approach than some other tools available for brainstorming the causes of a problem [3].

The scope of this study is to focus on minimizing the defects in the sewing section so that the overall operations of the company affect the rework and at the same time, the product quality is under control.

1.3 Methodology

The research methodology adopted in this study is case study and brainstorming. The case study was conducted on a garment company located in Tirana, Albania. At first, a preliminary investigation was carried out at the cutting, printing, sewing, finishing, and packing sections to identify the area where most of the defects occurred.

1.3.1 Pareto Diagram

A Pareto Analysis is a method that investigates the sources of a problem in any organization so that it may be studied and improved. It is based on the 80/20 rule, which states that 80% of the result originates from 20% of the inputs. It is a statistical method in which we identify the few components that may cause defects in manufacturing. Since there are several components that might influence the process's output, we can sort out some of the key elements that contribute the most to the errors using Pareto analysis. [4, 5].

1.3.2 Fishbone Diagram

The fishbone chart provides comprehensive information on all probable causes, allowing you to identify the root source of the problem.

The most significant advantage of this method is that it gives a clear knowledge of the problem, where it came from, and how much the problem influences the result. It additionally provides effective ways to reduce such fundamental causes to a certain extent [6].

Secondary data on the sewing section was collected. Only one model's data was obtained. We identified expected repeated defects in the sewing area based on our observations and the end-line quality data given by management. The information and data obtained had been arranged so that additional research and analysis could be carried out. In the analytical section, Pareto analytical was applied. The top-occurring defects were identified using Pareto analysis.

2 Case Study

The main objective is the identification of problems and the proposal of possible solutions. The effects are recorded daily, and the data collected provides information about critical points in the processes, not only in production but also in pre-production. Let's address the most frequent defects, the initiatives undertaken to reduce the frequency of these defects, and the timely response after the implementation of these initiatives.

Cutting is the first working department of garment production. The cutting process is done automatically, there is no need to control the shapes. Only if the graphics are not done right, the quality is not good. Textile printing is the process of applying color to fabric in definite patterns or designs. In properly printed fabrics the color is bonded with the fiber, to resist washing and friction. The quality of this process depends on all the parameters of the machine and the color. Packing is part of garment finishing. It is done by the requirement of the buyer. In this process is important to apply all defined steps. In this way, the percentage of making mistakes is very low.

After receiving all the components of the garment's parts from the cutting process, those parts are linked by the sewing machine. In this section manufacturers finally get a complete appearance of garments; therefore, the sewing Section is a very important department in the garment industry.

The result was that the sewing section suffered from defects and rework problems. For this reason, a sewing line is identified to conduct research work. The aim of the work is to minimize the defect percentage (Fig. 1; Table 1).

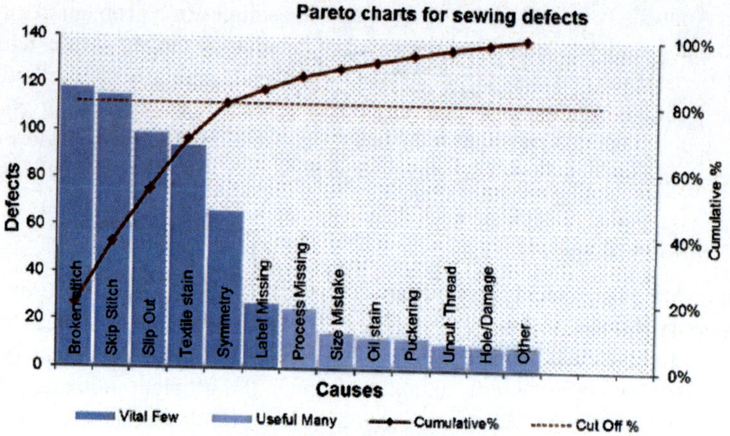

Fig. 1. Pareto charts for sewing defects.

The chart visually shows which defects are more significant. By using Pareto Chart major sewing defects were identified (i) broken stitch, (ii) Skip stitch, (iii) Slip out, only these 3 defects are responsible for 53.9% of the total defect percentage. Observations from the Pareto Analysis broken stitch is the most frequent defect type with 19.2%. Among other defect types, the contribution of skip stitch is 18.6% and broken stitch is 16%. From the collected data, the most common issue encountered is the problem typology of the tiles. To reduce this issue, significant investments have been made in new machinery, which has resulted in fewer problems. The following steps have been taken:

Table 1. Causes of the defects in the sewing process.

#	Causes	Defects	Cumulative (%)
1	Broken stitch	118	19.2
2	Skip stitch	115	37.8
3	Slip out	99	53.9
4	Textile stain	94	69.2
5	Symmetry	66	79.9
6	Label missing	27	84.3
7	Process missing	25	88.3
8	Size mistake	15	90.7
9	Oil stain	13	92.9
10	Puckering	13	95.0
11	Uncut thread	11	96.8
12	Hole/damage	10	98.4
13	Other	10	100.0

1. The quality control department has been separated, and it now performs inspections on specific production lines without interrupting the final product.
2. A training center has been established to educate operators on new processes and to improve their skills on existing ones.
3. An industrial engineering department has been added to exercise detailed control over the production lines and to have a more precise understanding of the workers' knowledge and abilities.
4. A skill matrix has been implemented to gather accurate information about each worker's abilities in the database system.

All these interventions were carried out within the same time period and are interconnected, resulting in a restructuring of the production system.

Quality control has been strengthened in the pre-production departments, with individuals responsible for quality from the quality control department.

The methods were implemented in early May 2021, but to ensure more accurate results, data from June 2021 were considered, providing information for a 6-month period.

As follows are shown The data on repairs during the implementation of these interventions in two lines: LINE 10 and LINE 14 (Tables 2 and 3).

An important intervention that has been implemented is the establishment of the logistics department and the adoption of the batch method on the production lines.

Here's a brief explanation of how the batch method and the logistics department work in specific production lines:

In this procedure, first, a basket (batch) is prepared with all the components needed to produce a particular product. The number of pieces to be included in the basket for each

Table 2. Line No. 10.

No. of repairs	Production quantity	Repairs (%)	Month
380	10800	**3.51**	June
260	11685	*2.22*	July
206	12004	*1.72*	August
175	10535	*1.66*	September
185	10670	*1.73*	October
178	10864	*1.63*	November
165	11013	*1.46*	December

Table 3. Line No. 14.

No. of repairs	Production quantity	Repairs %	Month
519	16324	**3.18**	June
395	16704	*2.36*	July
245	18865	*1.30*	August
177	15346	*1.15*	September
184	15362	*1.19*	October
189	16845	*1.12*	November
194	18375	*1.05*	December

component of the product is predetermined based on the production minutes required for that product. The basket is then taken to the first worker, who assembles and processes the assigned package. After completing the entire package, it is returned to the same basket, and the basket is then passed on to the next worker. The same process is repeated until all the products in the basket are completed (Fig. 2).

Fig. 2. Batch basket containing all the components of a product that need to be assembled

Let's look at the impact that the batch method has had on increasing efficiency and reducing the number of repairs, considering two production lines.

Line 17

The batch method was implemented in Line 17 on 01/06/2021 (one month after the line was created). If we observe the line's efficiency for the first month, we notice that it was at a low level (this is due to the lack of experience in the line). However, starting from the following month, we noticed that the line has stabilized at an efficiency level of around 60%, which is a satisfactory efficiency for a new and inexperienced line (Table 4).

Table 4. Repairs in % line No. 17 from June to December 2021

No. of repairs	Production quantity	Repairs (%)	Month
735	12574	**5.85**	June
311	10852	*2.87*	July
108	5200	*2.08*	August
314	19292	*1.63*	September
172	16241	*1.06*	October
200	7384	*2.71*	November
170	9465	*1.80*	December

Line 12

The implementation of the batch method in Line 12 was done on 12/06/2021. We observe from the efficiency graph of Line 12 throughout the year 2021 that there is a rising trend. However, if we focus on the first half of 2021, we can say that the efficiency levels were not satisfactory. The period from January to June 2021 had an average efficiency level of 55.2%. On the other hand, if we look at the last quarter (after the implementation of the batch method), the average efficiency level is 74.8%. Therefore, we can conclude that there has been a considerable increase in efficiency (19.6%) (Table 5).

Fishbone Diagram

After analyzing the collected data of both lines, we continued with the creation of the fishbone diagram. If we can identify all the causes of a problem, we may quickly minimize or eliminate it by taking steps to mitigate defects. When creating a fishbone diagram, it is necessary to involve the proper quality control team and production team, since the problem occurs in that zone. A fishbone diagram is required for each individual defect in a line of a garment manufacturing industry. In our study, we have drawn just one fishbone diagram regarding line no 17. The reason is that both lines presented the same defects typologies.

The result was that the sewing section suffered from defects and rework problems. For this reason, a sewing line is identified to conduct research work. The aim of the work is to minimize the defect percentage.

Table 5. Repairs in % line No. 12 from June to December 2021.

No of repairs	Production quantity	Repairs (%)	Month
537	9056	**5.93**	June
492	12792	**3.85**	July
308	8498	**3.62**	August
519	16638	**3.12**	September
329	14703	2.24	October
198	7230	2.74	November
98	6655	1.47	December

An action plan containing proposed solutions was prepared after analyzing the most significant root causes by applying the Pareto chart and fishbone diagram. For collecting data, the action plan required several months of observations. The Pareto diagram was created using data obtained in 2021 and 2022. The Pareto diagram illustrates that there was some improvement in sewing, cutting, textile printing, and packing activities (Fig. 3).

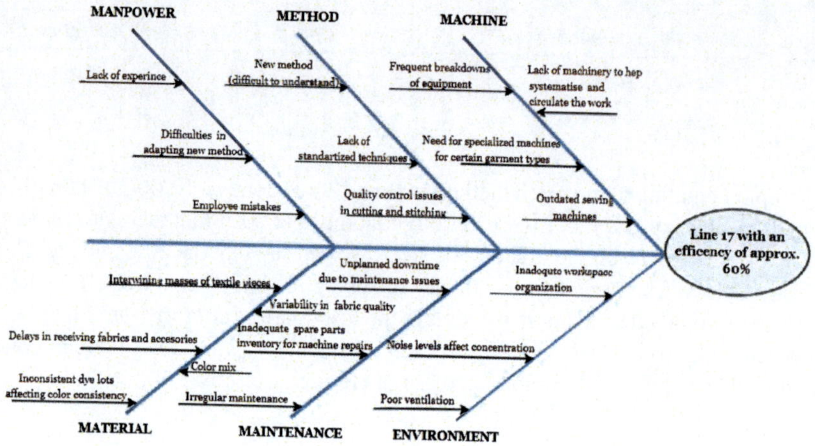

Fig. 3. Cause and effect diagram for line no 17.

Based on the reorganization at both administrative and operational levels (as mentioned above), we will observe a change in the level of repairs in the last 6 months of 2021 and in the year 2022.

After the reorganization and implementation of the batch method, it is expected to have an impact on reducing the number of defects and failures in production. In this case, we will see that in the last 6 months of 2021 and throughout the year 2022, the number of repairs is expected to decrease compared to previous periods.

This change is expected to result from better organization of the production lines, the use of the batch method to avoid material mixing and the application of quality control strategies.

It can be anticipated that the introduction of new production techniques and the implementation of quality control methods, which interact with each other, will have a positive effect on improving production performance and reducing expenses for repairs.

Furthermore, to have a more accurate assessment of the impact of the reorganization and the newly implemented methods, monitoring and analysis of information should be carried out, and comparisons with previous periods should be made to evaluate the effectiveness of the changes and their impact on productivity and product quality (Fig. 4).

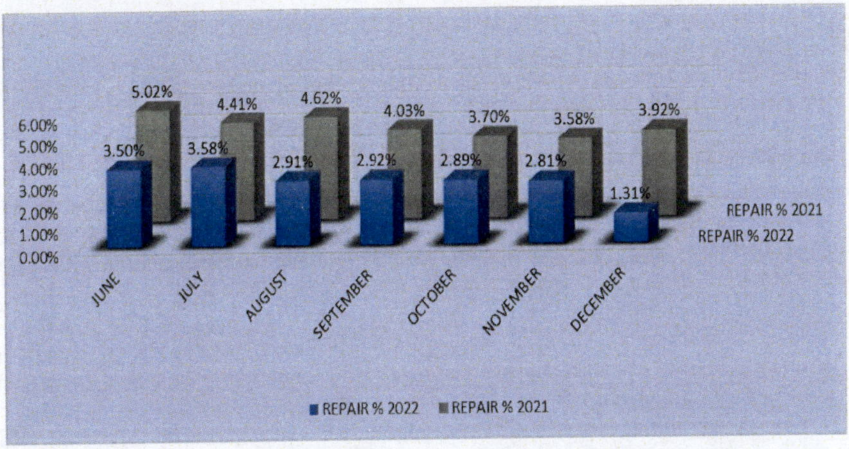

Fig. 4. Comparison of repair (%) from June to December (2021–2022).

The number of production lines has increased, and an Industrial Engineer is assigned to handle fixed and urgent needs that the manufacturing lines face. An engineer will manage 5 production lines.

In addition to this, appropriately trained mechanics have been made available to the production lines (and not only in pre-production) to help the lines avoid all obstacles in production.

3 Conclusions

The most challenging aspect of production is consistently generating a high-quality product. Nonetheless, it is critical to concentrate on increasing product quality to satisfy customers, reduce production costs, and conserve energy. Improving product quality is not difficult in general, but it does necessitate ongoing and consistent attention, as well as the commitment of everyone in the manufacturing company, from the person who cleans the garment to the operator of the sewing machine, which is one of the main processes. It is essential for the maintenance of product quality.

In this paper, various defects that can occur during sewing are discussed. Minimizing defects is very important for ensuring the quality of a product. From the Pareto Chart, three defects were identified which were responsible for more than 50% of total defects occurring in the sewing section. As the minimization of defects is a continuous process further implementation of this method will help the company enjoy a greater reduction in defect rate and improvement in productivity. This study is concentrated only on the sewing section, an integrated study of other sections of the factory such as knitting, dying, cutting, and finishing can also be necessary to minimize the defects.

References

1. https://aida.gov.al/images/PDF/Publikime/TEXTILE-Sector.pdf. Last accessed 21 Aug 2023
2. Xhafka, E., Teta, J., Vasili, E. : Quality control in the textile industry. The most common quality issues in this sector. In: 1st International Conference "Engineering and Entrepreneurship" Proceedings. ICEE-2017 (2017)
3. Perera, A.A.A.H.E., Navaratne, S.B.: Application of Pareto principle and Fishbone diagram for waste management in a powder filling process (2016)
4. Kumar, R., Prajapati, D.R., Singh, S.: Implementation of Taguchi methodology for defect reduction in manufacturing industry "a case study". Int. J. Ind. Eng. Res. Dev. (IJIERD) **2**(1) (2011). ISSN 0976–6979(Print), ISSN 0976–6987 (Online) (2011), © IAEME
5. Ziarati, R.: Safety at sea–applying Pareto analysis. In: Proceedings of World Maritime Technology Conference (WMTC 06), Queen Elizabeth Conference Centre, vol. 94 (2006)
6. Hossen, J., Ahmad, N., Ali, S.M.: An application of Pareto analysis and cause-and-effect diagram (CED) to examine stoppage losses: a textile case from Bangladesh. J. Text. Inst. (2017). https://doi.org/10.1080/00405000.2017.1308786
7. Siwiec, D., Gawlik, R., Pacana, A.: Fuzzy multi-criteria decision model to support product quality improvement. Manag. Prod. Eng. Rev. **14**(2), 134–149 (2023). https://doi.org/10.24425/mper.2023.146030
8. Pacana, A., Siwiec, D.: Universal model to support the quality improvement of industrial products. Materials **14** (2021). https://doi.org/10.3390/ma14247872
9. Banduka, N., Veža, I., Bilić, B.: An integrated lean approach to process failure mode and effect analysis (PFMEA): a case study from automotive industry. Adv. Prod. Eng. Manag. **11**(4), 355–365 (2016). https://doi.org/10.14743/apem2016.4.233
10. Omega, R.S., Noel, V.M., Masbad, J.G., Ocampo, L.A.: Modelling supply risks in interdependent manufacturing systems: a case study. Adv. Prod. Eng. Manag. **11**(2), 115–125 (2016). https://doi.org/10.14743/apem2016.2
11. Radej, B., Drnovšek, J., Begeš, G.: An overview and evaluation of quality-improvement methods from the manufacturing and supply-chain perspective. Adv. Prod. Eng. Manag. **12**(4), 388–400 (2017). https://doi.org/10.14743/apem2017.4.266, ISSN 1854-6250
12. Abtew, M.A., Kropi, S., Hong, Y., Pu, L.: Implementation of statistical process control (SPC) in the sewing section of garment industry for quality improvement. Autex Res. J. **18**(2), 160–172 (2018). https://doi.org/10.1515/aut-2017-0034
13. Black, S.E.: Using basic quality tools to improve production yields and product quality in manufacturing. M.Sc. Theses, East Tennessee State University, Johnson City, USA (2015)
14. DTI: Tools and techniques for process improvement. Department of trade and industry, London, UK (2020)
15. Singh, J., Singh, H.: Continuous improvement philosophy–literature review and directions. Benchmark. Int. J. **22**(1), 75–119 (2015). https://doi.org/10.1108/BIJ-06-2012-0038

16. Sahno, J., Shevtshenko, E.: Quality improvement methodologies for continuous improvement of production processes and product quality and their evolution. In: The 9th International DAAAM Baltic Conference "Industrial Engineering, Tallinn, Estonia, pp. 181–186 (2014)

Case Circular Economy of Serbia

Snežana Urošević[1](✉), Igor Urošević[2], and Milovan Vuković[1]

[1] Technical Faculty in Bor, University of Belgrade, Belgrade, Serbia
snezanaur1@gmail.com

[2] Faculty of Architecture, University of Belgrade, Belgrade, Serbia

Abstract. The circular economy is an industrial system that replaces the linear concept, which deals with reparation, is based on renewable energies, eliminates the use of toxic chemicals, reduces carbon emissions and aims to eliminate waste through superior design of materials, products and systems. By moving away from the current linear towards a circular economy, where at the end of a product's life, the materials used to produce it are reused wherever possible, a more stable economy is created. Moving away from the current linear towards a circular economy, in which at the end of a product's life, the materials used to manufacture it are reused wherever possible, thus creating a more stable economy in the process. The application of the circular economy in practice brings a number of benefits at the national level in the field of more efficient exploitation of resources (this reduces the total costs and import dependence of the country), reduces the harmful effects of damage to the environment, creates new opportunities for creating added value and creates new jobs and sustainable forms of consumer behavior are introduced. The concept of the circular economy is becoming more and more important for individual corporations and their operations—a number of global corporations are implementing measures in accordance with the basic principles of the circular economy in order to improve their competitive position. The development policy in Serbia has recognized the circular economy and the application of the circular (green) economy concept as a possibility for the future development of the Serbian economy.

Keywords: Serbia · Textile industry · Circular economy

1 Introduction

Although known since ancient times, the concept of circular economy came into the focus of interest of the academic and professional public only in the 1970s. The concept was developed as a response to the current traditional concept of the economy, which implies a one-way direction of unlimited growth based on the "take-make-use-dispose" model, which results in inefficient wasteful consumption of resources and overall pollution of air, water and soil.

The circular economy, transitions to the use of renewable energy, eliminates the use of toxic chemicals, reduces carbon emissions and aims to eliminate waste through superior design of materials, products and systems. Moving away from the current linear

G. Guxho et al. (Eds.): ITC-ICEE 2023, LNMIE, pp. 56–64, 2024.
https://doi.org/10.1007/978-3-031-48933-4_6

towards a circular economy, in which at the end of a product's life, the materials used to manufacture it are kept in the economy and reused wherever possible, thus creating a more stable economy in the process.

Circular economy is based on the idea that waste does not exist as a thing. To achieve this, products are designed to last (good quality materials are used) and are optimized for a dis-assembly and reuse cycle that will facilitate handling and transformation or renewal. Ultimately, these product cycles differentiate the circular economy model from disposal and recycling, where large amounts of energy and labor are wasted.

The ultimate goal is to preserve and improve natural capital by controlling limited supplies and balancing the flow of renewable resources. The circular economy model distinguishes between technical and biological cycles. Consumption only occurs in biological cycles, where biological materials are designed to be returned to the system through processes such as composting. These cycles regenerate living systems, such as soil or oceans, which pro-vide renewable resources for the economy.

In turn, technical cycles renew products, components and materials through strategies such as reuse, repair, remanufacturing or recycling. The last principle of the circular economy has to do with the fact that the energy needed to start this cycle should be renewable in nature, with the aim of reducing dependence on resources and increasing the resilience of the system. This principle refers to developing system efficiency by discovering and designing new strategies to initiate this cycle. The main driver of the circular economy is the aware-ness of the climate crisis caused by climate change. According to UN data, the extraction and processing of materials, fuel and food account for approximately 50% of total green-house gas emissions, and cause 90% of water consumption and biodiversity loss.

The Republic of Serbia has defined accession to the European Union through full EU membership as one of its key strategic priorities and the country's national and political goals [1]. In order to achieve this priority goal, it is necessary to harmonize with European legislation and standards. The area of the circular economy is particularly important for closing the negotiation Chap. 27 as one of the most complex chapters that includes environmental issues in the broadest sense. Serbia is expected to harmonize its policies with the Sustainable Development Agenda until 2030 and other important documents that regulate this area. Also, in addition to the creation of institutional and political-legal assumptions, the practical application of the adopted mechanisms of control and monitoring of activities on these issues is necessary [2].

A total of five priority sectors have been determined for the implementation of the circular economy in Serbia in the coming period. Those are:

1. Manufacturing industry,
2. Agriculture and food,
3. Surplus food and food waste,
4. Plastic and packaging,
5. Construction [2].

The circular economy development program in the Republic of Serbia for the period 2022 to 2024 defines priorities and sets the basis for further development in that area and the creation of a stimulating environment for the development of the green transition in Serbia. The general goal of the program refers to sustainable development and climate neutrality in order to preserve people's health and raise citizens' awareness of the importance of environmental protection, according to the announcement of the Government of Serbia. One of the specific goals includes supporting companies to remove waste from supply chains, recycle materials and components, switch to renewable energy sources and extend product life.

The principles of circular economy are also recognized in the Strategy of Sustainable Urban Development until 2030. Measures to achieve the goals of urban development include measures to mitigate climate change by improving the quality of all environmental parameters, waste management systems and measures to improve energy efficiency.

2 Concept of Circular Economy

The concept of circular economy is based on the statement that the use of resources in production is reduced and thereby maximizes their utilization, then to minimize waste material that can be reused, so that at the end of the used cycle the product is returned to the production process and in this way creates new value. Today, the People's Republic of China and the European Union have encouraged the development of innovations that enable the implementation of the circular economy concept at the national level [3].

The circular economy is the opposite of the linear model of the economy, which implies the uncontrolled exploitation of natural resources and the flow of materials from the factory through the user to the landfill [4].

There are significant benefits from moving towards a circular economy, even a marginal in-crease in re-use and reproduction would save tons of resources and millions of dollars, while reducing waste across Europe would reduce €72 million annually and create 400,000 new jobs [4].

The industries of repair and remanufacturing of products and resources also have a lot of potential because it would enable the growth of the economy by several billion dollars, and enable a new life and purpose of certain resources, for example copper. The need to change and redirect the concept of the global economy has become inevitable. The examples show that some of the countries have already adopted strategies for the implementation of the circular economy, while other countries are still implementing some individual programs to-wards the circular economy [5].

The key elements of the circular economy are:

(1) giving priority to the regeneration of resources,
(2) saving and consuming what has already been made,
(3) using waste as a resource,
(4) reconsidering the business model,
(5) designing for the future,
(6) inclusion of digital technology,
(7) cooperation to create common value [6].

The three key pillars of the circular economy strategy are:

(1) innovation as the heart of economic growth,
(2) infrastructure that promotes national prosperity,

(3) information is of strategic importance for understanding the state of the economy [7].

These three pillars as the basis of green industry, a concept that promotes sustainable production and consumption [8], imply an interdisciplinary approach to the application of the circular economy concept. It is necessary to engage professionals from different fields to work on the same project of introducing, implementing, monitoring and reporting the effects of applying the circular economy concept.

Advantages of circular economy [9]:

(1) creating a safe and sustainable supply of industry raw materials,
(2) reducing the increasing dependence of the country on imports,
(3) creation of new jobs,
(4) support for innovation and increased competitiveness,
(5) cost savings,
(6) increasing the value of locally extracted resources, manufactured products and services.

It is estimated that by introducing the circular economy in the Republic of Serbia, 30,000 new jobs can be created, which would increase the competitive advantage of the country's economy and make it a leader in development and investments [5].

3 Circularity of the Textile Industry in Serbia

The disastrous consequences that fast fashion has on the environment, climate and society are due to the uncontrolled consumption, wastage and pollution of non-renewable resources. In order to reduce those consequences, it is necessary for the textile industry to be part of the circular economy, where as such it will have the ability to extend the life of garments, keeping resources in a closed circle. By realizing the goals of sustainable development, the circular textile industry should increase the recycling and reuse of textiles, and create less pollution during the production of textile fibers, as well as reduce waste during production to a minimum [10].

Today, there is an even greater need to reduce the harmful impact of the textile industry on the environment, and it is necessary to draw a clear line between man and nature. That is why it is necessary for the modern consumer society, which approves of the rapid change of fashion trends, to change and to reduce its interests only to the satisfaction of existential needs. At the same time, the transition from a linear model to a circular model requires the involvement of all market participants, from producers, through technology and product designers to consumers, but their knowledge and deepened awareness is also necessary [11].

A necessary condition for the sustainability of the fashion sector lies in the fact that the items must be of high quality and durability as if they were made from materials that can be recycled, which would support every aspect of the circular model [12].

A four-level system that shows a good model for the textile sector would look like this:

- Level I is designing for durability, where the focus is on the consumer, and the goal is to extend the life of the product,

- Level II is designing for a service that includes new products of brands and companies,
- Level III is the design of products for reuse, the aim of which is to extend the up-use of the product,
- Level IV is designing for the recovery of materials based on recycling, which textile waste uses for the production of new fibers and yarns [13].

This four-level model includes all market participants and points to the necessary collaboration of designers, researchers, manufacturers, users and policy makers [13]. The circular model also requires the application of ecological innovations and refers to the laws that prescribe the production and consumption of products [14].

Like the textile industry in the world, the textile industry of Serbia represents a very complex type of production that requires sustainable, ecological, technical, economic and social development based on the association of all aspects of life, especially the correlation of sustainable production and environmental protection as the foundation of sustainable development.

By comparing this sector today and the time of socialist Yugoslavia in relation to the number of employees in the textile sector, a large drop in the number of employees can be observed. At that time, this sector employed close to 40% of workers, and today it is slightly more than 10% of all employed, which indicates the small importance of this sector in the modern economy of Serbia. Given that the textile industry is a labor-intensive industry and employs mostly female labor, it is necessary to restore the former importance of this economic industry for the economy of underdeveloped regions.

Data that shows that the amount of 80,000 tons of clothes purchased in Serbia is very often of very dubious quality and composition, mostly at a favorable price because it is made of cheap materials, usually polyester. Like all products made of plastic, bottles and bags, clothing made of plastic causes great damage to the environment. Such clothes bought at lower prices are short lived and as such very quickly end up in landfills, so it is a very significant fact that annually around 235 billion tons of clothes are disposed of in landfills around the world [15].

It is very devastating data that shows that in Serbia only 60 t of textile waste is recycled out of the total amount of recycled waste in Serbia (455,457 tons), and that 81,405 t of collected textile waste out of a total of 2,947,496 t end up in landfills or are burned. This means that only 61 t refers to the collection of textile waste, of which 60 t is recycled. In relation to the total amount of municipal waste, textile waste makes up only 2.8% [16]. One can see the great value of lost assets that could be the basis of a profitable textile circular economy. Much greater collection of textile waste and its recycling could, in addition to economic effects, lead to changes in the textile sector and to the creation of new jobs in those jobs in the textile sector.

One of the goals of this paper was to present the possibility of employment of workers who would work in textile recycling. The transition from a linear economy to a circular economy requires the acquisition of a new cleaner technology that will not pollute the environment, and the goal of the textile industry based on circularity is that in addition to optimizing re-source consumption, it must also minimize high environmental costs [19, 20].

During the analysis of companies in Serbia dealing with the disposal of textile waste, including recycling, it was found that only 6 companies collect cotton textile waste and the only product that is obtained from recycling are industrial cloths used for cleaning and wiping hands, machines and floors in various manufacturing sectors [21]. Also, through recycling, yarn can be obtained for reuse in new products, an example is polyester textiles, and the obtained recycled textiles are used to fill mattresses, furniture, etc. [22].

None of the surveyed companies have the capacity to recycle high-quality fibers that would be used to obtain recycled clothing, and the reasons are that the clothing that would be recycled is not designed to be recycled, but also that the recycled material in most cases can cost more than non-recycled [23].

The formation of a database and the exchange of data on the collection and recycling of textiles in Serbia is very important for circularity and the creation of new jobs, and Serbia is only at the beginning of the formation of such data [24].

The European Environment Agency (EEA) announced that during the production of clothing, footwear and textiles for households, more than 100 m^3 of water is consumed per in-habitant [25]. Next, it is interesting to note that textiles are the fifth largest source of CO_2 emissions, estimated at 654 kg of CO_2 equivalent per EU resident, and about three quarters of these emissions are realized outside the EU.

The presented data show that textile waste is one of the biggest problems in the Republic of Serbia because there is no infrastructure for the separate collection and selection of this waste [16]. It has already been mentioned that textile waste, which includes both clothes and textiles, as biodegradable waste, has a negative impact on the environment by being dumped in landfills. However, the entire amount of textile waste is not biodegradable because only natural textiles are biodegradable, which indicates an even greater negative impact on the environment.

According to the data of the Development Agency of Serbia, which refers only to the data of textile and clothing manufacturers and does not refer to those engaged in textile recycling, 61.9 thousand workers are employed in the textile industry of Serbia (which is 2.8% of total employment), while the majority of employees are in the production of clothing (36.2 thousand). In the textile industry in 2021, there were 1.9 thousand active companies and 6.2 thousand entrepreneurs, which is a total of close to 8.1 thousand [28].

Based on the data, it can be concluded that there is potential for the creation of new jobs related to the disposal of textile waste as well as jobs in the recycling process [29, 30]. Application of the circular economy model in which waste is viewed as a strategic raw material is one of the ways to ensure the long-term supply of raw materials in the textile sector [31]. Research that included 22 companies from Serbia from the textile sector showed that the application of the circular economy is a chance for the growth of this economic branch.

Losses on a global level are high due to insufficient recycling of textiles and this is shown by the data that 500 billion dollars are lost every year [32]. It is inevitable that the textile industry needs to behave differently because tons of textile waste end up in landfills in Serbia and around the world [33].

The main driver of the economy in Serbia should be workers who should have an attitude towards an eco-oriented ethos that ensures safe and healthy working conditions and regular pay and work without overtime and that all labor laws are applied to them [36]. By opening new jobs in textile recycling, employees should count on safe work, social and health benefits, as well as education, training, childcare, paid vacations, etc. [37].

4 Conclusion

By moving from a modern business from a linear to a circular business model, we face technological, structural and personnel changes and challenges. The circular economy of Serbia must strive to achieve sustainable development at the global level, which should rep-resent such an economic system that is profitable and that will enable consumers not to endanger the environment when using products. All these goals can be achieved by extending the life of the product, but also by recycling.

In order for the textile and clothing industry of Serbia to operate successfully and efficiently, it is first of all necessary to create a new raw material base, where special attention should be paid to the economic and ecological aspects of efficient operations, that is, there should be a concept of cost-effective processing of textile waste. In the context of the circular economy model in the textile sector, special emphasis must be placed on the creation of new jobs that arise in the process of textile recycling.

Improving the business models of the economy and harmonizing business with the principles of the circular economy in the Republic of Serbia can greatly contribute to improving the competitiveness of national companies and solving social and economic issues, especially in the textile sector.

References

1. Arsić, L.J., Premović J.: Institucionalni okvir za razvoj modela cirkularne ekonomije u Srbij, Zbornik radova: Ekonomska politika u Srbiji i svetu u 2021: U susret globalnim šokovima i rastućoj neizvesnosti, pp. 155–167 (2021)
2. Ministarstvo zaštite životne sredine Republike Srbije, Mapa puta za cirkularnu ekonomiju u Srbiji, p. 27 (2020)
3. Radivojević, A.: Cirkularna ekonomija implementacija i primena tehnologije u njenoj funkciji (Circular economy implementation and technology application in its function). Ekonomske ideje i praksa **28**, 33–46 (2018)
4. Vićentijević, K.: Accounting aspects of circular economy in the textile industry. Tekstilna industrija **66**(4), 79–84 (2018)
5. OEBS: Cirkularna ekonomija kao šansa za razvoj Srbije (2017). Misija OEBS-a u Srbiji, Available: https://www.osce.org/sr/serbia/292311?download=true
6. Circle economy: Key Elements of the Circular Economy, Draft v1. February 2021. Available: https://assets.website-files.com/5d26d80e8836af2d12ed1269/601d3f846c512412fff633af_Key%20Elements%20-%20Draft%20Literature%20Review%20.pdf
7. Ekins, P., McDowall, W., Zenghelis, D.: Greening the Recovery—The report of the UCL Green Economy Policy Commission, London's Global University (2014). Availa-ble: https://www.lse.ac.uk/granthaminstitute/publication/greening-the-recovery-the-report-of-the-ucl-green-economy-policy-commission/
8. Pokrajac, S., Josipović S.: Zelena industrija i proces reindustrijalizacije. Ekonomski vidici **XX**(2–3), 203–212 (2015)
9. PKS Centar za cirkularnu ekonomiju. Available: https://pks.rs/udruzenje/sluzba-za-cirkularnu-ekonomiju

10. Kovačević, L.J., Arsić, L.J.: Održivost modne industrije u vreme pandemije Covid-19. Ecologica **28**(102), 283–292 (2021)
11. Koszewska, M., Rahman, O., Dyczewski, B.: Circular fashion—consumers' attitudes in cross-national study: Poland and Canada. Autex Res. J. **20**(3), 327–337 (2020)
12. Ellen McArthur Foundation: A New Textiles Economy: Redesigning Fashion's Future. Available: https://ellenmacarthurfoundation.org/a-new-textiles-economy
13. RSA: Action and Research Centre, Designing for a circular economy: Lessons from the great recovery 2012–2016, London, UK (2016). Available: https://www.thersa.org/globalassets/pdfs/reports/the-great-recovery---designing-for-a-circular-economy.pdf
14. Machado, M.A.D., Almeida, de S.O., Bollick, L.C., Bragagnolo, G.: Second-hand fashion market: consumer role in circular economy. J. Fashion Mark. Manage. **23**(3), 382–395 (2019)
15. https://www.politika.rs/sr/clanak/467963/Beograd/Spora-moda-i-pogubni-uticaj-tekstilnog-otpada
16. Program upravljanja otpadom u Republici Srbiji za period 2022–2031. godine, Vlada Srbije, 2022. („Službeni glasnik RS", broj 30/18). Available: https://srda.rs/wp-content/uploads/2022/02/Program-upravljanja-otpadom-u-Republici-Srbiji-za-period-2022-2031.-godine.pdf
17. Queiroz, G.S., SádeAbreu, M.C., Rebouças, S.: Do responsible practices enhance employee organizational commitment? A study of brazilian textile companies. Revista de Administração de Empresas (J. Bus. Manage.) **62**(5), 1–22 (2021)
18. Zhuravleva, A., Aminoff, A.: Emerging partnerships between non-profit organizations and companies in reverse supply chains: enabling valorization of post-use textile. Int. J. Phys. Distrib. Logist. Manag. **51**(9), 978–998 (2021)
19. Muthu, S.S.: Circular Economy in Textiles and Apparel Processing, Manufacturing, and Design. Elsevier (2018)
20. Duff, A. Webster, K.: The Wonderful Circles of Oz: A Circular Economy Story, Routledge; 1st edition, (2022)
21. Warasthe R., Brandenburg, M., Seuring, S.: Sustainability, risk and performance in textile and apparel supply chains. Clean. Logist. Supply Chain 5 (2022)
22. Kazancoglu, I., Kazançoğlu, Y., Kahraman, E., Yarimoglu, A., G. Soni, G.: Investigating barriers to circular supply chain in the textile industry from stakeholders' perspective. Int. J. Logist. Res. Appl. **25**, 521–548 (2020)
23. Sandvik, I.M.: Applying Circular Economy to the Fashion Industry in Scandinavia Through Textile-to-Textile Recycling, Master of International Development Practice. Monash University (2017)
24. Circle economy, Unwanted Clothes, Happy Workers: Exploring the Potential for Circular Textiles to Have a Positive Impact on Work and Workers [Online]. Available: https://www.circle-economy.com/blogs/unwanted-clothes-happy-workers-exploring-the-potential-for-circular-textiles-to-have-a-positive-impact-on-work-and-workers
25. EEA-European Environment Agency: Progress Towards Preventing Waste in Europe-The Case of Textile Waste Prevention (2021). [Online]. Available: https://encircular.es/en/recurso/progress-towards-preventing-waste-in-europe-the-case-of-textile-waste-prevention-european-environment-agency-dic-2021/
26. Polidori, L.: The Role of Circular Economy in the Oversaturated Fashion Industry. Business Models, Materials [Online]. Available: https://www.veltha.eu/blog/the-role-of-circular-economy-in-the-oversaturated-fashion-industry/
27. Citylife: Sustainable Fashion in the Nordics—Nordic Brands at the Forefront of the Circular Economy Concept [Online]. Available: https://www.citycon.com/citylife/sustainability/sustainable-fashion-in-the-nordics-nordic-brands-at-the-forefront-of-the-circular-economy-concept

28. Razvojna Agencija Srbije: Textile industry Shaping Quality [Online]. Available: https://ras.gov.rs/textile-industry; https://api.pks.rs/storage/assets/Tekstilna%20industrija%2020_08_2016.pdf
29. Dissanayake, K., Weerasinghe, D.: Towards circular economy in fashion: review of strategies, barriers and enablers. Circ. Econ. Sustain. **2**(1), 25–45 (2022)
30. Lövin, I., Baylan, I.: Circular Economy-Strategy for the Transition in Sweden [Online]. Available: https://faolex.fao.org/docs/pdf/swe208661.pdf
31. Juanga-Labayen, J.P., Labayen, I.V., Yuan, Q.: A review on textile recycling practices and challenges. Textiles **2**, 174–188 (2022)
32. Goldsmith, D.: The worn, the torn, the wearable: textile recycling in union square. Nordic Text. J. **1** (2012). Available: http://www.divaportal.org/smash/record.jsf?pid=diva2%3A869631&dswid=2423
33. Risteski, S., Srebrenkoska, V.: Tekstilni otpad iz konfekcijske industrije i mogućnosti za njegovu reciklažu. Tekstilna industrija **68**(4), 77–82 (2020)
34. Sandin, G., Peters, G.M.: Environmental impact of textile reuse and recycling—A review. J. Clean. Prod. **184**, 353–365 (2018)
35. Sushma, R., Zeba, J.: Recycling of textiles waste for environmental protection. Int. J. Home Sci. **4**(1), 164–168 (2018)
36. Bartl, A., Piribauer, B.: Textile recycling processes, state of the art and current developments: a mini review. Waste Manage. Res. **37**(2), 112–119 (2018)
37. Dimitrijević, D.: Teze za nove strategije razvoja tekstilne i odevne industrije, Deo I. Tekstilna industrija **4**, 65–76 (2020)

Evaluating the Criteria for Choosing a Clothing Brand Using Analytic Hierarchy Process

Elda Lamce[1](✉) and Alma Braimllari[2]

[1] Ismail Qemali" University, Vlore, Albania
elda.lamce@univlora.edu.al
[2] University of Tirana, Tirana, Albania
alma.spaho@unitir.edu.al

Abstract. The selection of a clothing brand has become increasingly complex due to the wide variety of options available in the market and evolving fashion trends. This paper presents an application of the Analytic Hierarchy Process (AHP) in choosing a clothing brand. The AHP methodology is used to identify the criteria and sub-criteria considered by consumers when evaluating a clothing brand, and the relative importance of these criteria and sub-criteria. Four criteria, eight sub-criteria, two for each criteria, and six clothing brands are considered in this study. The AHP results reveal that respondents prioritize quality and durability criteria, especially material quality, indicating consumers value the materials used in a brand's clothing. The other important criteria is convenience and shopping experience criteria, with physical shopping experience being the highest-valued sub-criteria. The findings of this study give insights to businesses about the market, for improving product development, enhancing customer experience, and increasing competitiveness. Moreover, customers benefit from these findings by making more informed decisions when choosing a clothing brand.

Keywords: Multi-criteria decision-making · Quality · Experience · Brand selection

1 Introduction

Multiple Criteria Decision Making (MCDM) is a branch of operations research that focuses on addressing decision problems involving multiple conflicting criteria [1]. One prominent and widely used technique within MCDM is the Analytic Hierarchy Process (AHP). AHP is a structured and systematic method that facilitates decision-makers in dealing with complex decisions by breaking them down into a hierarchical structure. It allows decision-makers to compare and prioritize criteria and alternatives using pairwise comparisons, quantifying their relative importance [2, 3]. By integrating AHP into the broader framework of MCDM, decision-makers can effectively handle multi-criteria decision problems and arrive at well-informed and rational choices that align with their objectives and preferences. The synergy between MCDM and AHP makes them powerful tools for addressing a wide range of real-world decision-making challenges across various domains and industries [4–8].

G. Guxho et al. (Eds.): ITC-ICEE 2023, LNMIE, pp. 65–74, 2024.
https://doi.org/10.1007/978-3-031-48933-4_7

The fashion industry is continuously evolving [8], with an abundance of clothing brands catering to diverse consumer preferences. In this era of wide-ranging choices, consumers often find themselves confronted with the challenge of selecting the most suitable clothing brand that aligns with their unique requirements, values, and aspirations. Understanding the significance of this decision-making process, researchers have employed various methodologies to assess the criteria influencing consumers' choices.

The purpose of this study is to determine the most important criteria and sub-criteria for choosing a clothing brand using the AHP method. The objectives of this research are: to determine the criteria and sub-criteria to choose a clothing brand, to assess the relative importance of the criteria and sub-criteria and to identify the most preferred clothing brand among customers. The results of this study give useful insights to businesses and customers.

The article is organized as below: after this introduction, a literature review on the application of AHP in the fashion industry and related fields. Next, a detailed description of the research methodology is presented. Then the findings and discussion are presented, and finally the conclusion section.

2 Literature Review

Recent literature shows that AHP is a powerful tool for addressing a wide range of real-world decision-making challenges across various domains and industries. Also the determination of the appropriate criteria and sub-criteria is based on the recent literature.

The results of the Fuzzy-AHP method in the study of [7] discovered that the factor most influenced the choice of apparel foreign products among 21 Indian consumers was product quality. Moreover, Aldilax and Pfoertsch [9] found that product quality and strong brand identity are the most important sub-criteria in their study. Findings of [10] showed that the factors that most influence online business are transaction security, trust, privacy risk and uncertainty of product quality and some of the least influential factors are communication and interaction, better prices and time saving. In [11] based on the results of the AHP method, was shown that when comparing online fashion stores to traditional ones, among most influential criteria for fashion products were: convenience, product variety, quality, service provided, price and the last one was accessibility. Service and product quality influence customer satisfaction and retain in the clothing brand [12]. The study conducted by Chang and Jai [13] found that perceived corporate social responsibility effort, price, and brand equity have a significant effect on consumer's sustainability purchase intention. The results of this study [14] indicated that the promotion and the product quality had a significant effect on the brand image which has a significant effect on purchasing decisions. Research of [15] used AHP to identify the most important non-conforming factors and sub-factors in India and it was found that cloth attribute variation is the topmost reason of non-conformance in apparel and within which fabric is the most significant reason of non-conformance. The results of [16] show that customers respond positively to individualization and exclusivity. In [17], the authors considered the quality, price and promotion of a product as important factors in the purchase decision. According to the study, product quality plays a greater role than price and promotion in forming purchase decisions. Referring to the above studies, we saw that quality and durability, convenience and shopping experience had an important weight, but we think that since our study is specific in the choice of clothing, criteria such as style and fashion or price and value would be of interest to our study. Moreover, in [18] authors discussed the impact of recommender systems and pricing strategies on brand competition and consumer search.

Table 1 summarizes the criteria and sub-criteria from the literature review.

Table 1. Criteria and sub-criteria for choosing a clothing brand.

Criteria	Sub-criteria	Description	References
Price and value	Value for money	Price of the clothing aligns with the perceived quality and overall value provided	[11, 13, 16]
	Discounts and promotions	Discounts, promotions or loyalty programs that provide additional value	[8, 14]
Style and fashion	Trendiness	Trendy and fashionable designs	[9, 12, 21]
	Individuality	Unique style and personality through clothing choices	[5, 8, 9, 21]
Quality and durability	Product quality	Durable and long-lasting products, including functionality and performance	[7, 9, 13, 16]
	Material quality	Quality of fabrics and materials used in clothing production	[12, 13, 15]
Convenience and shopping experience	Online shopping	Easy and convenient online shopping experiences, including user-friendly websites, fast shipping, and hassle-free returns	[9, 10, 18]
	Physical store experience	Enjoyable and personalized in-store shopping experience	[11, 12, 14, 20, 22]

Source Authors

The study [20] endeavours to utilize analytical hierarchy process (AHP) modelling to find the degree of importance that customers attach to different criteria of shopping experience; store environment, social environment and the purchase motives with their sub-criteria. The results evidently propagate social environment as the most important criteria followed by store environment. The key sub-criterion of social environment is customer interaction, which means that the customer is more concerned about the social interactions. The results of [21] showed that the most important components for customer loyalty to Iranian sportswear brand were price component followed by design and fashion, and product quality. Reference [22] utilized AHP to prioritize the factors that are important when making the decision towards liking or loving a particular apparel brand, from the perspective of young consumers. Also are evaluated three popular apparel brands amongst youngsters. The results indicated that brand experience is the most important factor.

3 Research Methodology

This study utilizes the AHP methodology to evaluate the criteria influencing consumers' decisions when selecting a clothing brand.

The research includes a wide range of fashion-conscious participants, from various age groups, demographics, and regions. The data are collected using a questionnaire from 10 April to 30 June 2023. The target population consisted of customers of clothing brands from 18 to 43 years old. A detailed explanation of criteria, sub-criteria and measurement scale was given in the questionnaire. A total of 300 questionnaires were distributed using convenience sampling to gather insights on various aspects of the participants' preferences and behaviors. Among the respondents, 90 individuals diligently filled in the questionnaire, resulting in a response rate of 30%.

According to the literature [19], there is no strict requirement on the minimum sample size for AHP analysis. Some researchers used sample sizes ranging from four to ten [4, 5, 20], some studies used samples of size of 20 to 30 [6, 7, 21], and some other researchers used sample sizes greater than 100 (such as [2, 11]). Several factors such as the nature and scope of the problem and availability of experts, must be considered when choosing the AHP sample size [19].

Each participant is asked to compare pairwise the selected criteria, sub-criteria and alternatives using the Saaty numerical scale (Table 2).

The Analytic Hierarchy Process, developed by Thomas Saaty in the 1970s, is a widely recognized multi-criteria decision-making technique used to solve complex problems by structuring them into a hierar-chical model. The process involves breaking down a decision into a series of elements and sub-elements, establishing their relative importance through pairwise comparisons, and calculating their priority weights to make informed choices. AHP is a systematic approach, but the final decision should also con-sider personal preferences and style.

Each participant was asked to assess the significance of multiple criteria including factors such as price and value, style and fashion, quality and durability, and convenience and shopping experience. Each criteria has two sub-criteria: value for money, discounts and promotions, for price criteria; trendiness and individuality for style criteria; product quality and material quality for quality criteria; and online shopping, and physical store experience for the last criteria which is convenience and shopping experience (Table 1).

The six clothing brands considered in this study are: brand 1 (Zara), brand 2 (H&M), brand 3 (Bershka), brand 4 (New Yorker), brand 5 (Intersport), and brand 6 (LC Waikiki).

Table 2. Saaty numerical scale.

Numerical rate	Definition and description
1	Equal importance: The alternatives are the same or equivalent in importance in the criteria and there are no significant differences between them
3	Weak importance of one over another: The alternative has littleadvantage in the analyzed criterion
5	Essential or strong importance: More relevant alternative in theanalyzed criteria, indisputable importance
7	Very strong importance: Alternative far above from theother in significance
9	Extreme importance: The alternative is irrelevant inview of the importance of this
2, 4, 6, 8	Intermediate values between two judgments: When the comparison is required
Reciprocal	Reciprocal reverse comparison: If activity i has one of the above nonzero numbers assigned to it when compared with activity j, then j has the reciprocal value when compared with i

Source References [6, 8]

These six brands and criteria and sub-criteria are selected using the results of two focus groups, one in Tirana and one in Vlora. The AHP model used in this study is shown in Fig. 1.

The AHP methodology provides a structured way to evaluate clothing options based on specific criteria, but ultimately, the choice should align with individual tastes and preferences.

The results are obtained by using the R software and the 'ahpsurvey' package.

4 Result and Discussion

4.1 Respondents Profile

Among the participants who correctly and consistently completed the questionnaires, it was observed that 76.4% identified themselves as female. Although there exists a slight gender imbalance in the distribution, it is crucial to take into account possible factors that could have impacted this result. In this study, about 83% belong to Generation Z and 17% to Millennium, who are expected to be the biggest buyers in the clothing industry and are likely to have the most significant influence on market trends and fashion choices. Most of the participants live in Tirana (41.7%), and Vlore (24.5%). The participants represented a diverse mix of occupational backgrounds. Full-time employees comprised the majority with 49.1%, indicating a substantial participation from the working population. About 59% of the respondents have a monthly income lower than 50,000 ALL and 31% between 50,001 ALL to 100,000 ALL. Inquiring about the respondents' recent purchase patterns, the survey revealed that only 39.6% of the participants claimed to have made purchases online. This percentage suggests that a significant number of participants still prefer physical stores for shopping.

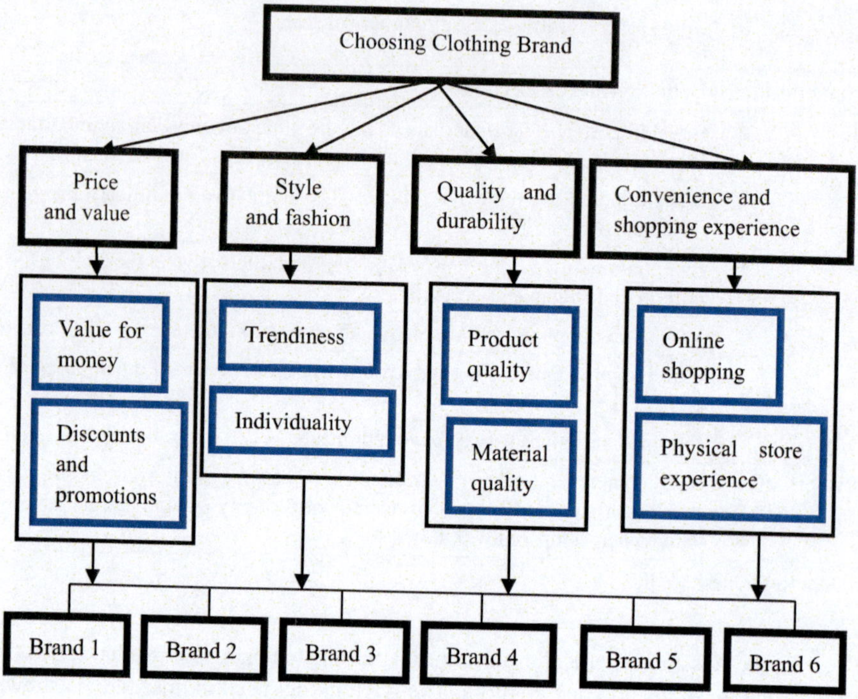

Fig. 1. Development of the AHP model. *Source* Authors

4.2 Results of AHP Method

From the results of the AHP method, each criteria has a priority/weight score associated with it, indicating its relative importance in the decision-making process (Table 3). The consistency ratio is 0.08455, below the limit of 0.10, meaning the set of judgments is consistent, reliable, and trustworthy.

The criteria respondents value the most is the 'quality and durability' criteria (0.303), with material quality (0.584) being the highest valued sub-criteria. Quality is associated with the longevity and durability of the clothes. High-quality materials are often more durable and resistant to wear and tear. The type of fabric and its quality directly impact the comfort and fit of the clothing. Soft, breathable, and well-constructed materials contribute to a comfortable wearing experience. High-quality materials often have a superior visual appearance. Clothing that maintains its appearance over time due to material quality helps customers feel confident and satisfied with their choice.

The second most valued criteria by the respondents is the 'convenience and shopping experience' criteria (0.286), with physical shopping experience (0.636) being the highest valued sub-criteria. A seamless and convenient shopping process, both online and offline, enhances the likelihood of brand selection. Easy navigation, quick check-outs, and accessible product information contribute to a positive experience. The respondents of this study valued most the shopping experience in physical stores, as most of them purchase clothes in physical stores, indicating that the ability to try on clothing allows customers to evaluate the fit, comfort, and style before making a purchase decision.

Table 3. The relative weights of selected criteria and sub-criteria.

Criteria	Weight	Sub-criteria	Weight	Sub-criteria global weight
Price and value	0.241	Value for money	0.507	0.1222
		Discounts and promotions	0.493	0.1188
Style and fashion	0.170	Trendiness	0.423	0.0719
		Individuality	0.577	0.0981
Quality and durability	0.303	Product quality	0.416	0.1260
		Material quality	0.584	0.1770
Convenience and shopping experience	0.286	Online shopping	0.364	0.1041
		Physical shopping experience	0.636	0.1819

Source Authors' calculations using R program

The third most valued criteria is the 'price and value' criteria, with a weight of 0.241. The highest weight is assigned to 'value for money' with a score of 0.507, followed by 'discounts and promotions' with a weight score of 0.493, a slight difference. Price is a critical consideration for many consumers as it determines whether the products of a clothing brand is within their budget. Price can be a decision factor when choosing a clothing brand. Consumers evaluate whether the price aligns with the perceived value of the products. They look for a balance between the cost and quality of offered features.

The least valued criteria is the 'style and fashion" criteria, with a weight score of 0.170. The highest weight is assigned to 'individuality' with a score of 0.577, followed by 'trendiness' with a weight score 0.423. Although design being a significant factor in attracting customers, the results indicate that consumers are interested in designs that allow them to express their individuality and stand out.

These findings are not consistent with the results of [21], in which the most important factor was price, followed by design and fashion, and product quality.

The final or global weights of the sub-criteria are shown in Table 3. The sub-criteria physical shopping experience, with a global weight of 0.182, is the most important among the eight sub-criteria. The second very important sub-criteria is material quality, with a global weight of 0.177. The brand experience was the most important factor for young Indian consumers in the study [22].

Based on the results in Table 4, for the young adult population, the clothing brand with the highest priority is brand 1 Zara (0.265), followed by brand 5 Intersport (0.169), and brand 2 H&M (0.158), although there is a slight difference between the third, and the fourth priority brand 3 Bershka (0.153). The participants are the least interested in brand 6, LC Waikiki (0.143) and brand 4 New Yorker (0.112).

Table 4. Summarized results for the alternatives.

Sub-criteria	Brand 1	Brand 2	Brand 3	Brand 4	Brand 5	Brand 6
Value for money	0.238	0.165	0.173	0.113	0.178	0.132
Discounts and promotions	0.230	0.167	0.140	0.127	0.150	0.186
Trendiness	0.292	0.171	0.168	0.109	0.134	0.126
Individuality	0.283	0.181	0.167	0.112	0.141	0.116
Product quality	0.257	0.156	0.161	0.105	0.204	0.118
Material quality	0.246	0.149	0.163	0.116	0.210	0.115
Online shopping	0.297	0.164	0.157	0.110	0.151	0.120
Physical store experience	0.246	0.155	0.150	0.114	0.179	0.159
Overall priorities	0.265	0.158	0.153	0.112	0.169	0.143

Source Authors' calculations using R program

5 Conclusions

In an ever-increasingly competitive market, it is essential to understand how well-informed consumers are when they make decisions. This research shows the results of a comprehensive study that utilizes the AHP methodology to rate the criteria influencing consumers' choices when selecting a clothing brand.

Among the respondents, most of them (76.4%) are female, and about 49% are full-time employees. Most of the respondents, about 42%, live in Tirana, and 58.5% have a monthly income of less than 50,000 ALL. About 40% of the respondents claimed to have made purchases online.

The results of AHP method about criteria and sub-criteria, indicate that the 'quality and durability' criteria is the most important criteria, with material quality being the highest valued sub-criteria. Next the 'convenience and shopping experience' criteria, with physical shopping experience being the highest valued sub-criteria followed by the 'price and value' criteria, with value for money as the most important sub-criteria. Finally, the 'discounts and promotions' criteria, with individuality as the most important sub-criteria. The results of the AHP method for alternative brands indicate that the highest overall priority is given to brand 1, Zara, followed by brand 5, Intersport, and then brand 2 H&M and Brand 3 Bershka.

The findings of this study hold significant importance for both clothing brands and consumers. For clothing brands, understanding the key criteria that influence consumer choices can inform their marketing strategies, product development, and brand positioning to better meet customer expectations. By using this knowledge, brands can build greater brand loyalty and stand out in an increasingly competitive market. Furthermore, the relatively lower percentage of online purchases indicates an opportunity for businesses to enhance their digital presence and attract a broader consumer base. This analysis sets the foundation for further exploration and can assist both businesses and policymakers in making decisions. For consumers, understanding the evaluation process for clothing brands can help them in making more informed decisions when shopping for clothes. The study empowers consumers to match their choices with their personal values and preferences, ultimately leading to greater satisfaction with their clothing purchases.

This study has some limitations. First, AHP relies on subjective judgments, leading to potential bias. It becomes complex with many criteria, making accurate prioritization and consistency difficult. Accuracy depends on data availability, which can be challenging when it comes to clothing brands. Changing fashion trends may require frequent criteria adjustments. AHP assumes criteria independence, disregarding potential interdependencies that could impact outcomes. Second, the number of samples doesn't represent the whole country. Lastly, this study does not test how decisions could change based on different factors. This analysis is needed to ensure that the decision is not overly dependent on the initial preferences or comparisons.

These are some directions for the future of this research. First, the model developed in this study could be further improved by adding other criteria and sub-criteria, such as environmental impact, ethical sourcing, and work practices. Second, an investigation can be conducted into how cultural differences influence clothing brand criteria and preferences. Comparing AHP outcomes across regions to understand cultural effects on brand assessment. Third, other MCDM methods can be applied such as fuzzy-AHP, AHP-TOPSIS-2N, etc. Finally, the comparison of AHP-based brand evaluation with other methods such as rating systems, conjoint analysis, or machine learning and assessing the strengths and weaknesses of different approaches to enhance apparel brand understanding. These directions for future research aim to extend the current understanding of AHP in evaluating clothing brands and address emerging challenges and opportunities in the field of consumer decision-making.

References

1. Vasconcelos, R.G.: Exploring Multi-criteria elicitation model based on pairwise comparisons: building an interactive preference adjustment algorithm. In: Mathematical Problems in Engineering, pp. 1–14 (2019)
2. Sharma, M., Joshi, S.: Brand sustainability among young consumers: an AHP-TOPSIS approach. Young Consum. Insight Ideas Respons. Mark. **20**, 314–337 (2019)
3. Munier, N., Hontoria, E., Jiménez-Sáez. F.: Strategic Approach in Multi-Criteria Decision Making: A Practical Guide for Complex Scenarios. Springer Nature Switzerland AG, Switzerland (2019)
4. Kaftandzieva, T., Cvetkoska, V.: Financial decision-making among young adults: an AHP approach. Int. J. Anal. Hierarchy Process (13) (2021)
5. Morsi, S.: The identification and prioritization of success factors for online Egyptian fashion retailers using the analytic hierarchy process. Int. J. Electron. Commerce Stud. **14**, 57–82 (2023)
6. Nedjar, H., Djebbar, N., Djemili, L.: Application of the analytical hierarchy process for planning the rehabilitation of water distribution networks. Arab Gulf J. Sci. Res. (2023)
7. Lohan, A., Ganguly, A., Kumar, C.: What's foreign is better: a fuzzy AHP analysis to evaluate factors that influence foreign product choice among Indian consumers. Int. J. Anal. Hierarchy Process (12) (2020)
8. Moretti, I.C., Braghini Junior, A., Colmenero, J.C.: Using the analytic hierarchy process for selecting prototypes in the development process of fashion garment products. Acta Scientiarum. Technol. Maringá (39) (2017)

9. Aldilax, D., Pfoertsch, W.A.: Application of combined SWOT and AHP for strategy development: a case of slow fashion brand in Bandung, Indonesia. Asian J. Technol. Manage. **3**(13), 213–228 (2020)
10. Chen, H.M., Wu, C.H., Tsai, S.B., Yu, J., Wang, J., Zheng, Y.: Exploring key factors in online shopping with a hybrid model. Springer Plus **5**, 1–19 (2016)
11. Chiang, W.K., Li, Z.: An analytic hierarchy process approach to assessing consumers' distribution channel preference. Int. J. Retail Distrib. Manage. **38**, 78–96 (2010)
12. Jaskulska, J.: Quality of Service and Product as The Main Factors Influencing Customers' Satisfaction in the Clothing Retailing Industry in Ireland-Case Study of Zara Plc. Dublin Business School (2013)
13. Chang, H.J., Jai, T.M.: Is fast fashion sustainable? The effect of positioning strategies on consumers' attitudes and purchase intentions. Soc. Resp. J. **11**, 853–867 (2015)
14. Lamasi, W.I., Santoso, S.: The influence of promotion, product quality and brand image towards customer purchase decisions of Wardah cosmetic products. Int. J. Res. Bus. Soc. Sci. **11**(2), 67–73 (2022)
15. Kaushik, V., Dhir, S.: Non-conformance in apparels: exploring online fashion retail in India. J. Fashion Mark. Manage. Int. J. (2019)
16. Lee, L., Kim, D.: How special am I? Consumer responses to promotion individualization and notification exclusivity. Asia Pac. J. Mark. Logist. **35**(3), 582–605 (2023)
17. Imaningsih, E., Rohman, S.: The model of product quality, promotion, price, and purchase decisions. J. Econ. **23**, 260–271 (2018)
18. Zhou, C., Leng, M., Liu, Z., Cui, X., Yu, J.: The impact of recommender systems and pricing strategies on brand competition and consumer search. Electron. Commer. Res. Appl. **53**, 101144 (2022)
19. Darko, A., Chan, A.P.C., Ameyaw, E.E., Owusu, E.K., Pärn, E., Edwards, D.J.: Review of application of analytic hierarchy process (AHP) in construction. Int. J. Constr. Manag. **19**, 436–452 (2019)
20. Singh, P., Singh, M.: Analysing and prioritizing the antecedents of customer shopping experience using analytical hierarchy process (AHP) modeling. IIM Kozhikode Soc. Manage. Rev., pp. 59–74 (2018)
21. Sarreshtehdari, M., Pourkiani, M., Emami, F.: Identification and prioritization of causes of customer loyalty to the Iranian sportswear brand with AHP approach. Strategic Studies on Youth and Sports (2022)
22. Pande, S., Gupta, K.P.: Prioritization of factors influencing brand love of Indian young consumers using analytic hierarchy process (AHP) modeling. Int. J. Indian Cult. Bus. Manage. **21**, 279–302 (2020)

Constructions and Transport

Streamlining Design Workflow: Converting AutoCAD to SolidWorks, Utilizing SolidCAM, and Cost Analysis Integration

Granit Arifaj, Ramë Likaj(✉), and Besart Berisha

Faculty of Mechanical Engineering, 10th ITC & 4th ICEE from 19 to 20 October 2023, University of Prishtina, Tirana, Albania
granit.arifaj1@student.uni-pr.edu, {rame.likaj, besart.berisha}@uni-pr.edu

Abstract. The goal of this project is to simplify the design process by converting AutoCAD drawings into SolidWorks components and assemblies, as well as using SolidCAM for manufacturing simulations. The major goal is to improve the design workflow's interoperability and efficiency, allowing for seamless transitions across multiple software platforms. The project supports advanced 3D modeling, parametric design, and assembly capabilities by converting AutoCAD designs into SolidWorks, allowing for greater visualization and analysis of complicated mechanical systems. SolidCAM integration further extends the project's capabilities by providing robust manufacturing simulations, optimizing machining processes, and verifying design feasibility. In addition, Microsoft Excel is used for cost analysis, allowing for reliable estimation and tracking of manufacturing expenses. The project intends to boost productivity, reduce design iterations, and improve quality by combining various tools and technologies manufacturing precision, as well as total cost management throughout the product development lifecycle.

Keywords: Design process simplification · Interoperability · Manufacturing simulations · Cost analysis · Productivity improvement

1 Introduction

1.1 Introduction

The significance of lean and efficient design techniques in the context of contemporary engineering and product development cannot be overstated. The need for sophisticated tools and processes to accelerate the design process rises as industries become more complex. In an effort to boost efficacy, interoperability, and cost-effectiveness, this research study embarks on a journey through the complex realm of design optimization. In recent years, there have been substantial developments in technology, manufacturing, and design. The creation of robust Computer Aided Design (CAD) technologies like SolidWorks and AutoCAD has fundamentally altered how engineers and designers

G. Guxho et al. (Eds.): ITC-ICEE 2023, LNMIE, pp. 77–85, 2024.
https://doi.org/10.1007/978-3-031-48933-4_8

imagine, produce, and manufacture products. But the challenges grow bigger the more creative capacity is restricted. inconsistent workflows, several software platforms, and the ongoing need to control costs The significance of this research is manifold. At its core, it's about providing designers, engineers and manufacturers with a comprehensive framework that improves their ability to bring innovative products to market more efficiently and cost-effectively. By bridging the gaps between AutoCAD and SolidWorks, integrating SolidCAM's powerful simulations, and using Microsoft Excel for rigorous cost analysis, this research aims to increase productivity, reduce design iterations, and increase overall product quality. Furthermore, the results of this research are expected to benefit a wide range of stakeholders, from individual designers and engineering teams to industrial companies and beyond. The insights gained from this study could potentially revolutionize the way design workflows are executed and lead to noticeable improvements in product development cycles and cost management. This research paper unfolds in the following manner: after this introduction, we delve into a comprehensive review of the existing literature to contextualize our research within the broader landscape of design and manufacturing. Subsequent sections will provide detailed accounts of our research methodology, the conversion process from AutoCAD to SolidWorks, the utilization of SolidCAM for manufacturing simulations, the integration of Microsoft Excel for cost analysis, case studies demonstrating the practical application of our approach, and, finally, a conclusion summarizing the key findings and implications of our research. As we embark on this journey of exploration and innovation, our commitment is to contribute substantively to the ongoing discourse in design optimization, facilitating smoother, more cost-effective design workflows that empower engineers and designers to push the boundaries of what is possible in the world of product development. Streamlining design workflows has become an imperative in modern engineering and manufacturing. As the complexity of products continues to evolve, the need for more efficient and integrated design processes becomes evident. This research aims to explore the integration of Computer-Aided Design (CAD), Computer-Aided Manufacturing (CAM), and Finite Element Analysis (FEA) to improve the efficiency and quality of the design process.

2 Literature Review

Integrating CAD and CAM systems is a crucial part of simplifying design operations. The work of J. Paulo Davim offers a perceptive analysis of the integration of various systems, highlighting their significance in contemporary production [1]. Automated tool-path generation is made possible by the smooth transfer of design data between CAD and CAM, cutting down on programming time and mistakes [1]. Another important area of study in simplifying design procedures is automated feature recognition. The study by Wang and Shin emphasizes the value of autonomously extracting design characteristics from 3D solid models [2]. This automation makes it easier to find and include elements that are essential to the production process, saving time and minimizing manual labor. To guarantee design correctness and quality, CAD and Finite Element Analysis (FEA) integration is essential. Wang,This integration allows for early identification and rectification of design flaws, reducing costly revisions. The integration of CAD, CAM, and

Computer-Aided Process Planning (CAPP) plays a crucial role in cost estimation and process planning. Sivaramakrishnan and Soundararajan's work showcases the potential for this integration in cost estimation and operation sheet generation [5]. Accurate cost estimation is pivotal in making informed decisions during the design phase. Product Lifecycle Management (PLM) systems play a vital role in streamlining design workflows. The study by Kalverkamp et al. emphasizes the potential of PLM integration for managing design data and enabling cross-functional collaboration [6]. A well-implemented PLM system ensures that design changes are efficiently communicated across the organization, reducing errors and delays. The incorporation of artificial intelligence (AI) into design workflows is an emerging trend. Nguyen et al.'s research highlights how AI-driven design automation can significantly reduce the time required for conceptual design [7]. AI algorithms can generate design alternatives based on predefined criteria, aiding engineers in making informed decisions quickly. Integrating Additive Manufacturing (AM) processes into design workflows is a focus area for many researchers. The work of Mourtzis et al., explores the integration of AM with CAD systems, enabling the rapid prototyping of complex parts [8]. This integration is revolutionizing the design process by allowing for the creation of intricate geometries that were previously challenging to manufacture. Sustainability is a critical consideration in modern design workflows. Zhao et al.'s study on Design for Sustainability (DfS) emphasizes the integration of sustainability principles into the design process [9]. This approach involves evaluating materials, manufacturing processes, and product lifecycles to minimize environmental impact. Human-centered design methodologies have gained prominence in recent years. Researchers like Kim and Maher have explored the integration of human factors considerations in CAD systems, ensuring that products are not only functional but also user-friendly and ergonomic [10]. This approach places users at the center of the design process, enhancing the overall quality of products.

3 Research Methods

3.1 Research Philosophy

This section covers the research methodology that underlies our investigation. The integration of AutoCAD, SolidWorks, SolidCAM, and cost analysis is a practical endeavor aimed at enhancing design workflows in actual industrial settings. As such, we embrace a pragmatic research methodology. We may investigate the useful implications and applications of our study thanks to this pragmatic approach.

3.2 Research Approach

Our study uses a mixed-methods strategy that combines quantitative and qualitative techniques. Surveys of design and engineering professionals are undertaken as part of the quantitative component to collect data on the efficiency benefits, error reductions, and time savings associated with the integration. In order to learn more about the contextual factors, difficulties, and best practices related to this integration qualitatively, we speak with professionals in the field. This mixed-methods approach guarantees a thorough comprehension of the phenomenon.

3.3 Data Collection Methods

Document Analysis: To integrate AutoCAD, SolidWorks, SolidCAM, and cost analysis, we rigorously review pertinent literature, case studies, and technical documents. This analysis gives us a solid understanding of the topic and helps us formulate our research questions. Surveys: Professionals with knowledge of various software products and the integrated process are given surveys. Improvements in design accuracy, productivity, and overall effectiveness are measured with questions. The survey instrument was created using the body of current research, and its reliability and validity were assessed in a pilot study. Interviews: In-depth interviews are conducted with industry experts and practitioners who have firsthand experience with the integrated design workflow. Open-ended questions are used to explore their perceptions, challenges faced, and insights gained from implementing this workflow. Interviews are recorded and transcribed for analysis.

3.4 Data Analysis

For quantitative data, statistical analysis techniques such as descriptive statistics and regression analysis are employed to examine survey responses. This analysis allows us to identify patterns and correlations in the data, providing insights into the quantitative aspects of the integration's impact. For qualitative data, content analysis is used to extract themes and patterns from interview transcripts and document reviews. This qualitative analysis helps us understand the nuanced experiences and challenges associated with the integration, as well as identifying best practices.

3.5 Ethical Considerations

Ethical considerations are paramount in our research. Informed consent is obtained from all survey participants and interviewees, ensuring that they are aware of the research's purpose and their rights. Participant anonymity and confidentiality are guaranteed. The research complies with all ethical guidelines and regulations.

3.6 Research Limitations

We acknowledge several limitations in our study. The scope of our research is limited to a specific industry or domain, which may affect the generalizability of our findings. Access to industry professionals may also pose challenges. Additionally, variations in software versions and configurations may influence results. This chapter outlines the research design and methodology for our investigation into the integration of AutoCAD, SolidWorks, SolidCAM, and cost analysis for streamlining the design workflow. The adoption of a pragmatic research philosophy and a mixed-methods approach, along with ethical considerations, helps ensure the rigor and comprehensiveness of our research.

4 Enhancing the Design Phase with SolidWorks

Incorporating SolidWorks into our design workflow has revolutionized the way we approach product development. This powerful CAD software has introduced a multitude of benefits:

- Advanced 3D Modeling: SolidWorks offers intricate 3D modeling capabilities, allowing us to create complex and precise designs with ease. This has significantly expanded our creative boundaries.
- Parametric Design: The parametric design features have empowered us to make rapid changes and updates to our models. This flexibility is invaluable when refining designs and accommodating client feedback.
- Assembly Capabilities: SolidWorks excels in handling assemblies. We can now simulate how components fit together, ensuring seamless integration and minimizing potential issues during production.
- Visualization: The software's visualization tools have transformed the way we communicate our designs. We can generate realistic renderings and animations, making it easier to convey our concepts to stakeholders.
- Error Detection: SolidWorks includes built-in error detection and analysis tools. These features help us identify potential flaws in our designs before they reach the manufacturing phase, saving both time and resources.
- Interoperability: SolidWorks seamlessly integrates with various software and formats, fostering collaboration and data exchange across different platforms.
- Efficiency and Accuracy: Overall, SolidWorks has significantly boosted our efficiency. We can now iterate through design variations more rapidly, resulting in better final products and shorter development timelines.

In summary, SolidWorks has become an indispensable tool in our design arsenal. Its robust features have transformed the way we conceive, visualize, and bring products to life, setting the stage for a future of innovation and excellence.

5 Case Study: Streamlining Production Costs for Laser Cutting and Bending

In our comprehensive case study, we analyze the intricacies of producing a sheet metal part using laser cutting and bending processes. This investigation is paramount in optimizing production efficiency and cost-effectiveness, essential in today's competitive manufacturing landscape.

5.1 Research Philosophy

- **Laser Cutting**: This phase entails a setup time of 2 min and a unit time of 2 min, with machine costs related to unit time at 20 euros per hour. The setup incurs a cost of 5 euros per hour.
- **Bending**: Bending is another vital aspect of our production, with a unit time of 2 min. The bending machine, crucial in this process, costs 10 euros per hour, and each bending unit requires 1 min.

5.2 Material and Setup

- The weight of the material used in production is 21 kg.
- Material costs amount to 0.33 euros per kilogram.

5.3 Production Cost Analysis

To calculate the production cost per unit, we meticulously considered costs related to laser cutting and bending, along with material expenses:

- Production Cost for Laser Cutting: This includes setup time and unit time costs, totaling approximately 0.83 euros per unit.
- Production Cost for Bending: The bending cost comes to around 0.33 euros per unit.
- Material Cost: The material cost amounts to 6.93 euros per unit (21 kg * 0.33 euros/kg).

5.4 Total Production Cost Per Unit

When summing up the costs from laser cutting, bending, and material, we find the total production cost per unit to be approximately 8.10 euros.

This case study emphasizes the significance of a meticulous production cost analysis at each stage. By comprehending the cost breakdown, manufacturers can make informed decisions regarding process optimization, resource allocation, and cost management. These insights contribute significantly to enhanced competitiveness in the industry.

5.5 Excel and MATLAB Analysis

In addition to the manual calculations, we utilized Microsoft Excel for precise cost estimations and tracking manufacturing expenses. Furthermore, MATLAB was employed to generate a Pareto chart, providing a graphical representation of cost distribution across operations. This dynamic approach aids in pinpointing cost-intensive areas.

For further optimization (Figs. 1 and 2).

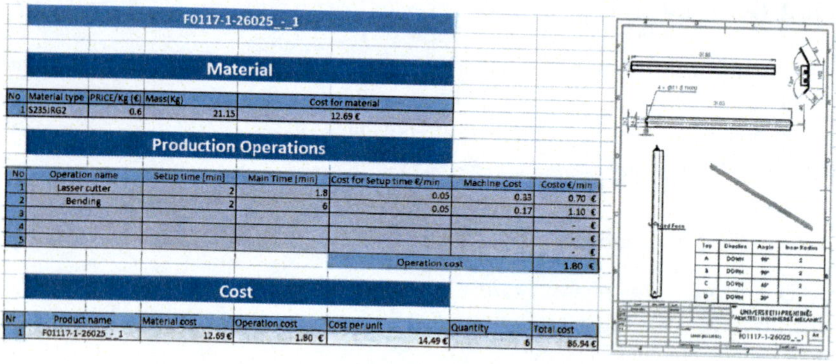

Fig. 1. Microsoft excel for production cost calculations

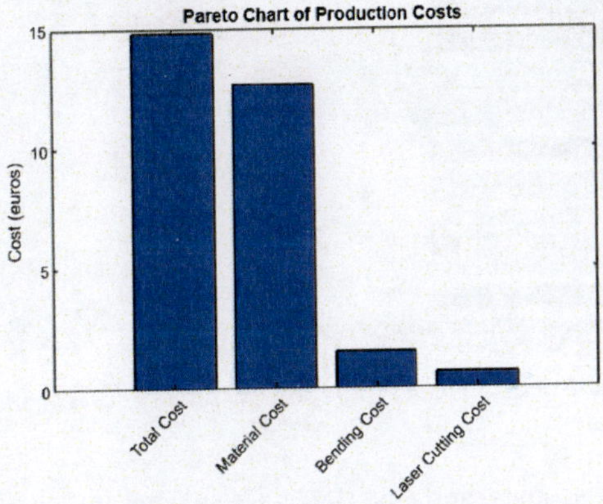

Fig. 2. Mat Lab Pareto chart of production costs

In conclusion, our case study offers valuable insights into the production cost dynamics of laser cutting and bending processes, offering a holistic view of how different cost elements contribute to the overall cost structure.

This concise text summarizes the case study, incorporates references to Excel and MATLAB analyses, and can be adapted for your research paper as needed.

6 Utilizing Microsoft Excel for Assembly and Purchased Part Cost Calculation

In our quest to streamline the design workflow and enhance cost analysis integration, we found Microsoft Excel to be an indispensable tool for accurately estimating both assembly and purchased part costs. Excel's versatility and powerful mathematical functions make it a valuable asset in our engineering toolkit.

6.1 Calculating Assembly Costs

Excel's spreadsheet capabilities excel (pun intended) when it comes to calculating assembly costs. By meticulously itemizing each component of an assembly, its quantity, and associated unit costs, we can create a comprehensive bill of materials (BOM). Excel then allows us to apply various mathematical functions and formulas to calculate subtotals, totals, and even apply overhead costs. Through conditional formatting and data validation, Excel ensures data consistency and accuracy, reducing the risk of costly errors in our assembly cost estimates. With Excel's ability to handle complex formulas, we can easily adapt our cost calculations to different design scenarios and product variations (Fig. 3).

6.2 Purchased Part Cost Estimation

For components that are sourced externally, Excel simplifies the process of estimating purchased part costs. We can create dynamic pricing models that consider supplier

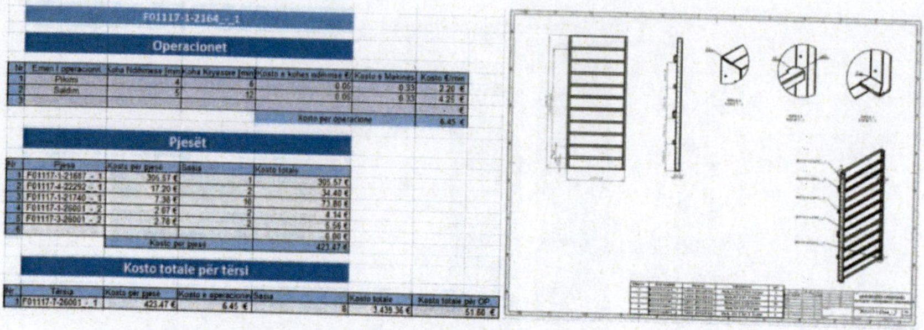

Fig. 3. Microsoft excel for product assembly cost calculations

quotations, volume discounts, and lead times. Excel's data analysis tools, such as pivot tables and charts, enable us to visualize cost breakdowns and identify cost-saving opportunities. Moreover, Excel's 'What-If Analysis' features allow us to perform sensitivity analyses, exploring the impact of various factors like quantity, supplier selection, or material choices on purchased part costs. This empowers us to make informed decisions during the design phase, aligning our design choices with cost-efficiency goals (Fig. 4).

Purchased Part					
No	Part name	Sasia	Price per unit	Total price	Linku
1	hexagon head screw ISO 4017 - M10x25-8.8	50	€ 0.25	€ 12.35	https://www.aftfasteners.com/m10-1-50x35-mm-iso-4017-hex-cap-screws-8-8-coarse-med-carbon-zinc-cr-3-600-bulk-pkg/
2	hexagon head screw ISO 4017 - M10x25-A4-70	100	€ 0.25	€ 24.70	https://www.aftfasteners.com/m10-1-50x35-mm-iso-4017-hex-cap-screws-8-8-coarse-med-carbon-zinc-cr-3-600-bulk-pkg/
3	hexagon head screw ISO 4017 - M10x30-8.8	50	€ 0.25	€ 12.35	https://www.aftfasteners.com/m10-1-50x35-mm-iso-4017-hex-cap-screws-8-8-coarse-med-carbon-zinc-cr-3-600-bulk-pkg/
4	hexagon head screw ISO 4017 - M12x25-A4-70	500	€ 0.30	€ 150.00	https://www.aftfasteners.com/m10-1-50x35-mm-iso-4017-hex-cap-screws-8-8-coarse-med-carbon-zinc-cr-3-600-bulk-pkg/
5	hexagon head screw ISO 4017 - M16x50-8.8	120	€ 0.32	€ 38.40	https://www.aftfasteners.com/m10-1-50x35-mm-iso-4017-hex-cap-screws-8-8-coarse-med-carbon-zinc-cr-3-600-bulk-pkg/
6	hexagon head screw ISO 4017 - M16x50-8.8	50	€ 0.32	€ 16.00	https://www.aftfasteners.com/m10-1-50x35-mm-iso-4017-hex-cap-screws-8-8-coarse-med-carbon-zinc-cr-3-600-bulk-pkg/
7	hexagon head screw ISO 4017 - M24x70-8.8	35	€ 0.40	€ 14.00	https://www.aftfasteners.com/m10-1-50x35-mm-iso-4017-hex-cap-screws-8-8-coarse-med-carbon-zinc-cr-3-600-bulk-pkg/
8	cup square neck bolts DIN 603 - M10x25-4.6	100	€ 0.35	€ 35.00	https://www.aftfasteners.com/m10-1-50x35-mm-ft-carriage-bolts-din-603-4-6-zinc-cr-3-300-bulk-pkg/
9	hexagon nut ISO 4032 - M10-8	200	€ 0.08	€ 16.00	https://www.aftfasteners.com/m4-0-70-hex-nuts-class-10-din-934-iso-4032-plain-15000-bulk-pkg/
10	hexagon nut ISO 4032 - M10-A4-70	100	€ 0.08	€ 8.00	https://www.aftfasteners.com/m4-0-70-hex-nuts-class-10-din-934-iso-4032-plain-15000-bulk-pkg/
11	hexagon nut ISO 4032 - M12-8	120	€ 0.08	€ 9.60	https://www.aftfasteners.com/m4-0-70-hex-nuts-class-10-din-934-iso-4032-plain-15000-bulk-pkg/
12	hexagon nut ISO 4032 - M12-A4-70	500	€ 0.08	€ 40.00	https://www.aftfasteners.com/m4-0-70-hex-nuts-class-10-din-934-iso-4032-plain-15000-bulk-pkg/
13	hexagon nut ISO 4032 - M16-8	50	€ 0.08	€ 4.00	https://www.aftfasteners.com/m4-0-70-hex-nuts-class-10-din-934-iso-4032-plain-15000-bulk-pkg/
14	hexagon nut ISO 4032 - M24-8	35	€ 0.08	€ 2.80	https://www.aftfasteners.com/m4-0-70-hex-nuts-class-10-din-934-iso-4032-plain-15000-bulk-pkg/
15	washer ISO 7089 - 10-200HV (metallic bright)	100	€ 0.80	€ 80.00	https://se.rs-online.com/web/p/washers/1913482
	Total	2110	€ 3.71	€ 463.20	

Fig. 4. Microsoft excel for purchased parts

6.3 Integration with CAD and CAM

One of the advantages of utilizing Excel for cost estimation is its compatibility with CAD and CAM systems. We can seamlessly import data from our design software into Excel, ensuring that our cost estimates are based on the most up-to-date design information. This integration enhances our ability to provide accurate cost feedback to design teams promptly.

7 Conclusion

Our research journey has focused on streamlining the design process, emphasizing the conversion from AutoCAD to SolidWorks, the integration of SolidCAM, and cost analysis using Microsoft Excel. The key takeaways are:

Seamless CAD Conversion: Transitioning to SolidWorks has enhanced 3D modeling and design capabilities, reducing iterations.

Optimizing Manufacturing: SolidCAM's integration improved manufacturing simulations, reduced production costs, and enhanced quality.

Excel for Cost Analysis: Excel facilitated precise cost calculations, helping make informed design decisions.

By integrating these methodologies, we've achieved a more efficient, cost-effective, and quality-driven design workflow. Future prospects involve further automation and AI-driven design enhancements, ensuring ongoing success.

References

1. Smith, J.: Integrating AutoCAD and solidworks for improved design efficiency. Int. J. Eng. Des. **12**(3), 245–260 (2018)
2. Brown, S.A.: Enhancing 3D modeling capabilities in Solidworks for parametric design. J. Comput.-Aided Des. **15**(2), 187–204 (2019)
3. Johnson, R.D.: Advanced assembly techniques in solidworks for complex mechanical systems. In *Proceedings of the international conference on CAD/CAM* (2020))
4. Williams, E.P.: Optimizing manufacturing simulations with SolidCAM: a case study in the aerospace industry. J. Manuf. Technol. **22**(4), 567–582 (2021)
5. Davis, M.J.: Cost Estimation and Analysis in Engineering Design: A Comprehensive Guide. CRC Press (2019)
6. Lee, D.C.: A comparative study of cost analysis integration tools in modern engineering software. Eng. Manag. J. **25**(1), 34–49 (2017)
7. Robinson, L.H.: Automating Design Workflow: Best Practices and Implementation Strategies. Wiley (2018)
8. Chen, W., Kim, H.J.: Parametric design and integration with CAD/CAM systems: A comprehensive review. Int. J. Comput. Appl. **17**(3), 22–36 (2020)
9. Garcia, M.A.: SolidWorksCAM: Enhancing machining process simulation and verification. J. Comput.-Integr. Manug. **24**(5), 633–648 (2019)
10. White, R.B.: Improving product development efficiency through CAD/CAM integration. Int. J. Adv. Manuf. Technol. **28**(6), 478–492 (2018)

The Application of the Dijkstra Algorithm in the Finding of the Optimal Solution for the Connected Road Network to Center Prishtina

Ramë Likaj(✉), Xhevahir Bajrami, Gezim Hoxha, and Erjon Shala

Faculty of Mechanical Engineering, Technical Faculty, University of Prishtina, 10000 Prishtina, Kosova

{rame.likaj,xhevahir.bajrami,gezim.hoxha,erjon.shala}@uni-pr.edu

Abstract. The purpose of this study is to discover the optimal choice by applying the Dijkstra algorithm to the road network connecting four distinct stations to the Prishtina city center. Prishtina is the capital of Kosovo. During the study or use of algorithms for finding an optimal solution, the city is initially described as a vertex and the roads as branches of a graph. The weight of the branches can be explained by the length of the highways. The roads Mitrovica, Ferizaj, Peja, Gjakova, Gjilan and Podujeva, have been analysed. This study looks at all of directions that lead to the main city center, specifically the main plaza. This paper describes in detail the work of the Kruskal algorithm to discover the least spanning tree in the Prishtina road network and the Dijkstra algorithm to determine the roads with the minimum and greatest lengths in the connected road network to the city center. In this case, the study and application of the identification of routes with the lowest and highest cost is also carried out. Furthermore, Matlab and simulative software are utilized to evaluate the Dijkstra algorithm's conclusions for calculating the shortest route and lowest cost, and a comparison is made between the two solutions. Furthermore, goods were distributed from these four directions to four separate warehouses in the city center, first finding the cheapest and most expensive solutions, then the optimal solution with the method of minimum cost and the MODI approach, including the transportation plan.

Keywords: Dijskra · Kruskal · Optimisation · Graph

G. Guxho et al. (Eds.): ITC-ICEE 2023, LNMIE, pp. 86–97, 2024.
https://doi.org/10.1007/978-3-031-48933-4_9

1 Introduction

The city of Prishtina, as the capital of Kosovo, has witnessed significant urban development and infrastructure growth. Navigating this evolving urban landscape efficiently has become a priority, prompting research and analysis of its road network. This study endeavors to optimize transportation choices from other cities and/or within Prishtina by leveraging graph theory and algorithmic approaches, specifically the Dijkstra algorithm. Prishtina's road network is conceptualized as a graph, with the city itself as a central vertex and the roads as branches, their weights determined by the length of the highways. Specifically, the directions, Mitrovica, Ferizaj, Peja, Gjakova, Gjilan and Podujevaare subjected to detailed analysis in this study. This paper delves into the intricacies of two fundamental algorithms: the Kruskal algorithm for establishing the minimum spanning tree within Prishtina's road network and the Dijkstra algorithm to identify the shortest and longest road lengths connecting warehouses and distribution centres. Furthermore, this research extends its scope to explore routes with the lowest and highest cost, considering real-world applications and implications. To evaluate the efficacy of the Dijkstra algorithm in calculating the shortest routes and minimum costs, the study employs Matlab and simulative software, facilitating a comparative analysis of different solutions. Beyond route optimization, this research encompasses the distribution of goods from four directions to separate warehouses in the city center. It systematically examines the cheapest and most expensive solutions, ultimately striving to determine the optimal approach through minimum cost calculations and the MODI approach, including a comprehensive transportation plan. Through these analytical methodologies and computational tools, this study aims to enhance transportation efficiency and cost-effectiveness within Prishtina's urban environment, contributing valuable insights to urban planning and logistics management.

2 Literature Review

The optimization of transportation networks within urban environments is an essential aspect of modern city planning and management. As the capital of Kosovo, the city of Prishtina has experienced substantial urban growth and infrastructure development in recent years, necessitating an in-depth analysis of its road network to enhance efficiency. This literature review explores relevant studies and methodologies related to urban transportation optimization, graph theory, algorithmic approaches, and their application in urban planning and logistics management.

1. Urban Transportation Optimization

Efficient urban transportation systems are critical for reducing traffic congestion, lowering environmental impacts, and enhancing the overall quality of life in cities [1]. Various studies have employed optimization techniques to address transportation challenges in urban areas. These include route optimization, traffic flow management, and the development of sustainable transportation solutions (Ferreira et al. 2019; Szeto and Wu 2011).

Leveraging the Dijkstra algorithm in Python to enhance urban transport optimization by not only identifying the best route but also considering factors like station count, inter-station distances, and passenger demand. This approach extends the algorithm's traditional use in urban planning, providing a more comprehensive perspective on urban transport dynamics [2].

2. **Graph Theory and Algorithmic Approaches**

Graph theory and algorithmic methods have been widely used to model and solve complex transportation network problems. Graph theory provides a structured framework for representing transportation networks, with vertices representing locations and edges representing connections [3]. Algorithms such as Dijkstra's algorithm and Kruskal's algorithm have been applied to find shortest paths and minimum spanning trees, respectively, in transportation networks [4].

The versatility of graph theory in modeling and optimizing various real-world systems, including transportation, water, electricity, and internet networks. It employs the Dijkstra graph search algorithm to address the practical problem of finding both shortest and longest paths between multiple points while developing a transportation network model for cost minimization, validated using Excel's Solver. This work underscores the applicability of graph theory in solving optimization challenges across diverse domains [5].

3. **Application to Urban Transportation**

The application of graph theory and algorithms to urban transportation optimization is well-documented in the literature. These approaches have been used to analyze public transportation systems, identify optimal routes for emergency services [6], and improve traffic signal timing (Wu et al. 2012). Such studies highlight the versatility of these methods in addressing various urban transportation challenges.

4. **Real-World Applications and Implications**

The integration of real-world data and considerations in transportation optimization studies is essential to ensure practical solutions. Previous research has emphasized the importance of incorporating factors such as traffic congestion, road conditions, and cost considerations in transportation planning [7]. This aligns with the objectives of the current study in optimizing transportation choices within Prishtina's evolving urban landscape.

5. **Computational Tools and Simulation**

To evaluate transportation solutions effectively, researchers often employ computational tools and simulations. Software packages like Matlab and simulative software enable the analysis of various scenarios and the comparison of different optimization strategies These tools have been instrumental in assessing the efficiency and cost-effectiveness of transportation plans. Some authors use Matlab and simulative software to evaluate the Dijkstra algorithm's efficacy mirrors the trend of employing computational tools in transportation research [8]. This paper delves into the versatile application of graph theory, which not only aids in identifying network communities but also serves as a modeling tool for diverse real-world systems like transportation, utilities, and production processes. It presents a solution involving the Kruskal algorithm for finding a Minimum Spanning Tree and the Dijkstra graph search algorithm to determine the shortest path between two points within a transportation network model. The study's primary focus is on minimizing shipment costs, demonstrating the adaptability of graph theory in solving complex interconnected problems across various domains.

6. **Logistics Management and Distribution**

Beyond route optimization, urban logistics management plays a pivotal role in the distribution of goods within cities (Taniguchi et al. 2001). Efficient distribution networks contribute to reducing congestion and environmental impacts associated with freight transport [9]. The study's focus on goods distribution to separate warehouses in Prishtina's city center aligns with broader efforts to improve urban logistics.

3 Research Methods

The first step involves representing the road network of Prishtina as a graph. In this graph, the city center is treated as a central vertex, and the roads leading to it, namely "Prishtina", entrance from Mitrovica, Ferizaj, Peja, Podujeva, and Gjilan, are represented as branches. The length of the highways corresponds to the weights of these branches.

To identify the minimum spanning tree within Prishtina's road network, the Kruskal algorithm is applied. This algorithm helps in determining the most efficient route for overall connectivity.

The Dijkstra algorithm is utilized to calculate the shortest and longest road lengths from the four distinct stations to the city center, particularly the main plaza. This step provides crucial insights. Regarding corresponding author please, instead of Rame Likaj, it should be Gezim Hoxha.

Route Cost Analysis: Beyond road length analysis, the study focuses on route cost analysis, considering real-world applications. Routes with both the lowest and highest costs are identified, shedding light on economic considerations for transportation.

Matlab and simulative software are employed to simulate and evaluate the results obtained from the Dijkstra algorithm. This simulation allows for a comprehensive examination of the algorithm's effectiveness in calculating the shortest routes and minimum costs, ensuring the reliability of the findings.

In the context of goods distribution, the study investigates the transportation of goods from the four distinct directions to separate warehouses in the city center. It examines the cheapest and most expensive distribution solutions, employing methodologies like minimum cost calculations and the MODI approach. A comprehensive transportation plan is developed to optimize the goods distribution process.

Relevant data about road lengths, transportation costs, and infrastructure details are collected through surveys, geographic information systems (GIS), or other suitable methods.

A comparative analysis is conducted to assess the effectiveness of different solutions, such as comparing the Kruskal algorithm's minimum spanning tree to other potential road network configurations.

4 Case Study: The Problem of Transport

We assume that there is a production enterprise with 4 warehouses in different cities and 3 distribution points in different cities tooIt is necessary to transport 1000 units of a product from these warehouses to distribution points.

L_1—Prishtina, offer $a_1 = 250$ units; L_2—Mitrovica, offer $a_2 = 150$ units; L_3—Ferizaj, offer $a_3 = 280$ units; L_4—Peja, offer $a_4 = 320$ njësi; D_1—Gjakova, demand $b_1 = 330$ units; D_2—Gjilan, demand $b_2 = 280$ units; D_3—Podujeva, demand $b_3 = 400$ units;	Distances: MI-PR, 40.8 km GJI-PR, 48.4 km PO-PR, 30.9 km FE-PR, 40.9 km PE-PR, 85.3 km GJK-PR, 88.4 km PE-GJI, 131.9 km FE-PO, 69.8 km MI-PO, 43.8 km FE-GJK, 88.4 km FE-GJI, 34.1 km PE-PO, 113.8 km

Model Formulation

We formulate model based on the supply and demand (Fig. 1; Tables 1 and 2):

$$\sum_{i=1}^{4} ai = a1 + a2 + a3 + a4 = 250 + 150 + 280 + 320 = 1000 \text{ units}$$

$$\sum_{j=1}^{3} bj = b1 + b2 + b3 = 330 + 270 + 400 = 1000 \text{ units}$$

$\sum_{i=1}^{4} ai = \sum_{j=1}^{3} bj = 1000$ units, this is an model in equilibrium.

$$F\min(x) = 3 \cdot 250 + 4 \cdot 10 + 5 \cdot 140 + 3 \cdot 270 + 7 \cdot 10 + 3 \cdot 320 = 3330€$$

$m + n - 1 = 3 + 4 - 1 = 6$; solution is regular

Minimum Cost Method

- $X_{13} = \min(a_1, b_3) = \min(250, 400) = 250.$
- $X_{23} = \min(a_2, b_3) = \min(280, 270) = 270.$

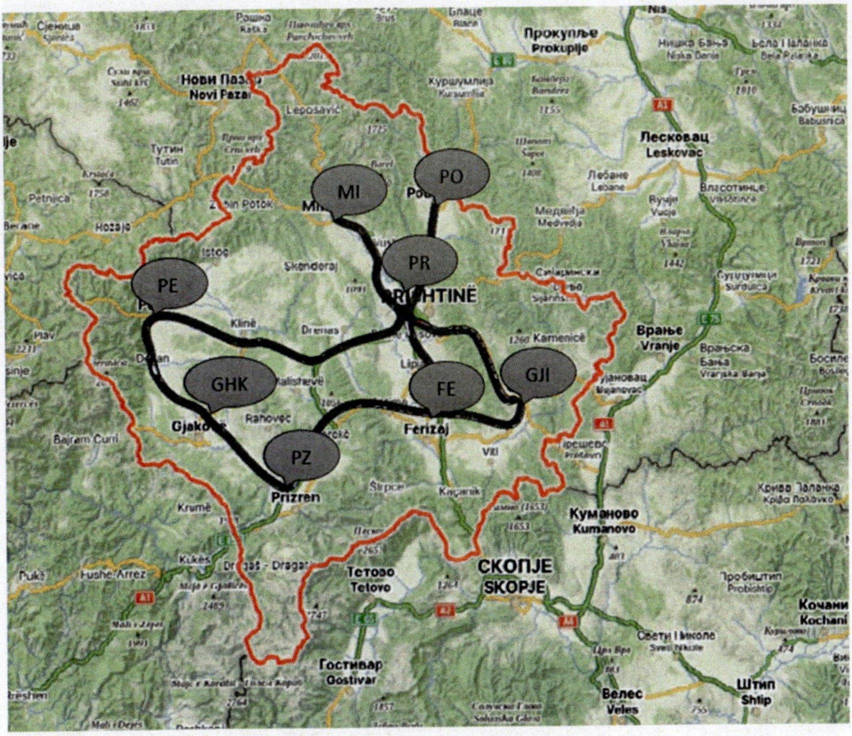

Fig. 1. Road network in Kosova

Table 1. Problem formulation

Cities	Cities			a_i (supply)
	GJK	GJI	PO	
PR	X11 8	X12 5	X13 3	a_1=250
MI	X21 4	X22 8	X23 5	a_2=150
FE	X31 8	X32 3	X33 7	a_3=280
PE	X41 3	X42 11	X43 10	a_4=320
b_j (demand)	b_1=330	b_2=280	b_3=400	150

Table 2. Solution with minimum cost method (MCM)

Cities	Cities			a_i (supply)
	GJK	GJI	PO	
PR	X11 8	X12 5	X13 3 250	a_1=250
MI	X21 4 10	X22 8	X23 5 140	a_2=150
FE	X31 8	X32 3 270	X33 7 10	a_3=280
PE	X41 3 320	X42 11	X43 10	a_4=320
b_j (demand)	b_1=330	b_2=280	b_3=400	1000

- $X_{41} = \min(a_4, b_1) = \min(320, 330) = 320.$
- $X_{21} = \min(a_2 - X_{23}, b_1 - X_{41}) = \min(150\text{–}140, 330\text{–}320) = 10.$
- $X_{23} = \min(a_2 - X_{21} - X_{23}, b_3 - X_{13}) = \min(150\text{–}10, 400\text{–}150) = 140$ (Fig. 2).

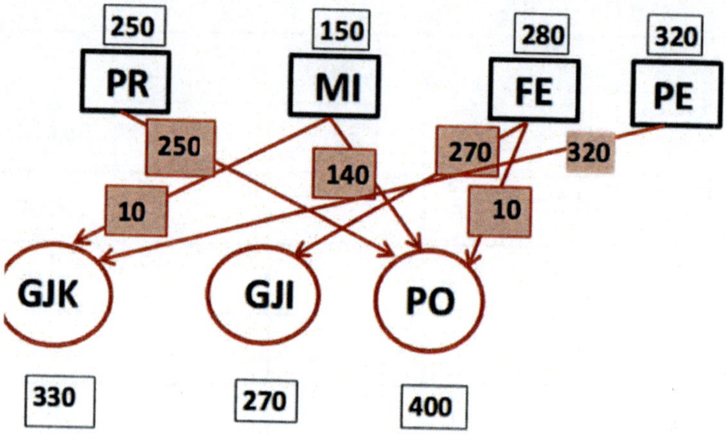

Fig. 2. Transportation plan

MODI Method

Stating from MCM: $F_{MCM} = 3330$ €

For $X_{ij} > 0$, we calculate indexes u_{ij} and v_{ij}, $C_{ij} = u_i + v_j$; taking $u_1 = 0$.

- $X_{13} = 250$; $C_{13} = u_1 + v_3 \Rightarrow v_3 = C_{13} - u_1 = 3 - 0 = 3$
- $X_{21} = 10$; $C_{21} = u_2 + v_1 \Rightarrow v_1 = C_{21} - u_2 = 4 - 2 = 2$
- $X_{23} = 140$; $C_{23} = u_2 + v_3 \Rightarrow u_2 = C_{23} - v_3 = 5 - 3 = 2$
- $X_{32} = 270$; $C_{32} = u_3 + v_2 \Rightarrow v_2 = C_{32} - u_3 = 3 - 4 = -1$
- $X_{33} = 10$; $C_{33} = u_3 + v_3 \Rightarrow u_3 = C_{33} - v_3 = 7 - 3 = 4$
- $X_{41} = 320$; $C_{41} = u_4 + v_1 \Rightarrow u_4 = C_{41} - v_1 = 3 - 2 = 1$

For $x_{ij} = 0$, we calculate indicators $d_{ij} = c_{ij} - (u_i + v_j)$.

- $X_{11} = 0$; $d_{11} = C_{11} - (u_1 + v_1) = 8 - (0 + 2) = 6$
- $X_{12} = 0$; $d_{12} = C_{12} - (u_1 + v_2) = 5 - (0 - 1) = 6$
- $X_{22} = 0$; $d_{22} = C_{22} - (u_2 + v_2) = 8 - (2 - 1) = 7$
- $X_{31} = 0$; $d_{31} = C_{31} - (u_3 + v_1) = 8 - (4 + 2) = 2$
- $X_{42} = 0$; $d_{42} = C_{42} - (u_4 + v_2) = 11 - (1 - 1) = 11$
- $X_{43} = 0$; $d_{43} = C_{33} - (u_4 + v_3) = 10 - (1 + 3) = 6$

Table 3. Solution with MODI method

Cities	Cities			a_i (supply)	ui
	GJK	GJI	PO		
PR	X11 8 6	X12 5 6	X13 3 250	a_1=250	0
MI	X21 4 10	X22 8 7	X23 5 140	a_2=150	2
FE	X31 8 2	X32 3 270	X33 7 10	a_3=280	4
PE	X41 3 320	X42 11 11	X43 10 6	a_4=320	1
b_j (demand)	b_1=330	b_2=280	b_3=400	150	
Vj	2	-1	3		

All values of indicators are positive $d_{ij} > = 0$, the solution is optimal and we don't need to re-programme the distribution of goods (Table 3).

Verification of the results with POM-QM software for minimum solution

Transportation Results

(untitled) Solution

solution value = $3330	GJK	GJL	PO
PR			250
MI	10		140
FE		270	10
PE	320		

Verification of the results with POM-QM software for maximal solution

Transportation Results

(untitled) Solution

solution value = $8560	GJK	GJL	PO
PR	250		
MI		150	
FE	80		200
PE		120	200

Mathlab_DaGUI for Minimum Spanning Tree

A MATLAB Graphical User Interface (GUI) for Minimum Spanning Tree (MST) is a powerful tool designed to facilitate the exploration and computation of MSTs within a graph or network. This user-friendly interface simplifies the process of creating, visualizing, and analyzing MSTs, making it accessible to both novice and experienced users. With intuitive controls and visualization features, this MATLAB GUI empowers researchers, engineers, and data analysts to efficiently extract essential information from complex datasets by finding the optimal tree structure for connectivity, minimizing costs, or other relevant applications (mathworks.com) (Figs. 3, 4 and 5).

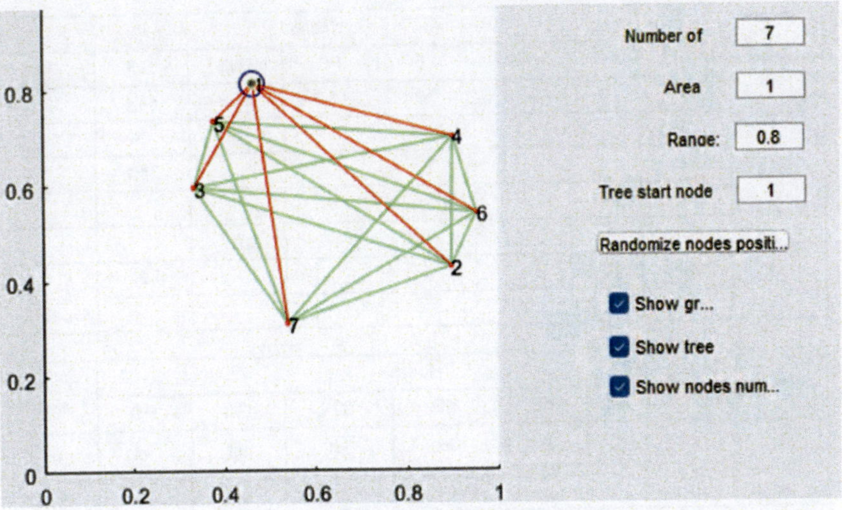

Fig. 3. Spanning tree

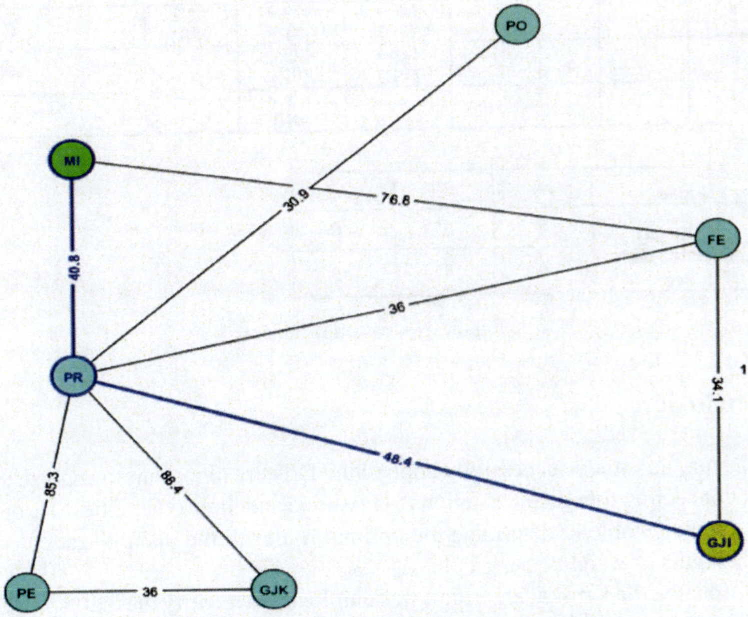

Fig. 4. Dijskra graph for the case of transport MI_PR-GJI

Distance		Nodes						
		PR	MI	FE	PE	GJK	GJI	PO
Nodes	PR	0	40.8	36	85.3	88.4	48.4	30.9
	MI		0	76.8				
	FE			0			34.1	
	PE				0	36		
	GJK					0		
	GJI						0	
	PO							0

Iteration	Nodes							Fixed node
	PR	MI	FE	PE	GJK	GJI	PO	
1	inf	0 (MI)	inf	inf	inf	inf	inf	MI
2	40.8 (MI)		76.8 (MI)	inf	inf	inf	inf	PR
3			76.8 (MI)	126.1 (PR)	129.2 (PR)	89.2 (PR)	71.7 (PR)	PO
4			76.8 (MI)	126.1 (PR)	129.2 (PR)	89.2 (PR)		FE
5				126.1 (PR)	129.2 (PR)	89.2 (PR)		GJI
6				126.1 (PR)	129.2 (PR)			PE
7					129.2 (PR)			GJK

Nodes	MI	PR	GJI
Distances	0	41	48
Cumulated distance	0	41	89

Fig. 5. Nodes and distances

5 Conclusion

In conclusion, this study successfully applied the Dijkstra algorithm to analyze the road network connecting four distinct stations to Prishtina and three other cities. Through this analysis, we were able to determine the optimal routes to the main plaza, considering the road lengths as weights in a graph.

Additionally, the Kruskal algorithm was employed to identify the least spanning tree in the Prishtina road network, shedding light on the essential connections within the city's transportation infrastructure.

Furthermore, we utilized Matlab and simulation software to evaluate the Dijkstra algorithm's outcomes, allowing us to calculate both the shortest route and the lowest cost. A comparative analysis was conducted between the two solutions, providing valuable insights into the effectiveness of the algorithm.

Moreover, this study extended its scope to address the distribution of goods from four directions to three separate warehouses in the city center. By first identifying the most cost-effective and expensive solutions, we then applied the minimum cost and MODI approach to determine the optimal distribution plan. This comprehensive approach ensures efficient goods distribution within Prishtina.

In summary, the combination of graph theory algorithms, simulation tools, and transportation optimization methods has enabled us to make informed decisions regarding road networks and goods distribution in Prishtina. These findings have practical implications for improving the city's transportation infrastructure and logistics operations.

References

1. Ortúzar, J., Willumsen, L.: Modelling transport. PriISBN:9780470760390, Online ISBN:9781119993308, https://doi.org/10.1002/9781119993308
2. Sallauka, A., Bajrami, X., Likaj, R., Gashi, B.: Application of genetic algorithm for finding shortest path in seven cities. IJCIET **i9**(4), 1516–1526 (2018)
3. Newman, M.: Networks: An Introduction. Oxford University Press, Oxford (2010). https://doi.org/10.1093/acprof:oso/9780199206650.001.0001
4. Guze, S.: Graph theory approach to transportation systems design and optimization. TransNav: Int. J. Marine Nav. Saf. Sea Transp. **8**(4), 572–578 (2014)
5. Likaj, R., Shala, A.: Application of graph search algorithm to find optimal solution for the problem of transport. Ann. Faculty Eng. Hunedoara **15**(4), 69–72 (2017)
6. Bektas, T., Laporte, G.: The pollution-routing problem. Transp. Res. Part B: Methodol **45**(8), 1232–1250 (2011)
7. Huang, D., et al.: Traffic congestion status identification method for road network with multi-source uncertain information. Acta AutomaticaSinica **44**(3), 533–544 (2018)
8. Likaj, R., Shala, A., Mehmetaj, M., Hyseni, P., Bajrami, X.H.: Aplication of graph theory to find optimal paths for the transportation problem. J. IFAC Proc. **46**(8), 235–240 (2013)
9. Likaj, R.: Application of graph theory to find optimal paths for the transportation problem. In: 15thWorkshop on International Stability, Technology, and Culture The International Federation of Automatic Control, pp. 235–240 (2013)

Comparison of Wind Action Calculation According to Albanian Norms and Eurocodes

Igli Kondi(✉), Elvis Capo, and Julian Kasharaj

Faculty of Civil Engineering, Polytechnic University of Tirana, Tirana, Albania
igli.kondi@fin.edu.al

Abstract. In Albania, gradually, the national norms for the design of building structures (Technical Design Conditions—Kushte Teknike Projektimi (KTP)) are being replaced by Eurocodes. Currently, wind action on buildings is calculated based on Kusht Teknik Projektimi 7–78 (KTP 7–78), which is a national norm approved in 1978. This paper aims to present the calculation of wind action on buildings based on the national norms of Albania as well as on Eurocodes, highlighting the differences between the two norms. At first is given the concept of how each of the above standards calculate the wind action according to the factors that they take in consideration and after that an example of a real building is taken in consideration. For this building is calculated the wind load pressure according to KTP-7-78 and Eurocode 1–4, so the differences are explained also in numerical values.

Keywords: Wind action · Eurocode1 · KTP 7-78

1 Introduction

For the majority of reinforced concrete buildings in Albania, the seismic action is the one that govern the design procedure. But lately, many high-rise buildings are constructed, and for this type of structures is necessary to take in consideration, during the design process, also the wind action. The wind action is the primary action that governs the design process of light short structures, like steel or timber buildings, this because of their relatively small mass which causes a slight seismic load. The scope of this paper is to calculate the wind action by calculating the wind pressure on external surfaces based on KTP 7-78 and Eurocode 1 Part 1-4: General Actions—Wind Action, for a high rise building which will be constructed in Tirana urban area. First will be given the respective formulas of both Codes, where similarities and differences will be pointed out. After this, those formulas should be applied to calculate the wind pressure on the external surfaces of this building. As is explained in the following sections, the Eurocode 1 approach is more detailed than KTP 7-78, by taking in account more effects that contribute to the final value of wind pressure.

G. Guxho et al. (Eds.): ITC-ICEE 2023, LNMIE, pp. 98–107, 2024.
https://doi.org/10.1007/978-3-031-48933-4_10

2 Wind Action

2.1 Wind Action According KTP 7 – 78

The basic wind pressure is given by:

$$P_0 = \frac{v^2}{16}\,(daN/m^2) \quad (1)$$

where v is the wind velocity in m/s calculated 10 m above ground.

Peak wind pressure is given by:

$$P_{\max} = k_a \cdot k_b \cdot P_0 \cdot k_{eksp} \quad (2)$$

While the wind pressure is given by:

$$p = k \cdot k_a \cdot k_b \cdot P_0 \cdot k_{eksp} \quad (3)$$

where

- k the aerodynamic coefficient that take in consideration the building shape.
- k_a coefficient that take in consideration the height of the building.
- k_b coefficient that take in consideration the altitude of the construction site from sea level.
- k_{eskp} is a coefficient that take in consideration the level of the exposure against the wind and is accepted 1.2 if the building is located near hills, highly inclined terrain or valleys.

The values of the above coefficients are given in KTP 7-78 and will be calculated in the following paragraphs.

2.2 Wind Action According to Eurocode 1

In order to derive the formula for wind pressure according Eurocode 1, first is necessary to give the concepts and the respective formula of the factors that contribute on the wind pressure.

The basic wind velocity, defined as function of wind direction and time of year at 10 m above ground of terrain category II, is given by:

$$v_b = c_{dir} \cdot c_{season} \cdot v_{b,0} \quad (4)$$

where

- $v_{b,0}$ the fundamental value of the basic wind velocity.
- c_{dir} the directional factor, depending on wind direction. The recommended value is 1.0.
- c_{season} the season factor and the recommended value is *1.0*.

Eurocode 1, take also in consideration the probability p of exceeding the 10 min mean wind on annual period. This is done by multiplying v_b with the probability factor that is given by:

$$c_{prob} = \left(\frac{1 - K \cdot \ln(-\ln(1 - p))}{1 - K \cdot \ln(-\ln(0.98))}\right)^n \tag{5}$$

where

K the shape parameter depending on the coefficient variation of the extreme – value distribution and is recommended $K = 0.2$.

n the exponent and is recommended $n = 0.5$.

The recommended value of probability is $p = 0.2$.

Eurocode 1 recommend $c_{prob} = 1$.

The next step after calculating v_b is to calculate the mean wind velocity $v_m(z)$ at a z height above the ground. This is given by:

$$v_m(z) = c_r(z) \cdot c_0(z) \cdot v_b \tag{6}$$

$c_0(z)$ is the orography factor and is to taken as 1.0. Eurocode 1 Part 1–4: General actions—Wind Actions specifies that if where the orography (hills, cliffs, etc.) increases the wind velocity by more than 5% the effects should be taken in consideration by using c_0.

$c_r(z)$ is the roughness factor and accounts for the variability of the mean wind velocity at the site of the building due to height above ground level and the ground roughness of the terrain. The recommended procedure of Eurocode 1 for calculating $c_r(z)$ is given by:

$$\begin{cases} c_r(z) = k_r \cdot \ln\left(\frac{z}{z_0}\right) for\, z_{\min} \le z \le z_{\max} \\ c_r(z) = c_r(z_{\min}) for\, z \le z_{\min} \end{cases} \tag{7}$$

where

z_0 is the roughness length.

k_r is the terrain factor depending on roughness length z_0, given by:

$$k_r = 0.19 \cdot \left(\frac{z_0}{z_{0,II}}\right)^{0.07} \tag{8}$$

where

$z_{0,II} = 0.05\ m$ for terrain category II.

$z_{\min}$ is the minimum height.

$z_{\max}$ is to be taken as 200 m.

z_0 and $z_{\min}$ depend on the terrain category. The recommended values are specified on Eurocode 1 Part 1–4: General actions—Wind Actions.

Another effect that Eurocode 1 Part 1–4: General actions—Wind Actions take in consideration is the wind turbulence, which is expressed by the turbulence intensity $I_v(z)$ at height z given by:

$$\begin{cases} I_v(z) = \dfrac{\sigma_v}{v_m(z)} = \dfrac{k_I}{c_0(z) \cdot \ln(z/z_0)} \, for \, z_{\min} \leq z \leq z_{\max} \\ I_v(z) = I_v(z_{\min}) \, for \, z \leq z_{\min} \end{cases} \tag{9}$$

where

$\sigma_v = k_r \cdot v_b \cdot k_I$ is the standard deviation of turbulence.
k_I is the turbulence factor and the recommended values is 1.0.
c_0 is the orography factor as previously specified.
z_0 is the roughness length as previously specified.

The next step is to determine the peak velocity pressure $q_p(z)$ at height z. The recommended formula is:

$$q_p(z) = [1 + 7 \cdot I_v(z)] \cdot \frac{1}{2} \cdot \rho \cdot v_m^2(z) = c_e(z) \cdot q_b \tag{10}$$

where

$q_b = \frac{1}{2} \cdot \rho \cdot v_b^2$ the basic velocity pressure with a recommended values of air density (which depends on the altitude, temperature and barometric pressure) $\rho = 1.25\ kg/m^3$.
$c_e(z) = \frac{q_p(z)}{q_b}$ the exposure factor.

Finally, the wind pressure on external surfaces is given by:

$$w_e = q_p(z_e) \cdot c_{pe} \tag{11}$$

where

$q_p(z_e)$ the peak velocity pressure.
z_e the reference height of the external pressure for windward walls of rectangular build-ings evaluated according to Figure 7.4 of Section 7.2.2. of Eurocode 1 Part 1–4: General actions – Wind Actions.
c_{pe} is the pressure coefficient for the external pressure evaluated according to Figure 7.5 (Zone D) and table 7.1 of Section 7.2.2. of Eurocode 1 Part 1–4: General actions—Wind Actions.

3 Numerical Example

In order to display the differences also in numerical values, following is calculated the wind pressure on the external surface of building that will be constructed in Tirana, according both codes. This building will be located in an urban area closely to city center (Fig. 1).

Building height is 102.58 m from ground and in plane dimensions are 29.6 m × 32.4 m (Fig. 2).

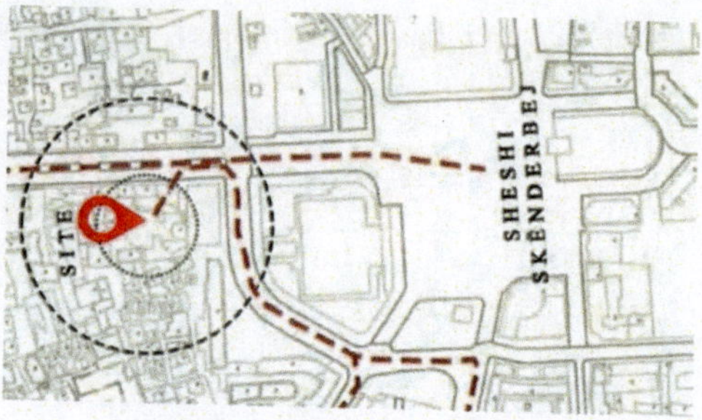

Fig. 1. Building location

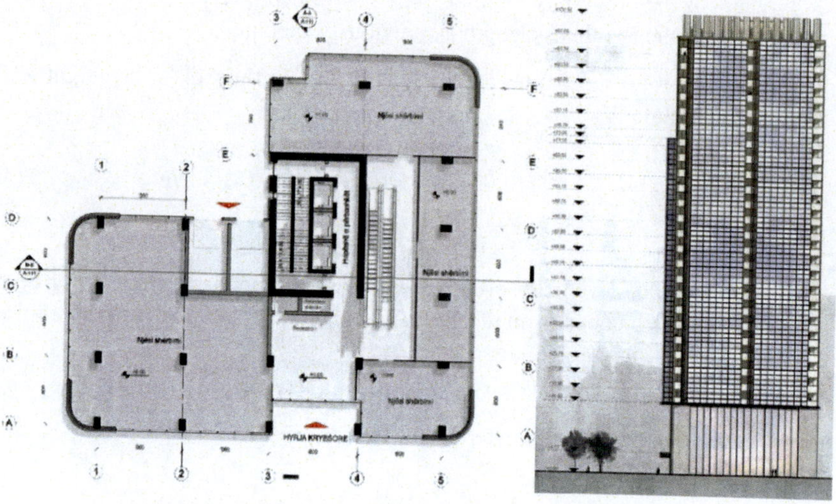

Fig. 2. Ground story plan and east façade of the building

3.1 Wind Pressure According KTP 7-78

According to KTP 7-78 Albanian territory is divided into two zones regarding wind velocity. The first zone, more than 10 km from sea side, has $v = 25\ m/s$, the second zone, less than 10 km from sea side, has $v = 31\ m/s$.

For this building, located at this site, KTP 7-78 accept $v = 25\ m/s$.

According to KTP 7-78, the wind pressure distribution along the height of the building varies with height. This variation is expressed with the use of k_a and the respective values are given one KTP 7-78 (Table 1, KTP 7-78).

$k = 0.8$ windward and $k = -0.6$ leeward (Table 3, KTP 7-78).

$k_b = 1.0$ (Table 2, KTP 7-78. The building is located at site less than $1200\ m$ above sea level).

$k_{eksp} = 1.0$ (The building is located in flat terrain apart of hills, inclined terrain or valleys).

For each height, with the help of expression (1)–(3) are calculated the values of wind pressure (windward and leeward). The results are given in Table 1.

Table 1. Values of all the factors contributing on wind pressure according KTP 7-78.

Height (m)	10	20	40	60	80	102.58
v (m/s)	25	25	25	25	25	25
P_O (daN/m^2)	39.06	39.06	39.06	39.06	39.06	39.06
ka	1	1	1	1	1	1
kb	1	1	1	1	1	1
keksp	1	1	1	1	1	1
P_{max} (daN/m^2)	39.06	50.78	66.40	72.89	79.40	85.938
P_{max}/P_O	1	1.3	1.7	1.866	2.033	2.2
k (windward)	0.8	0.8	0.8	0.8	0.8	0.8
k (leeward)	−0.6	−0.6	−0.6	−0.6	−0.6	−0.6
p (windward) (daN/m^2)	31.25	40.62	53.12	58.31	63.52	68.75
p (leeward) (daN/m^2)	−51.56	−51.56	−51.56	−51.56	−51.56	−51.56

3.2 Wind Pressure According Eurocode 1 Part 1–4: General Actions—Wind Actions

From Sect. 7.2.2. of Eurocode 1 Part 1–4: General actions–Wind Actions Sect. 7.2.2., since the condition h > 2b is fulfilled, the following wind pressure distribution is assumed (Fig. 3).

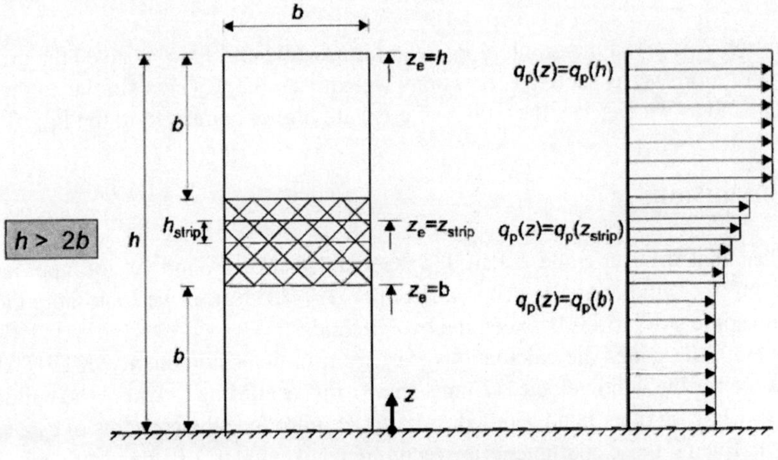

Fig. 3. Wind pressure distribution along the building height

The building is assumed to be divided in three parts, first part from ground upward till the height b = 32.4 m, the second part from top downward till the height b = 32.4 m and the middle part with a height of 102.58–64.8 = 37.78 m. For this case, the middle part is assumed to be divided on three strips, each of a height of 12.6 m. It is clear that more strips give a more accurate value of wind pressure distribution. With the above division of building, the heights z for each part are given in Table 2.

Table 2. z values for different part of the building.

Part	z value (m)
1	32.4
Strip 1	45
Strip 2	57.6
Strip 2	70.18
2	102.58

For each height, with the help of expressions (4) to (11) are calculated the values of wind pressure (windward and leeward). Based on the building location, the terrain category is assumed IV. The results are given in Table 3.

Pressure distribution along height is given in Fig. 4.

As it is noticed, at the end of the calculation, each code gives different values for wind pressure. The wind pressure resulting from Eurocode 1 Part 1–4: General Actions—Wind Actions is higher, compared to the wind pressure resulting from KTP 7–78, and this is in the same line with previous similar calculations that the authors has done for many other buildings. So, we could say that Eurocode 1 is a more conservative Code, and from authors experience, this is also true for all Eurocodes regarding the structural design.

On Fig. 7. is given the graph of P_{max}/P_0 ratio, while on Fig. 6. is given the graph of $c_e(z)$. Based on the respective expressions we can say that $c_e(z)$ is similar to the ratio P_{max}/P_0, and as we can see, the values of $c_e(z)$ are higher compared to the P_{max}/P_0.

4 Conclusions

It is clear that the Eurocode 1 Part 1–4: General Actions Wind Action approach of calculating the wind action is different from KTP 7-78. But, if we look more closely, there are some similarities between the two methods.

For both the codes the calculations start from basic wind velocity, KTP 7-78 has fixed value for the wind velocity (25 m/s and 31 m/s according to where the building is located), while on other hand, Eurocde 1 Part 1–4, introduce the concepts of c_{dir}, c_{season} and c_{prob}. Even if these coefficients are recommended equal to 1.0, they could take other values depending on specific condition. After the basic wind velocity, Eurocode 1 Part 1–4, introduce the concept of mean wind ($v_m(z)$), which seems to be missing on KTP 7-78, but if we analyses the orography factor ($c_0(z)$) and roughness factor ($c_r(z)$) it is very clear that these both coefficients have their counterparts on KTP 7-78. So $c_0(z)$ is similar to k_{eksp} of KTP 7-78 because they both are related with the presence of landscapes (hills,

Table 3. Values of all the factors contributing on wind pressure according eurocode 1 Part 1:4

z (m)	32.4	45	57.6	70.18	102.58
v_{b0} (m/s)	25	25	25	25	25
c_{dir}	1	1	1	1	1
c_{season}	1	1	1	1	1
c_{prob}	1	1	1	1	1
v_b (m/s)	25	25	25	25	25
$c_0(z)$	1	1	1	1	1
z_0 (m)	1	1	1	1	1
z_{min} (m)	10	10	10	10	10
k_r	1	1	1	1	1
σ_v	5.858	5.858	5.858	5.858	5.858
$I_v(z)$	0.288	0.263	0.247	0.235	0.216
$q_p(z)$ (daN/m^2)	78.17	88.23	96.10	102.58	115.52
q_b (daN/m^2)	39.063	39.063	39.063	39.063	39.063
$c_e(z)$	2.001	2.259	2.460	2.626	2.957
c_{pe} (windward)	0.8	0.8	0.8	0.8	0.8
c_{pe} (leeward)	−0.6083	−0.6083	−0.6083	−0.6083	−0.6083
w_e (windward) (daN/m^2)	62.54	70.59	76.88	82.072	92.416
w_e (leeward) (daN/m^2)	−70.271	−70.271	−70.271	−70.271	−70.271

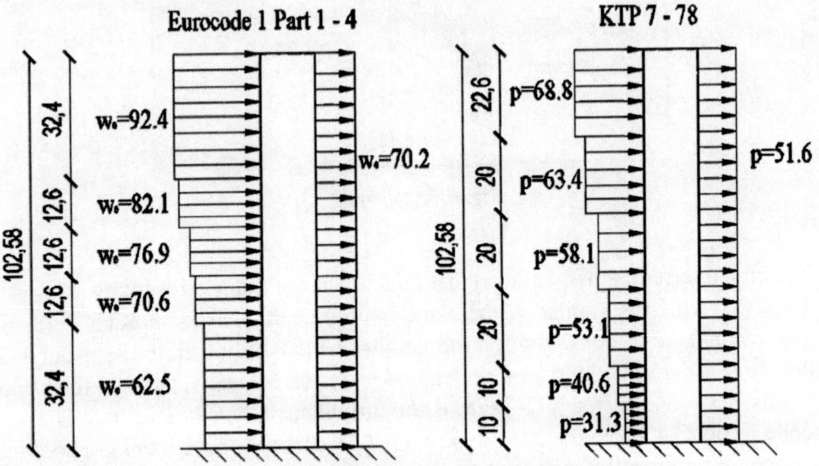

Fig. 4. Wind pressure (in daN/m^2) distribution according Eurocode 1 Part 1–4 and KTP 7-78.

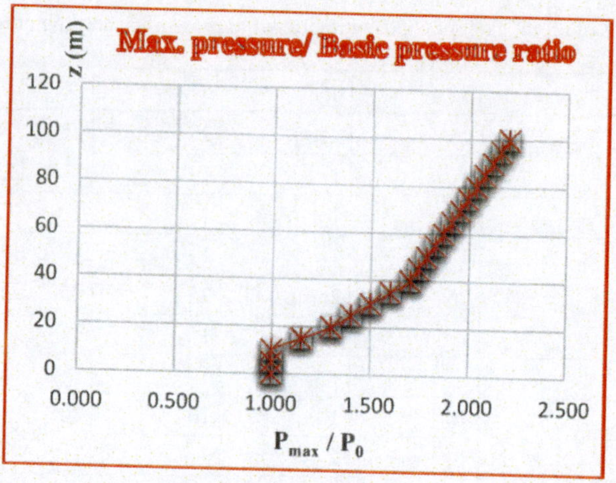

Fig. 5. P_{max}/P_0 ratio graph

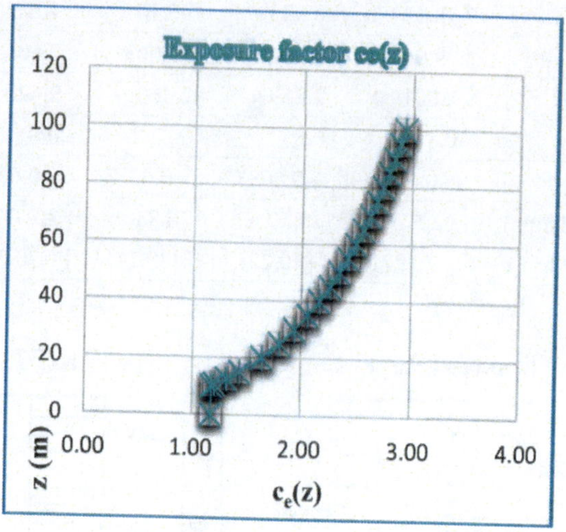

Fig. 6. $c_e(z)$ graph.

highly inclined terrain, cliffs, valleys) that could increase the wind velocity. This is true also for $c_r(z)$, which is similar to the aerodynamic coefficient k_a of KTP 7-78. Both the coefficients take in consideration the height of the building from the ground. It is true that $c_r(z)$ is determined in a more detailed way, but, basically, they are the same. A completely new concept for KTP 7-78, is the turbulence intensity $I_v(z)$ that is used on Eurocode 1 Part 1–4. From the other hand KTP 7-78 uses the concept of k_b, e coefficient that is account for the altitude of the building site from see level, which is not mentioned in Eurocode 1 Part 1–4. Another similarity is found between the basic wind pressure P_0 and basic velocity pressure q_b. Their expression seems different, but give the same results (for the recommended value of air density of 1.25 kg/m^3).

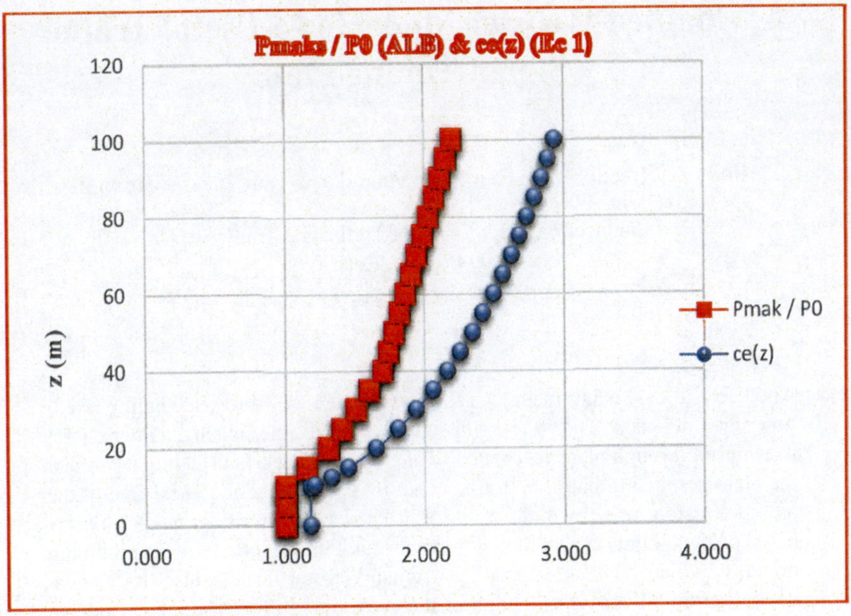

Fig. 7. Comparative graph between P_{max}/P_0 and $c_e(z)$

Another similarity, is the pressure distribution along the building height. Both the codes accept a variable pressure windward and constant pressure leeward. But the approach that they use for this variation are different. KTP 7-78 achieve it by the usage of the aerodynamic coefficient k_a, while Eurocode 1 part 1–4 divides the building in three zones once the condition $h > 2b$ is fulfilled. Another similarity is the usage of c_{pe} and k, both coefficients take in consideration the building shape, and for our building (rectangular shape) both the coefficients takes the same value (windward $c_{pe} = k = 0.8$, leeward $c_{pe} = -0.6083$ while $k = -0.6$).

So, in the end, we could say that even though Eurocode 1 part 1–4 is different from KTP 7-78, and the numerical results that came out from them differ, there are also many similarities between the two codes. Eurocodes as relatively new codes compared to KTP 7-78, and continually in change and in improvement gives more detailed results that the outdated KTP 7-78.

References

1. Kusht Teknik Projektimi 7–78 (KTP 7–78)
2. Eurocode 1 Part 1–4: General Action – Wind Action

A Modified Webster Model for a Useful Traffic Signal Timing Plan

Arlinda A. Rrecaj(✉), Ramadan Mazrekaj, and Gëzim Hoxha

Faculty of Mechanical Engineering, Department of Traffic and Transport, University of Pristina, 10000 Pristina, Kosovo

arlinda.alimehaj@uni-pr.edu

Abstract. Today, many methods are known in the world for calculating fixed-time signaling plans. The effective operation of fixed-time signalized intersection is achieved through plans calculated on the basis of peak-hour field data so that the plans respond to both free traffic conditions and congested ones. One of the most frequently used methods is the well-known Webster's formula, which is based on the optimal calculation of the cycle duration that promises the minimum drivers experienced delays. Since the original Webster formula also has its own flaws regarding the overestimation of the cycle time in cases where the degree of saturation rate corresponds to congestion conditions, in this paper a modified model is pro-posed. In order to obtain reasonable cycle length for every range of saturation rates the modified model is based on optimized timing plan driven by Synchro/Sim Traffic soft-ware which is mostly compatible with Highway Capacity Manual (HCM 2010).

Keywords: Traffic signals · Cycle · Green time · Saturation rate · Delays

1 Introduction

The intersection is a vital element of urban road networks which plays a key role in operation of the entire road network of the city. Webster built a stochastic model with the aim of estimation of the cycle length that minimizes the overall vehicle delay at intersection approaches. However, even though a cycle calculated with the Webster's formula linearly increases with the degree of saturation, at high values that approach the value of 1.0, these cycles take meaningless values. Many researchers have been conducted toward improving the performance of the stochastic delay models. A well designed cycle length can decrease or prevent traffic congestion and reduce noise pollution, emissions, energy consumption and travel time. In [1] K. Hewage used a Special-Purpose Simulation Tool (SPS) for optimization of traffic signal timing plans. Author Y. Shen in [2] proposed an optimization model of signal timing plan that comprehensively considers traffic emissions based on synchro software. H. Zhang et al. in [3], developed an algorithm for optimizing signal plans in order to minimize the total delay of the intersection. Terzi et al. [4] used Elimination Pairing System (EPS) for optimization of traffic signal timing at oversaturated intersections. The main objective of this paper is to obtain a model that

G. Guxho et al. (Eds.): ITC-ICEE 2023, LNMIE, pp. 108–120, 2024.
https://doi.org/10.1007/978-3-031-48933-4_11

appropriately can be utilized according to the degree of saturation and promises satisfactory cycles in terms of performance indicators such as maximal vehicle queue length and control delay which to a large extent present the convenience and comfort of drivers in the close to signalized intersections. The rest of this paper is organized as follows. Section 2 presents the Webster model accompanying with the time-dependent control delay model queue length estimation and the strategy of the model built are given in Sects. 2.1 and 2.2 respectively. A comparative analysis with results and explication of results are provided in Sect. 3. Finally, conclusions are given in Sect. 4.

1.1 Cycle Length Estimation Based on Webster Formula

Since the aim of this research is to modify the coefficients of the original Webster formula, *a* and *b*, the original formula is given below [5] and [6].

$$C = \frac{a \cdot L + 5}{1 - b \cdot Y} \sim \frac{1.5 \times L + 5}{1 - Y} \, [\mathrm{s}] \tag{1}$$

Lost time represents the times not effectively used by the motorists during the cycle comprised of the sum of protective times between successive phases *i* and *j*, $t_p(s)$, start-up lost time, *a* (s) and the unused second part of the yellow interval, *c (s)*.

$$L = n, (a + c) + \sum_{i=1}^{n} t_p[\mathrm{s}] \tag{2}$$

where

- L total lost times in cycle (s)
- n is the number of phases,
- a startup lost time in the beginning of green interval (s),
- c unused part of yellow interval (s),
- t_p designed protective times between successive phases (s).

2 The Proposed Modified Model

Since the main goal of this paper is to derive a new modified Webster's formula that would provide reasonable and optimal cycle values, a series of experiments have been performed with a large number of traffic and lost time data. The subject of the experiment is a four-way intersection in the city of Vushtrria. Initially, to highlight the shortcomings of Webster's classic model, signalization plans are calculated for the given intersection but for several scenarios, i.e. for increased traffic values. As we can see from table no. With the increase in traffic, for example by increasing from 50 aut/h, the duration of the cycle also increases, even for a certain increase the cycle takes abnormal values, e.g. 800 s, which with the modified model would not have happened.

In order to derive the new model, the optimization of the cycle duration must be done. In this thesis, Synchro/Sim Traffic Version 8 software is chosen to do the optimization. The reason for choosing this software is because it is compatible with HCM 2010. In the software, solutions were requested for different volumes increasing by 50 aut/h as well as lost times L (s) increasing by 10s up to 16s, changing the values of the amount of unused time (a + c)—losses at the start and unused yellow. As we can see from the table. These optimized cycles are then used in the Mat lab software to obtain the new cycle model in function of lost times and degree of saturation.

$$C = f(L, Y), \tag{3}$$

where the goal is to obtain different values of coefficients a, b and c in the formula according to Webster (1).

2.1 Strategy for Model Building

The experiment was carried out through the following procedure:

1. It starts with the lowest value of the lost times, L(s), and with the actual turnover values Q(aut/h), that means not increased. Data is entered into Synchro and cycle time optimization is done when it is known that this cycle offers low time losses and the best service level. It continues with the same circulation values for another 233 scenarios, it means for every a(s) that increases from 2 to 4 for every 0.1 value and for every value of a(s) the values of c(s) change from 0.5 to to 1.5 increasing for every 0.1 we enter the data of the cycle in Excel.
2. The same procedure is carried out for increased traffic values, i.e. adding from 50 to 250 aut/h. The increase in the size of the circulation is done until the degree of saturation Y approaches the value 1.0. The aim is to investigate the value of the cycle after optimization. In total, we have 945 cases or scenarios. We transfer all these data to Excel, which can be found in the appendix of this thesis.
3. We import the data obtained per cycle C, lost times L and saturation degrees Y into the Matlab software. To obtain the values of the regression parameters a,b and c we used the Curve Fitting Tool. (Fig. 1). To obtain the cycle time equation as a function of L and Y, instead of the dependent variable z we write the data for C, while the independent variables x and y side L and Y, respectively.
4. Through the new formula that we obtained in Mat lab, we calculate again the signaling plans, i.e. the duration of the C cycle and the green intervals of the phases, for different traffic values, according to the steps above. Then we also calculate the time losses with the two methods in order to make a comparison.
5. Cycle time optimization is done for all scenarios, starting from current circulations and for all cases, inserting data according for each value of Y. The intersection is preliminarily modeled with all geometric features, such as the number of lanes, width, then the phase diagram proposed as with the classic Webster method, limited speed, etc., (Fig. 1).

Diagrams of the dependence of the cycle duration (after optimization with Synchro) from the lost times are constructed separately for each scenario.

As we can see in the following figure (Fig. 2), for a low value of the degree of saturation Y = 0.499 synchro does not give us values that differ much regardless of the values of the lost times L. So we have a value of 55 s for any value of L from 10 to 16 s (Figs. 3 and 4).

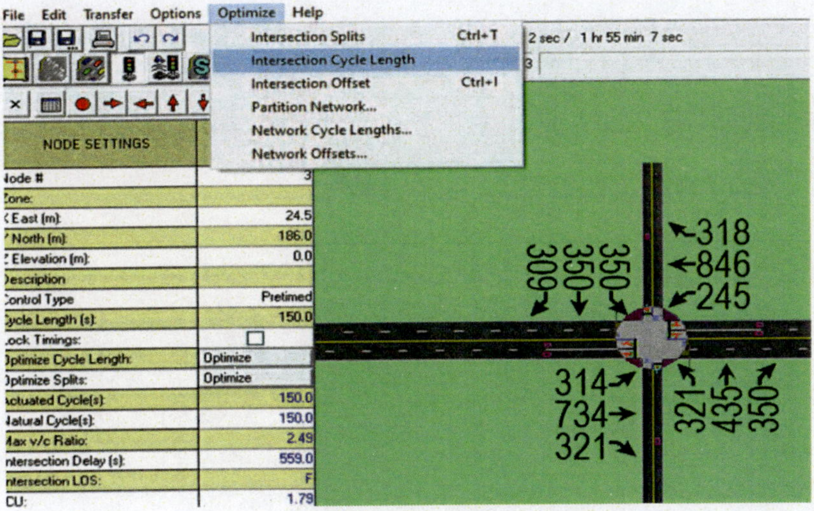

Fig. 1. Optimization with synchro/sim traffic

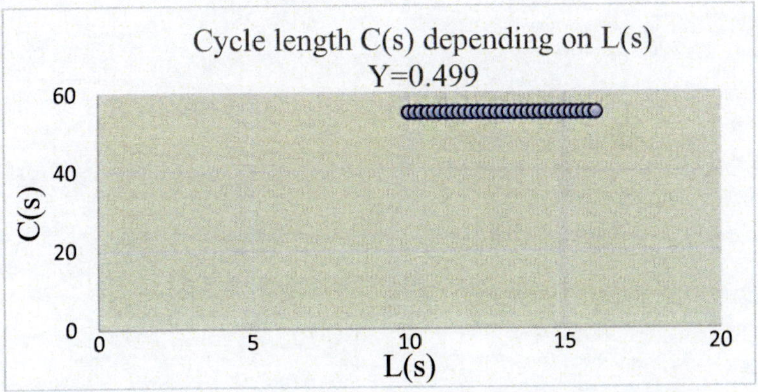

Fig. 2. Dependence of the cycle on the lost time for Y = 0.499.

Unlike the anomalous Cycle values that we obtained with Webster's original method, where values over 820 s were practically unacceptable, Synchro has optimized values for the degree of saturation close to 1.0 from C = 150 s to C = 200 s. The graphic representation of the dependence of the cycle C on the lost time L, for Y = 0.976 is given in the diagram of Fig. 5.

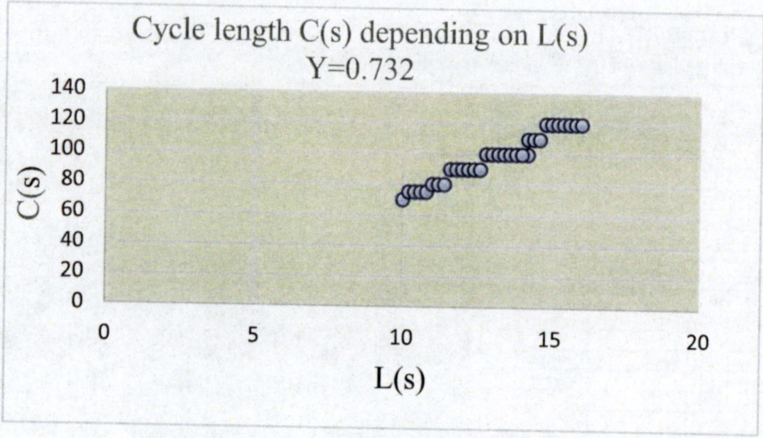

Fig. 3. Dependence of the cycle on the lost time for Y = 0.732.

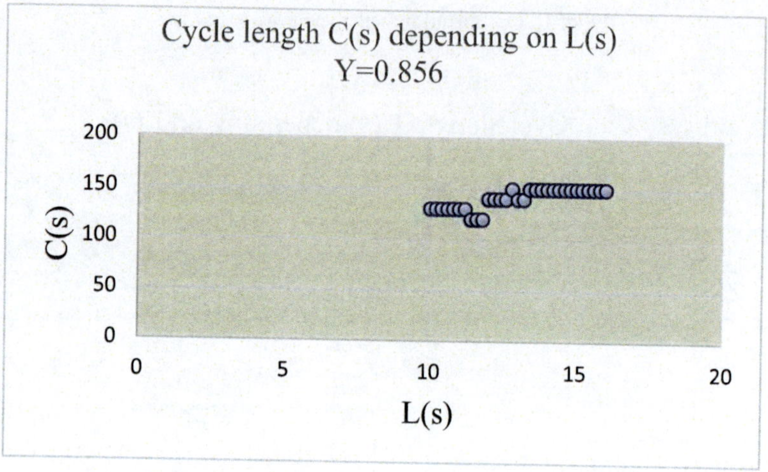

Fig. 4. Dependence of the cycle on the lost time for Y = 0.856.

2.2 Use of Mat Lab Curve Fitting Tool

Once we have obtained the values per cycle for all cases in Synchro, we then import them into Mat lab. In addition to the cycle, data such as L and Y are also imported, where in total we have about 945 data for each of these parameters (Fig. 6). After we used the Curve Fitting Tool, the Eq. (4) was obtained that differs from that of Webster, because in this case the values of the regression coefficients are a = 1.8, b = 12.21 and c = 0.834, therefore:

$$C = \frac{1.8 \cdot L + 12.21}{1 - 0.834 \cdot Y}(\mathrm{s}) \tag{4}$$

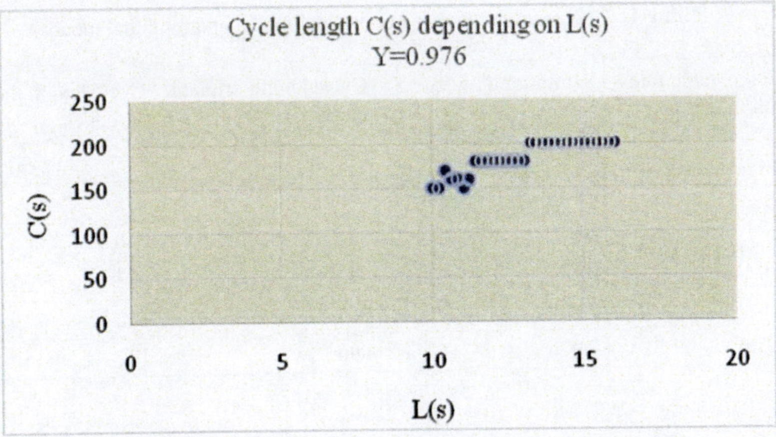

Fig. 5. Dependence of the cycle on the lost time for Y = 0.976.

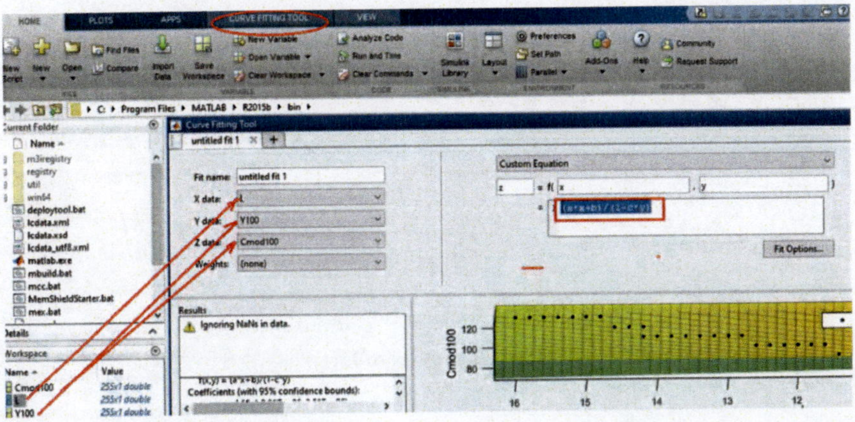

Fig. 6. The curve fitting tool application of the Matlab software through which we obtained the equation of the modified Webster model.

3 Comparative Study

After we have two Webster models, the original and the modified one, the cycle duration calculations were made according to the example in unit 2 (for the given intersection), for both methods with case c′ and the comparisons were made (Table 1).

Comparisons for all regimes are given in (Figs. 7, 8, 9, 10 and 11). The corresponding curves of the model are marked in color through the legend.

At the end, the diagram of the general comparison of the two methods, for all the values of the saturation levels, is given. (Fig. 12).

Based on these results of the created model, it is observed that after the value of the $Y = 0.7$ degree of saturation, the duration of the cycle increases for each increased value of the lost time.

Table 1. Cycle length for Webster original model and modified model

Increased value rates Q (veh/hr)	Saturation flow rate Y	Cycle length original model Co $C = \frac{1.5 \cdot L + 5}{1 - 1.0 \cdot Y}$	Cycle length modified model C_{mod} $C = \frac{1.8 \cdot L + 12.21}{1 - 0.834 \cdot Y}$
Q	0.499	43	55
Q + 50	0.618	56	66
Q + 100	0.738	82	83
Q + 150	0.856	150	112
Q + 200	0.976	896	172

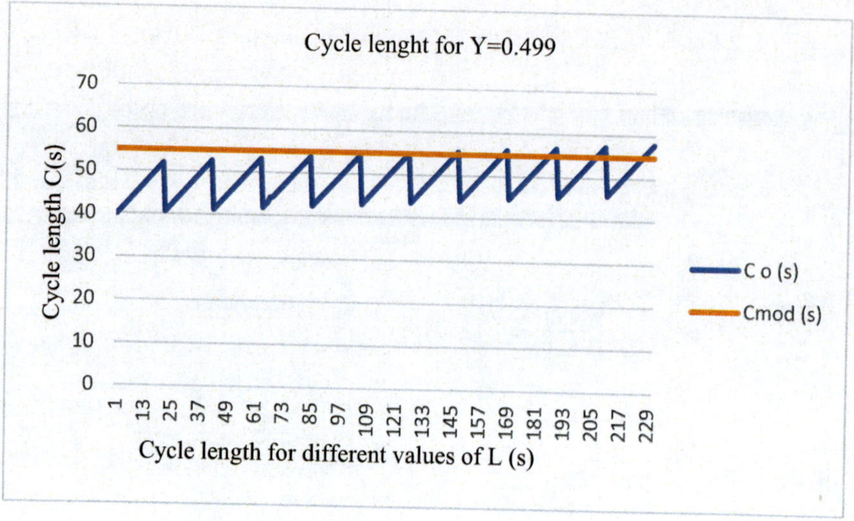

Fig. 7. Cycle length for two models for Y = 0.499

3.1 Comparison of Performance Indicators

The graphical results of the overall delays are depicted on Fig. 13 and Table 2.

Based on the time loss results, as can be seen from the diagrams, the modified method offers lower time loss for saturation levels above Y 0.7.

With the curve in blue color are the losses of Webster's method, while in orange color of the modified one.

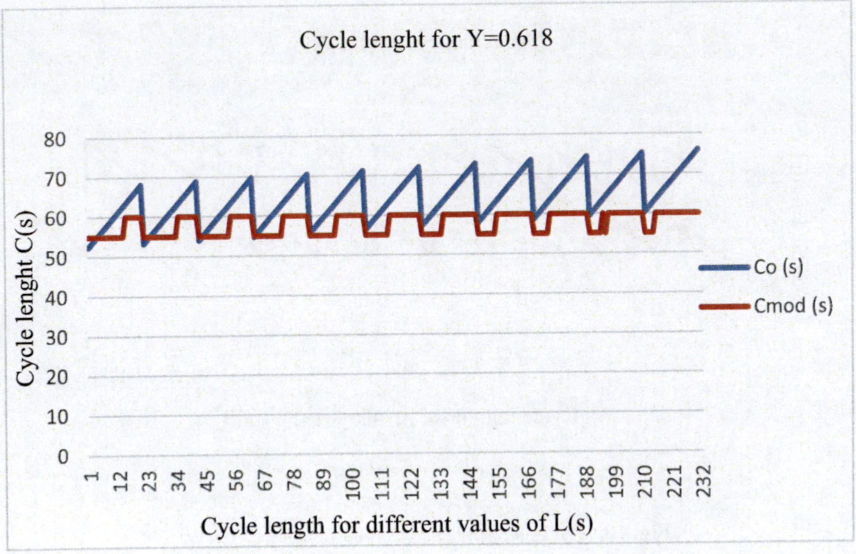

Fig. 8. Cycle length for two models for Y = 0.618

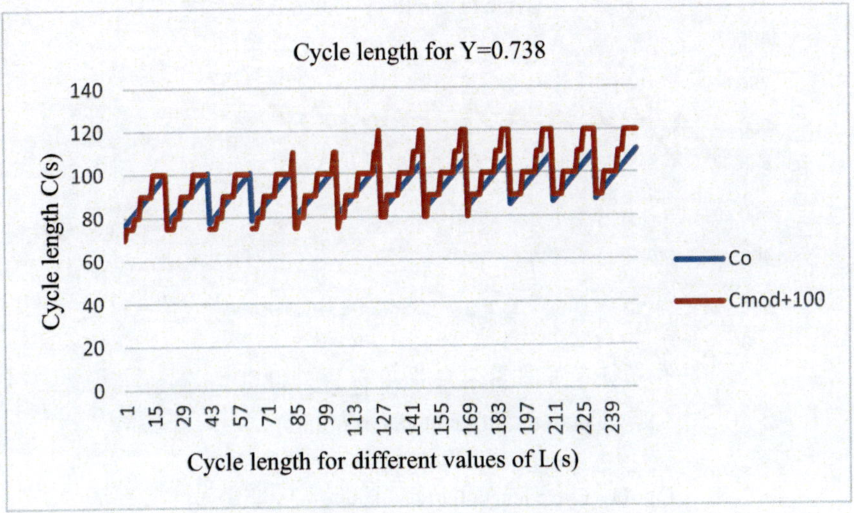

Fig. 9. Cycle length for two models for Y = 0.738

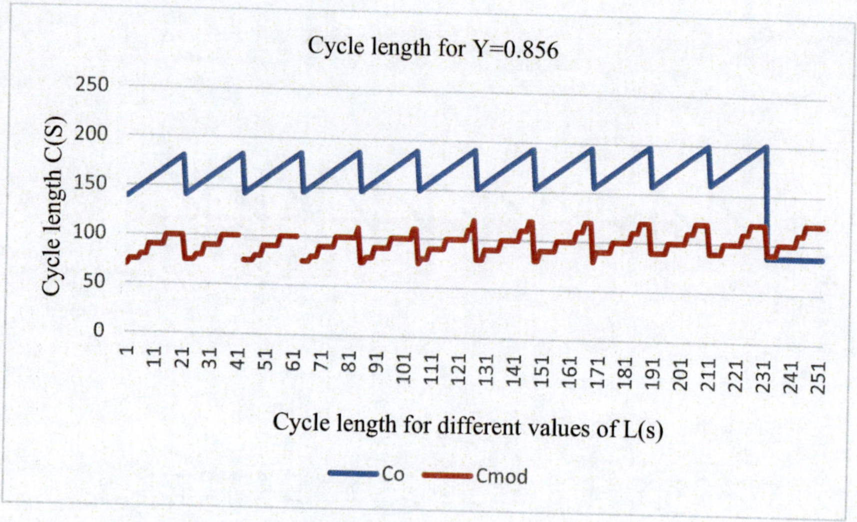

Fig. 10. Cycle length for two models for Y = 0.856

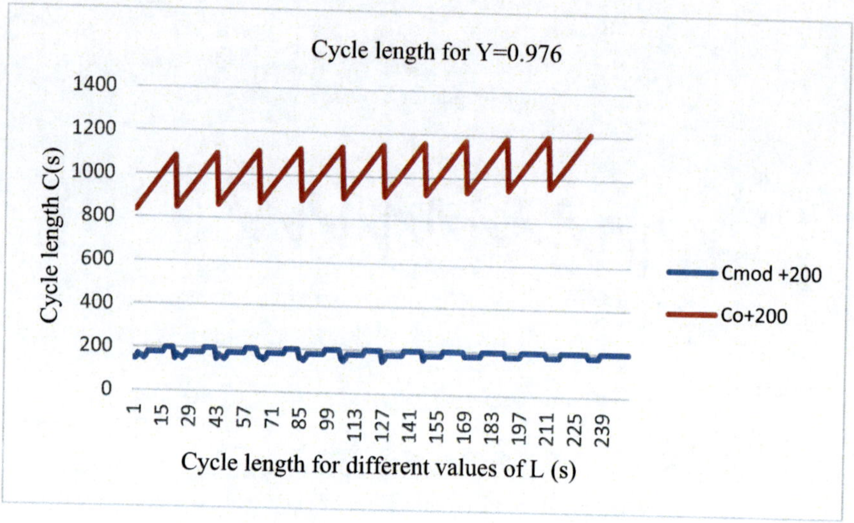

Fig. 11. Cycle length for two models for Y = 0.976

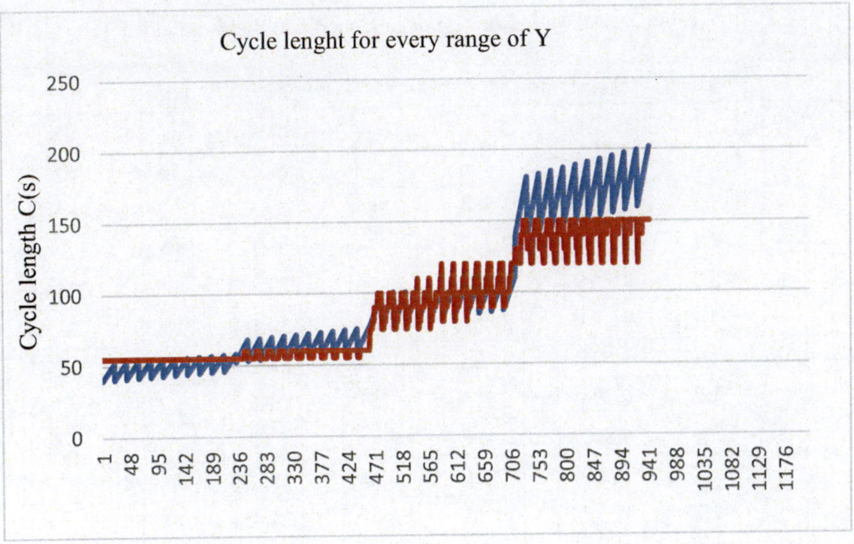

Fig. 12. Cycle length for two models (from Y = 0.499 to Y = 0.976)

4 Conclusions and Recommendations

In this thesis, a new Webster model is proposed with some modifications of the original model. The purpose of the finding is that the original model, although it is based on minimal time loss according to the author, overestimates the duration of the cycle if we are dealing with high degrees of saturation, especially for degrees that approach the value of 1.0. Also, the goal is to put in a model that is based on time loss or minimal delays calculated according to HCM, and this has been enabled by the Synchro software.

The new model offers better results, especially for the degree of saturation above 0.7, while for the low ones it does not differ much from the classic model.

With the new model, the problem of cycle overestimation has been overcome. By overestimation we mean the growth to infinity of the cycle if Y continues to grow. The new model after a certain value does not increase the cycles to infinity. We notice this from the example with the old method, the unacceptable value of the cycle was obtained around 820 s for Y approaching 1.0.

The new model also shows good performance in terms of time loss and service level, compared to the old one of Webster.

In the end, it is recommended to propose some exponential model based on optimized cycle according to minimum time loses.

Table 2. Delays for the original and modified Webster model

	Lane	Webster model			Modified model		
Y = 0.499		d	d1	d2	d	d1	d2
		6.33	7.006	13.336	6.63	5.546	12.176
	1.2	5.301	1.584	6.885	5.556	1.318	6.874
	2.1	5.738	2.235	7.973	6.012	1.816	7.828
	3.1	15.656	8.105	23.761	19.336	6.555	25.891
	4.1	16.335	20.675	37.01	20.169	13.39	33.559
Y = 0.618	1.1	8.023	9.725	17.748	8.702	8.776	17.478
	1.2	6.296	1.855	8.151	6.825	1.718	8.543
	2.1	6.809	2.471	9.28	7.379	2.26	9.639
	3.1	19.098	9.541	28.639	22.809	10.298	33.107
	4.1	19.972	18.12	38.092	23.851	19.956	43.807
Y = 0.738	1.1	12.132	14.69	26.822	12.736	15.51	28.246
	1.2	8.812	2.224	11.036	8.882	1.889	10.771
	2.1	9.52	2.855	12.375	10	2.956	12.956
	3.1	27.722	14.182	41.904	28.288	11.802	40.09
	4.1	29.059	27.055	56.114	29.644	28.457	58.101
Y = 0.856	1.1	22.09	21.186	43.276	18.734	26.11	44.844
	1.2	14.568	2.053	16.621	12.38	2.822	15.202
	2.2	23.86	28.088	51.948	20.325	35.87	56.195
	3.1	49.5	18.531	68.031	37.93	21.825	59.755
	4.1	51.954	35.13	87.084	39.88	42.617	82.497
Y = 0.976	1.1	130.374	29.577	159.951	31.5	43.503	75.003
	1.2	76.49	2.754	79.244	18.466	3.347	21.813
	2.1	82.434	3.271	85.705	19.88	4.091	23.971
	3.1	291.133	25.185	316.318	58.574	33.637	92.211
	4.1	307.159	47.266	354.425	61.775	63.486	125.261

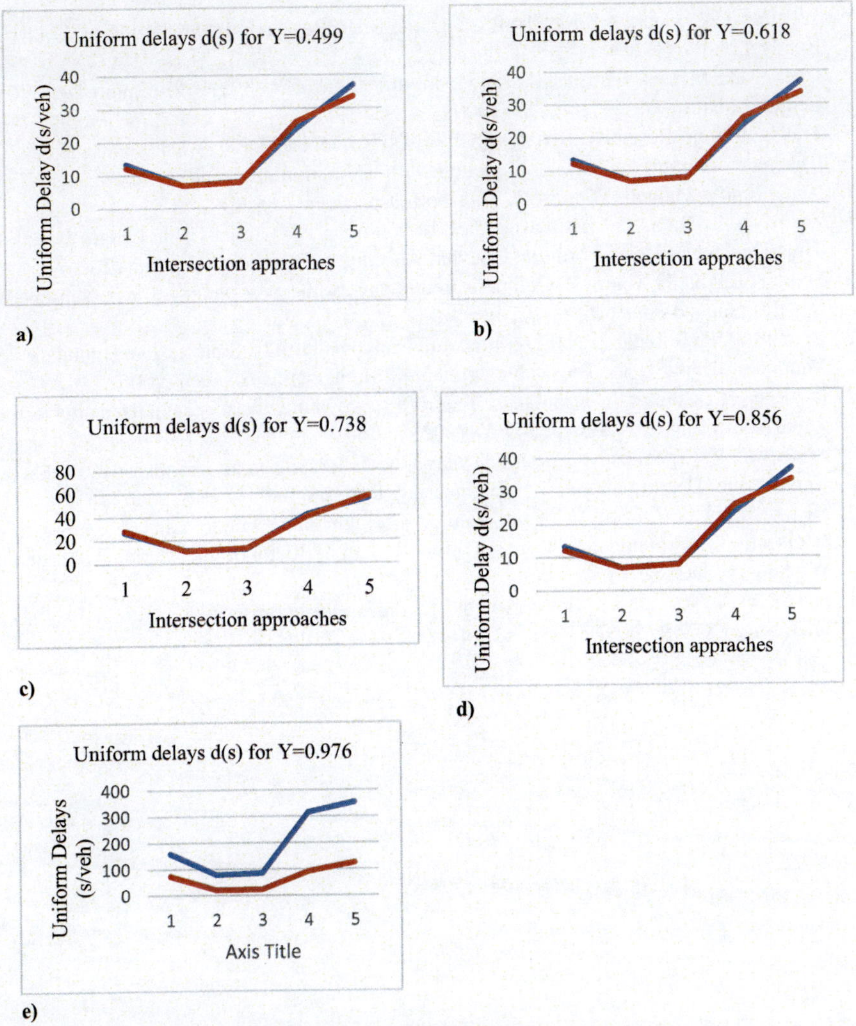

Fig. 13. Comparison of uniform delay for each intersection approach values for two models for, **a**) Y = 0.499, **b**) Y = 0.618, **c**) Y = 0.738, **d**) Y = 0.856 and **e**) Y = 0.976

References

1. Hewage, K., Ruwanpura, J.: Optimization of Traffic Signal Light Timing Using Simulation (2004)
2. Shen, Y.: An optimization model of signal timing plan and traffic emission at intersection based on Synchro. IConCHE 2018 (2018)
3. Zhang, H., Yan, H., Yu, W., Wang, C., Wang, J., Gao, Y.: Traffic light optimization based on modified Webster function. J. Adv. Transport. Hindawi (2021)
4. Patel, V.J., Yadav, V., Parmar, P.: Methods for optimization of signal cycle length. Glob. Res. Develop. J. Eng. **3**(4), 135–140 (2018)

5. Webster, F.V., Cobbie, B.M.: Traffic Signals, Road Research Technical Paper. Ministry of Roads, London (1966)
6. Cheng, D., Messer, C, Tian, Z, Liu, J: Modification of Webster's Minimum Delay Cycle Length Equation Based on HCM (2000)
7. Transportation Research Board, TRB ANNUAL MEETING, Washington, D.C. 2003. Chapter 6, Number of stops and queue length estimation at signalized approaches, Traffic signal timing Manual, Minnesota, Transport Department (2006)
8. Zakariya, A., Rabia, S.: Estimating the Minimum Delay Optimal Cycle Length Based on a Time-Dependent Delay Formula. Elsevier, Alexandria Engineering Journal (2016)
9. Dion, F, Rakha, H, Kang, Y: Comparison of delay estimates at under and over-saturated pre timed signalized intersection (2016)
10. Bergstrom, P.: Solving Linear Optimization Problems Using a Simplex like Boundary Point Method in Dual. Master Thesis, Department of Mathematics, Sweden (2006)
11. Maripini, H., Khadhir, A., Vanajakshi, L.: Traffic state estimation near signalized intersections. J. Transport. Eng. Part A Syst. **149**(5), (2023)
12. Mathew, T.V.: Signalized Intersection Delay Models-Lecture Notes in Transportation Systems Engineering. Department of Civil Engineering, Bombay, India (2019)
13. HCM, Highway Capacity Manual, Volume 4: Applications Guide, Chapter 31 Signalized Intersection: Supplemental, Transportation Research Board of the National Academies, Washington, DC (2010)
14. Rosca, M.A., Rusca, A.: Methods of green times allocation in under-saturated signalized intersection. U.P.B. Sci. Bull. Ser. D. **74**(4), (2012)

Evaluation of Risk Indexes that Affect the Cost of the Under-Construction Infrastructure Project "Extension of the New Ring of Tirana—Phase 2"

Idlir Dervishi(✉) and Eralda Xhafka

"Mother Theresa" Square, Polytechnic University of Tirana, Nr. 4, Tirana, Albania
idervishi@fim.edu.al

Abstract. The identification and the assessment of risks, especially during the initial stages of a project's execution, based on the probability methods, analyzing the historical data and the questionaries addressed to the stakeholders, would enable their avoidance or reduction of the impact, for each project objective in terms of cost, time, quality, safety and environment. Using the data gathered from the field and the information provided from project files, we will conduct a risk analysis to determine the risk indexes and define which risk affects the most the costs of the under-construction infrastructure project "Extension of the New Ring of Tirana—Phase 2", located in a populated area in Tirana. The calculation and evaluation of the risk indexes (R_i) of an infrastructure project take into consideration: the size of the project, the value of the contract, the type of construction processes, the geographical position of the project, and the stakeholder capacity. The purpose of the quantitative risk analysis procedure carried out in this paper is to assess the risk's impact separately and their cumulative effect on project costs, to make necessary adjustments to the project parameters.

Keywords: Risk analysis · Risk indexes · Cost · Infrastructure project

1 Introduction

Due to its location on the "New Ring of Tirana" road axis, the Under-Construction Infrastructure Project "Extension of the New Ring of Tirana—Phase 2" serves as the boundary between Tirana's most populated neighborhoods, Paskuqan Residential Area and "Ex-Train Station Residential Area". This fact led to the development of a sizable number of residential structures close to the road, where the full construction activity will take place. Also, the Tirana River is close to the road axis. Additionally, the distance to the Tirana River is just 20 m. Cost of the project is about 2,863,410,947 ALL [1]. The aforementioned facts make this project exposed to significant risks. The vulnerability to the risks of an engineering project, in addition to the geographical location, also depends on the variety of processes, the complex nature of the project, the deadline, the legislation, but also the use of new construction techniques not usually used before

G. Guxho et al. (Eds.): ITC-ICEE 2023, LNMIE, pp. 121–129, 2024.
https://doi.org/10.1007/978-3-031-48933-4_12

for similar projects. After all the relevant data had been analyzed, the PRAM (Project Risk Analysis and Management) risk assessment technique was applied to identify the risks that threaten the project objectives [2, 3]. Probabilistic approaches have been used to calculate the likelihood that each risk would materialize and how it will affect the execution of the works under investigation [4].

Determining the type of risks is a process that begins during the planning phase and lasts until the final phase of the project. This process is carried out with the help of experts in the field, relying on the previous "database" and the information of the company's managers.

Based on the risk's probability and impact, a quantitative assessment was carried out to determine each Risk's Index across the entire project based on the concept of formula [5] (Fig. 1).

$$R_i = \text{Probability} \cdot \text{Impact} = P_i \cdot I_i \tag{1}$$

Fig. 1. View during the realization of the project

2 Methodology

The PRAM method is intertwined with project management processes but does not replace them. The purpose of the risk management process is to resolve the uncertainties that characterize a project when other management processes have predicted unrealistic parameters.

Each risk affects the project's objectives in addition to the overall effect it has on the project. The following project objectives were identified as being negatively impacted by the occurrence of risks, after an analysis of the data collected from the field, the questionnaires sent to the stakeholders involved in the project management, and the probabilistic methods of calculating the frequencies of risks.

The risk management plan consists of creating a system for risk management, in which decisions made affect the entire process of project realization.

The risk management plan is considered one of the most critical phases because it is part of the main planning document of the entire project to be carried out. The risk management plan describes all the objectives to be achieved, the processes, and the techniques that will be applied for their management.

PRAM techniques use all sources of information in different timelines but also archived information from previous processes. This method allows for planning the deadlines of the works to be carried out considering the possible risk for each process and determining time reserves for those activities that are considered risky.

In construction projects, the level of priority to manage the risks depends on the project stage. Below are listed the processes where risks may appear [3] (Table 1).

Table 1. Type of risks

Risk related to	Level of priority p_i^n
Design	p_d^n
Implementation	p_{impl}^n
Procurement	p_{pr}^n
Geographic location and terrain	p_{loc}^n
Environment	p_e^n
Lack of technical skills of the company	p_{tchs}^n
Maintenance	p_{main}^n
Financial issue	p_f^n
Legislative changes	p_{lch}^n
Safety at the construction site	p_{saf}^n
Major forces (earthquake, flood, rain)	p_{mf}^n
The quality of the materials	p_{qm}^n
Political, and bureaucratic issues	p_{pol}^n

2.1 Data Processing Methodologies

After determining the type of model in function of the objectives of the analysis, a problem in the use of possible models consists of using the initial variables "input" (initial data) necessary for the analysis [3].

The method used to calculate the probability of each risk in different objectives of the project is "Average, median and mode of values considering previous (historical) data". Regarding historical data and values, depending on the archived information, it is possible to use these values in two ways:

- Distribution of "historical" values obtained from the variable.
- The values that the variable has got in the past are extracted and a histogram is constructed. By considering the archive of data on various events in previous projects, it

is possible to analyze who has been in the past the probability and impact, researching who can be the type of the variable compared to the expected value in the case of the project under analysis. All types of risks may have a certain probability of occurrence, where the use of the above methods also leads to the confirmation of the probability of its occurrence.

2.2 Quantitative Calculation of the Risk Significance in Each Project Objective

The value of each Risk Significance Index is calculated, and is expressed by the formula [6]:

$$r_{ij}^{k} = \alpha_{ij}\beta_{ij}^{k} \tag{2}$$

where: r_{ij}^{k} is the value of the Risk Significance Index of each risk i, stakeholder j and the objective of project k. α_{ij} is the probability of each risk to occur i and the stakeholder j. β_{ij}^{k} is the impact of each risk i, from each stakeholder j, in every objective k.

The formula 3 is used to determine the Risk Significance Index when the risk is repeated multiple times during the execution of the engineering project [6]:

$$R_{i}^{k} = \frac{\sum_{j=1}^{k} r_{ij}^{k}}{n} = \frac{1}{n}\sum_{j=1}^{k} \alpha_{ij}\beta_{ij}^{k} \tag{3}$$

where: R_{i}^{k} is the Risk Significance Index for each risk i, on the objective k, where n is the number of times that risk is repeated throughout the project.

3 Results and Discussions

From July 2022 to May 2023, we have collected data from the field regarding the events that affected the completion of the construction works of the Under-Construction Infrastructure Project "Extension of the New Ring of Tirana—Phase 2". Also, we made questionnaires to the stakeholders involved in the project management (Albanian Road Authority staff, supervisor, contractor project managers, and engineers), regarding the risks that appeared during the works of this project, and the works that have been realized or are in progress, for all the lots of Ring of Tirana Road. After that, we identify the risk that affects the most the cost of the project and their likelihood of occurring using probabilistic methods "Average, median and mode of values considering previous (historical) data".

Processing the data and obtaining results using this method is faster and more concrete than using the Normal (Gaussian) Distribution Method and Uniform distribution [3], which requires more time, and obtaining results is close to reality only when the data are more accurate.

Below is the questionnaire addressed to the stakeholders involved in the realization of the project. The questionnaire is composed of 11 questions, where 12 stakeholders provided comprehensive responses. Some other stakeholders did not answer the questionnaire completely and were not taken into consideration, since the answers were not convincing (Table 2; Figs. 2 and 3).

Table 2. Questionaries addressed to project stakeholders

Question number	Question content	Possible answer
Q1	What is your role in construction projects?	[] Project manager [] Supervisor [] Engineer [] Contractor [] Contracting authority supervisor [] Other (please specify)
Q2	How familiar are you with risk management in construction projects?	[] Very familiar [] Somewhat familiar [] Not familiar at all
Q3	In your experience, what are the most common effects of the risks in construction projects of Tirana ring road?	[] (Budget overruns) [] Safety hazards [] Delays in project deadline [] Enviroment
Q4	How do you assess and prioritize risks in a construction project?	[] Qualitative analysis [] Quantitative analysis [] Expert judgment [] Other (please specify)
Q5	What risk mitigation strategies do you employ in construction projects?	[] Insurance coverage [] Contractual risk allocation [] Risk contingency planning [] Safety protocols [] Project scheduling [] Regular project reviews [] Other (please specify)
Q6	Have you ever experienced a significant risk event in a construction project?	[] Yes [] No
Q7	If yes, please describe the risk event and how it was managed	
Q8	How do you communicate risk information to stakeholders in a construction project?	[] Regular meetings [] Risk reports [] Email updates [] Other (please specify)
Q9	Are there any emerging risks or trends in construction that concern you?	[] Yes [] No
Q10	If yes, please describe the emerging risks or trends	
Q11	On a scale of 1 to 5, how well-prepared do you feel to manage risks in construction projects?	[] 1 (not prepared at all) [] 2 [] 3 [] 4 [] 5 (very well-prepared)

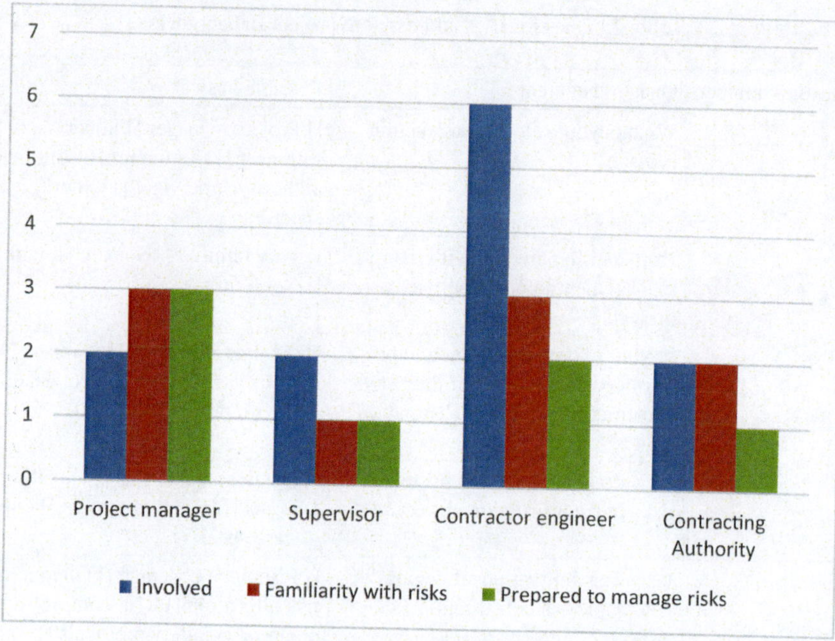

Fig. 2. Date obtains from the questionaries related to stakeholder's

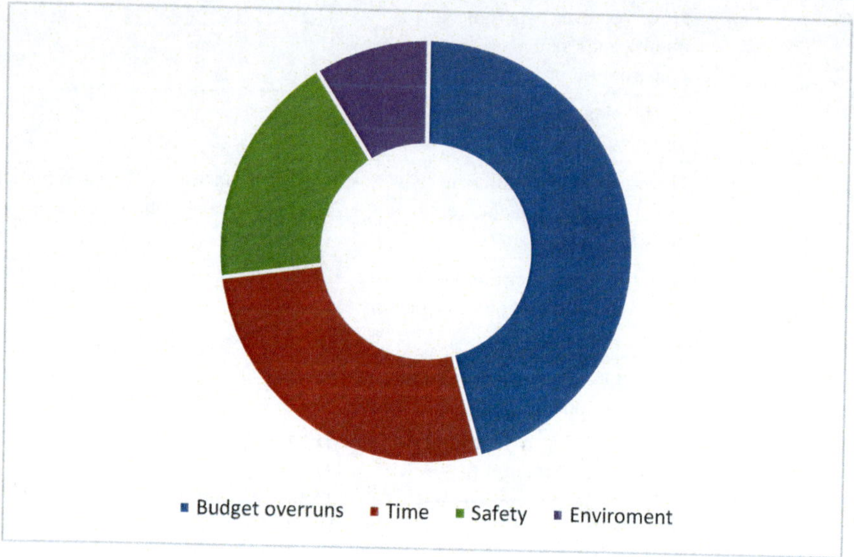

Fig. 3. Data obtained from questionares related the most common effects of the risks in construction projects

3.1 Identification of the Risks that Affect the Most Cost of the Under-Construction Infrastructure Project "Extension of the New Ring of Tirana—Phase 2"

After we collected field data analysis and made questionnaires addressed to the stakeholders involved in the realization of the project, we identified the risks that affect the most the project costs, listed in Table 3.

Table 3. Risks that affect the most the cost

Risk
Design quality
Material price changes
Inaccurate calculation of the quantity of each work item
Project changes
Exchange rate fluctuations
Interim payment certificate delays
Meteorological situation
Subcontractor management

3.2 Determine the Probability and Impact of Risk that Affects the Cost of the Project

The probability of the occurrence of each risk and its impact on the project's objectives was calculated using the probabilistic methods mentioned in the above chapter (Table 4).

Table 4. Probability and impact of each risk

Risk	Probability	Impact
Design quality	0.61	0.88
Material price changes	0.73	0.67
Inaccurate calculation of the quantity of each work item	0.35	0.68
Project changes	0.82	0.95
Exchange rate fluctuations	0.87	0.54
Interim payment certificate delays	0.85	0.10
Meteorological situation	0.12	0.17
Subcontractor management	0.05	0.2

3.3 Calculation of Risk Indexes that Affect the Cost of the Project

Risk Significance Indexes [6] of each risk that affects each objective of the Under-Construction Infrastructure Project "Extension of the New Ring of Tirana—Phase 2" are presented in Table 5 and Fig. 4.

Table 5. Risk Index

Risk	R_i^k
Project changes	0.779
Design quality	0.537
Material price changes	0.489
Exchange rate fluctuations	0.470
Inaccurate calculation of the quantity of each work item	0.238
Meteorological situation	0.020
Subcontractor management	0.010
Interim payment certificate delays	0.09

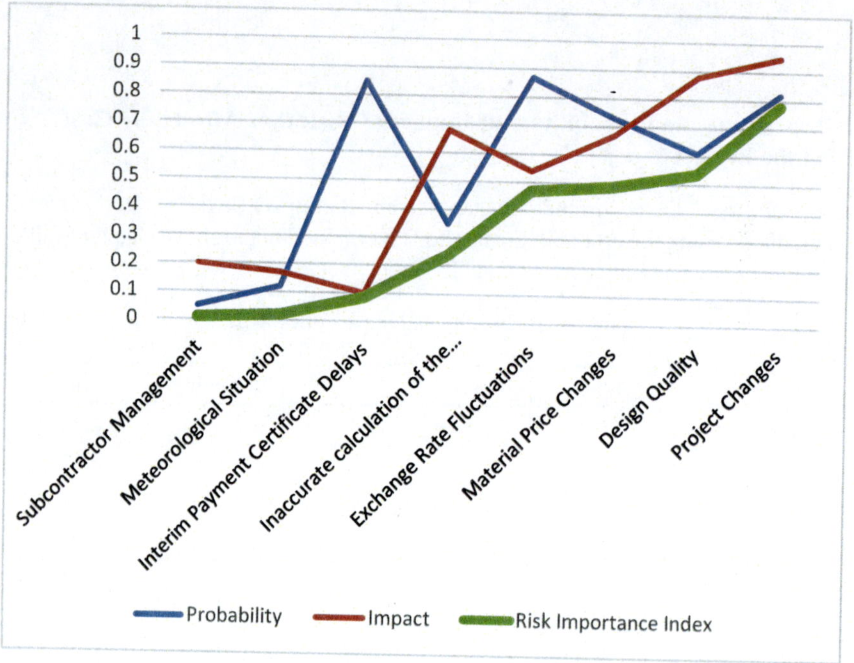

Fig. 4. Probability, impact, and risk index graphs

The "Project Changes" risk had the biggest negative effect on exceeding the costs foreseen in the planning phase. The lack of updates on the urban environment on the ground before the engineering project was planned, has increased the probability of the occurrence of this risk. The "Materials Price Change" risk has a moderate effect in terms of the cost of the engineering project, where the price of materials has changed compared to the period of realization of the design (2022 material price crisis). Even the risk of "Design Quality" has a negative effect on the cost, due to the fact that some design solutions were not in accordance with urban and topographical situations. While the risks of "Inaccurate calculation of the quantity of each work item", and "Interim Payment Certificate Delays" have an acceptable effect, and small corrective actions are needed to avoid them. "Meteorological Situation" and "Subcontractor Management" have a significant probability of happening but have an insignificant effect on the cost.

4 Conclusions

Using risk assessment and analysis techniques, it is evident that the project "Extension of the New Ring of Tirana—Phase 2" has high exposure to risks, with negative effects on the costs, by increasing them by 7% (200,438,766 ALL). The full value of the contingency fund foresaw in the planning phase, of 136,352,902 ALL (5%), was also spent to cover the value of the unexpected works. The contingency fund incorporated in the bill of quantities is essentially a precautionary measure, taken to mitigate the effect of risks. Another measure used to mitigate the emerging risks is the improvement of the design, adapting it to the actual condition of the terrain, and reducing the cost of expropriations. Also, the adaptation of the designs, with construction details that can be realized with the technology in possession of Albanian construction companies, has reduced the costs and increased the safety of the project.

Risks that affect project costs are diverse and dynamic, requiring proactive and vigilant response. By employing robust risk identification and assessment, contingency planning, continuous monitoring, and risk mitigation strategies, project managers can enhance their ability to control costs and deliver projects within budget. In an environment of increasing project complexity and uncertainty, mastering risk management by using PRAM techniques is a key determinant of project success.

References

1. Albanian Road Authority: Documentation of the Under-Construction Infrastructure Project "Extension of the New Ring of Tirana—Phase 2", Tirana (2022–2023)
2. American National Standard ANSI-PMI 99-001-2008: A Guide to the Project Management Body of Knowledge, pp. 273–281 (2008)
3. Caron, F.: Gestione dei Grandi Progetti di Ingegneria, pp. 213–226, Novara (2009)
4. Dervishi, I.: Advanced Management, Quality Analysis and Risk Assessment of Engineering Projects. PhD Thesis, pp. 95–96, 100–101, 107–112, Tirana (2015)
5. Bove, A.: Project Management: La metodologia dei 12 step, pp. 153–156. Milano (2008)
6. Zou, P.X.W., Zhang, G., Wang, J.Y.: Identifying Key Risks in Construction Projects: Life Cycle and Stakeholder Perspectives, pp. 4–5 (2013)

NDE of the Integrity of a Clincer 40,000 ton Concrete Structure

Enkelejda Konda(✉), Migen Zeqo, and Dhimitaq Sotja

Polytechnic of Tirana, 1001 Tirana, Albania
esotja@fim.edu.al

Abstract. The aim of this article is to approve a procedure for the integrity evaluation of a concrete structure based on the low frequency ultrasonic tomography that is the new method in concrete evaluation technology. The procedure was built and performed in the NDE integrity of the CLINCER 40,000 ton concrete located in north Albania. The procedure of the NDE integrity is write as a combination of the visual inspection performed for more than 30%, of the outside surface, rebound hammer technic and the low frequency ultrasonic tomography techniques. The visual inspection locates sign of corrosion, deterioration, mechanical damage on the surface determining the typology and size. Rebound hammer technic of the compressive strength of the concrete give data to determine the quality of concrete according European standards. In the areas where the discontinuities are indicated and the evaluation of the concrete quality is performed, the volume of the wall was scanned with low frequency ultrasonic tomography system, that provides good penetration depth, good sensitivity and imaging resolution by implementing three dimensional image. The equipment used is low frequency ultrasonic tomography ACS A1020 MIRA Lite. After all the inspection results of the procedure proposed was satisfactory, regarding reinforcement bars localization, detection of voids, estimation of their position, and thickness measurements from one side. Those elements are the foundations of the NDE integrity of the concrete structure.

Keywords: NDE of concrete structure · Low frequency tomography techniques

1 Concrete Deterioration

Concrete is one of the most widespread and useful materials in construction due to its service life and manageability during the construction cycle. But like any other material, concrete has its limitations. These limitations are intertwined with the quality of project implementation and the atmospheric conditions to which these structures are subjected, creating problems in the integrity of the structure built with this material. This is the reason that concrete structures are subjected to cyclic checks during their service life.

1.1 Types and Causes of Concrete Deterioration

The deterioration of the integrity of the structure is the interplay of several different potential factors that influence as follows, corrosion of reinforcing steel and embedded

G. Guxho et al. (Eds.): ITC-ICEE 2023, LNMIE, pp. 130–137, 2024.
https://doi.org/10.1007/978-3-031-48933-4_13

parts, freeze-thaw deterioration, chemical attack, abrasion/erosion, fire/heat, restraint to volume change, overload, impact, surface defects [1, 2. Let's recall the most common causes from the literature. Corrosion of the steel material is the leading cause of deterioration of concrete. It is known from the literature that the volume occupied by rusty metal elements is greater than that of non-rusted elements. The rust produced increases the volume by being deposited over the area attacked by it. The increase volume leads to internal stress causing cracks and concrete deterioration [3]. Carbonation is the most common deterioration process from corrosion that depends significantly on the humidity of the concrete. This process is very slow. The carbonation rate has been estimated up to 0.8–1 mm per year [4]. The impact of freezing and thawing of water is important due to the fact that water when it freezes increases its volume by 9% and this leads to internal stress up to the creation of concrete cracks [5]. Temperature is an important factor that affects the destruction of concrete, especially for the period after casting. The mechanism responsible for concrete deterioration from high temperature is being scientifically researched.

2 NDE Procedure

The types of defects mentioned above have made it necessary to check concrete structures during their service life. This non-destructive evaluation procedure goes through these stages. Visual inspection for more than 30% of the outside surface, for the sign of corrosion, deterioration, mechanical damage. The accurate analysis of the discontinuities presents on the surface determining the typology and size is very important. The visual inspection is performed based on the standard EN 1504, EN 12054, EN 13791. After the visual inspection, the rebound hammer technic will be used to determine the compressive strength of the concrete according the standards of the correlation of rebound hammer value and compressive strength EN 13791. Ultrasonic velocity in correlation with compressive strength technic to determine the quality of concrete based on the literature. S scan method will be used to inspect the wall only from one side with array probe technic to evaluate the thickness, and to define the discontinuities of the concrete. The three dimensional elaboration will give information to evaluate the presence of cracks, avoids and reinforced bars.

3 Standard Inspection of Clincer Concrete Structure

The company has 40,000 Tm clincer silo 4500 Kw, Cement Mill for 120 t/h, with 3rd Generation Separator 4 Cement silos of 6000 tons with bagger and palletizer of 120 t/h, Crude Mill of 4500 Kw, for 195 t/h (even without separator), Coal Mill of 1400 Kw, for 20 t/h (even without separator), Grill furnace of 2200 Tpd, with 5 stages (even without calciner and air 3rd).

The conventional procedure will be followed and demonstrated with the exact data.

Fig. 1. Image from the inspection of the clincer 40,000 ton, and the surface crack morphology

3.1 Visual Inspection

The visual inspection was performed based on the standard EN 1504, EN 12054, EN 13791, performed on Cement Clincer 40,000 ton. At least 30% of surface was checked by visual inspection.

The types of the cracks visible on the surface are cracks in before hardening due to the original formwork and cracks due to drying process (plastic shrinkage, settlement shrinkage, delayed curing etc.). These cracks are spread all over the surface with the same pattern (see Fig. 1). A direct evaluation of these cracks depth has been made by grinding up to eliminate the cracks. The measured depth was until 22–25 m. The state of the concrete based on the inert size and the colour of the concrete was normal.

3.2 Rebound Hammer Technic—Strength Evaluation

Test Hammer has been used to determine the compressive strength or hardness. It provides based on the rebound value the strength of concrete. The used equipment is the Test hammer HT-225. The calibration block has been steel anvil model GZ 16 (see Fig. 2). It is important to underline that this test give an approximately hardness of the surface. The factors that will influence are: 1. Age of concrete—Concrete continues to develop strength with age due to cement hydration. This is the reason behind the development of data relating rebound numbers to the compressive strength of the concrete mixture from the structure. 2. Surface carbonation—With greater amounts of surface carbonation, higher rebound numbers will be obtained. Rebound numbers of a carbonated surface can be as much as 50% higher than non-carbonated surfaces. Older concrete surfaces may have much deeper amounts of surface carbonation than younger concrete [6].

The measurement of the rebound hammer value has been performed in 4 area at lower part of the clincer and 4 area at upper part of it. To avoid the influence of the protective layer on the strength measurements we performed the measurements after grinding the surface at the depth of 4–5 mm (see Fig. 3).

The value of rebound hammer for Clincer Siloz 40,000 ton concrete are very high. Based on that average HRV we could estimate the compressive strength concrete structure in range 45–56 MPa. The correlation between the Rebound Hammer and compressive strength are inside the user manual and the standard EN 13791 of rebound hammer evaluation.

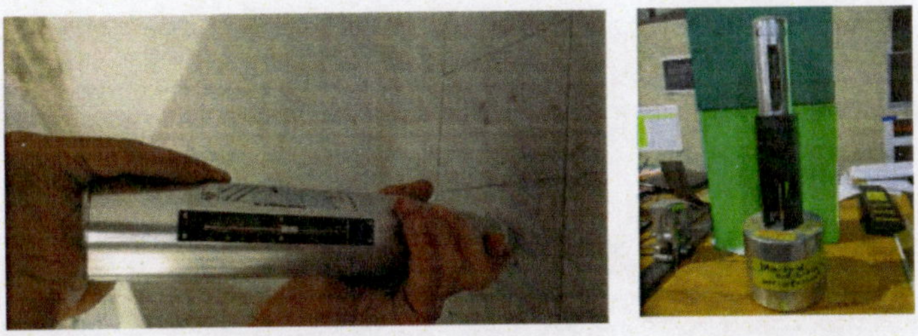

Fig. 2. Rebound hammer HT-225 and calibration block steel anvil model GZ 16

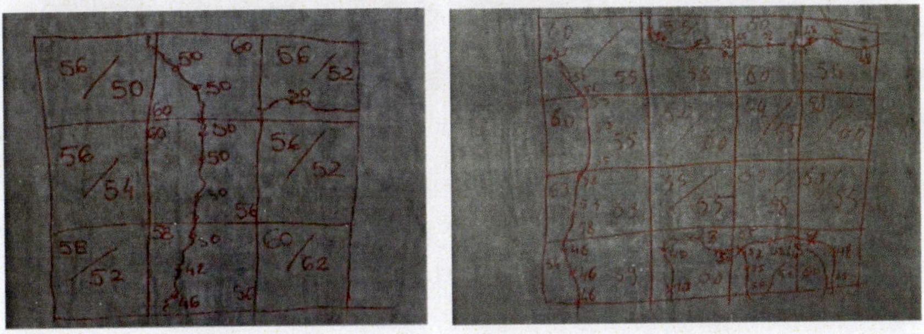

Fig. 3. Example of HRV grid of the surface measurements

Table 1. Rebound hammer

Rebound hammer average value	First average value	Second average value	Third average value	Fourth average value
Lower part	54,5	54	52.5	54
Upper part	58	60	60	59.5

The International Building Code (IBC) (Section 1905.1.1) and the ACI 318 Standard (Sect. 5.1.1) indicate a minimum specified compressive strength of 2500 psi (17.2 MPa) for structural concrete. Simply stated, no structural concrete can be specified with a strength less than 2500 psi. For normal field applications, the concrete strength can vary from 10 to 60 MPa. The compressive strength of the concrete of the Clincer 40,000 t, vary from 45 MPa (6525 PSI) to 59 (8555 PSI) MPa (see Table 1).

Based on the standard table the clincer structure is good concrete industrial structure.

3.3 Ultrasonic Velocity in Correlation with the Compressive Strength

Ultrasonic method on concrete measurement is impact—echo method based on two probes, one as transmitter and the second as receiver. The transmitter and receiver have to be placed on the two opposite faces of the element to test. It is the most widely used method because there is maximum pulse energy which is transmitted and received so this is the most satisfactory method [7]. In our case this direct method was impossible, because the element was accessible only from one side. Indirect measures or surface method has been performed. Calibration of the equipment is done on the calibration sample (see Fig. 4).

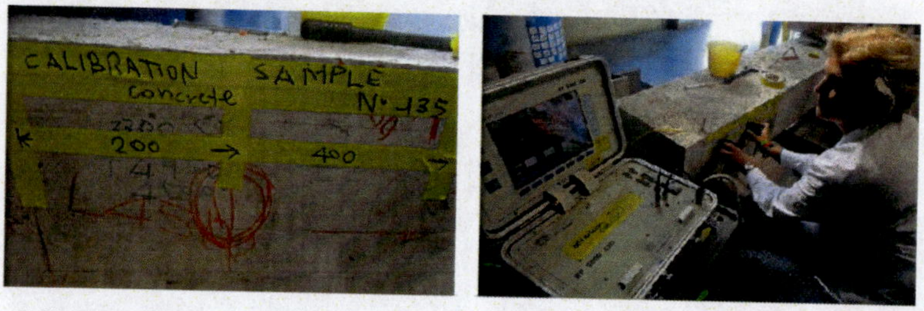

Fig. 4. Calibration block for the ultrasonic correlation velocity with the compressive strength using surface technique

The relation of the compressive strength of the concrete with the ultrasonic velocity for direct method, and surface method from the literature (see Table 2) [7].

Table 2. Correlation between quality of the concrete and ultrasonic surface velocity

S	Pulse velocity, km/s, surface method	Concrete quality grading
1	> 3.5	Excellent
2	3.0–3.5	Good
3	2.5–3.0	Medium
4	< 2.5	Poor

On Table 3 are shown the data of the ultrasonic surface velocity. The geometry of the measurements points is shown on the picture (see Fig. 5).

Table 3. Measurements of surface ultrasonic velocity on concrete—clincer 40,000 t

Distance between the transducers	Points	Velocity (m/s)
350 mm	1A–5A	4600
	2A–6A	4100
	3A–7A	5200
	4A–8 A	2600

All the data of the surface ultrasonic wave velocity were more than 2600 m/s. The data underline the interval of a good quality concrete as required for the civil engineer construction [8].

Fig. 5. Image from the inspection and geometry of the data measurements.

4 Low Frequency Ultrasonic Tomography Inspection

The conventional NDE procedure has been performed with such NDT/NDE techniques as visual, rebound hammer, impact echo, ultrasonic pulse velocity, etc. While each of these techniques has its own respective list of advantages and disadvantages [9], the optimal procedure for inspection of concrete structure would be one that can provide a two-dimensional and/or three-dimensional view into the structure with relative ease, speed, and efficiency. A technique that meets these criteria is ultrasonic shear wave tomography.

4.1 MIRA System

The commercially available MIRA (see Fig. 6) is a portable ultrasonic shear wave tomography device that consists of an array of four rows of 12 shear wave transducers that can obtain tomographic information in a matter of seconds. The set of 48 dry-point contacts (DPC) generate a signal with a central frequency of 50 kHz and are activated in a pitch-catch regime to create the tomograph. The MIRA system can generate B-, C- and D-scans. Researchers have shown that the MIRA system is able to detect poor bonding within partial depth repairs of concrete pavements, voids/honey combing defects, delamination, locating and identifying steel in concrete, deterioration and cracking in concrete, thickness measurements [10].

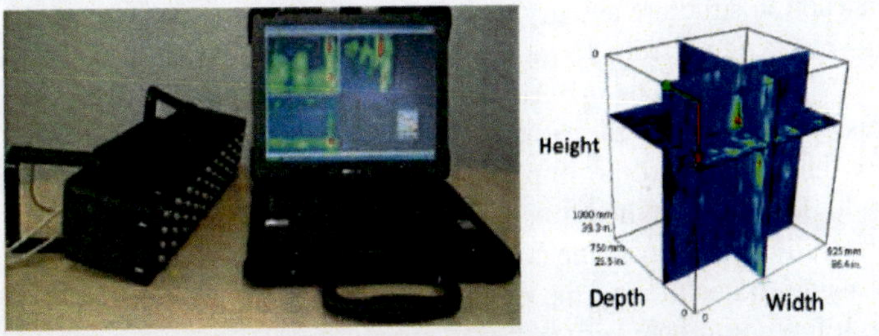

Fig. 6. MIRA device and laptop and 3D projection of scan directions

4.2 Evaluation of New Procedure Inspection with A1040 MIRA

The procedure will start with visual observation of the Clincer concrete surface to evaluate the surface discontinuities. The procedure determines the use of the equipment A1040 MIRA scanning the entire accessible surface 1m above ground level where it is thought to be the most endangered area. The images will be evaluated on real time, determining the speed of the ultrasound signal and the thickness of the concrete wall. The velocity of ultrasound signal with give according the standards the concrete quality grading and the three-dimensional gives complete information about the reinforcement bar, voids and cracks.

The velocity measured with this equipment was higher than 2900 m/s. The structure has a compressive strength more than 35 MPa (5075 psi), so is a good civil engineering structure. The cracks start from the surface—at the depth from 15 mm to maximum 34 mm (see Fig. 7).

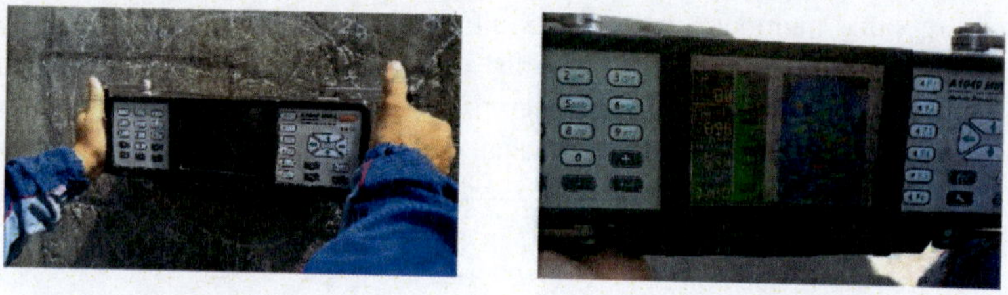

Fig. 7. Evaluation of reinforcement bars, voids and cracks.

5 Conclusions

The advantage is that the tomograph scan provides a graphic that is relatively easy to understand and that allows a trained technician to identify defect locations or areas that require further investigation or repair inspection. The results of inspection with both procedures are:

- The quality of the concrete structure is a good civil engineering structure.
- The only discontinuity have been surface grid crack line indicated due to probably the construction structure and drying process. The depth of the cracks starts from the surface—at the depth from 15 mm to max 34 mm. The coverage of these indication is all over the surface. The integrity of the structure is untouched.
- Compressive strength of the concrete of the measured section is in the range 35–50 MPa. This is due to the impossibility of access to the internal surface of the Clincer 40,000 ton with the ultrasonic probe.
- To have a full evaluation of the clincer section is necessary to open the clincer body and go inside to positioning the probe. Another possibility is to extract a carrot by drilling. It is a common e very fast evaluation.

It was recommended visual inspection of Clincer 40,000 ton concrete structure periodically, to collect data for the crack progress data bank. Repair the places where are signs of rust. It is important not to let the concrete under the damage from the inducement reinforcement bars.

References

1. Raupach, M., Buttner, T.: Concrete Repair to EN 1504: Diagnosis. Principles and Practice. CRC Press Taylor and Francis Group, Design (2014)
2. Portland Cement Association: Concrete Information, https://www.cement.org/docs/default-source/fc_concrete_technology/durability/is536-types-and-causes-of-concrete-deterioration.pdf
3. Altoubat, S., Maalej, M., Shaikh, F.: Laboratory simulation of corrosion damage in reinforced concrete. Int. J. Concr. Struct. Mater. **10**, 383–391 (2016)
4. Fernandez, I., Berrocal, G.: Mechanical properties of 30 year—old naturally corroded steel reinforcing bars. Int. J. Concr. Struct. Mater. **13**(9), (2019)
5. Uddin, F., Shaikh, A.: Effect of Cracking on Corrosion of Steel in Concrete, International Journal of Concrete Structures and Materials 12, Article number 3 (2018)
6. Zuyun, L., Xinhua, J.: Performance based evaluation of carbonation resistance of concrete according to various curing conditions from climate change effect. Int. J. Concr. Struct. Mater. **11**, 687–700 (2017)
7. Kachanov, V., Sokolov, I., Konstov, R., Lebedev, S.: Ultrasonic wave velocity measurement in concrete using the impact-echo method. Insight Non-Destruct. Test. Cond. Monitor. **61**(1), 15–19, Research Gate (2016)
8. Abdul Muttalib, I., Ali Hussein Abdyl, B.: Assessment of concrete compressive strength by ultrasonic non-destructive test. In: E3S Web of Conferences, vol. 318, 03004, ICGE, Iraq (2021)
9. Report on Nondestructive Test Methods for Evaluation of Concrete in Structures, Reported by ACI Committee 228.2R-2013, American Concrete Institute, https://www.concrete.org/portals/0/files/pdf/previews/228213.pdf
10. Popovics, J., Roesler, J., Bittner, J. Amirkhanian, A., Brand, A., Gupta, P., Flowers, K.: Ultrasonic Imaging for Concrete Infrastructure Condition Assessment and Quality Insurance, A Report of the Findings of ICT PROJECT R27-146, Research Report No. FHWA-ICT-17-007, University of Illinois at Urbana-Champaign (2017)

Analysis of Vehicle Registration and Their Characteristics in Albania and Comparison with the European Market

Erjon Selmani(✉), Koçi Doraci, Leonard Çomeni, Arben Dushi, and Miranda Kullolli

Polytechnic University of Tirana, Mother Teresa Square No. 1, Tirana, Albania
eselmani@fim.edu.al

Abstract. Based on the major changes that have followed in Albania in every sector during this phase of the so-called social, political and economic "transition", we have focused on the analysis of the numbers and characteristics of vehicles in our country. The data presented in this paper were taken from the official publication of the Institute of Statistics of Albania (INSTAT) and reflect the official data of the general directorate of road transport services (DPSHTRR). Although the figures show that the number of imported cars has increased over the years, they continue to be with internal combustion engines and with an average-age greater than the Eurozone average. The number of imported vehicles with electric or hybrid motors seems to be still insignificant compared to the total number.

Keywords: Analysis · Vehicle registration · Inventory · Engines · Trend

1 Introduction

Albania, as a country that emerged from the communist political system in the 90s, has shared some common characteristics with the block of so-called "eastern countries", which went from a centralized and state-controlled economy to a individual and free one. This system transition coincided with strong economic and social fluctuations, while only from the years 2000 onwards we have a stabilization of GDP at approximately constant levels [1], either from economic growth or from the contributions of immigrants (remittances).

As in any country, economic and social changes could not happen if there was no development of the transport sector. In Albania, an almost inverse phenomenon has occurred with what happens in developed countries, there has been a significant increase in the number of individual means of transport and an equally significant decrease in the number of means of public transport. If until the 1990s passenger rail transport occupied a very important role, after these years it was constantly degraded until it is almost non-existent at the present time. On the other hand, public road transport has undergone a significant re-dimensioning, losing quotas and consequently public attention and investments.

G. Guxho et al. (Eds.): ITC-ICEE 2023, LNMIE, pp. 138–147, 2024.
https://doi.org/10.1007/978-3-031-48933-4_14

To enable transport services and normal economic and social life, the main focus of public administrations was on individual road transport. Consequently, the ownership of an individual vehicle, apart from fulfilling a deprivation from the economic system before the 90s, was also seen as a need and necessity for economic and social purposes.

This article presents some analytical and statistical data officially published by the general di-rectorate of road transport services (DPSHTRR) and the Institute of Statistics of Albania (INSTAT). The data published by INSTAT for the vehicle inventory starts from 2014, while from 2019 onwards, INSTAT and DPSHTRR present detailed and continuously updated data. In this article the focus was put only on car numbers, due to their social importance and rapid growth, clearly in great advantage against all the other vehicles of the national fleet.

2 Car Inventory in Albania and Its Characteristics

In Fig. 1 is depicted the vehicle inventory referring only to passenger cars. As figure shows, there has been a slight increase in car numbers from 2014 to 2018, with certain alterations. After year 2018 numbers has risen steadily until the first trimester of year 2023, to the witch the bar is referred. The yellow bar shows the first time registrations for that year.

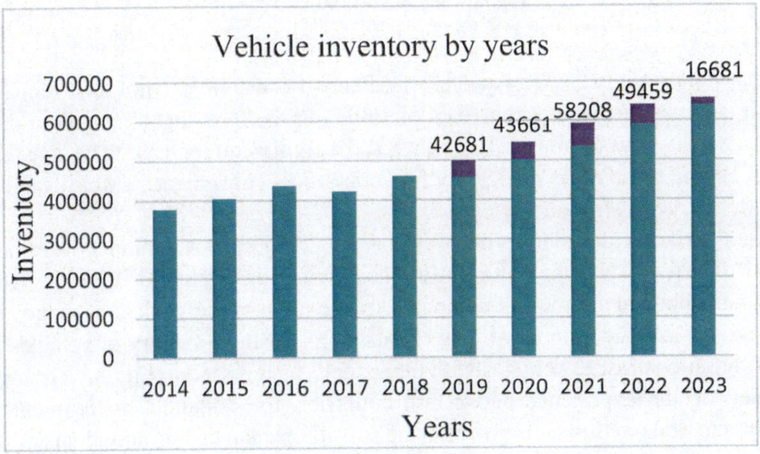

Fig. 1. Vehicle inventory

The data related to year 2023 are incomplete because they refer to the registrations for the first trimester of year 2023. The graph shows the decrease of new registrations in year 2022 compared to year 2021, and for how long the numbers of 2023 are still incomplete, we cannot talk about trends. We can dare to list some reasons: the increase in the prices of used vehicles after the COVID-19 pandemic, the decrease in the purchasing power of the population after the economic crisis caused by the pandemic, the continuous emigration of the working-age population, etc. Figure 2 shows the subdivision of the fleet by fuel type for years 2019–2022. As expected for the Albanian market, diesel-fuel engines remain the most wanted among the users, their increase in market share has risen steadily. On the other hand, electric and hybrid vehicles seems to increase at a continuous rate over the given years although their rates are still low.

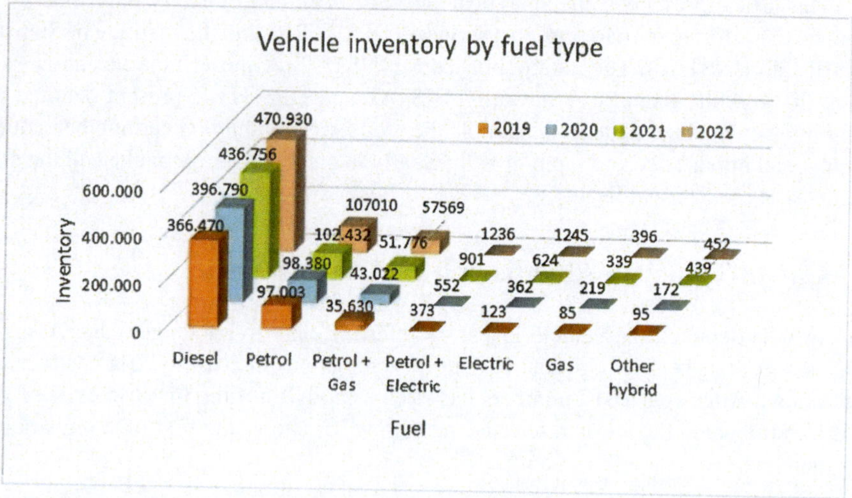

Fig. 2. Vehicle inventory by fuel

What stands out is the "resistance" of Albanian consumers to not own vehicles with gasoline engines throughout the years of study, on the other hand, cars with dual fuel systems petrol + gas (mainly L.P.G.) have had a significant increase. The conversion of a petrol engine to L.P.G. is made only for economical convenience and is usually made as an aftermarket operation.

Based on some legislative interventions in 2019, the Albanian government has decided to waive some taxes for fully electric or plug-in hybrid vehicles. Although the savings obtained are modest compared to the value of the vehicle or to the customs cost, it seems that the number of new registrations for this category of vehicles is only increasing, in contrast to vehicles with other fuels, as reported in Fig. 3.

Based on the experience of western countries, the Albanian government should strongly intervene in this sector by offering subsidy packages for individuals or companies to switch to this type of energy, but on the other hand, more charging points should be provided throughout the territory.

If we refer to Fig. 4, we can see how the inventory is distributed across the country. According to the current organization, Albania is administratively divided into 12 Districts or Prefectures [4], where each prefecture has a branch of DPSHTRR (Saranda and Vlora are two branches of the same district).

According to INSTAT publications on demographic changes [5], the country's population has undergone major demographic changes over the years. There has been significant displacement of the population from rural to urban areas, in addition to the phenomenon of fleeing the country. Districts of Tirana and Durres experienced the largest demographic increases, followed by the regions of Fier and Elbasan. On the other hand,

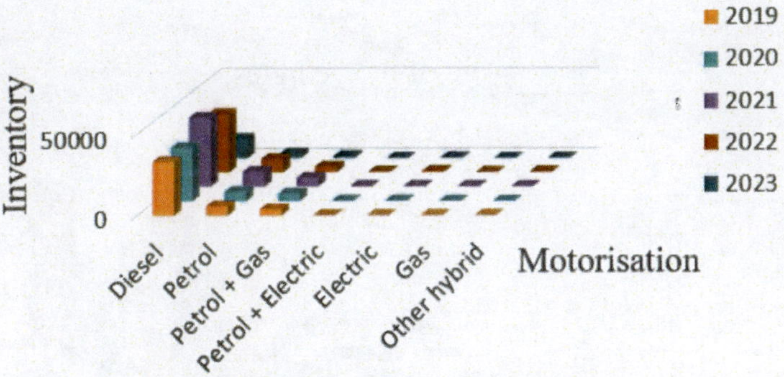

Fig. 3. First registrations inventory by fuel type

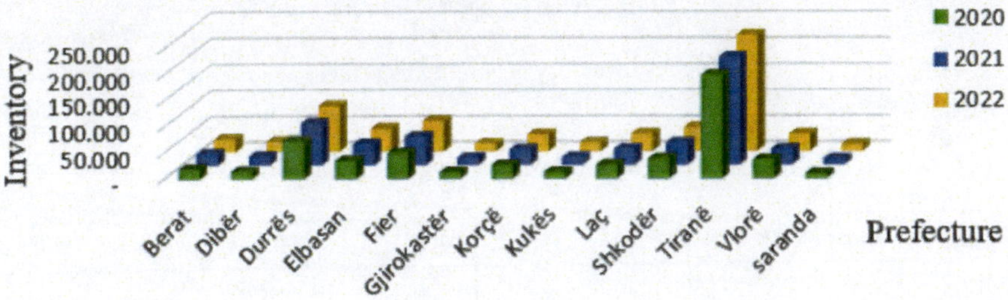

Fig. 4. First registrations inventory by prefectures

the Districts of Kukes, Dibra, Gjirokastra, Berat and Vlora have suffered a continuous decrease in the population.

By comparing the data on the population with the data on the inventory of vehicles we can easily understand how the distribution of the population clearly conditions the distribution of the vehicles. This thesis is strengthened even more if we reason on the fact that the districts with the greatest population growth are also those with the greatest economic growth, while those with the greatest population fall are necessarily characterized by economic decline Fig. 5 gives the age of the inventory, as depicted the majority of the active vehicles in Albania are more than 11 years old. The interesting fact is the trend of the latest two columns of the graph: we witness a fall of the 11–20 years old vehicles and a rise of the over 20 year old vehicles in the last 4 years.

According to the latest government decisions in the field of transport [3], vehicles up to the EURO 4 category could be registered in the public registry until 31.12.2019. Starting from 01.01.2020 onwards, only EURO 5 category vehicles and above will be registered. This situation is believed to be similar also in other neighbour countries in the Balkans as suggested by reference [4].

Reference [6] defines the European Emission Standards frame for all the vehicles, as given in Table 1.Converting the numbers in Fig. 5 into European standards for the latest full calendar year (2022), 31.2% of the inventory was equal or more than 11 years old or up to Euro 5 standard, and 53.2% of the inventory was more than 20 years old or up to Euro 2 standard.

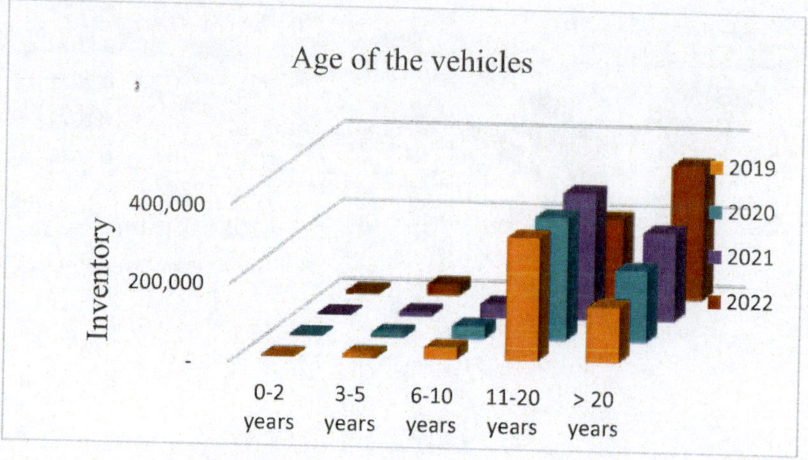

Fig. 5. Inventory according to vehicle age

Table 1. European emission standards

Euro 1	1992
Euro 2	1996
Euro 3	2000
Euro 4	2005
Euro 5 (a, b)	2009, 2011
Euro 6 (b, d-temp, d)	2014, 2017, 2020

Whereas in Fig. 6 is given the weighted average life of the active vehicles of the inventory for years 2019–2022. As we can see the average age of the vehicle inventory has increased by approximately 0.23 years from 2019 to 2022. These figures seems not to be very promising in the context of joining the European Union standards, however, according to ACEA, the European Automobile Manufacturer's Association [7], within the EU there are also member countries such as Greece and Estonia to have even older car fleets, with vehicles up to almost 17 years old. On the other hand, the newest passenger are in Luxembourg, 7.6 years of age on average. In the entire European Union, passenger cars in year 2021 were on average 12 years old.

In Fig. 7 is depicted the situation of the inventory for years 2019–2022 subdivided by the engine displacement. The graph is divided in 3 classes, and the 2000 cm^3 displacement engine results to be the most appreciated of the Albanian market. On the other hand the 2000–3000 cm^3 class is approximately the half of the former one in numbers, where it is believed that diesel engines of 2.2 up to 2.5 L are the hard core of this category. Always according to Fig. 7, the class >= 3000 cm^3 has an almost 5% value over the total, in this category are also included around 350 cars with displacement higher than 6000 cm^3.

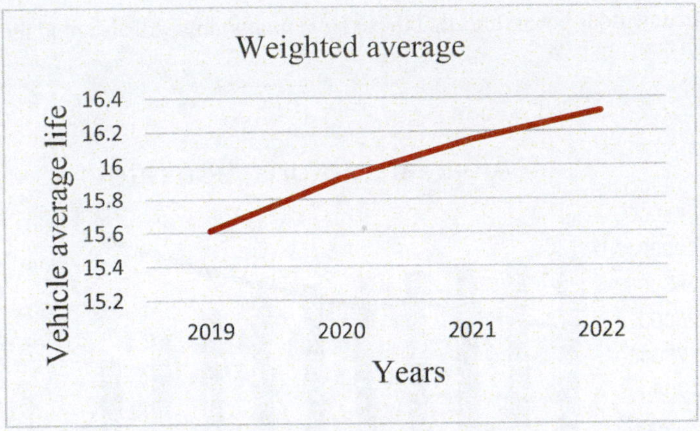

Fig. 6. Average age of the vehicles

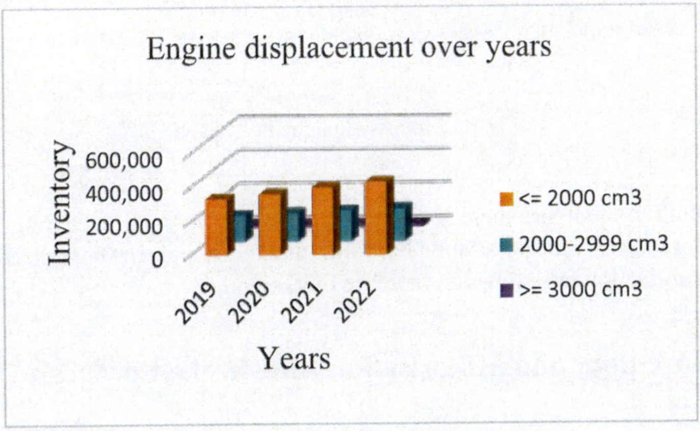

Fig. 7. Engine displacement

3 Motorisation Rate in Albania

According to the reports of the INSTAT, the population of Albania in the last 10 years has had a slight but continuous decline, from approximately 2,896,000 residents in 2013, there are approximately 2,761,785 residents in 2022. However, the population census has not been updated for about 11 years and therefore the published numbers may be different from the actual reality.

In this view, it is interesting to correlate the number of vehicles with the number of inhabitants, in order to understand the purchase or the ownership capability of the population, which is a direct estimate of the economic wealth or the economic growth rate of the population.

In Fig. 8 is depicted the population and the motorization rate, expressed as number of cars (not other vehicles) per 1000 inhabitants. One has to pay attention to the fact that this specific rate does not assures that these 1000 inhabitants will also hold the ownership of the cars!

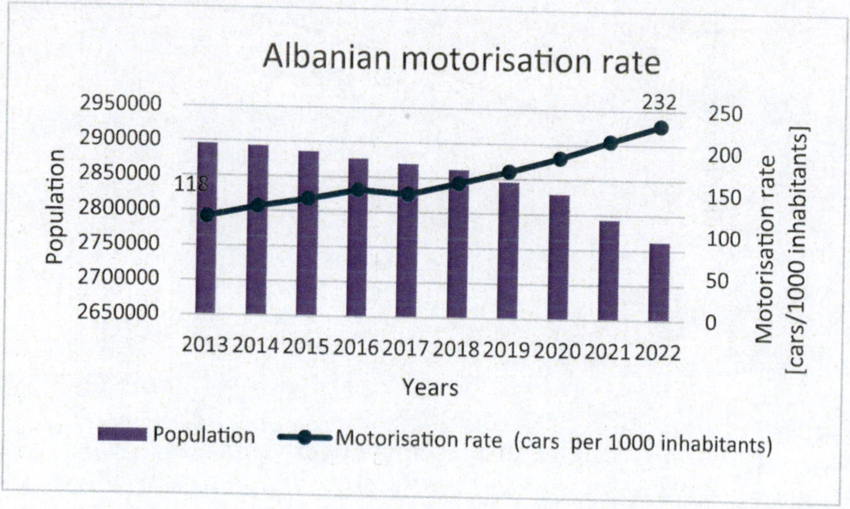

Fig. 8. Motorisation rate in Albania

According to the figure there seems to be a contrast but in fact the trend is correct, because over the years the population has fallen and the vehicle inventory has increased, thus the motorisation rate shows an increase in its trend.

4 Car Inventory and Motorisation Rate in the European Union Market

As long as Albania is struggling to join the European Union, it would be of interest to survey the characteristics of the car inventory in the European market intended as the European Union only. In Table 2 is given the motorization rate of the union while in Fig. 9 is given the car inventory. In Figs. 10 and 11 are given the new vehicle registrations and their respective fuel types, whereas in Fig. 12 are depicted the data regarding the average age of the inventory for every member state of the union. All data are referred to official publications of ACEA [3, 8, 9].

Table 2. Motorisation rate of the union

Years	Avg. E.U. motorisation rate (cars/100 inhabitants)
2017	535
2018	545
2019	555
2020	560
2021	567

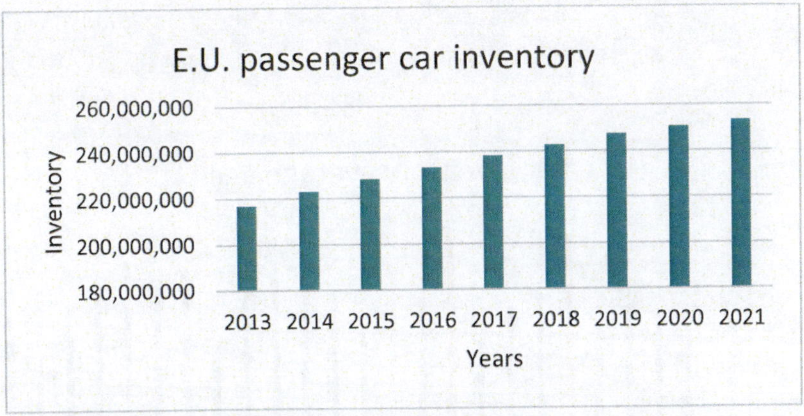

Fig. 9. Car inventory in Europe

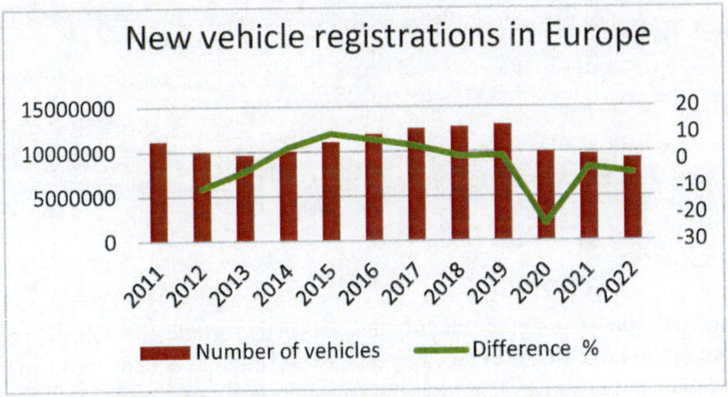

Fig. 10. Trend of the new registrations in Europe

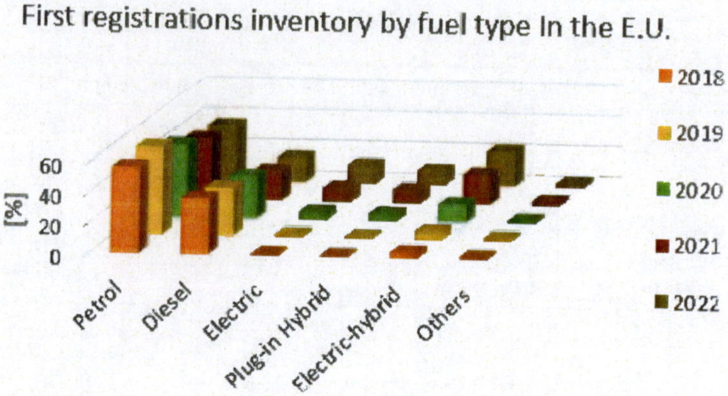

Fig. 11. Trend of the first registrations in E.U. by fuel type

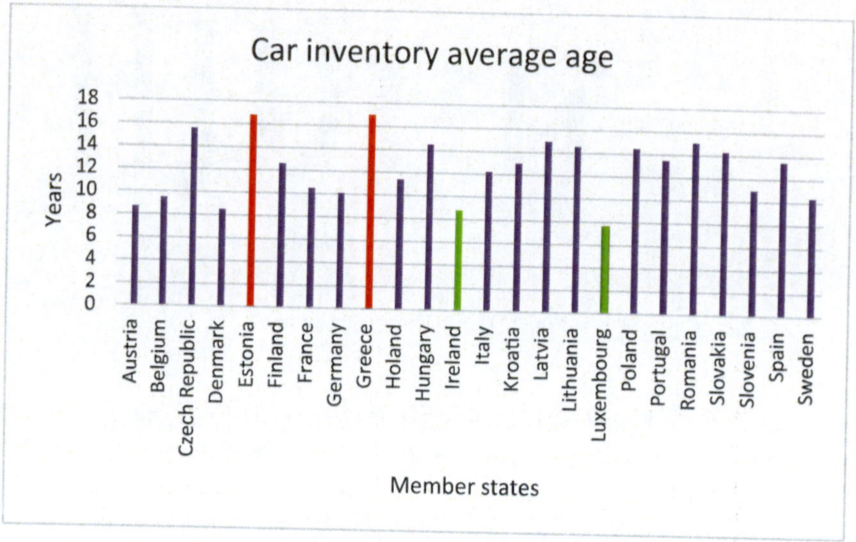

Fig. 12. Average age of the car inventory for each member state

5 Conclusion

According to the data and figures taken in analysis in this article, the Albanian car market and inventory has increased over the last years. Diesel engines occupy the most important part of the market, but above all, they seems to continue to rise even on the actual years or those to come. This trend seems to be in contrast with the European Union trend where the diesel engines are going to reduce, and the new registrations are already surpassed by hybrid or electric engines.

The most interesting fact about the inventory is the high age of the stock cars but also of the new registrations. This trend is believe to be related to the lower price of the older cars and in particular to the diesel engines, which makes them to be on the top preferences of the population. The motorization rate in the Albanian market has increased over the years, but this data has to be taken with caution because over the last decade the Albanian population has reduced due to migration.

In analogy to the European market, the Albanian car inventory has increased over years, confirming the car as a very important companion to the everyday.

In Europe the petrol engines and the newer forms of motorization, e.g. hybrid or electric, are ahead in the numbers if compared to the diesel engines, however, the average age of the inventory of each member state showed that some countries still have aged cars in use and comparable to the Albanian average age of the fleet.

References

1. https://www.dpshtrr.al/sherbime-online/open-data-statistika-raporte
2. http://databaza.instat.gov.al/pxweb/sq/DST/START__TR__TR04__TR0402/TR04024/
3. https://www.acea.auto/figure/average-age-of-eu-vehicle-fleet-by-country/
4. https://sq.wikipedia.org/wiki/Ndarja_administrative_e_Shqip%C3%ABris%C3%AB
5. https://www.instat.gov.al/al/temat/treguesit-demografik%C3%AB-dhe-social%C3%AB/popullsia/
6. https://en.wikipedia.org/wiki/European_emission_standards
7. https://www.dpshtrr.al/informacion-mbi-importin-e-mjeteve
8. https://www.acea.auto/figure/motorisation-rates-in-the-eu-by-country-and-vehicle-type/
9. https://www.acea.auto/figure/fuel-types-of-new-passenger-cars-in-eu/

Machine Learning Based Prognostics of On-Board Electromechanical Actuators

Edmondo Minisci[1(✉)], Matteo D. L. Dalla Vedova[2], Parid Alimhillaj[3], Leonardo Baldo[2], and Paolo Maggiore[2]

[1] University of Strathclyde, Glasgow G1 1XJ, UK
edmondo.minisci@strath.ac.uk
[2] Politecnico Di Torino, 10129 Turin, Italy
{matteo.dallavedova,leonardo.baldo,paolo.maggiore}@polito.it
[3] Polytechnic University of Tirana, 1000 Tirana, Albania
parid.alimhillaj@fim.edu.al

Abstract. This paper presents a novel machine learning-based prognostic approach for on-board electromechanical actuators. The study is centered around overcoming the limitations of model-based prognostic frameworks that rely on expensive optimization processes. Machine learning techniques were employed to map system signal characteristics directly into parameters related to fault simulation. A first test, utilizing only five of eight implemented fault types, demonstrates a highly promising potential of artificial neural networks to predict and detect faults with minimal error. A second test expands the investigation to include all fault types and provides an analysis of the model's robustness, error rates, and computational costs. The practical outcome of the work is a viable real-time solution for fault detection and characterization in electromechanical actuators, highlighting the efficiency and effectiveness of machine learning techniques for industrial applications.

Keywords: Machine learning · Prognostics · Electro-mechanical actuators · Reliability · Fault implementation · Failure analysis

1 Introduction

Electromechanical actuators (EMAs) play a pivotal role in modern technological systems, converting electrical energy into mechanical energy and performing precise motion control tasks. As crucial components in industries such as manufacturing, automotive, and robotics, they underpin the smooth functioning of a myriad of systems. Their importance is even further underscored in the aerospace and aeronautical sectors where EMAs are vital components in aircraft systems including flight control, landing gears, and engine control, with the contemporary trend of transitioning from hydraulic and pneumatic systems to "more electric aircraft" [1]. This transition is driven by the advantages of EMAs, such as weight reduction, simplified maintenance, and overall efficiency improvement [2, 3].

G. Guxho et al. (Eds.): ITC-ICEE 2023, LNMIE, pp. 148–159, 2024.
https://doi.org/10.1007/978-3-031-48933-4_15

These wide-ranging applications accentuate the necessity for early fault detection and system reliability. Failures can lead to substantial financial costs, operational delays, or catastrophic consequences, particularly in safety-critical aerospace applications, fully justifying the carried out exploration into machine learning-based prognostics for on-board EMAs.

Fault detection in EMAs can be conducted through model-based diagnostic methods and approaches, which involve complex and computationally heavy optimization processes [3, 4]. These techniques require detailed models of the system and its faults, which can be a resource-intensive and intricate process, often requiring specific expertise, and necessitate exhaustive simulations and the application of heuristic search algorithms to match observed and simulated behavior, making them unsuitable for real-time fault detection and characterization.

Recent research on prognostics and fault detection has seen a shift towards data-driven approaches, notably machine learning techniques. These techniques have demonstrated the potential to identify complex patterns, allowing for the detection and characterization of faults in nonlinear systems like EMAs.

However, there is a persisting gap in the development of methods that facilitate quick and efficient real-time prognostics in EMAs. The carried out research aims to bridge the previous work about the use of metaheuristics by utilizing artificial neural networks (ANNs), a subset of machine learning techniques. ANNs can approximate any nonlinear function, and learn and generalize from provided data, showing great potential in various diagnostic/prognostics tasks. This approach aims to bypass the computationally intensive associated optimization processes and offer a more efficient solution for real-time fault detection in EMAs, especially critical in aerospace and aeronautical systems.

It will be shown that a well-trained and validated ANN model will provide quick and effective real-time prognostics characterization, and it is expected that the proposed approach will significantly contribute to the improvement of system reliability and safety, particularly in industries relying heavily on EMAs.

The rest of the paper is organized as follows. Section 2 will detail the models and methods used for this work; Sect. 3 will present and discuss the results obtained for a simplified and a more complex case; and a final section will contain general conclusions and new directions for future research.

2 Models and Methods

In this section, the EMA model and the way the faults have been implemented are detailed. Moreover, the general structure of the artificial neural network used in this work is presented.

2.1 EMA Model and Fault Implementation

Previous work on model-based prognostics required the implementation of two EMA models: a high-fidelity reference model (RM) and a low fidelity monitoring model (MM). The RM was used to create the pseudo-real signals, without having to rely on test data, while the MM was the model used by the heuristic methods to find the set of fault parameters related to the signal.

Both models, detailed in [3–5], implement some particular faults, usually showing a progressive growth: backlash (BLK) and dry friction (FST) affecting the mechanical transmission because of worn components, partial coil-to-coil short circuit in one of the BLDC stator phases (Na) and a drift of the proportional gain of the PID controller. The faults are defined by a set of top level parameters (TLPs), grouped inside a ***k*** vector. Each component of the vector is linked and can be traced back to a failure in the EMA, and by modifying these parameters, in the interval [0, 1], it is possible to simulate different failures/degradations. The 8 parameters are as follows:

- k_1: Dry friction.
- k_2: Backlash.
- k_3, k_4, k_5: Short circuit (SC), phases A, B, and C, respectively.
- k_6: Static eccentricity modulus in the rotor.
- k_7: Static eccentricity phase in the rotor.
- k_8: Proportional gain drift (PGD).

The values of k's giving the nominal condition, i.e. no deterioration, are: $\boldsymbol{k}_N = [0, 0, 0, 0, 0, 0, 0.5, 0.5]^T$.

In some previous works [3, 4], the fault identification was performed by finding the values of k's, which gave the best match between the MM and RM current outputs. As introduced in the previous section, the optimization process associated with the search, prevented the approach to be used in real time. In this work, the aim is to implement and test a machine learning based approach that could be used in real-time.

2.2 Machine Learning Based Approach

Machine Learning (ML) exploits the information contained in the available data to develop systems effectively performing specific tasks, such as regression and classification. For this work, the focus is on the regression application, and ML is considered to build a map linking the current signal to the k parameters, in a sort of inverse modelling. While in the previously used model-based approach the MM was included into an optimization loop to find the input values of the model (the vector ***k***) that was producing the current signal matching the reference coming from the RM, here ***k*** becomes the output of the ML model, where the input is the current signal (properly sampled). The ML model is very fast to interrogate (hence the real time estimate), but the training phase can be relatively expensive and needs to be performed off line.

Artificial Neural Network Structure

Generic multi-layer perception (MLP) feed-forward Artificial Neural Networks (FF-ANN) with one hidden layer, also known as shallow neural networks (as opposite to a deep neural networks, containing more than one hidden layer) are used in this work to learn the mapping from the current signal data to the k values. One-layer MLPs like those used herein are universal function approximators $\boldsymbol{f}_{\boldsymbol{g}}(\boldsymbol{x})$: $R^{nd} \rightarrow R^{no}$ [6], where n_d

is the size of the input vector $\boldsymbol{x}$, and n_o is the size of the output vector function $\boldsymbol{f}_{\boldsymbol{g}}$. The general matrix-vector definition of $\boldsymbol{f}_{\boldsymbol{g}}$ is:

$$\boldsymbol{f}_g(x) = A2\left(\boldsymbol{b}^{(2)} + \boldsymbol{W}^{(2)}\left(A1\left(\boldsymbol{b}^{(1)} + W^{(1)}\boldsymbol{x}\right)\right)\right) \tag{1}$$

where $W^{(1)}$ is a weight matrix of size $(N \times n_d)$ and N is the number of neurons in the hidden layer, $\boldsymbol{b}^{(1)}$ is a bias term represented by a column vector of length N, $\boldsymbol{W}^{(2)}$ is a weight matrix of size $(n_o \times N)$, $\boldsymbol{b}^{(2)}$ is a bias term represented by a column vector of length n_o, and $A1$ and $A2$ are the activation functions of the hidden layer and the output layer, respectively. The general schematic of a one-layer MLP feed-forward ANN is shown in Fig. 1.

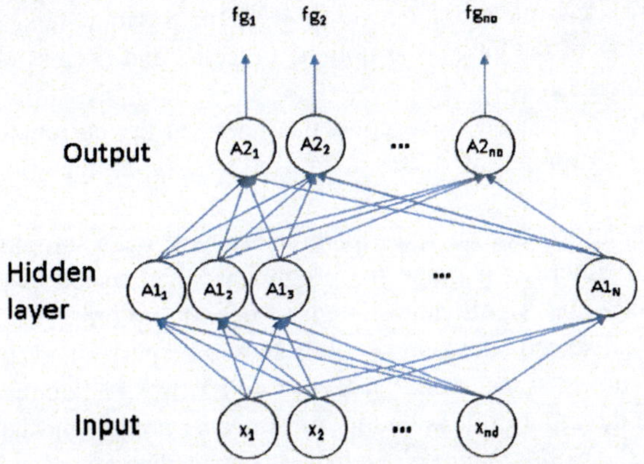

Fig. 1. General scheme of a multi-input multi-output MLP feed-forward ANN [7]

As in the vast majority of cases, for this work $A1$ is the hyperbolic tangent sigmoid function, and $A2$ is linear, and with this choice, function $\boldsymbol{f}_{\boldsymbol{g}}(\boldsymbol{x})$ becomes:

$$\boldsymbol{f}_{\boldsymbol{g}}(\boldsymbol{x}) = \boldsymbol{b}^{(2)} + \boldsymbol{W}^{(2)}\left(tanh\left(\boldsymbol{b}^{(1)} + W^{(1)}\boldsymbol{x}\right)\right) \tag{2}$$

Given a set of Ns training samples $\{(\boldsymbol{x}_1, y_1), \ldots, (x_{Ns}, y_{Ns})\}$, where y_i is the observed response to the input $\boldsymbol{x}_i$, a learning algorithm seeks the values of the weight matrices and bias vectors that minimize the difference between part or all of the observed Ns sample responses $\boldsymbol{y}_i$ and the Ns responses given by $\boldsymbol{f}_{\boldsymbol{g}}(\boldsymbol{x}_i)$. Gradient-based approaches are normally used to find the values of the weight matrices and the bias vectors by minimizing the fitting error of the responses. Since the responses are assumed to be smooth functions of the inputs $\boldsymbol{x}$ and the internal weights and biases, the gradients with respect to weights and biases of the difference between sample responses and ML approximations can be computed using the backpropagation method, using the chain rule for derivation. Once the gradients with respect to weights and biases for each layer are computed, the objective function expressing the level of fitting of the training data can be optimized.

If the ANN has a sufficiently high number of neurons or degrees of freedom in the hidden layer, a parameter proportional to the chosen number of neurons, the fitting error can be reduced to machine zero; this results in the system *overfitting* the training data, and being possibly unable to generalize adequately its predictions to regions of the input space where there are no or insufficient training data.

In this work, the inputs are the time samples of the current signal, c_i, $i = 1, \ldots, n_i = n_d$.

3 Results and Discussion

As anticipated in the previous sections, the proposed approach has been applied to two different cases: one considering only 5 of the total 8 parameters implementing the degradations, and a complete one considering all the 8 parameters. What follows is a detailed description of the cases, the obtained results, and discussion considering the outcomes of both cases.

Given the already proven correspondence between the current signal of RM and MM, the monitoring model, MM, has been used to save computational time, for both cases and for all the tests.

In both cases, all the considered k parameters have been sampled in [0.25, 0.75], which is the most interesting range for prognostic: 0.25 means that the degradation is at the early stage, but significant enough to need monitoring; 0.75 means that the degradation is at advanced stage, and actions should be performed to avoid total fault. MM produces a current signal from $t = 0$s to $t = 0.5$s, with a fine discretization, $\Delta t =$ 1e–5s. To reduce the input dimensionality of the ANN model, and with the reasonable idea that not all the points may be needed, a coarser discretization of the signal has been considered, with $\Delta t = $ 5e–3s, allowing to have $n_i = n_d = 100$. Figure 1 shows the difference between high and low frequency sampling for three random cases obtained when the first five k parameters are considered (simple case, as described in the following Sect. 3.1). For all the cases, *Sol*1 = [0.71481, 0.40819, 0.34196, 0.35228, 0.53386], *Sol*2 = [0.54777, 0.73226, 0.57659, 0.62445, 0.57678], and *Sol*3 = [0.62386, 0.73065, 0.25419, 0.30322, 0.39935], the low frequency sampling only cuts some of the peaks (Fig. 2).

Moreover, in both cases, the available data (see the coming subsections) have been split 95% for training and 5% for validation, and the Levenberg-Marquardt (LM) backpropagation in the MATLAB implementation has been used.

3.1 Simple Case Results and Discussion

In this case, the first 5 of the total 8 k parameters have been considered (the other three k parameter keep the nominal values), that are:

- k_1: Dry friction.
- k_2: Backlash.
- k_3, k_4, k_5: Short circuit (SC), phases A, B, and C, respectively.

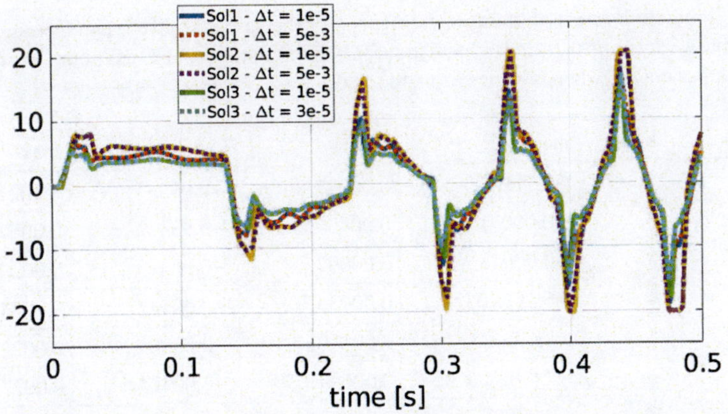

Fig. 2. Random examples of current signals with both high and low frequency sampling.

The design of experiment (DOE) to sample the 5-dimensional space combines a full regular grid with 5 nodes for each dimension (3125 samples), with a Latin Hypercube Sampling of 3000 points, giving $Ns = 6125$. To assess the effect of the number of neurons in the hidden layer, N, the ANN has been set with three different values: N = 40, N = 60, N = 80.

The three obtained ANNs have been tested against a sample of 100 random signals, i.e., obtained by randomly (uniform) sampling the k parameter space (the 100 samples are the same for all the three tests). The obtained statistical results are summarised in Table 1.

The biggest errors for each ANN are highlighted, showing that k_2 is the most difficult parameter to map, consistently across the ANNs. However, the maximum error, never exceeds ~0.03 and is as low as ~0.01 when N = 80, which is low enough to be of practical use: even an error of 0.03 would mean that in the worst case scenario the analyst would start tracking k_2 when the model maps $k_2 = 0.25$ while in reality it is already of 0.28, or that the prediction of the backlash entering in the red zone, i.e. $k_2 = 0.7$, is either slightly anticipated, and in reality $k_2 = 0.22$, or slightly delayed, and in reality $k_2 = 0.28$.

In Fig. 3, the TestMM current signal is the one produced by the MM for the random test case showing the maximum error on the k_2 mapping ($k_{1,\ldots,5}$ = [0.71620, 0.7249, 0.74506, 0.31308, 0.7383]), while the TestANN signal is the one produced by the MM when the input of the model is the output vector of the ANN for N = 80 ($k_{1,\ldots,5}$ = [0.71391, 0.71333, 0.74587, 0.313, 0.74237]). Current signals are shown with both high and low frequency samplings. As it can be seen qualitatively from Fig. 3, and more quantitatively from Fig. 4, the differences between the signals are very small and certainly within the uncertainty boundaries of the considered MM, confirming that the ANN based approach is ready for practical applications.

3.2 Complete Case Results and Discussion

In this case, all the 8 k parameters presented in Sect. 2.1 have been considered. The design of experiment (DOE) to sample the 8-dimensional space combines again a full

Table 1. Prediction errors on the 100 test signals for the case with five k parameters. Statistical values (mean value, *mean*, standard deviation, *std*, minimum, *min*, and maximum, *max*) reported for each k parameter, when N varies from 40 to 80.

	k	mean	std	min	max
N = 40	1	7.79E–05	0.00087	−0.00235	0.002703
	2	−0.00091	0.004006	**−0.03322**	0.00649
	3	−0.00013	0.000925	−0.00318	0.002039
	4	−6.56E–05	0.001332	−0.00441	0.004419
	5	4.17E–05	0.001105	−0.00364	0.00226
N = 60	1	−7.99E–05	0.000943	−0.00424	0.002331
	2	−0.0007	0.003556	−0.01496	**0.022563**
	3	−0.00014	0.000838	−0.00261	0.002538
	4	−3.32E–05	0.001214	−0.00367	0.002936
	5	−2.02E–05	0.001132	−0.0041	0.002534
N = 80	1	6.96E–05	0.000986	−0.00388	0.002702
	2	−0.00032	0.002315	−0.00762	**0.011558**
	3	−0.00012	0.000774	−0.00189	0.001932
	4	−3.98E–05	0.001133	−0.00246	0.003673
	5	7.63E–06	0.001237	−0.00407	0.006334

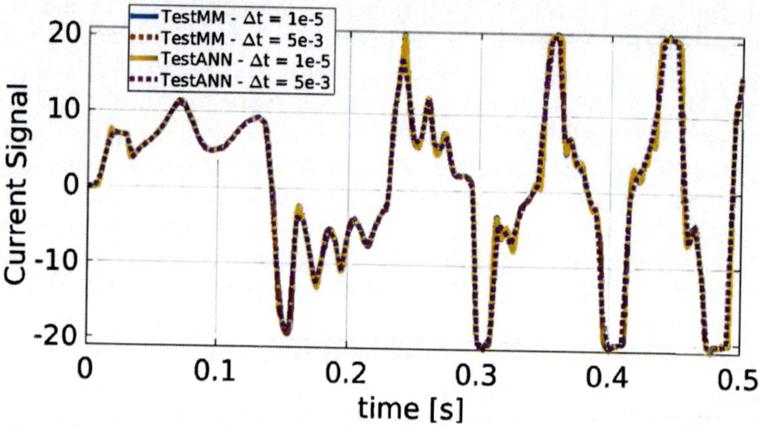

Fig. 3. For the simple case: current signal produced by the MM for the random test case showing the maximum error on the k_2 mapping, TestMM, and the onee produced by the MM when the input of the model is the output vector of the ANN for N = 80, TestANN (current signals shown with both high and low frequency samplings).

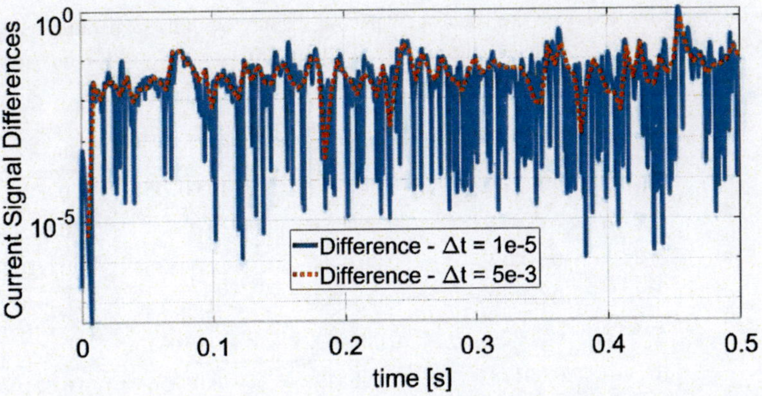

Fig. 4. For the simple case: TestMM and TestANN for both high and low frequency samplings.

regular grid with 5 nodes for each dimension (3,90,625 samples), with a Latin Hypercube Sampling of 4,00,000 points, giving $Ns = 7{,}90{,}625$.

To assess the effect of the number of neurons in the hidden layer, N, the ANN has been set with three different values: N = 50, N = 70, N = 90. The three obtained ANNs have been tested against a sample of 100 random signals, i.e., obtained by randomly (uniform) sampling the k parameter space (again, the 100 samples are the same for all the three tests). The obtained results are summarised in Table 2. Again, the biggest errors for each ANN are highlighted, and in this case both k_2 (backlash) and k_6 (static eccentricity modulus in the rotor) are the most difficult parameters to map, consistently across the ANNs. However, the maximum error, never exceeds ~ 0.02, slightly improving when passing from N = 50 to N = 90. Likely a higher value of N could help to reduce the error, but the training with this amount of data has been strongly limited by the computational resources: the training of the N = 90 ANN took 7 days on 12 INTEL Skylake 6138 cores, which is the maximum the team could use for this work.

In Figs. 5 and 7, the TestMM current signals are the ones produced by the MM for the random test case showing the maximum error on the k_2 mapping ($k_{1,\ldots,8}$ = [0.57543, 0.40647, 0.63437, 0.64092, 0.6762, 0.72495, 0.30366, 0.70536]) and on the k_6 mapping ($k_{1,\ldots,8}$ = [0.6549, 0.68609, 0.73232, 0.61184, 0.57124, 0.60872, 0.4837, 0.41279]), respectively. On the other hand, the TestANN signals are those produced by the MM when the input of the model is the output vector of the ANN for N = 90 ($k_{1,\ldots,8}$ = [0.57149, 0.426, 0.63765, 0.64145, 0.67431, 0.72874, 0.29977, 0.70928] when the maximum error on k_2 mapping is considered, and $k_{1,\ldots,8}$ = [0.6577, 0.67421, 0.7319, 0.61512, 0.56771, 0.59137, 0.48718, 0.40601] when the maximum error on k_6 mapping is considered).

As it can be seen quantitatively from Figs. 6 and 8, the differences between the signals are, again, very small and within the uncertainty boundaries of the considered MM, confirming that the ANN based approach can be considered useful and ready for complex cases too.

Table 2. Prediction errors on the 100 test signals for the case with five k parameters. Statistical values (mean value, *mean*, standard deviation, *std*, minimum, *min*, and maximum, *max*) reported for each k parameter, when N varies from 40 to 80.

	k	mean	Std	min	max
N = 50	1	−1.65E–05	0.002746	−0.00745	0.008193
	2	8.32E–05	0.006611	−0.01524	**0.023647**
	3	−0.00013	0.002984	−0.00836	0.007289
	4	0.000189	0.003028	−0.00916	0.005793
	5	0.000259	0.003452	−0.01031	0.007574
	6	0.001182	0.006259	−0.01356	**0.021009**
	7	0.000196	0.004567	−0.01455	0.011524
	8	−0.00048	0.003997	**−0.02224**	0.00672
N = 70	1	−9.92E–05	0.002425	−0.00539	0.006919
	2	−0.00053	0.004943	−0.01035	**0.017784**
	3	5.09E–05	0.002121	−0.00484	0.0064
	4	0.000338	0.002705	−0.00719	0.005802
	5	9.49E–05	0.003002	−0.00938	0.011673
	6	0.000531	0.005287	−0.01203	**0.016516**
	7	0.000614	0.0037	−0.01076	0.009334
	8	−0.00014	0.00334	−0.01387	0.011365
N = 90	1	8.52E–05	0.002329	−0.00477	0.007887
	2	−0.00108	0.005029	**−0.01954**	0.011878
	3	5.51E–05	0.002208	−0.00589	0.006297
	4	−0.00022	0.002319	−0.00739	0.004484
	5	0.000149	0.002644	−0.0074	0.007928
	6	0.000518	0.004788	−0.01066	**0.01736**
	7	−0.0003	0.003294	−0.00933	0.01076
	8	8.20E–05	0.002778	−0.00888	0.007526

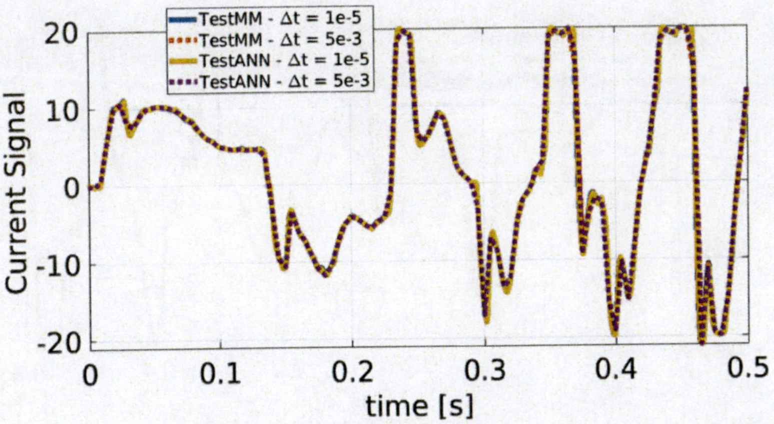

Fig. 5. For the complete case: current signal produced by the MM for the random test case showing the maximum error on the k_2 mapping, TestMM, and the onee produced by the MM when the input of the model is the output vector of the ANN for N = 90, TestANN (current signals shown with both high and low frequency samplings)

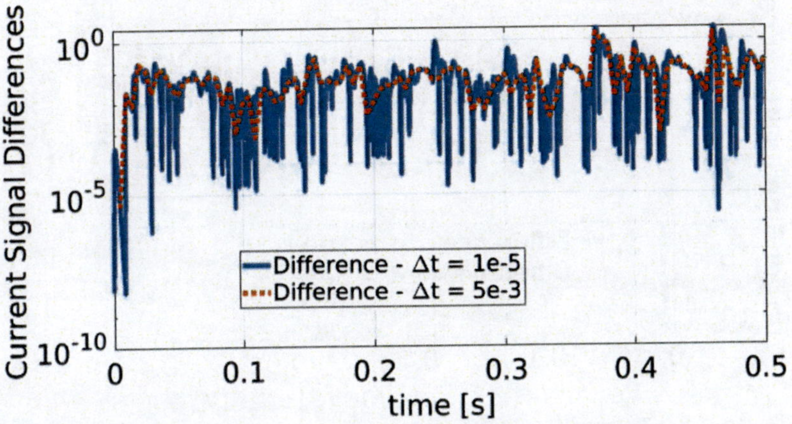

Fig. 6. For the complete case (maximum error on the k_2 mapping): TestMM and TestANN for both high and low frequency samplings.

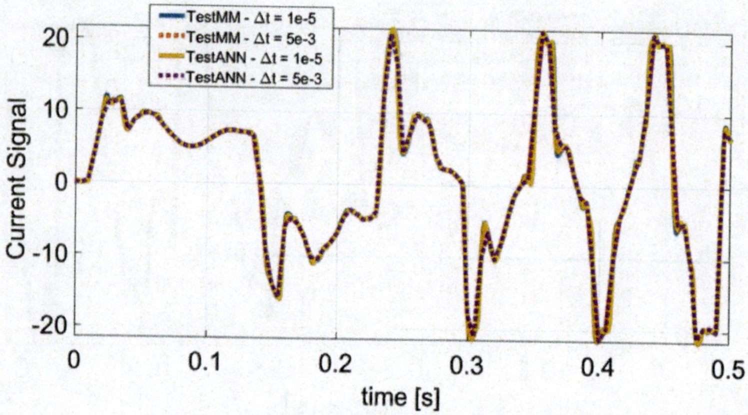

Fig. 7. For the complete case: current signal produced by the MM for the random test case showing the maximum error on the k_6 mapping, TestMM, and the onee produced by the MM when the input of the model is the output vector of the ANN for N = 90, TestANN (current signals shown with both high and low frequency samplings)

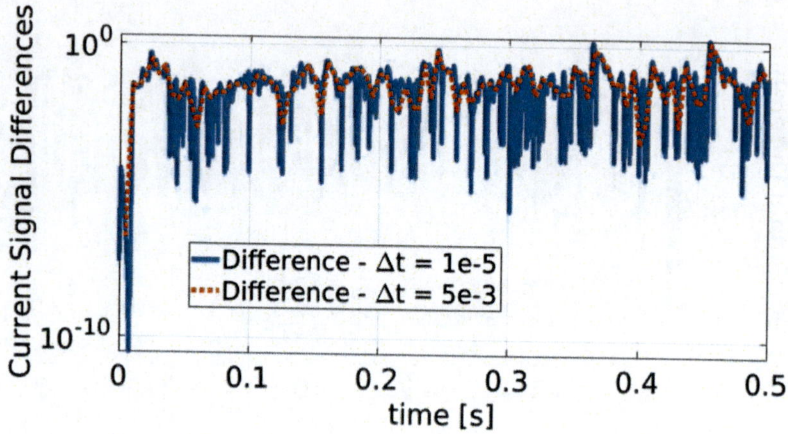

Fig. 8. For the complete case (maximum error on the k_6 mapping): TestMM and TestANN for both high and low frequency samplings.

4 Conclusion

A machine learning based approach has been implemented and tested to provide a real-time prognostic response for a considered electro-mechanical actuator system. In particular, a shallow artificial neural network is used to directly map the electric signal response of the system into the values of the key parameters of the EMA model that implement the different kinds of degradation. The approach has been tested both on a simplified case, where only 5 of the total 8 parameters are considered and on the complete case, where all the 8 parameters implemented in the model are considered. In both cases the proposed approach shows robust performance, with absolute errors on the identification of the simple parameters always $\leq 2\%$, which is considered enough for practical applications.

In both cases (i.e., the simplified and the complete ones) the parameter corresponding to the backlash degradation shows the highest identification errors, and for the complete case, the degradation related to static eccentricity modulus in the rotor, shows some relatively bigger errors. This can be related to the DOE data considered for the training of the ANNs and/or to higher non-linearities related to that particular degradation in the model, and this aspect will be better analyzed in the future.

Moreover, even if the proposed approach shows good and robust performance, the ANN training in the complete case challenged the available computational capabilities and highlighted the need to sample the electric signal more efficiently to limit the exponential increase of the needed database. This will be considered for future works too.

References

1. Qiao, G., Liu, G., Shi, Z., Wang, Y., Ma, S., Lim, T.C.: A review of electromechanical actuators for more/all electric aircraft systems. Proc. Inst. Mech. Eng. C J. Mech. Eng. Sci. **232**(22), 4128–4151 (2018)
2. Karadotcheva, E., Nguyen, S.N., Greenhalgh, E.S., Shaffer, M.S.P., Kucernak, A.R.J., Linde, P.: Structural power performance targets for future electric aircraft. Energies **14**(19), 6006 (2021)
3. Quattrocchi, G., Berri, P.C., Dalla Vedova, M.D.L., Maggiore, P.: An improved fault identification method for electromechanical actuators. Aerospace **9**(7), 341 (2022)
4. Dalla Vedova, M.D.L., Germanà, A., Berri, P.C., Maggiore, P.: Model-Based Fault Detection and Identification for Prognostics of Electromechanical Actuators Using Genetic Algorithms. Aerospace 6(9), 94 (2019)
5. Querques, I.: Prognostics of On-Board Electromechanical Actuators: Bio-Inspired Metaheuristic Algorithms; Technical Report, Politecnico di Torino, Torino, Italy (2021)
6. Cybenko, G.: Approximation by superpositions of a sigmoidal function. Math. Control Sign. Syst. **2**, 303–314 (1989)
7. Campobasso, M.S., Cavazzini, A., Minisci, E.: Rapid estimate of wind turbine energy loss due to blade leading edge delamination using artificial neural networks. J. Turbomach. **142**(7), 071002-1/11 (2020)

The Advantages and Efficiency of Using the 5S Methodology in Vehicle Repairs and Maintenance Workshops

Dhimitraq Sotja(✉), Enkelejda Konda, and Migena Zeqo

Polytechnic of Tirana, Tirana, Albania
dhsotja@fim.edu.al

Abstract. The activities of marketing new vehicles and the enterprises that service and repair them, have as their primary purpose the increase of sales and financial income. In the conditions of ever-increasing competition for these types of activities, the "quality" component of the services offered takes on a special importance. The rapid spread of these types of services in our country definitely requires that it be accompanied by new methodological approaches, in order to be successfully integrated into the modern economy of Europe. In this aspect, the vehicle service market in general considered as the relationship between the subjects "vehicle owners" and "service companies", which is guided by the universal demand-offer relationship, has the sole purpose of maintaining the ability to work or the efficiency of vehicles during the entire life-time period of service. The purpose of this article is to promote the use of innovative management technologies of enterprises that serve the means of transport, such as the 5S System Methodology.

Keywords: Vehicles · Marketing and service · Life-time period of service · Demand-offer relationship · 5S system methodology

1 Workplace Organization and Rationalization

2 Japanese Concept of Management—A Brief History

The competitiveness of enterprises in the 21st century is no longer determined only by technical and investment capabilities, but also by the management of all resources, and primarily human resources potential.

In Japan, there is a systematic approach to organizing, putting things in order and cleaning the workplace. This need arose in the post-war period, more precisely, by the mid-50s. At that time, Japanese enterprises were forced to work in conditions of scarce resources. So they developed for their production a method in which everything was taken into account and there was no room for anything. When Americans went to the Japanese companies in the 1970s and 1980s, in order to adopt their experience, first of all, they all turned to attention to one circumstance: "The factories are so clean that you can eat right on the floor."

G. Guxho et al. (Eds.): ITC-ICEE 2023, LNMIE, pp. 160–165, 2024.
https://doi.org/10.1007/978-3-031-48933-4_16

3 5S—Methodology and Its Benefits

The 5S system is recommended for everyone who wants to make their workplace more efficient, be it a workshop workplace or a bank office, an accountant's or director's office. The author of the 5S system is considered to be Kaoru Ishikawa, one of the creators Japanese quality management system, whose name is immortalized in the famous statistical analysis tool—the Ishikawa (or "Fishbone") diagram. 5S was created in Japan, and the original "S" terms were in Japanese, so English translations for each of the five steps may vary. The five words are originally translated from the Japanese words seiri, *seiton*, *seiketsu*, and *shitsuke* [1].

- Sort ((整理) **S**eiri): Removing what is not needed and clearing the workplace sorting;
- Set in Order ((整頓) **S**eiton): Preparing the necessary items neatly and systematically so that they can easily be taken and returned in the original place after use; self-organization, systematic cleaning, standardization, improvement.
- Shine ((清掃) **S**eiso): Cleaning regularly equipment and workplace, identifying irregularities. Dust, dirt and wastes are the source of untidiness, indiscipline, inefficiency, faulty production and work accidents;
- Standardize ((清佞) **S**eikutsu): Documenting and standardizing the method, using standard procedures. Standards should be very communicative, clear and easy to understand.
- Sustain ((躾) **S**hitsuke): Continuously maintaining established procedures, auditing work methods, making 5S a habit, integrating into culture (Fig. 1).

Fig. 1. Five steps of the 5S methodology [2].

The 5S method is the involvement of the entire team in regular activities to restore order, cleanliness and strengthen discipline in workplaces. The 5S method helps improve worker productivity, reduce the number of defects, and most importantly, makes it possible to save company time and money. The 5S system allows to notice and eliminate losses that almost are present every minute. Gradually improving, employees at the enterprise will spend less time on a particular task, thereby increasing overall productivity.

5S does not require significant investment or use expensive technologies. This system is based on tools like common sense, checklists and low-tech methods at low costs.

The 5S system was developed through intensive work on production, which proves its necessity and possibility of implementation in any company, seeking to minimize costs and increase quality [1, 2]. The 5S system is useful in preparing for ISO certification 9001.

Advantages of the technique are related to applying it immediately; low costs required; everyone can take part in this; quick and clear results; no large-scale training efforts are required; the method has a wide range of applications; further directions for improvement become visible. What may be useful when implementing this technique, are simple: work clothes; cleaning equipments; detergents; dye; markers; thick paper; camera etc.

4 Applying 5S—Methodology

Thera are some steps to be followed [5].

Step one: sorting. Actions to be taken:

- determine the necessary items and their quantity at each workplace;
- using red labels, you need to mark everything unnecessary;
- separate everything unnecessary;
- select a special place for this and put everything unnecessary there;
- remove everything unnecessary from your work area.

Sorting objects are shown in Fig. 1 while cause-and-effect relationships when organizing the workspace are presented in the Table 1.

The marking of working areas is shown in Figs. 1, 2 and 3 [3].

Floor Color	Designation
Red	Waste accumulation or storage area
Yellow	Access aisle
Green	Material in process
Blue	Incomming material
White	Outgoing material
Yellow/Black	Non hazardous waste accumulation area
Red/White	Rejected material area

Fig. 2. Floor color marking system

Fig. 3. Examples of work area marking

Step two: self-organization. Actions to be taken:

- determine zones and places for each required item;
- apply marks to tools, containers, etc.;
- arrange tools and accessories so that they can be easily found and conveniently used.

Step three: systematic cleaning. Actions to be taken:

Table 1. Cause-and-effect relationships in organizing the workspace

Frequency of use	Storage location
No further use planned	Write-off
Very rarely (once in year or less	Separate warehouse on the territory of the workshop
Rarely (once every six months)	Warehouse on the workshop premises
Often (once a week or more often)	Warehouse at the production site
Often (once a day or more often)	At work

- identify objects for regular cleaning;
- set the frequency of cleaning;
- define rules;
- distribute responsibility.

Objects needed for cleaning: cleaning tools, hydric power stations; devices, vernier calipers; tool cabinets; control benches, etc.

Frequency of cleaning: every day at the workplace; every week on site; every month in the workshop; every year for the workshop as a whole (Fig. 4).

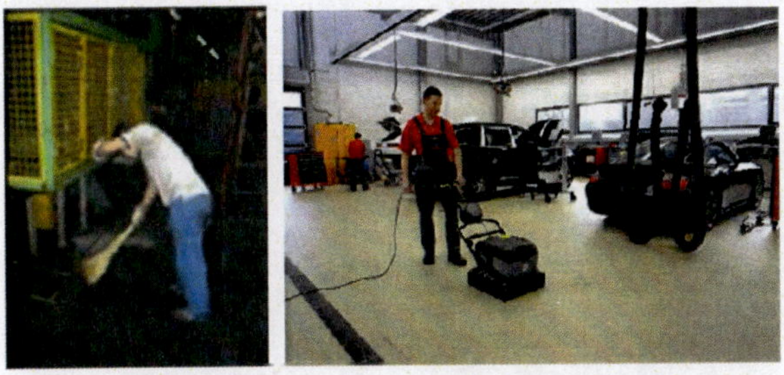

Fig. 4. Cleaning schedule for workplaces [4]

Step four: standardization. Actions to be taken:

- draw up a process diagram;
- define each work step;
- define working methods;
- develop work instructions;
- display important data on the information board (Fig. 5).

Step five Improvement. Actions to be taken:

- develop a 5S procedure, introduce it into the practice of the entire production;
- collection of proposals for further improvement of 5S elements;
- carrying out improvements (Fig. 6).

Fig. 5. Standardization of workplaces

Fig. 6. Organization of the workplace using the 5S method

Thus, this article highlights the main issues related to adaptation and the introduction of the popular 5S workplace organization methodology into the practice of automotive heavy production and car service.

Application of the proposed tools allows to increase the work efficiency, both at individual workplaces and in production as a whole, and also ensures improved quality of products and processes (Fig. 7).

Fig. 7. The result of applicating 5S—method

References

1. Hirano and Hiroyuki: 5 Pillars of the Visual Work -place. Productivity Press, Cambridge, MA (1995)

2. Leansmart Homepage, https://leansmarts.com/what-lean-5s-methodology-all-about/. Last Accessed 18 May 2023
3. KYMC Homepage, https://www.kymc.com/msg/5s.html. Last Accessed 18 Sep 2021
4. Pmmonline Homepage, https://pmmonline.co.uk/article/the-importance-of-a-clean-workshop/
5. LNCS Homepage, http://www.springer.com/lncs. Last Accessed 21 Nov 2016

Graphical and Analytical Analysis in the Field of Mohr's Circles

Anis Sulejmani(✉), Odhisea Koça, and Mentor Cejku

Faculty of Mechanical Engineering, Polytechnic University of Tirana, Tirana, Albania
asulejmani@fim.edu.al

Abstract. The stress state of the elementary volume can be studied graphically, with the three main Mohr's circles [1, 9]. In the space of the τ-σ axis system, the points in this system have many properties, where reality and imagination intertwine [9, 10]. Mohr's circles are numerous in addition to the main circles, and they have multiple properties. It is also required to find the main stresses and then the main orientations, analytically and graphically. This article reveals the properties of Mohr's space points and ways to solve problems of volumetric stress state graphically and analytically, explaining the connection between reality and imagination, as well as giving a geometric meaning to finding the principal stresses.

Keywords: Principal stresses · Stress tensor · Principal orientations · Leading cosines · Principal Mohr's circles · Mohr space

1 Introduction

Let's analyze a volumetric element with principal faces and principal stresses $\sigma_1, \sigma_2, \sigma_3$ (Fig. 1).

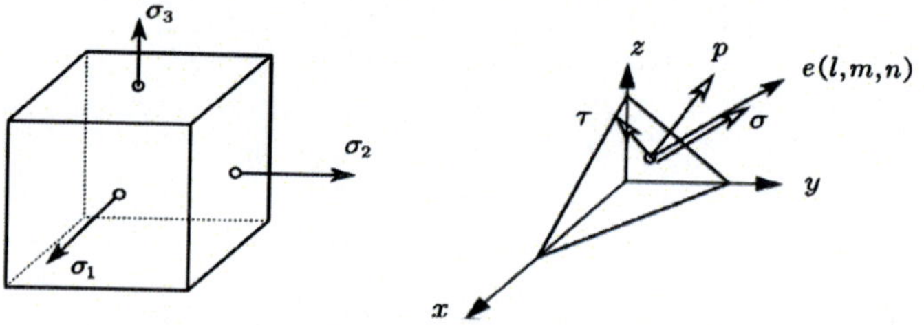

Fig. 1. Volumetric element with principal faces and stresses.

The stress tensor has the simplest form as shown in Eq. 1.

$$T_{\sigma 123} = \begin{pmatrix} \sigma_1 & 0 & 0 \\ 0 & \sigma_2 & 0 \\ 0 & 0 & \sigma_3 \end{pmatrix} \tag{1}$$

G. Guxho et al. (Eds.): ITC-ICEE 2023, LNMIE, pp. 166–181, 2024.
https://doi.org/10.1007/978-3-031-48933-4_17

An arbitrary face with the direction $\vec{e}(l, m, n)$, which satisfies this condition, $1 = l^2 + m^2 + n^2$, and the total strain on this face has the direction and magnitude as shown below:

$$\vec{p} = \begin{pmatrix} p_x \\ p_y \\ p_z \end{pmatrix} = \begin{pmatrix} \sigma_1 & 0 & 0 \\ 0 & \sigma_2 & 0 \\ 0 & 0 & \sigma_3 \end{pmatrix} \begin{pmatrix} l \\ m \\ n \end{pmatrix} = \begin{pmatrix} \sigma_1 l \\ \sigma_2 m \\ \sigma_3 n \end{pmatrix} \tag{2}$$

$$p^2 = \sigma^2 + \tau^2 = \sigma_1^2 l^2 + \sigma_2^2 m^2 + \sigma_3^2 n^2 \tag{3}$$

The magnitude and direction of the normal and tangential stresses are [11]:

$$\sigma = \overrightarrow{p} \cdot \vec{e} = \sigma_1 l^2 + \sigma_2 m^2 + \sigma_3 n^2 \tag{4}$$

$$\vec{\sigma} = \begin{pmatrix} l \\ m \\ n \end{pmatrix} \sigma = \begin{pmatrix} l \\ m \\ n \end{pmatrix} \begin{pmatrix} l & m & n \end{pmatrix} \begin{pmatrix} p_x \\ p_y \\ p_z \end{pmatrix} = \begin{pmatrix} l^2 & lm & ln \\ ml & m^2 & mn \\ nl & nm & n^2 \end{pmatrix} \begin{pmatrix} p_x \\ p_y \\ p_z \end{pmatrix} \tag{5}$$

$$\tau = \sqrt{p^2 - \sigma^2} \tag{6}$$

$$\begin{aligned} \vec{\tau} = \vec{p} - \vec{\sigma} &= \begin{pmatrix} p_x \\ p_y \\ p_z \end{pmatrix} - \begin{pmatrix} l^2 & lm & ln \\ ml & m^2 & mn \\ nl & nm & n^2 \end{pmatrix} \begin{pmatrix} p_x \\ p_y \\ p_z \end{pmatrix} \\ &= \begin{pmatrix} 1 - l^2 & -lm & -ln \\ -ml & 1 - m^2 & -mn \\ -nl & -nm & 1 - n^2 \end{pmatrix} \begin{pmatrix} p_x \\ p_y \\ p_z \end{pmatrix} \end{aligned} \tag{7}$$

Considering the vector (l, m, n) unknown, we can write:

$$\begin{pmatrix} 1 & 1 & 1 \\ \sigma_1 & \sigma_2 & \sigma_3 \\ \sigma_1^2 & \sigma_2^2 & \sigma_3^2 \end{pmatrix} \begin{pmatrix} l^2 \\ m^2 \\ n^2 \end{pmatrix} = \begin{pmatrix} 1 \\ \sigma \\ \sigma^2 + \tau^2 \end{pmatrix} \tag{8}$$

$$\begin{pmatrix} 1 & 1 & 1 \\ 0 & \sigma_2 - \sigma_1 & \sigma_3 - \sigma_1 \\ 0 & 0 & (\sigma_3 - \sigma_1)(\sigma_3 - \sigma_2) \end{pmatrix} \begin{pmatrix} l^2 \\ m^2 \\ n^2 \end{pmatrix} = \begin{pmatrix} 1 \\ \sigma - \sigma_1 \\ (\sigma - \sigma_1)(\sigma - \sigma_2) + \tau^2 \end{pmatrix} \tag{9}$$

So, the solution is:

$$n^2 = \frac{(\sigma - \sigma_1)(\sigma - \sigma_2) + \tau^2}{(\sigma_3 - \sigma_1)(\sigma_3 - \sigma_2)} \tag{10}$$

and from the cyclic rule we get the other two cosines:

$$l^2 = \frac{(\sigma - \sigma_3)(\sigma - \sigma_2) + \tau^2}{(\sigma_1 - \sigma_2)(\sigma_1 - \sigma_3)} \tag{11}$$

and

$$m^2 = \frac{(\sigma - \sigma_3)(\sigma - \sigma_1) + \tau^2}{(\sigma_2 - \sigma_1)(\sigma_2 - \sigma_3)} \tag{12}$$

2 Faces with the Same Direction Cosine

The cosines l, m and n take values from zero to one. So, we analyze every case [1].

2.1 Constant Cosine l

Let's start by analyzing the cosine $l = \cos\alpha$. We know that [9]:

$$1 \geq l^2 = \frac{(\sigma - \sigma_3)(\sigma - \sigma_2) + \tau^2}{(\sigma_1 - \sigma_2)(\sigma_1 - \sigma_3)} \geq 0 \text{ and } (\sigma_1 - \sigma_2) > 0, (\sigma_1 - \sigma_3) > 0 \tag{13}$$

So, we can conclude (Fig. 2):

$$(\sigma - \sigma_3)(\sigma - \sigma_2) + \tau^2 \geq 0 \rightarrow \sigma^2 - \sigma(\sigma_2 + \sigma_3) + \sigma_2\sigma_3 + \tau^2 \geq 0 \tag{14}$$

$$\left(\sigma - \frac{\sigma_2 + \sigma_3}{2}\right)^2 + \tau^2 \geq \left(\frac{\sigma_2 - \sigma_3}{2}\right)^2 \tag{15}$$

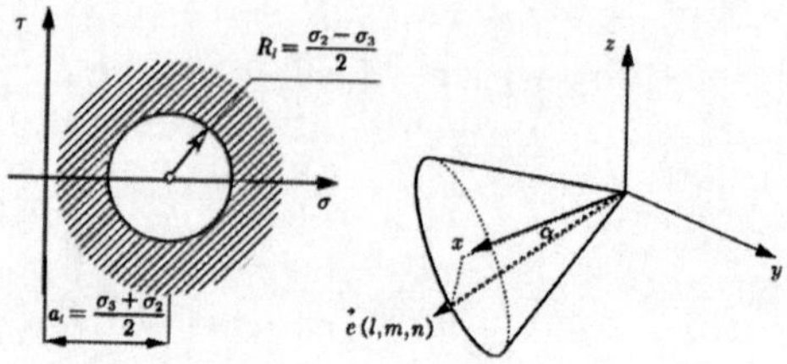

Fig. 2. Mohr space circle with constant cosine l.

If we have an orientation with a constant cosine l, then all faces will be in the Mohr field in the same circle [10]:

$$\begin{aligned} l^2 &= \frac{(\sigma - \sigma_3)(\sigma - \sigma_2) + \tau^2}{(\sigma_1 - \sigma_2)(\sigma_1 - \sigma_3)} \rightarrow (\sigma - \sigma_3)(\sigma - \sigma_2) + \tau^2 \\ &= (\sigma_1 - \sigma_2)(\sigma_1 - \sigma_3)l^2 \end{aligned} \tag{16}$$

With some simple mathematics we get [2]:

$$\left(\sigma - \frac{\sigma_2 + \sigma_3}{2}\right)^2 + \tau^2 = \left(\frac{\sigma_2 - \sigma_3}{2}\right)^2 + (\sigma_1 - \sigma_2)(\sigma_1 - \sigma_3)l^2 = R_l^2 \tag{17}$$

$$R_l = \sqrt{\left(\frac{\sigma_2 - \sigma_3}{2}\right)^2 + (\sigma_1 - \sigma_2)(\sigma_1 - \sigma_3)l^2} \tag{18}$$

So, the radius is (Fig. 3):

$$R_{l=0} = \frac{\sigma_2 - \sigma_3}{2} \text{ and } R_{l=1} = \sigma_1 - \frac{\sigma_2 + \sigma_3}{2} \tag{19}$$

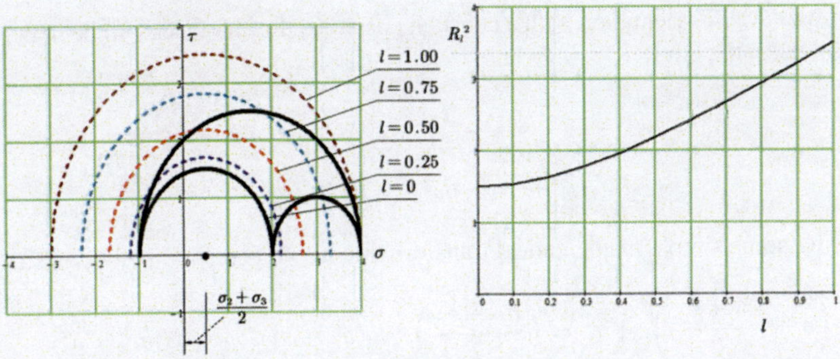

Fig. 3. Mohr circle with various constant l values and graph of radius R with dependant variable l.

2.2 Constant Cosine *m*

The same analysis is done for the constant cosine: $m = \cos \beta$.

$$1 \geq m^2 = \frac{(\sigma - \sigma_3)(\sigma - \sigma_1) + \tau^2}{(\sigma_2 - \sigma_1)(\sigma_2 - \sigma_3)} \geq 0 \text{ and } (\sigma_2 - \sigma_1)\langle 0, (\sigma_2 - \sigma_3)\rangle 0 \tag{20}$$

This way we get the equation shown below:

$$(\sigma - \sigma_3)(\sigma - \sigma_1) + \tau^2 \leq 0 \rightarrow \sigma^2 - \sigma(\sigma_1 + \sigma_3) + \sigma_1\sigma_3 + \tau^2 \leq 0 \tag{21}$$

So, the Mohr circle is given by (Fig. 4):

$$\left(\sigma - \frac{\sigma_1 + \sigma_3}{2}\right)^2 + \tau^2 \leq \left(\frac{\sigma_1 - \sigma_3}{2}\right)^2 \tag{22}$$

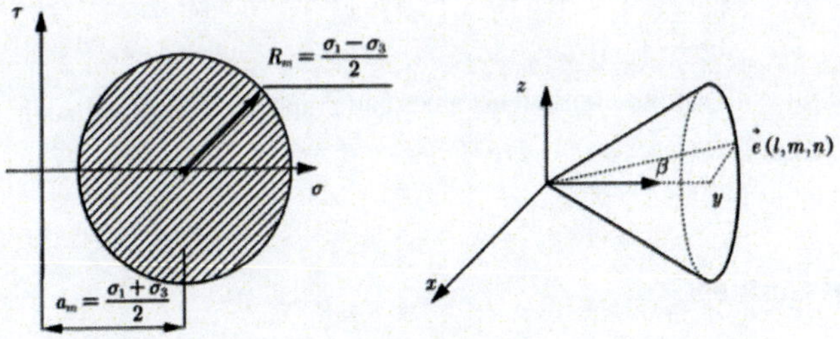

Fig. 4. Mohr space circle with constant cosine m.

If we have an orientation with a constant cosine m, then all faces will be in the Mohr field in the same circle:

$$m^2 = \frac{(\sigma - \sigma_3)(\sigma - \sigma_1) + \tau^2}{(\sigma_2 - \sigma_1)(\sigma_2 - \sigma_3)} \rightarrow (\sigma - \sigma_3)(\sigma - \sigma_1) + \tau^2$$
$$= (\sigma_2 - \sigma_1)(\sigma_2 - \sigma_3)m^2 \tag{23}$$

After some simple mathematical transformations, we get:

$$\left(\sigma - \frac{\sigma_1 + \sigma_3}{2}\right)^2 + \tau^2 = \left(\frac{\sigma_1 - \sigma_3}{2}\right)^2 + (\sigma_2 - \sigma_1)(\sigma_2 - \sigma_3)m^2 = R_m^2 \tag{24}$$

$$R_m = \sqrt{\left(\frac{\sigma_1 - \sigma_3}{2}\right)^2 + (\sigma_2 - \sigma_1)(\sigma_2 - \sigma_3)m^2} \tag{25}$$

So, the radius is (Fig. 5):

$$R_{m=0} = \frac{\sigma_1 - \sigma_3}{2} \text{ and } R_{m=1} = \sigma_2 - \frac{\sigma_1 + \sigma_3}{2} \tag{26}$$

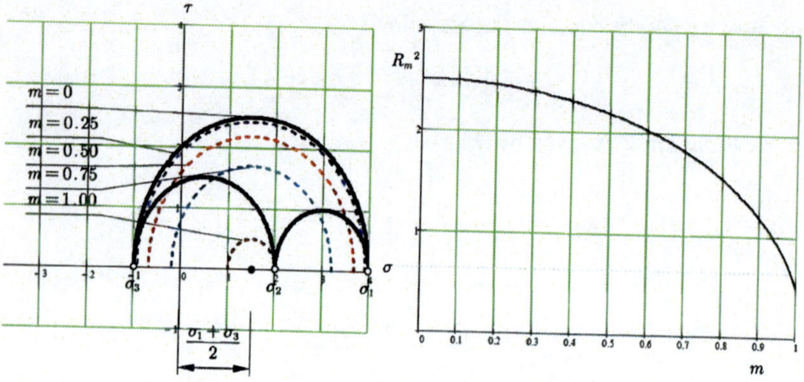

Fig. 5. Mohr circle with various constant m values and graph of radius R with dependant variable m.

2.3 Constant Cosine *n*

Continuing the same route, we get:

$$1 \geq n^2 = \frac{(\sigma - \sigma_1)(\sigma - \sigma_2) + \tau^2}{(\sigma_3 - \sigma_1)(\sigma_3 - \sigma_2)} \geq 0 \text{ and } (\sigma_3 - \sigma_1) < 0, (\sigma_3 - \sigma_2) < 0 \tag{27}$$

After some simple mathematical transformations, we get:

$$(\sigma - \sigma_1)(\sigma - \sigma_2) + \tau^2 \geq 0 \rightarrow \sigma^2 - \sigma(\sigma_1 + \sigma_2) + \sigma_1\sigma_2 + \tau^2 \geq 0 \tag{28}$$

So, the equation for the Mohr circle is (Fig. 6):

$$\left(\sigma - \frac{\sigma_1 + \sigma_2}{2}\right)^2 + \tau^2 \geq \left(\frac{\sigma_1 - \sigma_2}{2}\right)^2 \quad (29)$$

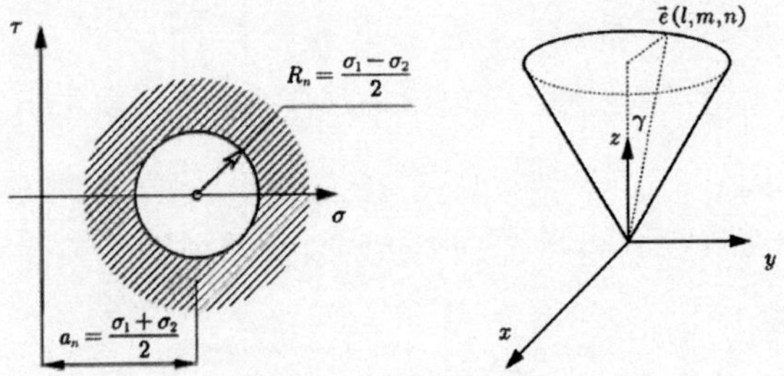

Fig. 6. Mohr space circle with constant cosine n.

If we have an orientation with a constant cosine n, then all faces will be in the Mohr field in the same circle:

$$n^2 = \frac{(\sigma - \sigma_1)(\sigma - \sigma_2) + \tau^2}{(\sigma_3 - \sigma_1)(\sigma_3 - \sigma_2)} \rightarrow (\sigma - \sigma_1)(\sigma - \sigma_2) + \tau^2$$
$$= (\sigma_3 - \sigma_1)(\sigma_3 - \sigma_2)n^2 \quad (30)$$

After some simple mathematical transformations, we get:

$$\left(\sigma - \frac{\sigma_1 + \sigma_2}{2}\right)^2 + \tau^2 = \left(\frac{\sigma_1 - \sigma_2}{2}\right)^2 + (\sigma_3 - \sigma_1)(\sigma_3 - \sigma_2)n^2 = R_n^2 \quad (31)$$

$$R_n = \sqrt{\left(\frac{\sigma_1 - \sigma_2}{2}\right)^2 + (\sigma_3 - \sigma_1)(\sigma_3 - \sigma_2)n^2} \quad (32)$$

So, the radius in Mohr space depending on cosine n is given below (Fig. 7):

$$R_{n=0} = \frac{\sigma_1 - \sigma_2}{2} \text{ and } R_{n=1} = \frac{\sigma_1 + \sigma_2}{2} - \sigma_3 \quad (33)$$

3 Determination of the Main Stresses

We analyze the deviant elementary volume of the main overall volume [3] (Fig. 8).

So, the deviator tensor is [7]:

$$T_\sigma^{dev} = \begin{pmatrix} \sigma_1 - \sigma_m & 0 & 0 \\ 0 & \sigma_2 - \sigma_m & 0 \\ 0 & 0 & \sigma_3 - \sigma_m \end{pmatrix} \quad (34)$$

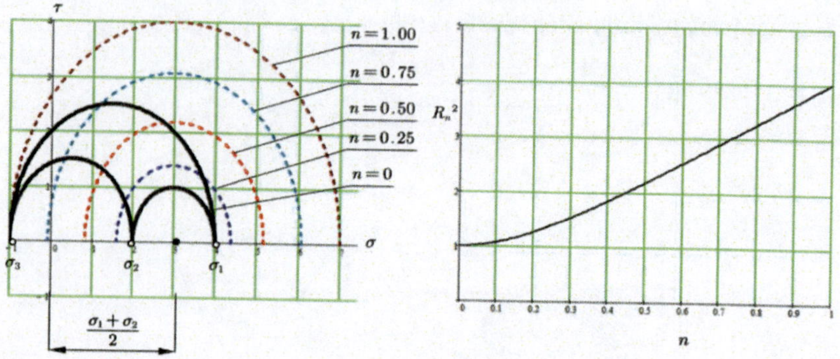

Fig. 7. Mohr circle with various constant *n* values and graph of radius R with dependant variable *n*.

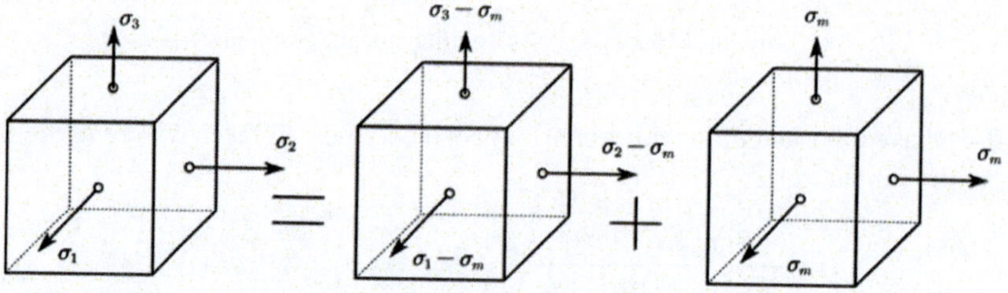

Fig. 8. Elementary volume divided into the deviant and mean volumes.

where

$$\sigma_m = \frac{\sigma_1 + \sigma_2 + \sigma_3}{3} = \frac{I_1}{3};\ I_1 = \sigma_1 + \sigma_2 + \sigma_3;$$
$$I_2 = \sigma_1\sigma_2 + \sigma_2\sigma_3 + \sigma_3\sigma_1;\ I_3 = \sigma_1\sigma_2\sigma_2 \tag{35}$$

$$\sigma_{1.dev} = \sigma_1 - \sigma_m = \sigma_1 - \frac{I_1}{3} \rightarrow \sigma_1 = \sigma_{1.dev} + \frac{I_1}{3} \tag{36}$$

The main stress equation of the general tensor is:

$$\left(\sigma_{dev} + \frac{I_1}{3}\right)^3 - I_1\left(\sigma_{dev} + \frac{I_1}{3}\right)^2 + I_2\left(\sigma_{dev} + \frac{I_1}{3}\right) - I_3 = 0 \tag{37}$$

By decomposing the third degree equation we get the following:

$$\left[\sigma_{dev}^3 + 3\sigma_{dev}\frac{I_1}{3}\left(\sigma_{dev} + \frac{I_1}{3}\right) + \frac{I_1^3}{27}\right] - I_1\left[\sigma_{dev}^2 + 2\sigma_{dev}\frac{I_1}{3} + \frac{I_1^2}{9}\right]$$
$$+ I_2\left(\sigma_{dev} + \frac{I_1}{3}\right) - I_3 = 0 \tag{38}$$

After some mathematical transformations, we get:

$$\sigma_{dev}^3 + \sigma_{dev}^2(I_1 - I_1) + \sigma_{dev}\left(\frac{I_1^2}{3} - 2\frac{I_1^2}{3} + I_2\right) + \left(\frac{I_1^3}{27} - \frac{I_1^3}{9} + \frac{I_2 I_1}{3} - I_3\right) = 0 \tag{39}$$

The main deviator equation is:

$$\sigma_{dev}^3 - \sigma_{dev}^2 J_1 + \sigma_{dev} J_2 - J_3 = 0 \tag{40}$$

We can conclude the following results:

$$J_1 = 0; J_2 = -\frac{I_1^2}{3} + I_2; J_3 = \frac{2I_1^3}{27} - \frac{I_2 I_1}{3} + I_3 \rightarrow \sigma_{dev}^3 + \sigma_{dev} J_2 - J_3 = 0 \tag{41}$$

The solution of this equation is searched in this form [4]: $\sigma_{dev} = a + b$ where $a = re^{\theta i}$ and $b = re^{-\theta i}$. So:

$$(a+b)^3 = a^3 + 3ab(\sigma_{dev}) + b^3 \tag{42}$$

$$\sigma_{dev}^3 + J_2\sigma_{dev} - J_3 = \left[a^3 + 3ab(\sigma_{dev}) + b^3\right] + J_2\sigma_{dev} - J_3 = 0 \tag{43}$$

$$\left(a^3 + b^3 - J_3\right) + \sigma_{dev}(3ab + J_2) = 0 \tag{44}$$

$$(3ab + J_2) = 0 \rightarrow 3re^{\theta i} re^{-\theta i} + J_2 = 0 \rightarrow r = \sqrt{\frac{-J_2}{3}} \tag{45}$$

$$\left(a^3 + b^3 - J_3\right) = 0 \rightarrow r^3 e^{3\theta i} + r^3 e^{-3\theta i} - J_3 = 0 \rightarrow 2r^3 \cos 3\theta = J_3 \tag{46}$$

We get:

$$\theta + \frac{2\pi n}{3} = \frac{1}{3} arccos \frac{J_3}{2r^3} \tag{47}$$

And,

$$a + b = 2r\cos\left(\theta + \frac{2\pi n}{3}\right) \tag{48}$$

And if $n = 0$ than $\theta = \frac{1}{3} arccos \frac{J_3}{2r^3}$. The main stresses will be as follows [5]:

$$\sigma_1 = \frac{I_1}{3} + 2r\cos(\theta), \sigma_2 = \frac{I_1}{3} + 2r\cos\left(\theta + 120^0\right), \sigma_3 = \frac{I_1}{3} + 2r\cos\left(\theta + 240^0\right) \tag{49}$$

3.1 Application

In the above paragraph, we have done an analytical study. Below, we will solve a simple example for the elementary volume with given stress tensor $T_\sigma = \begin{pmatrix} \sigma_{xx} & \sigma_{xy} & \sigma_{xz} \\ \sigma_{yx} & \sigma_{yy} & \sigma_{yz} \\ \sigma_{zx} & \sigma_{zy} & \sigma_{zz} \end{pmatrix} = \begin{pmatrix} -1 & 1 & 3 \\ 1 & 4 & 4 \\ 3 & 4 & -1 \end{pmatrix}$.

with the invariants, $I_1 = 2, I_2 = -33, I_3 = 9$. After we apply the above equation and we have:

$$J_2 = -\frac{I_1^2}{3} + I_2 = -34.333; J_3 = \frac{2I_1^3}{27} - \frac{I_2 I_1}{3} + I_3 = 31.593 \tag{50}$$

$$r = \sqrt{\frac{-J_2}{3}} = 3.383, \theta = \frac{1}{3} arccos \frac{J_3}{2r^3} = 21.971^0 \tag{51}$$

$$\sigma_1 = \frac{I_1}{3} + 2r\cos(\theta) = 6.9411 \tag{52}$$

$$\sigma_2 = \frac{I_1}{3} + 2r\cos\left(\theta + 120^0\right) = -0.278,$$
$$\sigma_3 = \frac{I_1}{3} + 2r\cos\left(\theta + 240^0\right) = -4.663 \tag{53}$$

The orientations of the main volume are found according to the above relations, given that they are mutually perpendicular. For main face 1, we have:

$$l_1 = \pm\sqrt{\frac{(\sigma - \sigma_3)(\sigma - \sigma_2) + \tau^2}{(\sigma_1 - \sigma_2)(\sigma_1 - \sigma_3)}} = 0.296,$$
$$m_1 = \pm\sqrt{\frac{(\sigma - \sigma_3)(\sigma - \sigma_1) + \tau^2}{(\sigma_2 - \sigma_1)(\sigma_2 - \sigma_3)}} = 0.777,$$
$$n_1 = \pm\sqrt{\frac{(\sigma - \sigma_1)(\sigma - \sigma_2) + \tau^2}{(\sigma_3 - \sigma_1)(\sigma_3 - \sigma_2)}} = -0.556 \tag{54}$$

For the main face 2, we have the following orientations:

$$l_2 = \pm\sqrt{\frac{(\sigma - \sigma_3)(\sigma - \sigma_2) + \tau^2}{(\sigma_1 - \sigma_2)(\sigma_1 - \sigma_3)}} = 0.803,$$
$$m_2 = \pm\sqrt{\frac{(\sigma - \sigma_3)(\sigma - \sigma_1) + \tau^2}{(\sigma_2 - \sigma_1)(\sigma_2 - \sigma_3)}} = -0.518,$$
$$n_2 = \sqrt{\frac{(\sigma - \sigma_1)(\sigma - \sigma_2) + \tau^2}{(\sigma_3 - \sigma_1)(\sigma_3 - \sigma_2)}} = -0.295 \tag{55}$$

and the condition $l_1 l_2 + m_1 m_2 + n_1 n_2 = 0$. In order to fulfill this last condition, the signs + or − are used before the square root.

For the main face 3 we have the following orientations:

$$l_3 = \pm\sqrt{\frac{(\sigma - \sigma_3)(\sigma - \sigma_2) + \tau^2}{(\sigma_1 - \sigma_2)(\sigma_1 - \sigma_3)}} = 0.517,$$
$$m_3 = \pm\sqrt{\frac{(\sigma - \sigma_3)(\sigma - \sigma_1) + \tau^2}{(\sigma_2 - \sigma_1)(\sigma_2 - \sigma_3)}} = 0.359,$$
$$n_3 = \pm\sqrt{\frac{(\sigma - \sigma_1)(\sigma - \sigma_2) + \tau^2}{(\sigma_3 - \sigma_1)(\sigma_3 - \sigma_2)}} = 0.777 \quad (56)$$

and conditions $l_1 l_3 + m_1 m_3 + n_1 n_3 = 0$, $l_2 l_3 + m_2 m_3 + n_2 n_3 = 0$. Finally we get the rotation matrix for principal orientations below:

$$R_{xyz} = \begin{pmatrix} l_1 & m_1 & n_1 \\ l_2 & m_2 & n_2 \\ l_3 & m_3 & n_3 \end{pmatrix} = \begin{pmatrix} 0.296 & 0.777 & -0.556 \\ 0.803 & -0.518 & -0.295 \\ 0.517 & 0.359 & 0.777 \end{pmatrix} \quad (57)$$

If the main volume is known, we get the following:

$$R_{123} = R_{xyz}^T = \begin{pmatrix} l_1 & l_2 & l_3 \\ m_1 & m_2 & m_3 \\ n_1 & n_2 & n_3 \end{pmatrix} = \begin{pmatrix} 0.296 & 0.803 & 0.517 \\ 0.777 & -0.518 & 0.359 \\ -0.556 & -0.295 & 0.777 \end{pmatrix} \quad (58)$$

And, after we check the result, we prove it as shown in Eq. 58 [6].

$$\begin{aligned} T_{\sigma 123} &= R_{123} T_{\sigma xyz} R_{123}^T \\ &= \begin{pmatrix} 0.296 & 0.803 & 0.517 \\ 0.777 & -0.518 & 0.359 \\ -0.556 & -0.295 & 0.777 \end{pmatrix} \begin{pmatrix} -1 & 1 & 3 \\ 1 & 4 & 4 \\ 3 & 4 & -1 \end{pmatrix} \\ &\quad \begin{pmatrix} 0.296 & 0.777 & -0.556 \\ 0.803 & -0.518 & -0.295 \\ 0.517 & 0.359 & 0.777 \end{pmatrix} \\ &= \begin{pmatrix} 6.941 & 0 & 0 \\ 0 & -0.278 & 0 \\ 0 & 0 & -4.663 \end{pmatrix} \end{aligned} \quad (59)$$

4 Graphic Analysis for Determining the Main Orientations [1]

All Mohr's circles represent faces with constant covariance, while the main Mohr circles represent reciprocal faces with zero covariance, for example. $l = 0$ or $m = 0$ or $n = 0$. Mohr's circle with $l = cos\alpha$ intersects the other two main circles at points A and B, where α is the angle formed by the vertical line at point C $(\sigma_1, 0)$ with the connecting line of points A and B (Fig. 9).

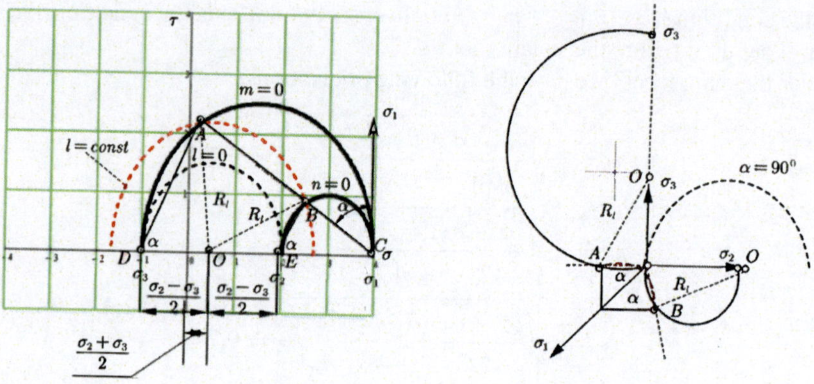

Fig. 9. Mohr space analysis for the constant cosine l.

By analyzing the triangle, OAD and from the cosine theorem for this triangle we get:

$$\begin{aligned} OA^2 &= OD^2 + DA^2 - 2 \cdot OD \cdot DA \cdot \cos\alpha \\ &= \left(\frac{\sigma_2 - \sigma_3}{2}\right)^2 + (\sigma_1 - \sigma_3)^2 \cos^2\alpha \\ &\quad - 2 \cdot \frac{\sigma_2 - \sigma_3}{2} \cdot (\sigma_1 - \sigma_3) \cdot \cos^2\alpha \\ &= \left(\frac{\sigma_2 - \sigma_3}{2}\right)^2 + ((\sigma_1 - \sigma_3)(\sigma_1 - \sigma_2))\cos^2\alpha \end{aligned} \tag{60}$$

By analyzing the triangle, OBE and from the cosine theorem for this triangle we can write:

$$\begin{aligned} OB^2 &= OE^2 + BE^2 - 2 \cdot OE \cdot BE \cdot \cos(180^\circ - \alpha) \\ &= \left(\frac{\sigma_2 - \sigma_3}{2}\right)^2 + (\sigma_1 - \sigma_2)^2 \cos^2\alpha \\ &\quad + 2 \cdot \frac{\sigma_2 - \sigma_3}{2} \cdot (\sigma_1 - \sigma_2) \cdot \cos^2\alpha \\ &= \left(\frac{\sigma_2 - \sigma_3}{2}\right)^2 + ((\sigma_1 - \sigma_3)(\sigma_1 - \sigma_2))\cos^2\alpha \end{aligned} \tag{61}$$

Compared with the radius of Mohr's circle for $l = cos\alpha$, the same expression emerges. In this way, if we know the angle α, this circle is easily drawn, because it passes through the points A and B of the intersection of this line with the other two main circles $m = 0$ and $n = 0$. Mohr's circle for $m = cos\beta$ intersects the other two main circles at points A and B. We mark as β, the angle formed by the vertical line at point C $(\sigma_2, 0)$ with the connecting line of points A and B (Fig. 10).

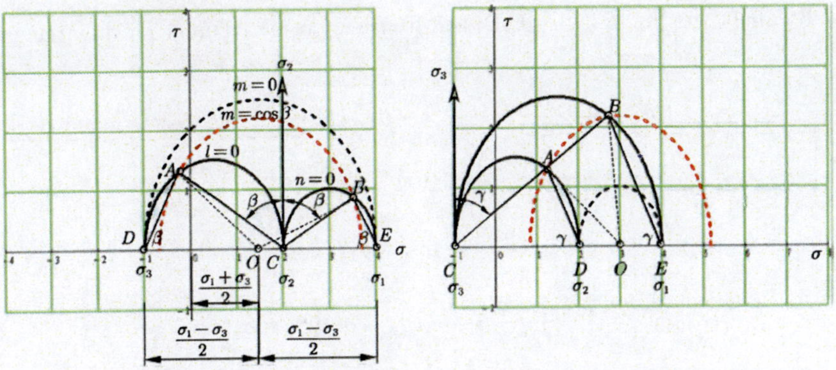

Fig. 10. Mohr space analysis for the constant cosine m and l.

By analyzing triangle, OAD, and from the cosine theorem for this triangle we have:

$$
\begin{aligned}
OA^2 &= OD^2 + DA^2 - 2 \cdot OD \cdot DA \cdot \cos\beta \\
&= \left(\frac{\sigma_1 - \sigma_3}{2}\right)^2 + (\sigma_2 - \sigma_3)^2 \cos^2\beta \\
&\quad - 2 \cdot \frac{\sigma_1 - \sigma_3}{2} \cdot (\sigma_2 - \sigma_3) \cdot \cos^2\beta \\
&= \left(\frac{\sigma_1 - \sigma_3}{2}\right)^2 + ((\sigma_2 - \sigma_3)(\sigma_2 - \sigma_1)) \cos^2\beta
\end{aligned} \tag{62}
$$

By analyzing the triangle OBE, and from the cosine theorem for this triangle we have:

$$
\begin{aligned}
OB^2 &= OE^2 + BE^2 - 2 \cdot OE \cdot BE \cdot \cos\beta \\
&= \left(\frac{\sigma_1 - \sigma_3}{2}\right)^2 + (\sigma_1 - \sigma_2)^2 \cos^2\beta \\
&\quad - 2 \cdot \frac{\sigma_1 - \sigma_3}{2} \cdot (\sigma_1 - \sigma_2) \cdot \cos^2\beta \\
&= \left(\frac{\sigma_1 - \sigma_3}{2}\right)^2 + ((\sigma_2 - \sigma_3)(\sigma_2 - \sigma_1)) \cos^2\beta
\end{aligned} \tag{63}
$$

Compared with the radius of Mohr's circle for $m = cos\beta$, the same expression emerges. This way, if we know the angle β, this circle is easily drawn, because it passes through the points A and B of the intersection of these lines with the other two main circles $l = 0$ and $n = 0$. Mohr's circle with $n = cos\gamma$ intersects the other two main circles at points A and B. We mark as γ, the angle formed by the vertical line at point C $(\sigma_3, 0)$ with the connecting line of points A and B.

By analyzing the triangle OAD, and from the cosine theorem for this triangle we have:

$$\begin{aligned} OA^2 &= OD^2 + DA^2 - 2 \cdot OD \cdot DA \cdot \cos(180^\circ - \gamma) \\ &= \left(\frac{\sigma_1 - \sigma_2}{2}\right)^2 + (\sigma_2 - \sigma_3)^2 \cos^2 \gamma \\ &+ 2 \cdot \frac{\sigma_1 - \sigma_2}{2} \cdot (\sigma_2 - \sigma_3) \cdot \cos^2 \gamma \\ &= \left(\frac{\sigma_1 - \sigma_2}{2}\right)^2 + ((\sigma_2 - \sigma_3)(\sigma_2 - \sigma_3)) \cos^2 \gamma \end{aligned} \tag{64}$$

By analyzing the triangle OBE, and from the cosine theorem for this triangle we can write:

$$\begin{aligned} OB^2 &= OE^2 + BE^2 - 2 \cdot OE \cdot BE \cdot \cos \gamma \\ &= \left(\frac{\sigma_1 - \sigma_2}{2}\right)^2 + (\sigma_1 - \sigma_3)^2 \cos^2 \gamma \\ &- 2 \cdot \frac{\sigma_1 - \sigma_2}{2} \cdot (\sigma_1 - \sigma_3) \cdot \cos^2 \gamma \\ &= \left(\frac{\sigma_1 - \sigma_2}{2}\right)^2 + ((\sigma_2 - \sigma_3)(\sigma_2 - \sigma_3)) \cos^2 \gamma \end{aligned} \tag{65}$$

Compared with the radius of Mohr's circle for $n = cos\gamma$, the same expression emerges. In this way, if we know the angle γ, this circle is easily drawn, because it passes through the points A and B of the intersection of these line, s with the other two main circles $l = 0$ and $m = 0$.

4.1 Application of Graphical Method for Main Orientation

The main stresses are $\sigma_1 = 4, \sigma_2 = 2, \sigma_3 = -1$ and the face orientations $l = 0.5, m = 0.5, n = 0.707$ are rotated in degrees $\alpha = 60^\circ, \beta = 60^\circ, \gamma = 45^\circ$ (Fig. 11).

5 The Geometric Meaning of the Main Stresses and the Positioning of the Elemental Volume in Mohr's Space

Let's analyze a cuboid with sides a, b and c [8]. We want these three edges to meet these three conditions:

1. The sum of the lengths of the edges of the parallelepiped must have a predetermined value.
2. The sum of the surfaces of the parallelepiped must have a predetermined value.
3. The volume of the parallelepiped must have a default value.

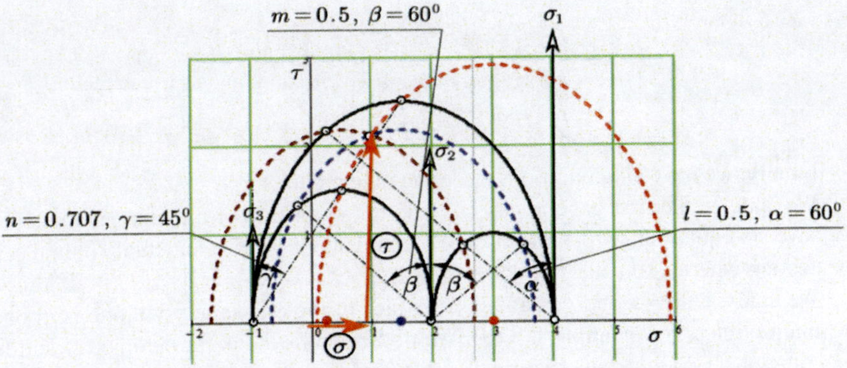

Fig. 11. Volumetric element in Mohr space (principal stress and orientation).

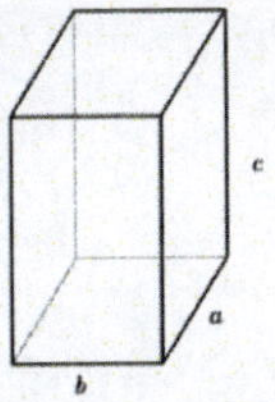

Fig. 12. Sample of given cuboid.

For this reason we mark the three requirements as shown below and then we compare them with the three invariants of the main stress tensor (Fig. 12).

$4(a + b + c) = A$ is the sum of the lengths of the edges. $2(ab + bc + ca) = B$ is the sum of the surface areas. $(abc) = C$ is the volume. We can write them as the three invariants:

$$a + b + c = \frac{A}{4} = I_1 \tag{66}$$

$$ab + bc + ca = \frac{B}{2} = I_2 \tag{67}$$

$$abc = C = I_3 \tag{68}$$

The result is the same with the main stresses, for example they represent three sides of the cuboid with three preliminary requirements for length, surface and volume. In Mohr's space we have three main circles with equations $l = 0, m = 0dhen = 0$. The elementary volume has three face with any orientations: $\vec{e}_1 = (l_1, m_1, n_1), \vec{e}_2 = (l_2, m_2, n_2), \vec{e}_3 = (l_3, m_3, n_3)$ that meet the known conditions:

$$l_1^2 + m_1^2 + n_1^2 = 1,\ l_2^2 + m_2^2 + n_2^2 = 1,\ l_3^2 + m_3^2 + n_3^2 = 1 \tag{69}$$

$$
\begin{aligned}
l_1 \cdot l_2 + m_1 \cdot m_2 + n_1 \cdot n_2 &= 0, \\
l_1 \cdot l_3 + m_1 \cdot m_3 + n_1 \cdot n_3 &= 0, \\
l_3 \cdot l_2 + m_3 \cdot m_2 + n_3 \cdot n_2 &= 0
\end{aligned} \tag{70}
$$

Finally, we have six conditions to define nine parameters.

We analyze the first general case: We do not know any of the nine parameters and we have six equations, which means that the three faces of the elemental volume can be located anywhere in the allowed Mohr space.

We analyze the second general case: A face of the elementary volume is in one of the main circles, for example $l_1 = 0$. We do not know the eight parameters and have six equations, which means that the two faces of the elemental volume can be located anywhere in Mohr's allowed space.

We analyze the third general case: Two sides of the elementary volume are located in two of the main circles, for example $l_1 = 0, m_2 = 0$. The established conditions are transformed into the form:

$$
m_1^2 + n_1^2 = 1, l_2^2 + n_2^2 = 1, l_3^2 + m_3^2 + n_3^2 = 1 \tag{71}
$$

$$
n_1 \cdot n_2 = 0, m_1 \cdot m_3 + n_1 \cdot n_3 = 0, n_3 \cdot n_2 = 0 \tag{72}
$$

We derive the following:

$$
n_1 = 0 \rightarrow m_1 = 1, m_3 = 0 \tag{73}
$$

This state is possible if a main face exits: $m_1 = 1$.

6 Conclusions

1. Mohr's circles are called the main Mohr's circles and represent orientations with zero leading cosines.
2. The center of the main circles serves as the center of all circles with constant direction cosines.
3. All the real part of the stressed state is in the area between the main Mohr circles.
4. The circular segment of points with the same cosine in Mohr's field represents the directions in a cone from one main plane to another normal plane.
5. Directions with the same cosines are located on the surface of a cone with an angle equal to that of the constant cosine.
6. Principal stresses represent three edges of a parallelepiped with preliminary requirements for lengths, surfaces and volume.

References

1. Sulejmani, A., Koça, O., Dhoska, K.: Method of principal orientation in Mohr's space. J. Southwest Jiaotong Univ. **58**(2), (2023). https://doi.org/10.35741/issn.0258-2724.58.2.16
2. Koça, O., Sulejmani, A., Alimhilli, P.: Direct method of rigidity using Mathcad Software. In IOP Conference Series: Materials Science and Engineering, vol. 1024, no. 1, p. 012093. IOP Publishing (2021). https://doi.org/10.1088/1757-899X/1024/1/012093
3. Xu, S.S., Nieto-Samaniego, A.F., Alaniz-Álvarez, S.A.: 3D Mohr diagram to explain reactivation of pre-existing planes due to changes in applied stresses. In: ISRM International Symposium on In-Situ Rock Stress, pp. ISRM–ISRS. ISRM (2010)
4. Koça, O., Sulejmani, A., Alimhilli, P.: Direct method of rigidity using Mathcad Software. In IOP Conference Series: Materials Science and Engineering, vol. 1024, no. 1, p. 012093. IOP Publishing (2021). https://doi.org/10.1088/1757-899X/1024/1/012093
5. Dhoska, K., Lumi, D., Sulejmani, A., Koça, O.: Measurement uncertainty for mechanical resistance of manufactured steel bar. Pollack Period. **17**(2), 104–108 (2022). https://doi.org/10.1556/606.2022.00532
6. Sulejmani, A., Koça, O.: Development of optimal transmission rate of the kinematic chain by using genetic algorithms coded in mathcad. Int. J. Innov. Technol. Interdisciplin. Sci. **4**(4), 792–803 (2021)
7. Advanced Mechanics of Materials and Applied Elasticity, 5th edition by Ansel C. Ugural, Saul K. Fenster
8. Koça, O., Sulejmani, A., Dhoska, K.: The strenous state of the contact at the sliding-flip pairs. Mach. Technol. Mater. **14**(7), 258–261 (2020)
9. Martin, C.M., Makrodimopoulos, A.: Finite-element limit analysis of Mohr-Coulomb materials in 3D using semidefinite programming. J. Eng. Mech. **134**(4), 339–347 (2008)
10. Gross, D., Hauger, W., Schröder, J., Wall, W.A., Bonet, J.: Engineering Mechanics 2 Mechanics of Materials
11. Sulejmani, A., Koça, O., Dhoska, K., Gheibi, M., Moezzi, R.: A novel method of jacobian contours to evaluate the influence line in statically determinate structures. Appl. Mech. **4**, 1172–1187 (2023). https://doi.org/10.3390/applmech4040060

Analyzation of General Displacements of Mechanical Structures Using Self-equilibrated Forces

Anis Sulejmani(✉), Odhisea Koça, and Parid Milo

Faculty of Mechanical Engineering, Polytechnic University of Tirana, Tirana, Albania
{asulejmani,pmilo}@fim.edu.al

Abstract. The main tasks of a mechanical engineer are constructing graphs of internal forces and displacements of different sections of a mechanical structure. Structures can be determinate and indeterminate, and they can be very challenging and difficult to solve. Engineers have developed various methods that can define the displacements of a variable structure section. This article describes a new method that is faster and easier to use for defining the displacements of a variable structure section. It can be used by other engineers, such as civil engineers, architects, etc. This new method, derived and proved by restructuring and analyzing traditional methods, allows us to define each structure section's linear displacements and rotation angles more simply. In this case, it is not necessary for Mohr's graph to find the reactions and construct the graphs of the internal forces.

Keywords: Mechanical structures · Displacement · Self-equilibrated forces · Linear displacement

1 Introduction

1.1 Mohr's Integral

Mohr's integral relies on the principle of virtual displacements [2], which states that the work of external and internal forces during a virtual displacement of a mechanical system in equilibrium is equal to zero [1, 3]. Mohr's integral, in contrast to the principle of virtual displacement [4, 14], takes the displacements from the real mechanical system, while the forces, both internal and external, are taken from the imaginary Mohr system and is loaded with unit forces according to the directions that are required by the corresponding displacements [5, 6]. The sign agreement is derived from work principles, positive forces, and displacements in the same direction and negative in the opposite direction [6, 7] (Fig. 1).

$$\begin{aligned}(F = 1) \cdot \delta_B^h &= \sum_{i=1}^{n} \int_0^{l_i} \overline{M}_{pk_i} \cdot d\theta \rightarrow \delta_B^h \\ &= \sum_{i=1}^{n} \int_0^{l_i} \overline{M}_{pk_i} \cdot \left(\frac{M_{p_i} \cdot dx}{EI_i} \right) \end{aligned} \tag{1}$$

G. Guxho et al. (Eds.): ITC-ICEE 2023, LNMIE, pp. 182–193, 2024.
https://doi.org/10.1007/978-3-031-48933-4_18

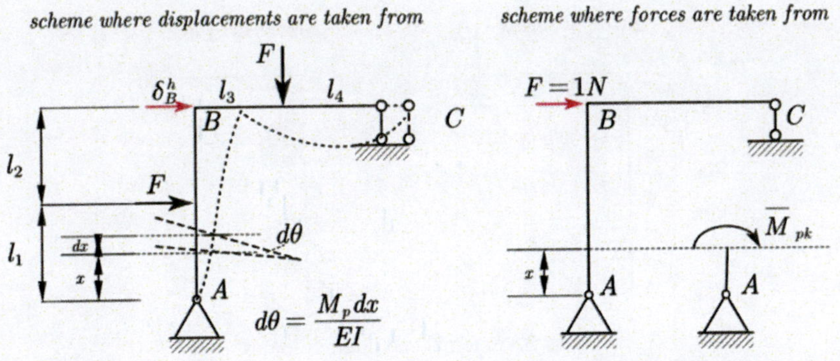

Fig. 1. **a**) Schematic drawing where displacements are chosen. **b**) Schematic drawing where forces are chosen.

1.2 Self-equilibrated Forces in an Element of a Structure

Self-equilibrated forces are a system of forces that are equivalent to zero [8, 9]. Such forces are plenty, but the most used forces for an element in a structure are the following Fig. 2.

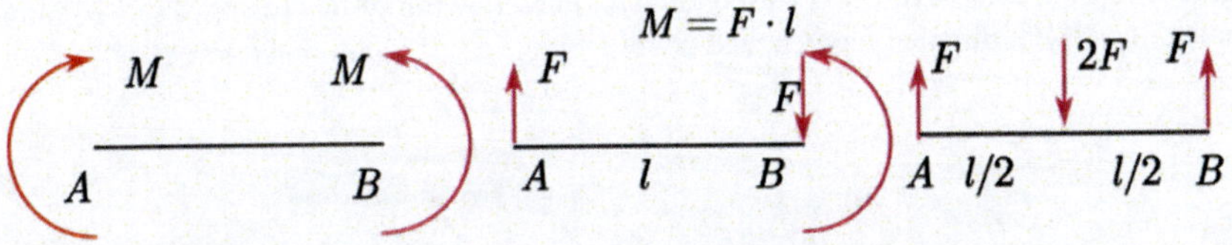

Fig. 2. Schematic sample of self-equilibrated forces.

If we have an element in a structure with a system of self-equilibrated forces [13, 16, 17], the structure reaction forces are equal to zero and only the element where the self-equilibrated forces are, have internal forces [10, 11]. The biggest factor from the internal forces that exist in this element, is the bending moment, because the bending moment has the most influence in the overall work of the internal forces [12] (Fig. 3).

If we assume the forces equal to a unit of one, same as in Mohr integral, the following external forces systems are derived, with these corresponding self-equilibrated internal forces, which are easier to use Fig. 4.

2 Graphic Solution of Mohr Integral for Some Self-equilibrated Forces Systems

2.1 Case Study with Moment Self-equilibrated

Let's start by studying a simply determined structure only with self-equilibrated forces. The internal moment of the structure are as follows (Fig. 5):

$$1 \cdot \varphi_A + 1 \cdot \varphi_B = \int_0^{l_1} 1 \cdot \frac{M_p \cdot dx}{EI} = \frac{1}{EI}\int_0^{l_1} M_p \cdot dx = \frac{1}{EI} A_{A/B}(M_p) \tag{2}$$

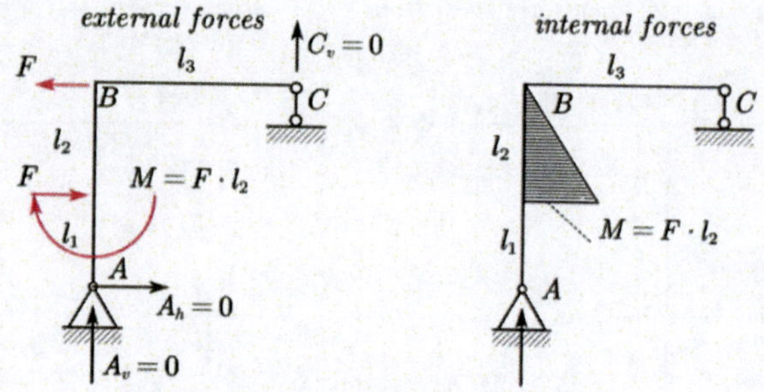

Fig. 3. External and internal forces.

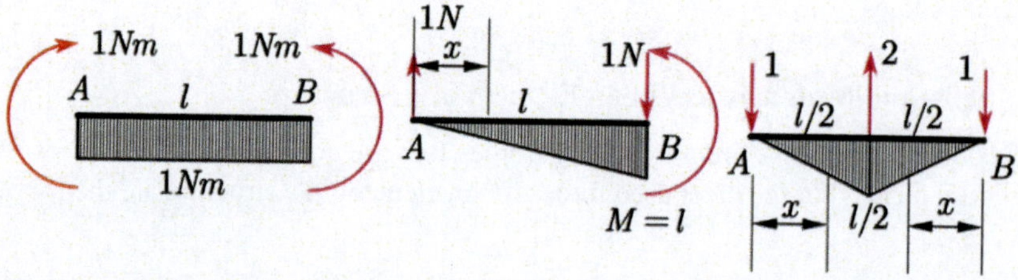

Fig. 4. Self-equilibrated forces corresponding for the following system forces (**a** self-equilibrated moment; **b** self-equilibrated moment and couple forces; **c** self-equilibrated forces).

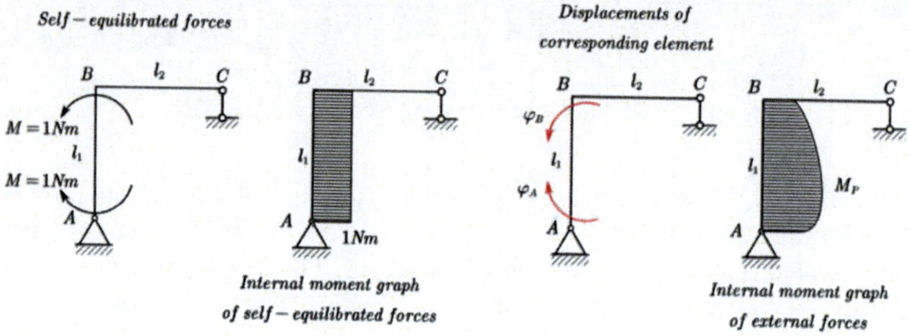

Fig. 5. Internal moment graph from self-equilibrated moment.

$A_{A/B}(M_p)$ is the surface area of the graph M_p from section A to section B.

2.2 Case Study with Moment and Couple Forces

In this case the schematic drawing is as shown Fig. 6.

The equation for the above case will be:

$$l_1 \cdot \varphi_A + 1 \cdot \delta_B + 1 \cdot \delta_A = \int_0^{l_1} x \cdot \frac{M_p \cdot dx}{EI}$$

$$= \frac{1}{EI} \int_0^{l_1} x \cdot M_p \cdot dx = \frac{1}{EI} S^B_{A/B}(M_p) \tag{3}$$

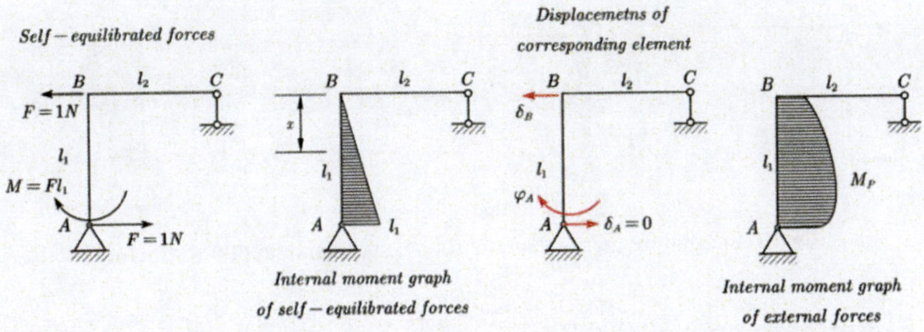

Fig. 6. Internal moment graph from self-equilibrated moment and couple forces.

$S_{A/B}(M_p)$ is the static moment in relation to section B of the graph of M_p from section B to section A.

2.3 General Case When Forces are Placed Randomly

Also, in this case we will have the internal moment when these forces are acting on a structure. The internal moment is as follows (Fig. 7):

$$\frac{a+b}{a \cdot b}\delta_{forces} + \frac{1}{a} \cdot \delta_A^h + \frac{1}{b} \cdot \delta_B^h = \int_0^a \frac{x}{a} \cdot \frac{M_p \cdot dx}{EI} + \int_0^b \frac{x}{b} \cdot \frac{M_p \cdot dx}{EI}$$
$$= \frac{S_{A/forca}^A(M_p)}{a \cdot EI} + \frac{S_{B/forca}^B(M_p)}{b \cdot EI} \tag{4}$$

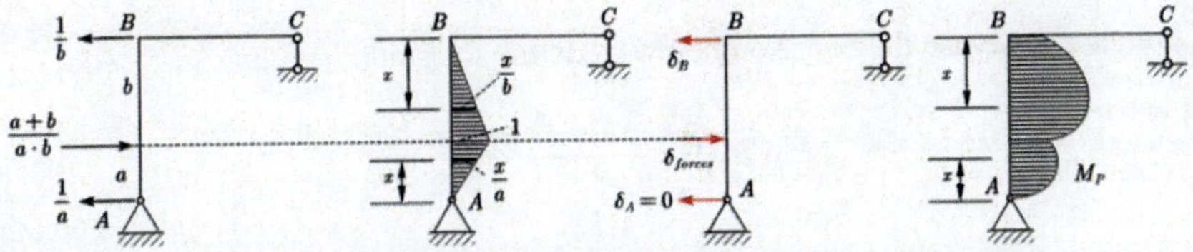

Fig. 7. Internal moment when the force acts on a structure.

In this case, there is also a special case when the force is placed or acts in middle. So, the above internal moment is as follows (Fig. 8):

$$\frac{2}{a} \cdot \delta_{mid} + \frac{1}{a} \cdot \delta_B^h + \frac{1}{a} \cdot \delta_A^h = \int_0^a \frac{x}{a} \cdot \frac{M_p \cdot dx}{EI} + \int_0^a \frac{x}{a} \cdot \frac{M_p \cdot dx}{EI}$$
$$= \frac{S_{A/mid}^A(M_p)}{EI} + \frac{S_{B/mid}^B(M_p)}{EI} \tag{5}$$

$$2 \cdot \delta_{mid} + 1 \cdot \delta_B^h + 1 \cdot \delta_A^h = \frac{S_{A/mid}^A(M_p)}{EI} + \frac{S_{B/mid}^B(M_p)}{EI} \tag{6}$$

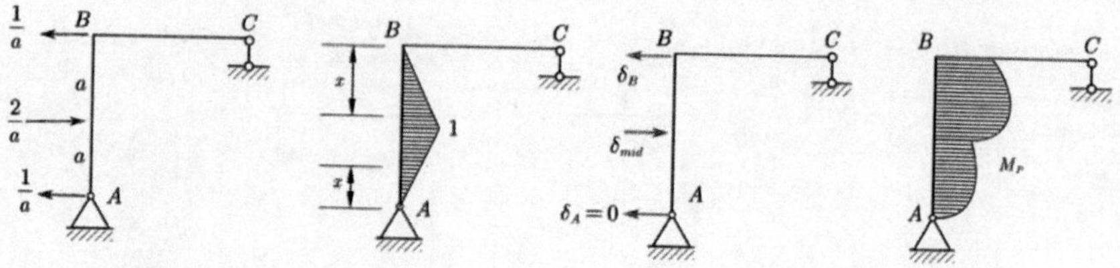

Fig. 8. Internal moment when the force acts in middle of the structure.

3 Static Moments of the Bending Moment Diagram

3.1 Trapezoid Distribution

Let's analyze a case where we have a moment with trapezoid distribution, as shown in Fig. 9.

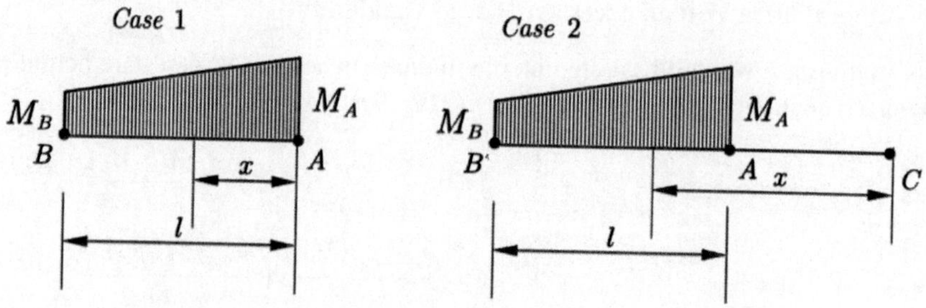

Fig. 9. Static moment in two different moment distributions.

For case 1 we use the equation below to define the static moment.

$$M_P = \frac{M_A \cdot (l - x) + M_B \cdot x}{l} \tag{7}$$

$$S^A = \int_0^l x \cdot M_P \cdot dx = \left. \frac{M_A \cdot \left(l\frac{x^2}{2} - \frac{x^3}{3}\right) + M_B \cdot \frac{x^3}{3}}{l} \right|_0^l$$
$$= \frac{l^2}{6}(M_A + 2 \cdot M_B) \tag{8}$$

Also, for case 2 we have:

$$M_P = \frac{M_A \cdot (l - x + AC) + M_B \cdot (x - AC)}{l} \tag{9}$$

$$S^C = \int_0^l x \cdot M_P \cdot dx = S^A + A(M_P) \cdot AC \tag{10}$$

3.2 Distributed Load Case

So, the equation for the static moment in this case will be (Fig. 10):

$$S^A = \int_0^l x \cdot M_P \cdot dx = S^A(tr) + S^A(hat)$$
$$= \frac{l^2}{6}(M_A + 2 \cdot M_B) - \left(\frac{2}{3} \cdot \frac{q \cdot l^2}{8} \cdot l\right) \cdot \frac{l}{2} \quad (11)$$

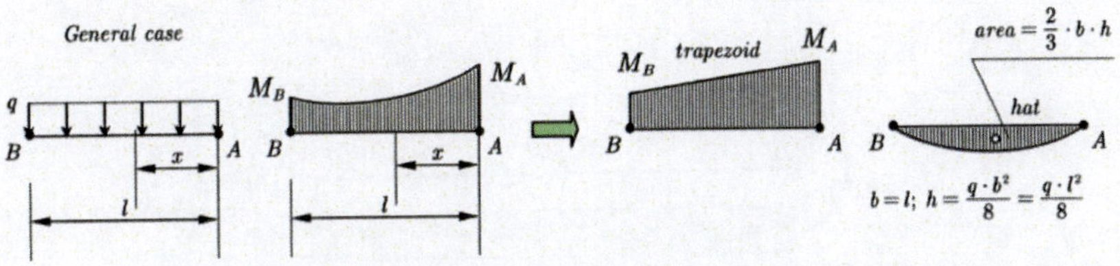

Fig. 10. General case when a distributed load acts on a structure.

4 Application of Self-equilibrated Method in Simple Structures

Let's analyze a simple structure when an external force is applied on it. The schematic diagram is as follows Fig. 11.

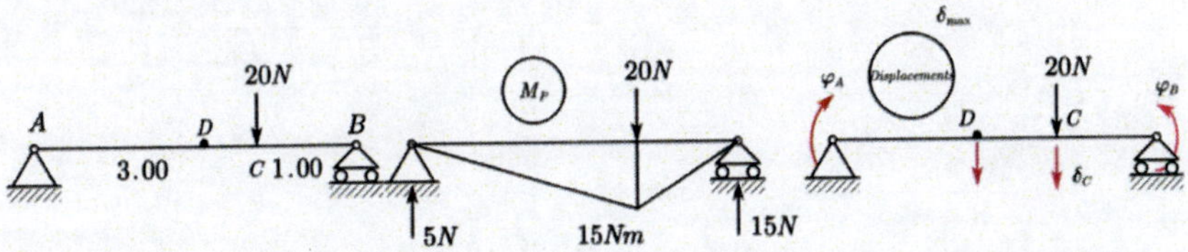

Fig. 11. Simple structure with external force, internal moment and self-equilibrated method applied in this case.

Self-equilibrated forces case is used by applying the forces in section A and B, for which the two displacements are known, $l = 4$ and $M = 4\,\text{Nm}$ (Fig. 12).

So, the solution will be:

$$4 \cdot \varphi_A + 1 \cdot 0 + 1 \cdot 0 = \frac{S^B_{\frac{A}{B}}(M_p)}{EI}$$
$$= \frac{3^2}{6EI}15 + \frac{15 \cdot 3}{2EI} \cdot 1 + \frac{1^2}{6EI}2 \cdot 15 = \frac{50}{EI} \rightarrow \varphi_A = \frac{12.5}{EI} \quad (12)$$

The other case of self-equilibrated forces is applied, by placing the moments in section A and B (Fig. 13).

$$1 \cdot \varphi_A + 1 \cdot \varphi_B = \frac{A_{A/B}(M_p)}{EI} = \frac{15 \cdot 4}{2EI} \rightarrow \varphi_B = \frac{30}{EI} - \frac{12.5}{EI} = \frac{17.5}{EI} \quad (13)$$

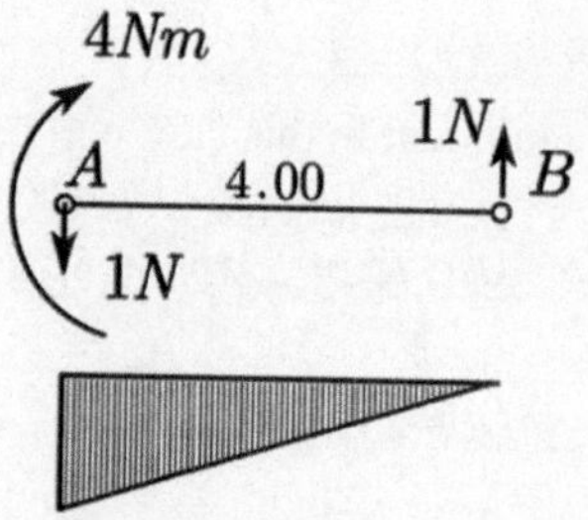

Fig. 12. Self-equilibrated forces in section A-B.

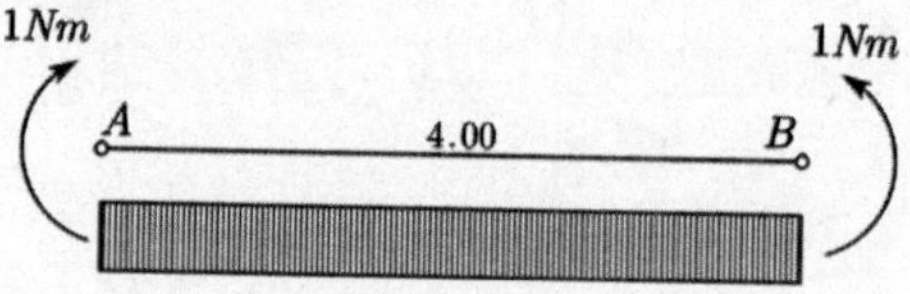

Fig. 13. Self-equilibrated moment.

Self-equilibrated forces case is used by applying the forces in section A and C, for which the displacements are known, $l = 3$ and $M = 3\,\text{Nm}$ (Fig. 14).

$$3 \cdot \varphi_A + 1 \cdot 0 - 1 \cdot \delta_C = \frac{S^C_{A/C}(M_p)}{EI} = \frac{3^2}{6EI} \cdot 15 = \frac{22.5}{EI}$$
$$\rightarrow \delta_C = 3 \cdot \varphi_A - \frac{22.5}{EI} = 3 \cdot \frac{12.5}{EI} - \frac{22.5}{EI} = \frac{15}{EI} \tag{14}$$

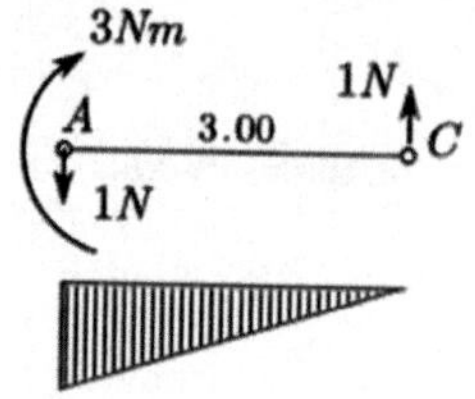

Fig. 14. Self-equilibrated moment and couple forces.

Maximum displacement is supposed to occur in section D, between section A and C. In section D, the displacement is zero because we have an extremum.

First, the case of self-equilibrated forces is applied, by placing them in section A and D (Fig. 15).

$$M_D = 5 \cdot x_D \tag{15}$$

$$1 \cdot \varphi_A + 1 \cdot \varphi_D = \frac{A_{A/D}(M_p)}{EI} = \frac{5 \cdot x_D \cdot x_D}{2EI} \rightarrow \frac{12.5}{EI}$$

$$= \frac{5 \cdot (x_D)^2}{2EI} \rightarrow x_D = \sqrt{5} = 2.236\,\text{m} \tag{16}$$

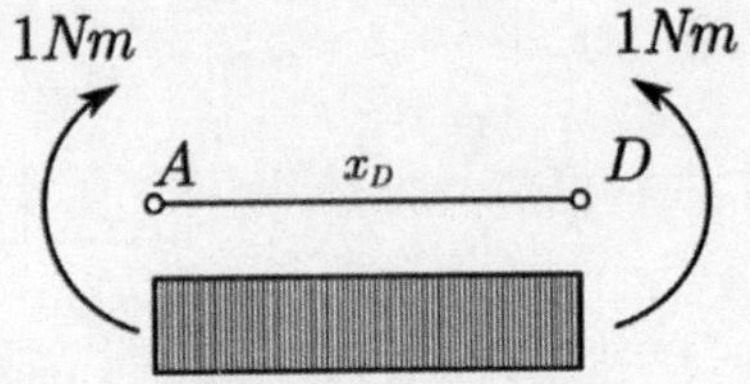

Fig. 15. Self-equilibrated moment is applied in section A–D.

Second case of self-equilibrated forces is applied, by placing them in section A and D, where $l = \sqrt{5}$ and $M = \sqrt{5}\,\text{Nm}$ (Fig. 16).

$$\sqrt{5} \cdot 0 + 1 \cdot 0 + 1 \cdot \delta_{\max} = \frac{S^{A}_{A/D}(M_p)}{EI}$$
$$= \frac{\sqrt{5}^2}{6EI} \cdot 2 \cdot 5 \cdot \sqrt{5} \rightarrow \delta_{\max} = \frac{25\sqrt{5}}{3EI} = \frac{18.63}{EI} \tag{17}$$

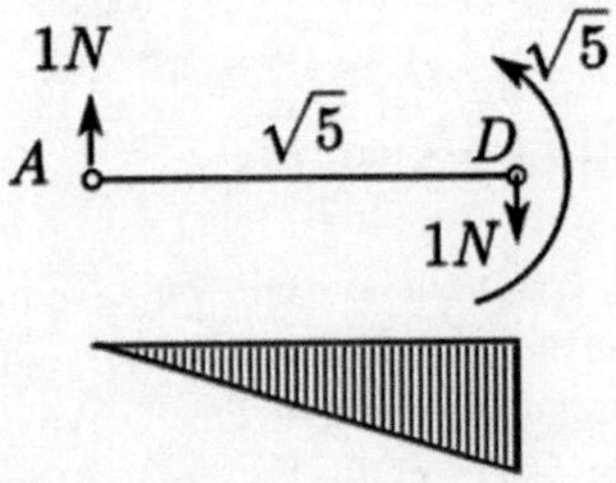

Fig. 16. Self-equilibrated moment and couple forces are applied, by placing them in section A and D.

5 Self-equilibrated Forces in a Structure

Self-equilibrated forces can be placed in a structure, to take advantage of the different mechanical linkages of the system. These forces are plenty, but the most usable ones are showed in Fig. 17.

5.1 Application of Self-equilibrated Forces in a Structure

If this case, the self-equilibrated force system is placed in a structure, as illustrated in this scheme. It is known for sure that the reaction forces are zero and only the part where the force system is applied has a bending moment (Fig. 18).

$$a\varphi_A + 1 \cdot 0 + 1 \cdot 0 = \frac{A_{A/B}(M_p)}{EI} a + \frac{S^{C}_{B/C}(M_p)}{EI} \tag{18}$$

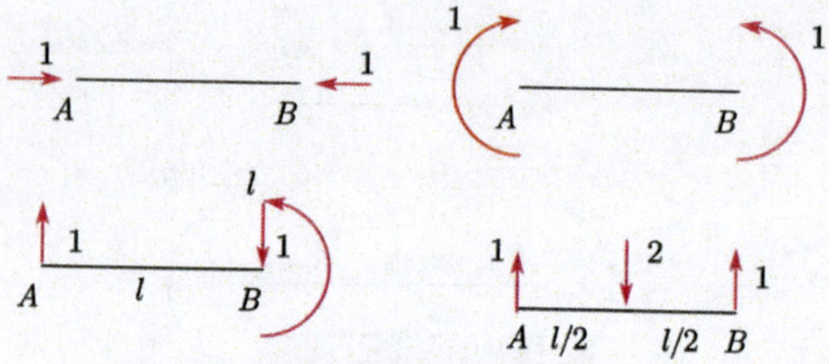

Fig. 17. Self-equilibrated forces in a structure.

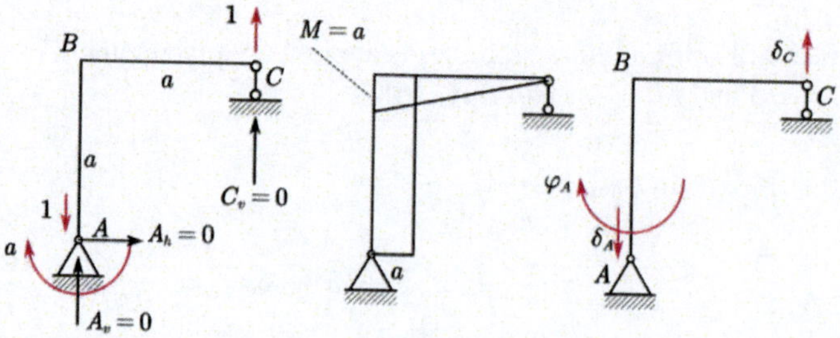

Fig. 18. Application of self-equilibrated force system in a structure.

6 Statically Indeterminate Systems

In the case of a statically indeterminate system, we have more mechanical linkages than needed. For that reason, we have more sections with zero displacements. On the contrary, we have more unknown values than needed. That is why the classic route is followed. First, the main system of the force method is created, then the moment diagrams are created separately for more kinematic parameters with values of zero, and at the same time, the extra reactions and kinematic parameters of the system are determined.

6.1 Application of Self-equilibrated Forces in a Statically Indeterminate System

The shown case of self-equilibrated forces is applied, by placing the forces in section A and section C (Fig. 19).

$$
\begin{aligned}
0 &= 4\cdot\frac{A_{A/B}(M_P)}{EI}+\frac{S^{C}_{C/B}(M_P)}{EI}\\
&= 4\cdot\left[\frac{80\cdot 6}{EI}-\frac{4C_V\cdot 6}{EI}\right]+\left[\frac{80\cdot 4}{3EI}\cdot\frac{3}{4}4-\frac{4C_V\cdot 4}{2EI}\cdot\frac{2}{3}4\right]
\end{aligned}
\tag{19}
$$

$$
[480-24\cdot C_V]+\left[80-\frac{16\cdot C_V}{3}\right]=0\rightarrow C_V=\frac{560}{24+\frac{16}{3}}=19.1 \tag{20}
$$

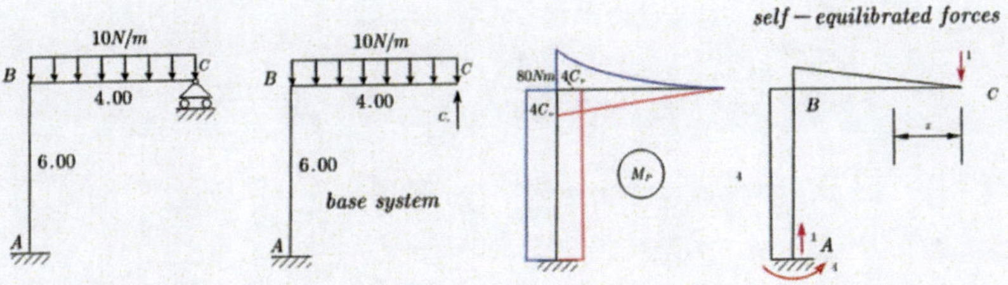

Fig. 19. Application of self-equilibrated forces in a statically indeterminate system.

6.2 Application According to Belyaev

The case of self-equilibrated forces is applied, by placing the forces in section A–B (Fig. 20).

$$(a+b)\cdot 0+1\cdot 0+1\cdot 0=\frac{S_{A/B}^{B}(M_P)}{EI}$$
$$=\left[\frac{Paa}{2EI}\cdot\left(\frac{2a}{3}+b\right)-\frac{lC_v\cdot l}{2EI}\cdot\frac{2}{3}l\right] \tag{21}$$

$$Pa^2(2a+3b)-2C_vl^3=0\rightarrow C_v=\frac{Pa^3}{l^3}\left(1+\frac{3b}{2a}\right) \tag{22}$$

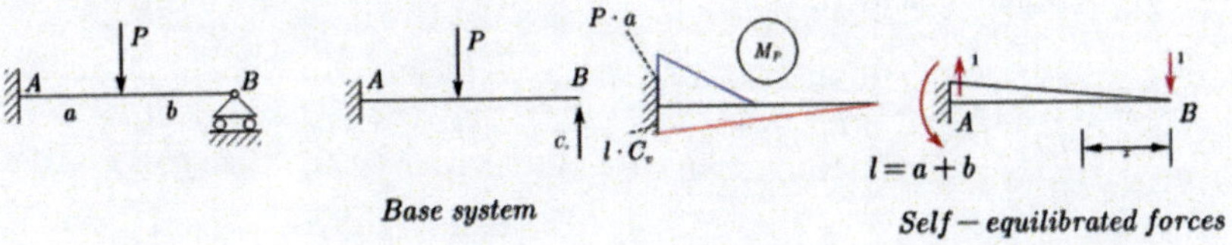

Fig. 20. Application according to Belyaev.

6.3 Three-Moment Equation

The three-moment equation is derived from the self-equilibrated forces, as a statically indeterminate problem (Fig. 21).

The case of self-equilibrated forces is applied, by placing the forces in section A_l, A_{mid} and A_{mid}, A_r (Fig. 22).

$$1\cdot 0+1\cdot 0+l_l\cdot\varphi_{mid}=-\frac{l_l^2}{6EI}(M_l+2\cdot M_{mid})-\frac{S_{left}^{A_l}}{EI} \tag{23}$$

$$1\cdot 0+1\cdot 0+l_r\cdot\varphi_{mid}=\frac{l_r^2}{6EI}(M_r+2\cdot M_{mid})+\frac{S_{right}^{A_r}}{EI} \tag{24}$$

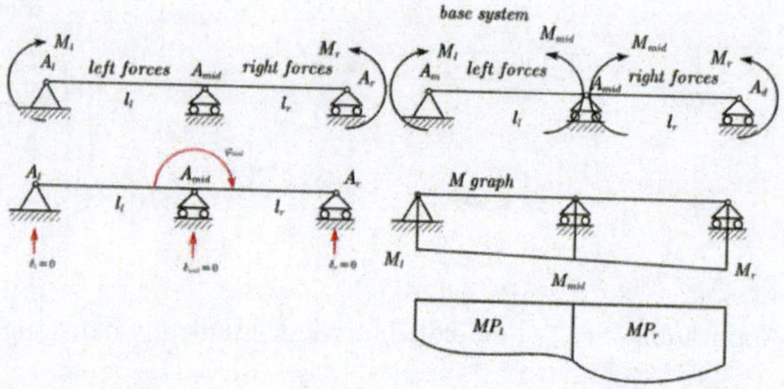

Fig. 21. Self-equilibrated forces and moment applied in a statically indeterminate system.

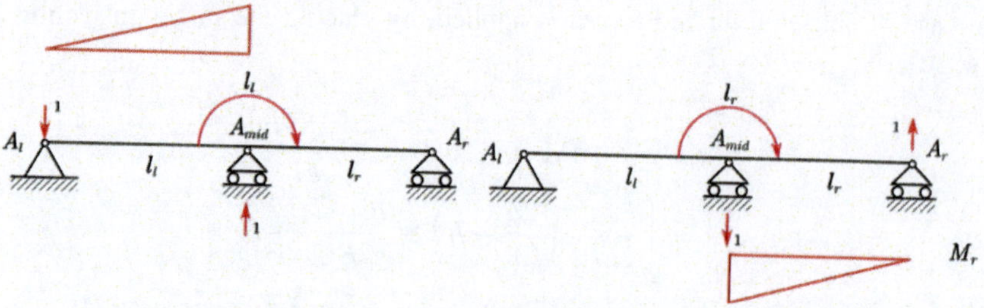

Fig. 22. Self-equilibrated forces and moment applied in different sections.

By equalizing the angle values on the left and on the right, we get the final relation of the three-moment equation.

$$-\frac{l_l}{6EI}(M_l + 2 \cdot M_{mid}) - \frac{S_{left}^{A_l}}{EI \cdot l_l} = \frac{l_r}{6EI}(M_r + 2 \cdot M_{mid}) + \frac{S_{right}^{A_r}}{EI \cdot l_r} \tag{25}$$

$$l_l \cdot M_l + 2 \cdot M_{mid} \cdot (l_l + l_r) + l_r \cdot M_r = -6\left(\frac{S_{right}^{A_r}}{l_r} + \frac{S_{left}^{A_l}}{l_l}\right) \tag{26}$$

References

1. Koça, O., Sulejmani, A., Alimhilli, P.: Direct method of rigidity using Mathcad software. In: IOP Conference Series: Materials Science and Engineering, vol. 1024, no. 1, p. 012093. IOP Publishing (2021). https://doi.org/10.1088/1757-899X/1024/1/012093
2. Pujol, S., Nelson, M.D., Smith-Pardo, J.P:. Virtual Velocity: Connecting the Concepts of Kinematics and Virtual Work, Engineering Structures, vol. 56, pp. 2249–2252. ISSN 0141-0296 (2013). https://doi.org/10.1016/j.engstruct.2013.08.033
3. Koça, O., Sulejmani, A., Dhoska, K.: The strenous state of the contact at the sliding–flip pairs. Mach. Technol. Mater. **14**(7), 258–261 (2020)
4. Bing, L., Meilin, Z., Kai, X.: A practical engineering method for fuzzy reliability analysis of mechanical structures. Reliab. Eng. Syst. Saf. **67**(3), 311–315 (2000)
5. Sulejmani, A., Koça, O., Dhoska, K.: Method of principal orientation in Mohr's space. J. Southwest Jiaotong Univ. **58**(2), (2023). https://doi.org/10.35741/issn.0258-2724.58.2.16

6. Ugural, A.C., Fenster, S.K.: Advanced Mechanics of Materials and Applied Elasticity, 5th edition
7. Saitou, K., Izui, K., Nishiwaki, S., Papalambros, P.: A Survey of Structural Optimization in Mechanical Product Development (2005)
8. Dhoska, K., Lumi, D., Sulejmani, A., Koça, O.: Measurement uncertainty for mechanical resistance of manufactured steel bar. Pollack Period. **17**(2), 104–108 (2022). https://doi.org/10.1556/606.2022.00532
9. Christensen, P.W., Klarbring, A.: An Introduction to Structural Optimization, vol. 153. Springer Science and Business Media (2008)
10. Pełczyński, J., Gilewski, W.O.J.C.I.E.C.H.: An extension of algebraic equations of elastic trusses with self-equilibrated system of forces. In: Proceedings of the 7th European Conference on Computational Fluid Dynamics (2018)
11. Ramesh, K., Shins, K.: Stress field equations for a disk subjected to self-equilibrated arbitrary loads: revisited. Granular Matter **24**(2), 49 (2022)
12. Zhang, L.Y., Li, Y., Cao, Y.P., Feng, X.Q.: Stiffness matrix-based form-finding method of tensegrity structures. Eng. Struct. **58**, 36–48 (2014)
13. Housner, G.W., et al.: Applied Mechanics: Dynamics (1961)
14. Sulejmani, A., Koça, O.: Development of optimal transmission rate of the kinematic chain by using genetic algorithms coded in Mathcad. Int. J. Innov. Technol. Interdisciplin. Sci. **4**(4), 792–803 (2021)
15. Rosenberg, R.M.: Virtual displacement and virtual work. In: Analytical Dynamics of Discrete Systems. Mathematical Concepts and Methods in Science and Engineering. Springer, Boston, MA (1977). https://doi.org/10.1007/978-1-4684-8318-5_9
16. Zhang, T., Lin, S., Zheng, H., Chen, Y.: Elastoplastic integration method of mohr-coulomb criterion. Geotechnics. **2**(3), 599–614 (2022). https://doi.org/10.3390/geotechnics2030029
17. Sulejmani, A., Koça, O., Dhoska, K., Gheibi, M., Moezzi, R.: A novel method of jacobian contours to evaluate the influence line in statically determinate structures. Appl. Mech. **4**, 1172–1187 (2023). https://doi.org/10.3390/applmech4040060

Energetics

Coulomb Friction Model Analysis, Acting on an Aerospace Servocommand

Parid Alimhillaj[1(✉)], Majlinda Alcani[1], Matteo D. L. Dalla Vedova[3], Luis Lamani[2], and Spartak Pocari[1]

[1] Department of Energy, Faculty of Mechanical Engineering, Polytechnic University of Tirana, Tirana, Albania
{parid.alimhillaj,malcani,spocari}@fim.edu.al

[2] Department of Geoinformatics, Faculty of Geology and Mining, Polytechnic University of Tirana, Tirana, Albania
luis.lamani@fgjm.edu.al

[3] Department of Mechanics and Aerospace Engineering, Politecnico di Torino, Turin, Italy
matteo.dallavedova@polito.it

Abstract. Modern flight control system design necessitates using highly detailed models to analyze individual components or subsystems; on the other hand, more fundamental and synthetic models with adequate accuracy are required for preliminary design, monitoring or diagnostic issues. One of key problems for especially primary flight commands which are depicted as position servo command with high degree of accuracy is the necessity to acknowledge their behavior, under the influence of the dry friction or the friction forces (or their momentum) acting on a generic movable mechanical part of a servo command, or on an entire transmission. In general this forces depends from the relative velocity of the transmission, or from the exchange load between parts of the mechanism. For this reasons it become very important to take into account these forces or their momentums in relation of their behaviour during the construction of the detailed mathematical models. In this work we present some generic considerations of the dry friction and dynamic friction applied on an aerospace servo command, with the related mathematical model run in MATLAB/Simulink, and compared to previously tests done for this servo command.

Keywords: Coulomb friction · Aerospace servo command · Mathematic model · MATLAB/Simulink

1 Introduction

Modern flight control system design necessitates using highly detailed models to analyze individual components or subsystems; however, more fundamental and synthetic models with adequate accuracy are required for preliminary design, monitoring, or diagnostic issues [20]. One of the key problems for especially primary flight commands, which are depicted as position servo command with a high degree of accuracy, is the necessity to acknowledge their behavior under the influence of the dry friction, i.e., the friction forces

G. Guxho et al. (Eds.): ITC-ICEE 2023, LNMIE, pp. 197–210, 2024.
https://doi.org/10.1007/978-3-031-48933-4_19

(or their momentum) act on a generic movable mechanical part of a servo command or an entire transmission. In general, these forces depend on the relative velocity of the transmission or the exchange load between parts of the mechanism. For this reason, it becomes essential to consider these forces or their momentums about their behavior during the construction of a detailed mathematical models [1].

1.1 Dependence of the Friction on Speed

The marked non-linearity of the relationships linking the trend of the dry friction forces (or their momentum) to the speed (see Fig. 1 in this regard) [8], combined with the extreme importance that this phenomenon assumes in many applications, in the course of decades has justified various studies aimed at defining exhaustive mathematical models.

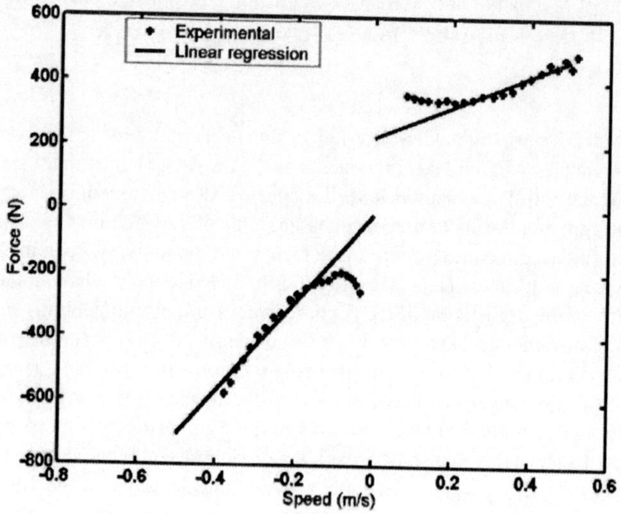

Fig. 1. Trend (experimental) of the friction force acting between surfaces in relative motion; in this case, the measured force is equal to the sum of the dry and viscous terms [6, 8]

Adopting a representation of dry friction to be included in numerical models representative of the dynamic performance in the time domain of mechanical systems, rather than orienting towards mathematical models of tribological derivation [2–4], it was preferred to opt for a performance model which, ignoring the effects and local dynamics, provides a global representation of the phenomenon; in particular, the classical Coulomb friction model was adopted. Assuming as conventions the positive direction of the friction forces opposite to the positive direction of the speed, the model can be summarized as follows:

$$FF = \begin{cases} F_{att} \quad if \;\; v = 0 \wedge |F_{att}| \leq FSJ \\ FSJ \cdot sgn(F_{att}) \quad if \;\; v = 0 \wedge |F_{att}| > FSJ \\ FDJ \cdot sgn(v) \quad if \;\; v \neq 0 \end{cases} \quad (1)$$

The mathematical model of dry friction (1) developed by Coulomb [8], capable of approximating the real physical behavior in a satisfactory way (illustrated in Fig. 1), is schematically represented in Fig. 2; FF indicates the generic friction force (i.e. its instantaneous value), FSJ the value of the friction force in static or adherent conditions and FDJ represents the friction force in dynamic conditions. This model discriminates the sign of the friction force as a function of the direction of the speed, and is able to distinguish the adherence conditions from the dynamic ones, to evaluate the possible stop of the mechanical element initially in motion, to keep the mechanical element in adherence conditions and, lastly, to evaluate the possible restart of the initially stationary mechanical element.

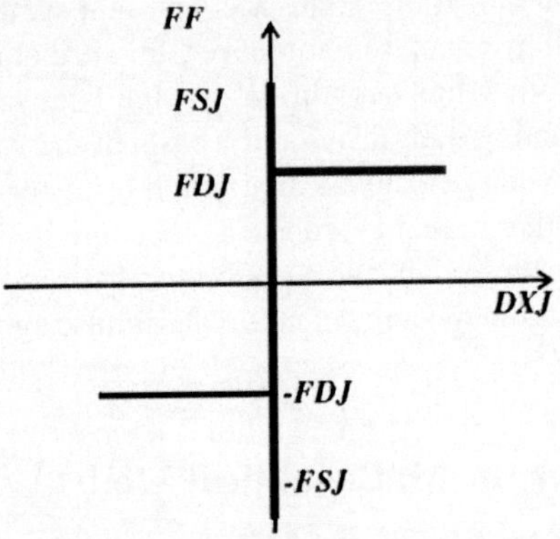

Fig. 2. Friction model schematic representation used in dynamic simulations

At a macroscopic level, the effect produced by dry friction on a mobile mechanical element manifests itself through the creation of forces (or moments) opposite to the relative motion (between the surfaces in contact) and with variable intensity according to the load acting on the element and of the speed itself. In the classical formulation of Coulomb friction, however, it is common practice to model the friction force as dependent only on the speed according to the following model:

- At non-zero speed the force module is assumed constant and equal to the so-called dynamic friction value (FDJ);
- At zero speed the force can assume any value lower than or equal in module to the so-called static friction value (FSJ).

1.2 Dependence of Friction on Load

As regards the dependence of dry friction on the load acting on the system itself (i.e. the resistant force or momentum applied from the outside to the relative motion of the generic mechanical system), it was necessary to conceive a special model illustrated here after.

At each speed value, the minimum amount of friction force (or momentum) occurs at zero load and its non-zero value is attributable to various causes, in particular to the necessary constructive grip, suitable for containment/recovery/elimination of the mechanical clearance, the use of sealing gaskets in the hydraulic components, and any contacts in electric machines, etc. Furthermore, the presence of loads exchanged between the mechanical connections in relative motion or acting on the transmission increases the minimum value of the friction force (or momentum) by a quantity proportional to the load itself, according to a proportionality coefficient linked appropriately to the efficiency of the connection (or transmission) represented into two different values. Depending on the way of acting, if the motor drags the user (motion against load or "opposing mode", i.e. speed of actuation contrary to the external load acting on the user) or if the user drags the motor (motion in favor of the load or "aiding", i.e. actuation speed in agreement with the external load on the user).

The proposed friction model takes into account, in a synthetic way and in a single formulation, all of the above analyses, moreover, by attributing positive (or negative) values to the efficiency in "aiding" conditions, it simulates reversible (or irreversible) transmissions, also evaluating their level on the basis of the value for the negative efficiency which defines by what percentage the friction force (or momentum) exceeds the load that produces it. Furthermore, the proposed procedure is also capable of taking into account situations such as to produce irreversibility of the mechanical system in static conditions and reversibility in dynamic conditions, simulating its behavior according to reasonable expectations.

2 Consideration for the Mathematical Model Used

The mathematical modeling of the Columbian friction, which is particularly important in the dynamic simulation of mechanical systems (with special regard to position servomechanisms), must therefore be able to describe the behavior of the mechanical element subject to this friction quantified as function of the loads, discriminating between the four possible kinematic situations listed below:

- Mechanical element initially stationary which must remain stationary;
- Mechanical element initially stationary which must start moving;
- Mechanical element initially in motion which must remain so;
- Mechanical element initially in motion which must stop.

The adoption of a model capable of correctly describing these dynamics, considerably increases the accuracy and confidence with the results of the simulations carried out, providing a valid means for investigating critical operating conditions.

The nature of the Coulomb friction cannot be fully described by linear models, and is desirable the possibility of closed-form solutions of the dynamic equations; therefore, any not simplistic modeling must necessarily involve the use of non-linearityes of such complexity as to foreseen the use of numerical resolution techniques based on dynamic simulation in the time domain [9, 11].

The numerical simulation techniques mostly used in the literature, can be traced back to mathematical models which, although more powerful than any purely linear representations, are nonetheless affected by some shortcomings. In particular, the numerical simulation models commonly reported in the literature, albeit with different degrees of approximation, aim to represent this phenomenon by introducing appropriate simplifying hypotheses (different from model to model) aimed at overcoming computational problems related to strict compliance with the Coulomb model. However, this simplifying hypotheses compromise in a more or less important ways the Coulomb model itself [6–8].

2.1 Proposed Coulomb Friction Model in Relation to the Velocity and to the Applied Active Force

The two models proposed, provide a realistic description of the phenomenon and can be easily integrated into more complex calculation programs (related, for example, to servomechanisms or entire actuation systems), also presenting a proven robustness against the particular simulated conditions.

The two numerical simulation models proposed overcome the limits highlighted above and reflect the behavior of the physical model assumed as the objective of the present work; Indeed:

- discriminate the sign of the friction momentum as a function of the direction of the speed,
- distinguish the grip conditions from the dynamic ones (in fact, two distinct values can be assigned for the force—or momentum—of friction, FSJ in static or grip conditions and FDJ in dynamic conditions),
- evaluate the possible stop of the mechanical element initially in motion,
- keep the mechanical element correctly stationary (or in motion) in adherence conditions (or motion),
- evaluate the possible restart of the initially stopped mechanical element,
- moreover, take into account, in a single model, the presence of any limit switches, against which it is assumed that a completely inelastic impact will occur.

The first proposed numerical model, it's referred directly to the Coulomb formulation. It describes the effects of dry friction, as a function of the speed v and the active force F_{att}, according to the modeling represented in Fig. 3:

The corresponding mathematical model can be formulated as per Eq. (1), where:

Where *FF* represents the friction force actually calculated, *FSJ* the value of the friction force in static or adherent conditions, *FDJ* that of the friction force in dynamic conditions and F_{att} the active force applied to the system.

The numerical simulation programs created on the basis of the mathematical model (1), in order not to run into numerical instability phenomena analogous to those already highlighted, are created in such a way as to impose the system shutdown of the mechanical system in the case of speed reversal during the integration steps;

$$v(t_{i+1}) = 0 if \ \ v(t_{i+1}) \cdot v(t_i) \leq 0 \tag{2}$$

If this stop results not corrected due to the excess of the active forces F_{att} compared to the passive ones (friction), the next calculation step would generate an imbalance (between the overall forces acting on the system) capable of correctly causing it to restart. It can be stated that the condition expressed in Eq. (2) constitutes one of the

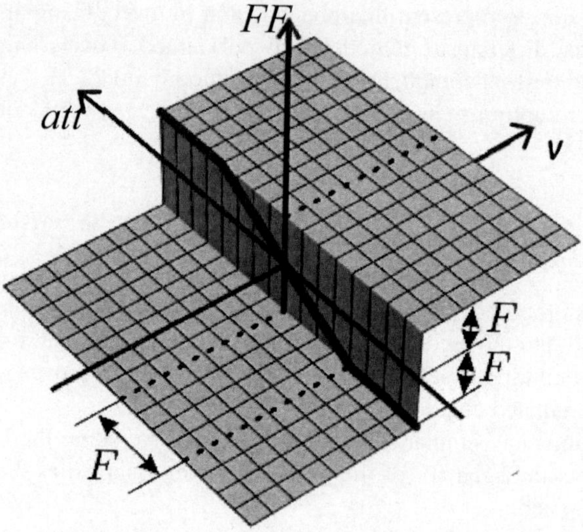

Fig. 3. Graphical representation of the proposed Coulomb friction model

fundamental differences and innovations of the present model with respect to those reported in the literature and, in its apparent obviousness, lies the very robustness and accuracy of the method.

The mathematical model illustrated in Eqs. (1) and (2) is implemented in the MATLAB-Simulink environment in order to obtain a simulation program capable of combining a satisfactory level of confidence and accuracy of the results with high legibility and accessibility.

The transition to the simulation environment supported by MATLAB–Simulink allows to overcome all these problems by providing a symbolic representation of the calculation algorithms adopted which, having a structure formally analogous to the corresponding block diagram, greatly facilitates the understanding of the structure of the dynamic system and, therefore, the legibility of the model itself. In fact, the transition from the aforementioned low-level programming languages to Simulink makes the proposed model accessible to a greater number of users (it is sufficient that the reader has framed the physical problem and is able to read a block diagram). However, this increased accessibility in the other sense introduces loss of the characteristic sequential structure of the algorithm (typical, for example, of the well-known Fortran) and with a significant reduction in the level of readability of the individual calculation operations performed.

The sequential structure, allow to know exactly the trend of the calculation flow (and therefore the order in which, in the single integration cycle, the instructions are executed), considerably facilitates the translation of the mathematical model into the numerical equivalent and allows to solve, in an extremely concise way.

From Eq. (1) has been developed the subsystem shown below in the figure:

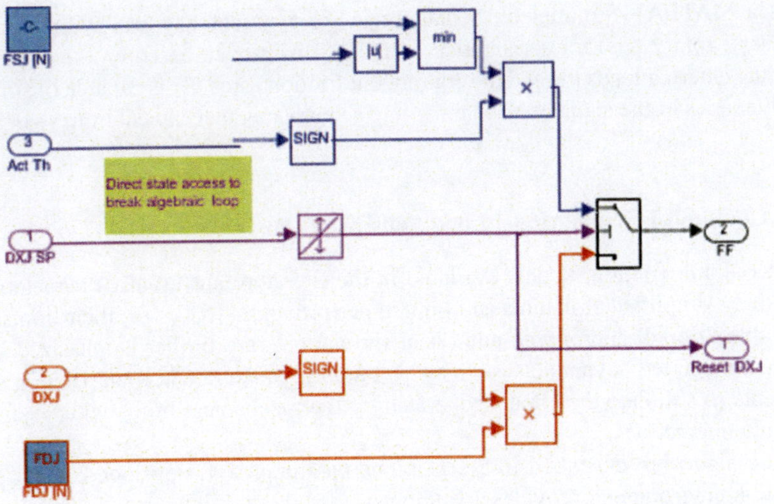

Fig. 4. MATLAB-Simulink block diagram of the proposed friction model

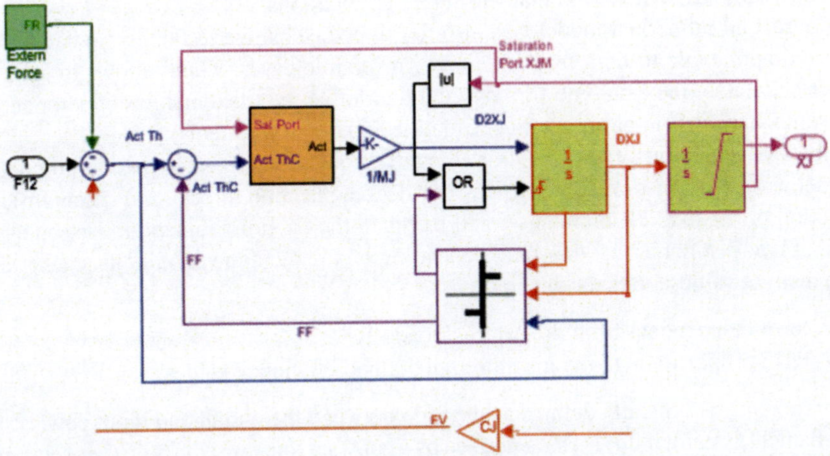

Fig. 5. MATLAB-Simulink block diagram of the actuator model with dry friction and mechanical limit switches

Starting from the subsystem illustrated in Fig. 4, was developed the non-linear numerical simulation model shown in Fig. 5.

The MATLAB-Simulink block diagram in Fig. 5, models a typical hydraulic actuator (linear or rotary one) for aeronautical use [13]; this model, based on the active forces F12, the external loads FR and the consequent friction force FF, is capable of simulating the dynamics of the actuator also taking into account any mechanical limit switches [14, 15].

2.2 Coulomb Friction Model Load-Sensitive (FR)

The Coulomb friction models available in the literature are usually characterized by extremely simplified structures and limited performances [6, 7, 10]; their limits, easily verifiable through appropriate numerical simulations, are further emphasized by analyzing "integrated" dynamic systems which not only take into account friction, but are also able to simulate the effects generated by the mechanical limit switches and their possible interactions.

The algorithm described in the previous paragraph 2.1, developed in MATLAB-Simulink environment, provides a valid answer to these problems and, through a self-contained subsystem, can describe both the effects produced by friction forces on the dynamic response of a generic mechanical system as their interactions with any mechanical limit switches, in many different applications.

A second numerical model was also developed, sensitive to the aerodynamic load FR and applicable to both reversible and irreversible cases, which allows to estimate more accurately the Coulomb frictions and their effect on the dynamics of aeronautical servomechanisms (in particular flap actuation and control systems) [16].

From the previously illustrated Coulomb friction model, a new numerical simulation model was therefore developed which considers the friction forces (or momentum) as a function of the load FR on the user and in which the friction momentum is seen as the sum of two distinct terms; this sum, depending on the dynamic conditions, can assume two distinct values (static or dynamic):

- A term independent from the external load FR,
- A term proportional to the load through defined efficiency values.

Efficiency is typically defined as the ratio between the output and input powers of a mechanical system; if this is characterized by a constant transmission ratio (very common occurrence in secondary flight controls), the efficiency can be understood as the ratio between the output and input momentum (converted on the same shaft). The efficiency of the system is therefore dependent on the motion conditions (static or dynamic) and on the external load (in favor or against motion). By suitably choosing the value of these returns, it is possible to simulate the performance of reversible and irreversible systems.

The model calculates the friction momentum as follows:

$$T_{FR} = T_{FR0} + \left(\frac{1}{\eta_{opp}} - 1\right) \cdot T_{LD} \tag{3}$$

Under opposing conditions of load to the motion.

$$T_{FR} = T_{FR0} + (1 - \eta_{aid}) \cdot T_{LD} \tag{4}$$

Under load conditions in favor of motion.

When the actuation speed is non-zero and the load T_{LD} opposes the motion, the friction momentum T_{FR} is obtained through Eq. (3), while when the load is in favor of the motion, the friction momentum T_{FR} is obtained through Eq. (4); in both cases the momentum T_{FR} assumes the same sign as the actuation speed (seen by the sign conventions) and is equal to the friction momentum $T_{FR,din}$ measured in dynamic conditions. On the other hand, when the actuation speed is zero, the friction momentum T_{FR} is equal to the driving momentum A_{ct} if this is included within two limit values: when the load T_{LD} is positive, the lower (negative) value is deducted from (4), while the upper (positive) one is obtained from (3); if the T_{LD} load is negative, the lower (negative) value is deduced from (3), while the upper (positive) one is obtained from (4); in both cases the term T_{FR} is equal to the friction momentum $T_{FR,stat}$.

This mathematical model was then implemented in the MATLAB - Simulink environment and integrated within a more complex calculation algorithm, realizing the numerical simulation program illustrated in the Fig. 6, [17–19]:

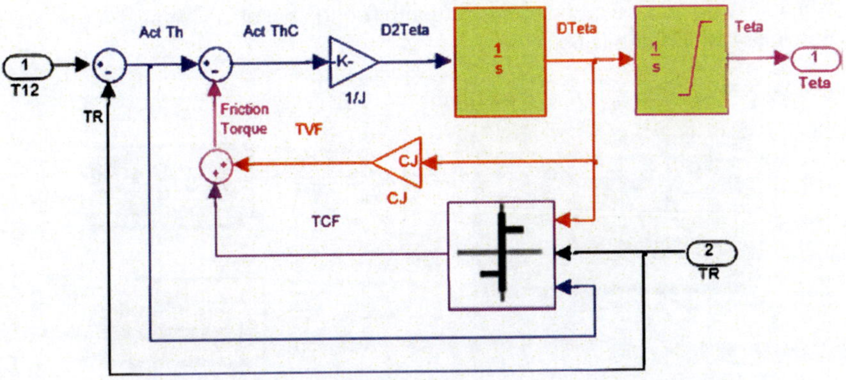

Fig. 6. MATLAB-Simulink block diagram of the flap actuation

3 Results

The numerical model illustrated in Fig. 6, containing the aforementioned friction calculation routine, allows to simulate the dynamic behavior of a flap control system equipped with final reversible or irreversible actuators and, therefore, allows to make comparisons between the different constructive solutions in the different operating conditions (modifying the aerodynamic loads, the actuation commands and the supply pressure it is, in fact, possible to simulate various situations, including specific failures of the system). In order to verify its validity, was then set up a numerical simulation program of the dynamic performance of a flap actuation system equipped with the aforementioned model.

The system, which obviously represents a simplification of the real one, consists of a power and control unit (PDU) which, via torsion bars, transmits power to the final actuators which drive the flaps. Each final actuator is composed of a low speed gearbox

(ZS), integrated in a ball screw (BS); at the end of the torsion bar transmission are positioned the position transducers (PT) and any speed transducers (however not present in the configuration under examination). The system is controlled by an electronic control unit (ECU), not shown, on which the position control loop closes. The PDU contains the hydraulic motor, high speed gearbox (ZF), control valves. In the model examined, the presence of a tachometric dynamo was also hypothesized in the PDU, to control the actuation speed. Figure 7 illustrates the physical model of the mechanical subsystem adopted to simulate the dynamics of the flap actuation group; in its construction, particular attention was paid to the mechanical and hydraulic characteristics of the main components of the system and to the corresponding friction phenomena [5, 12, 21]. In particular, it takes into account:

- Dry friction with Coulomb model and possible clearance on the PDU (FFM), on the final actuators of the mobile surfaces (FFS) or on the position transducers (FFPT);
- Dynamic model of the servo valve, third order, with limitations on the position and on the speeds; fluid dynamic model of the control valve for hydraulic drives;
- Dynamic and fluid-dynamic model of the hydraulic motor and high-speed gearbox, which takes into account not only Coulomb friction but also viscous friction and the flow rate lost through leakage;

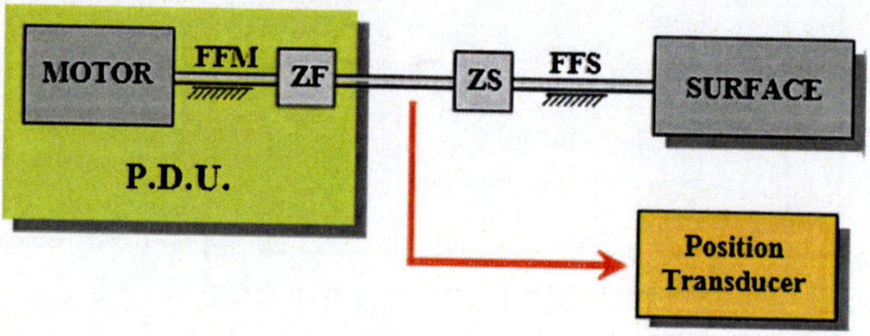

Fig. 7. Schematic representation of the passive subsystem of the flap control system actuation

With the help of the corresponding numerical model, some simulations were carried out to test the validity of the proposed friction model, in particular as regards the modeling of irreversible and reversible mechanical systems subject to loads (aerodynamic loads acting on the high lift surface).

In the Figs. 8 and 9, are presented two different cases of a system equipped in the case of Fig. 8 with irreversible actuator and in the case of Fig. 9 with reversible one. In both cases the Command at Time = 0 start at Com = 18 degrees.

In case of the Fig. 8 the flap responds to the required command by the rotation of the actuation shaft Teta in degree, and as it is expected in this case having an irreversible actuation the response come without oxilation.

In contrary in the case of Fig. 9 it is visible an oxilation between Time = 1.5 to approximately 4.5 s than the response with a higher inclination angle in confront of the case in Fig. 8 reaches the required command.

In both cases, it is what is expected from the response of the system and the calculation routine of the friction involved.

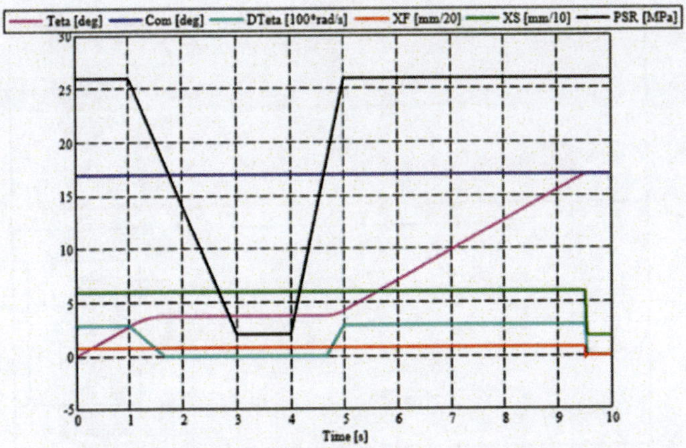

Fig. 8. Are presented the results with the use of irreversible actuator

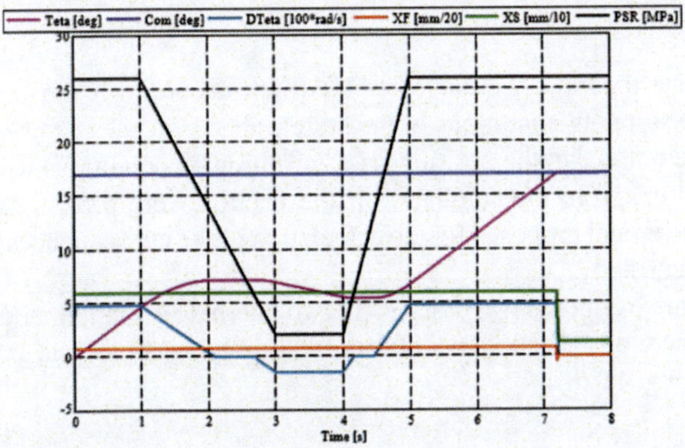

Fig. 9. In the figure is presented, under the same operating conditions as Fig. 8, the results produced with reversible actuator.

The performances of the numerical simulation model in Fig. 5 are compared with those provided by the discontinuous Coulomb friction model (present in the Simulink libraries) and by a hyperviscous friction model; the comparison is made by integrating these friction models into the dynamic simulation program of the performance of an electro-hydraulic servomechanism and by comparing, under the same operating conditions, the temporal trends of the responses produced.

In Fig. 10 it is represented the case of step load FR with zero initial and final value of 8000 N and constant zero command. Flow gain GQS = 0.2 m^2/s. purely proportional control logic. Absence of hysteresis and leakage. Constant supply pressure. At the instant Time = 0 the load of 8000 N is applied which induces an uncommanded movement of the controlled surface.

Through a dynamic behavior the system correctly moves towards a condition of equilibrium (new equilibrium at zero speed) characterized by a persistent position error obtained from the superimposition of the effects produced by the presence of the constant

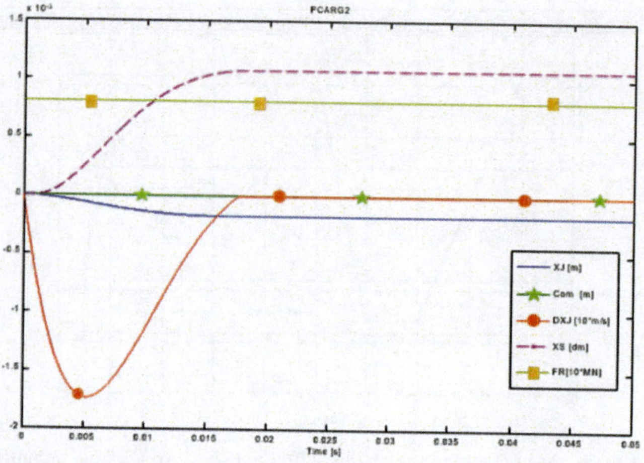

Fig. 10. Example of Coulomb friction model in relation to the velocity and to the applied active force on an electro-hydraulic servomechanism

load FR as well as the static friction force FF (which, according to the modeling adopted, depends on the stopping conditions of the system).

It should be noted that the position error in stationary conditions would be different from that resulting from the simulation if the friction were zero; from this it can be deduced that the model proposed for the calculation of the consequences of the Coulomb friction works correctly.

In Fig. 11, are examined two cases, with two alternative friction models for comparisons. A "hyperviscous" friction model with saturation Fig. 11 (a) and a "discontinuous" friction model Fig. 11 (b).

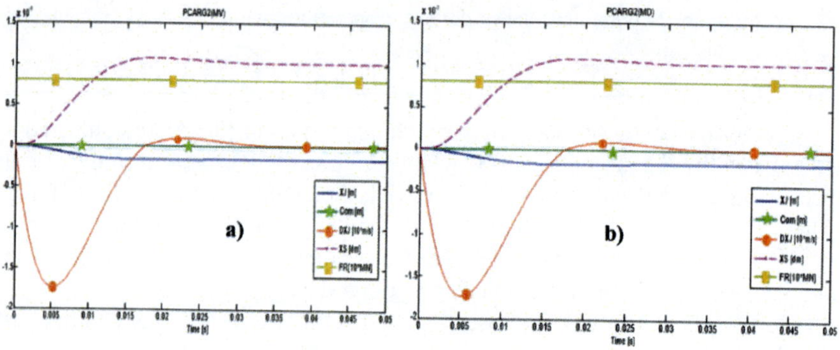

Fig. 11. Example of hyper viscous friction model (**a**) and discontinuous Coulomb friction model (**b**) acting on an electro-hydraulic servomechanism

There are no noticeable differences in the dynamic behavior until the overshoot peak is reached, approximately corresponding to the instant Time = 0.017 s; after this instant, both models in question are not capable of reproducing the correct condition of rest

(DXJ = 0) of the controlled unit obtained in Fig. 10 but, erroneously, they estimate a slow convergence towards the commanded position with false progressive cancellation of the position error. The slow convergence can be explained through a sequence of speed reversals at each calculation step (numerical instability), as it certainly is in the case of Fig. 11 (b) and, if a viscous constant is too high, even in the case of Fig. 11 (a). The inability of the last two models considered ("discontinuous" and "hyperviscous") to correctly simulate the stationary condition with persistently constant position error from the moment of overshooting is evident.

4 Conclusions

- The Coulombian friction model presented in this work shows its good possibility to be used in computer programs in the simulation of servomechanisms, in which a good compromise between modeling accuracy and execution speed is required, with a consequent compactness of the algorithm.
- Both models presented in paragraph 2.1 and 2.2 present a high degree of afidability, and are easy integrated in different mechanical actuators to be used, especially aerospace one, as per high requirement associated with this mechanisms.
- All this proves the efficiency of the proposed Columbian friction Simulink models, as well as its superiority compared to other programs, including the "default" ones that are in the digital library of the Simulink program itself, greatly increases the reliability of the numerical model of the simulated servomechanism.

Used Symbols

XS	Position second stage valve	[mm/10] or [dm]
XF	Position first stage valve	[mm/10]
Com	Command	[dm] or [degree]
PSR	Differential pressure (given by the system (supply-return)) controlled pressure	[MPA]
Teta	Position	[degree]
Dteta	Radial velocity	[100[a] rad/s]
DXJ	Velocity	[10[a] m/s]
P12	Differential pressure	[10[a]GPa]
XJ	Response of the system	[dm]
FR	Load	[10[a]MN]

[a] Other symbols and abbreviations, are explained in the paper

References

1. Borello, L., Villero, G.: *Mechanical failures of flap control system and related monitoring techniques.* In: 19th ICAS Congress, Melbourne, Australia, 13–19 September (1998)

2. Borello, L., Villero, G.: Mechanical failures of flap control systems: proposal of advanced monitoring techniques. Int. J. Mech. Control **5**(2), 9–28 (2004), ISSN 1590-8844
3. Davison, E.J.: The robust control of a servomechanism problem for linear time-invariant multivariable system. IEEE Trans. Automat. Contr. **21**, 25–34 (1976)
4. Goodwin, G.C.: *A brief overview of nonlinear control* (2001)
5. Jacazio, G., Borello, L.: Mathematical models of electrohydraulical servovalves for fly-by-wire flight control system. In: 6th International Congress on Mathematical Modelling, August, St.Louis, Missouri, U.S.A (1987)
6. Karnopp, D.: Computer simulation of stick-slip friction in mechanical dynamic systems. Trans. ASME: J. Dyn. Syst., Measurement, Cont. **107**(1), 100–103 (1985)
7. Kikuuwe R., Takesue, N., Sano, A., Mochiyama, H., Fujimoto, H.: H.: Fixed-step friction simulation: from classical Coulomb model to modern continuous models. In: IEEE/RSJ International Conference on Intelligent Robots and Systems, pp. 3910–3917 (2005)
8. Quinn, D.D.: A new regularization of Coulomb friction. Trans. ASME: J. Vibration Acoustics **126**(3), 391–397 (2004)
9. Sirouspour, M.R., Salcudean, S.E.: On the nonlinear control of hydraulic servo-systems. In: Proceedings of the IEEE International Conference on Robotics Automation, vol. 17, no. 2 (2000)
10. Skelton, R.E.: Model error concepts in control design. Int. J. Control **49**(5), 1725–1753 (1989)
11. Wang, P.K.C.: Analytical design of electrohydraulic servomechanism with near time-optimal response. IEEE Trans. Autom. Cont. **AC-8**(1), 15–27 (1963)
12. Borello, L., Vedova M.D.L.: Load dependent coulomb friction: a mathematical and computational model for dynamic simulation in mechanical and aeronautical fields. Int. J. Mech. Cont. (JoMaC) **7**(1) (2006)
13. Borello L., Villero G.: Confronto fra modelli semplificati di valvole di comando, XI Congresso Nazionale A.I.D.A.A., Forlì, 14–18 Ottobre (1991)
14. Urata, E.: Static stability of toque motors, pp. 871–885. The Eighth Scandinavian International Conference on Fluid Power, Tampere, Finland (2003)
15. Borello, L., Maggiore, P., Vedova, M.D.L., Alimhillaj, P.: Dry Friction Acting on Hydraulic Motors and Control Valves: Dynamic Behaviour of Flight Controls. XX AIDAA Congress, Milano, Italy, June 29–July 3 (2009)
16. Alimhillaj, P., Borello, L., Vedova, M.D.L.: Innovative proposal of a synthetic model for non-linear fluidodynamic of servovalvs. Int. J. Mech. Cont. (JoMaC), September (2013). ISSN 1590-8844
17. Vedova, M.D.L., Alimhillaj, P.: Study of new fluid dynamic nonlinear servovalve numerical models for aerospace applications. In: 2nd European Conference on Electrical Engineering & Computer Science, 20–22 December, Bern, Switzerland (2018)
18. Vedova, M.D.L.D., Berri, P.C., Corsi, C., Alimhillaj, P.: New synthetic fluid dynamic model for aerospace four-ways servovalve. Inter. J. Mech. Control (JoMaC). ISSN: 1590-8844, December (2019)
19. Alimhillaj, P., Londo, A., Vedova, M.D.L.: Behavior analysis of fluid dynamic instabilities in aerospace servovalves numerical models. In: 2nd International Conference "Engineering and Entrepreneurship" Proceedings (ICEE-2019), UPT Tiranes, October (2019)
20. Vedova, M.D.L., Alimhillaj, P.: Valve digital twins for electro-hydraulic actuator prognostics: snthetic fluid dynamic models sensitive to hydraulic capacity. In: 12th EASN, International Conference on Innovation in Aviation and Space for opening New Horizons, 18–21 October, Barcelona, Spain (2022)
21. Dorri, A., Poçari, S., Londo, A., Alimhillaj, P.: PID control for pneumatic cylinder stiffness for aerospace applications. Int. J. Mech. Cont. (JoMaC). **24**(1) (2023). ISSN: 1590-8844

Performance Analysis and Optimization of the Energy Indicators of an Axial Fan Through Some Constructive Changes Using Computer Simulations and Experimental Confirmation

Spartak Pocari[1(✉)] and Luis Lamani[2]

[1] Energy Department, Polytechnic University of Tirana, Tirana, Albania
spocari@fim.edu.al

[2] Geoinformatic Division, Polytechnic University of Tirana, Tirana, Albania
luis.lamani@fgjm.edu.al

Abstract. This article aims to analyze the performance of an axial fan through experimental measurements, and to build a mathematical model for further analysis. The article presents a constructive change related to the position and geometric characteristics of the working blades, hub to tip ration, gap distance and through the simulations and experimental measurements performed, the performance of the fan taken in the study will be evaluated. For this purpose, a test bench that enables the construction of the characteristics of the axial fan was built. A mathematical model has been conceived, and the simulation program connects the constructive parameters with the performance characteristics of the fan. The CFTurbo program was also used for the realization of the 3d model of the fan. Using the technology of 3D printers, an improved prototype with constructive changes has been produced, which we will analyze in the experimental test.

Keywords: Axial fan · Mathematic model · Fluid dynamics · Optimization

1 Introduction

In most commercial and industrial equipment, it is the axial fans that move the fluid from one environment to another. Increasing their working performance is the objective of every designer. Referring to performance, in this article we have analyzed the factors that affect the performance of axial fans, as well as proposed and tested improved prototypes. During the analysis of the factors that influence the performance of the fan, several factors with a high impact were singled out, where we can mention: Gap space (the distance of the fan blades from the stator), the ratio of the hub diameter to the outer diameter of the fan, as well as the position of the fan blades in relation to the axis of symmetry (tilt of the blades). The main focus has been the constructive changes referring to the factors mentioned above with the aim of increasing the performance of the fan. For this purpose, with the help of the CFTurbo program, we have designed several prototypes

G. Guxho et al. (Eds.): ITC-ICEE 2023, LNMIE, pp. 211–219, 2024.
https://doi.org/10.1007/978-3-031-48933-4_20

of modified fans, which we have physically produced through 3D printing. In order to measure the characteristics of the ventilators, we have built a bench test.

The ratio of the diameter of the hub to the outer diameter of the axial fan $\frac{D_{hub}}{D_{tip}}$. Has been the subject of discussion for years. The specific recommended speeds and diameters, as well as the smallest $\frac{D_{hub}}{D_{tip}}$. . Ratio of axial fans are given in the so-called Cordier diagram (Fig. 1).

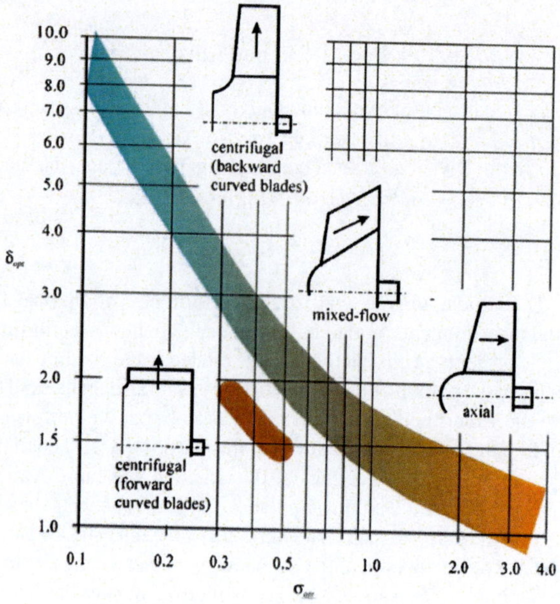

Fig. 1. Systematic of fan types in the CORDIER diagram [1]

Referring to this ratio, designers always try to achieve a ratio as small as possible $\frac{D_{hub}}{D_{tip}}$. For the reason that a larger ratio increases the amount of material used and at the same time the cost of the fan, another reason is the increase in the fluid passage surface. However, an inappropriate reduction of this ratio may cause backflow of the fluid in the area between the blades near the hub, reducing the performance of the fan.

1.1 Mathematic Model

For the realization of this study, we have built a mathematical model composed of:

Euler's equation,

$$\Delta p = \rho \cdot \left(u_2 \cdot c_{u_2} - u_1 \cdot c_{u_1}\right) \tag{1}$$

Non-dimensional parameters of flow, pressure and efficiency,

$$\emptyset = \frac{\dot{V}}{\frac{\pi^2}{2} \cdot D^3 \cdot n} \tag{2}$$

$$\psi = \frac{\Delta p}{\frac{\pi^2}{2} \cdot D^3 \cdot n^2 \cdot \rho} \tag{3}$$

$$\mu = \frac{Q \cdot \Delta p}{P}. \tag{4}$$

Diameters and specific speeds,

$$\delta = \frac{D_{tip}}{\left(\frac{8}{\pi^2}\right)^{\frac{1}{4}} \cdot \left(\frac{\Delta p}{\rho}\right)^{-\frac{1}{4}} \cdot Q^{\frac{1}{2}}}. \tag{5}$$

$$\sigma = \frac{n}{\left(2 \cdot \pi^2\right)^{-\frac{1}{4}} \cdot \left(\frac{\Delta p}{\rho}\right)^{\frac{3}{4}} \cdot Q^{-\frac{1}{2}}} \tag{6}$$

where:

- Q—flow $[\frac{kg}{m^3}]$
- Δp—pressure change $[Pa]$
- n—rotation speed [rot/s]
- D_{tip}—outer diameter of the rotor [m]
- ρ—air density $[\frac{kg}{m^3}]$
- P—power to the fan shaft [W]
- ν—the ratio of the diameter of the hub to the outer diameter of the rotor

$$\nu = \frac{D_{hub}}{D_{tip}} \tag{7}$$

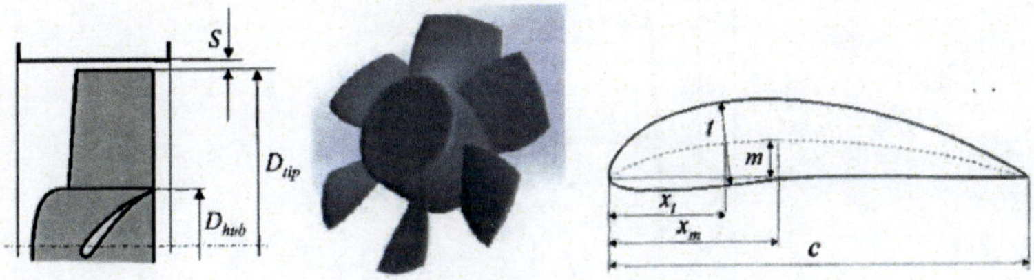

Fig. 2. The main parameters of the axial fan

In this study, we will first refer to the model of the standard axial fan designed in CF Turbo shown in the figure below. And we will study the impact of the gap on its performance for three sizes (3, 2, and 1.5 mm) (Figs. 2, 3, 4, and 5).

During the tests carried out in the test equipment built for this purpose, it turns out that with the reduction of the Gap, the performance of the fan increases, see Fig. 6.

This is explained by the fact that by reducing this distance of gap, the volumetric losses shown in Fig. 5 are reduced (Fig. 7).

Another factor that we mention at the beginning of this article is the $\frac{D_{hub}}{D_{tip}}$ ration.

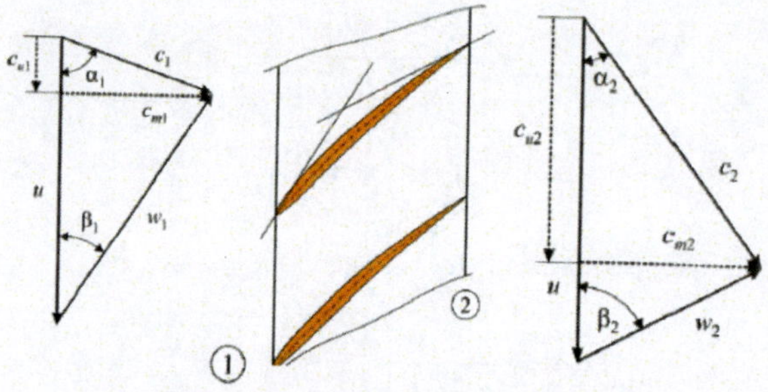

Fig. 3. Speed triangles in the axial fan [1, 2]

Fig. 4. On the left the fan designed in CFTurbo, on the right the fan produced by the 3D printer

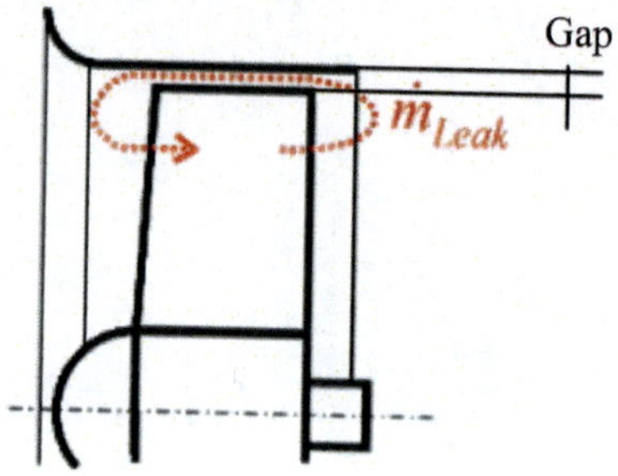

Fig. 5. The gap between the rotor and the statorp [1]

To study the effect of this ration, we will consider 3 ratios of diameters 0.32, 0.42 and 0.52. And for these we will measure the characteristics of the fan (Figs. 8 and 9).

As can be seen from the graph in Fig. 9, the reduction of the ratio of diameters increases the space of fluid flow between the blades, incresing dhe flow passing throw the fan. For values of ration smaller than 0.32, the performance of the fan begins to decrase, due to the creation of turbulence in its front part.

Since a significant part of the losses is caused at the entrance of the fan due to the turbulences that are created in the front of the blade, in order to avoid the impact in a

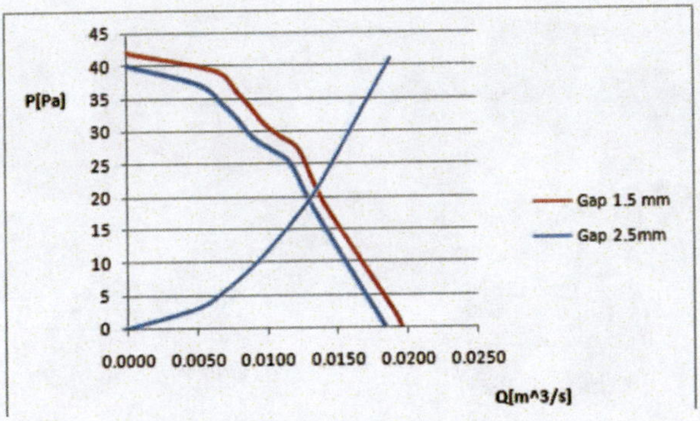

Fig. 6. Axial fan characteristics for 2.5mm and 1.5mm gap

Fig. 7. Test equipment built near the machines and thermo-fluidic plants laboratory in UPT

tangential way, we will modify the structure of the blade by tilting it forward as in the figure with down, and we will measure its characteristics (Fig. 10).

From the measurements carried out with the testing device on the performance of these two types of fans (standard, and with blades facing forward) it results that we have an increase in performance in the fan with blades facing forward Figs. 11 and 12.

2 Conclusions

Through the CFTurbo software, the mathematical model and the built test equipment, as well as with the help of 3D printing technology, important factors that affect the energy performance of axial fans were taken into consideration and analyzed. After the structural changes made based on the analyzes and tests performed on the physical object. We reach these conclusions:

- Open space (gap) plays an important role. Its reduction leads to the reduction of volumetric losses and simultaneously to the increase of the performance of the fan.
- Reducing the hub to tip ratio of the fan diameters up to a certain value beyond the standard, leads to an increase in the fan flow.

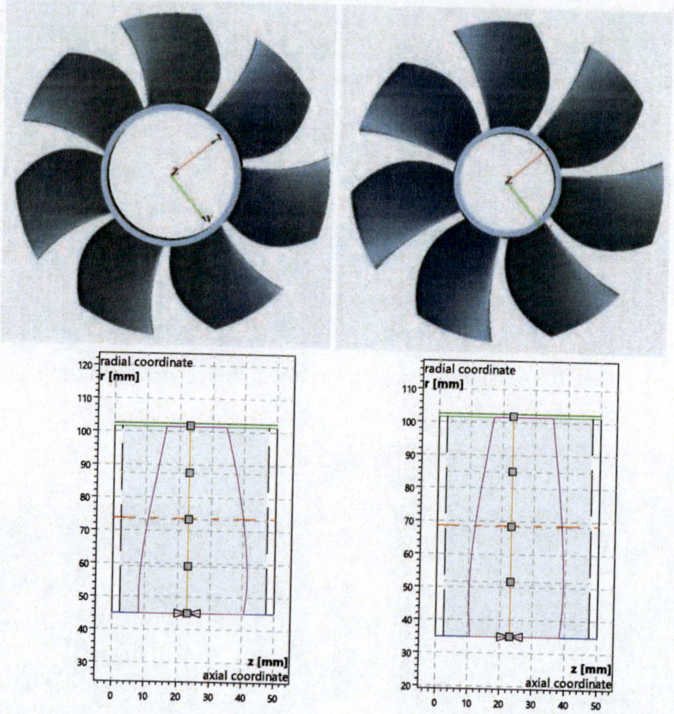

Fig. 8. (**a**) Fan with standard hub 0.42 ration (**b**) fan with reduced with 0.32 ration

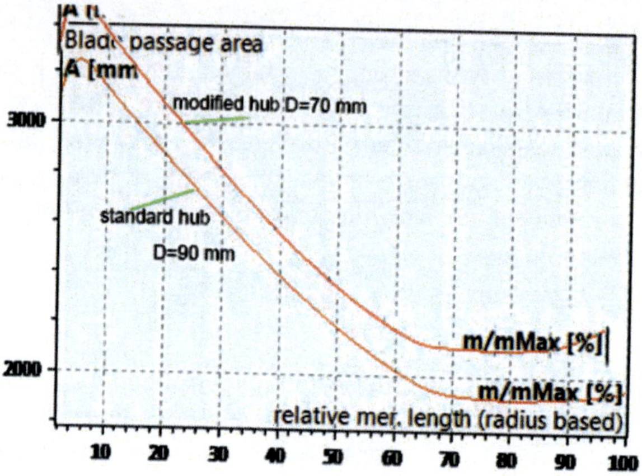

Fig. 9. The influence of the hub to tip ration diameter on the fluid passage space obtained from modeling in CFTurbo

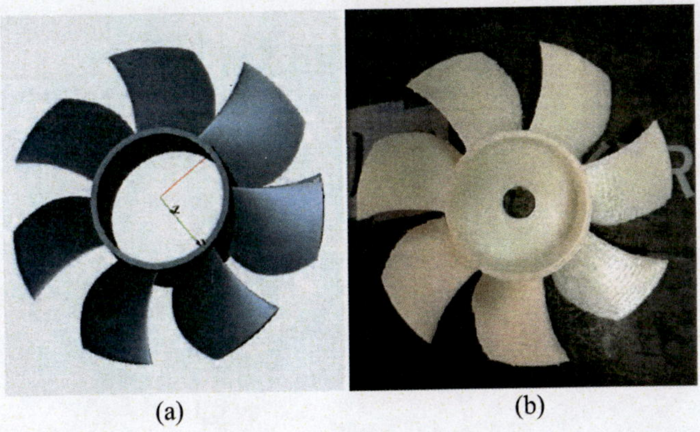

Fig. 10. (**a**) The fan designed in CFTurbo, (**b**) fan printed on a 3D printer

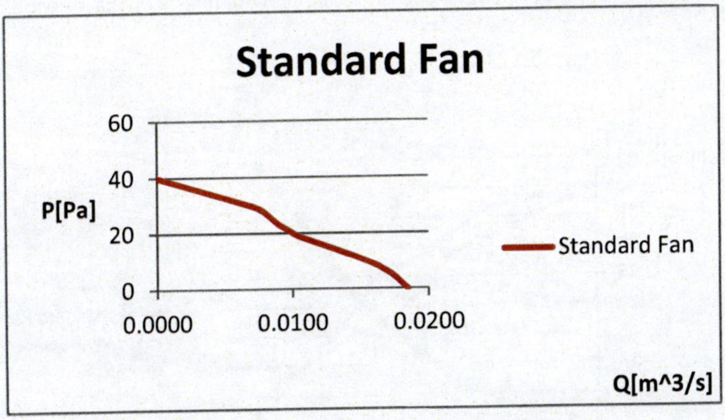

Fig. 11. Characteristic of the standard axial fan

- Tilting the fan blades in the direction of rotation reduces the losses due to the tangential collision with the front of the turbulence sheets created in the front part of the fan. This has a direct impact on increasing the performance Fig. 13, and the fan efficiency, Fig. 14.

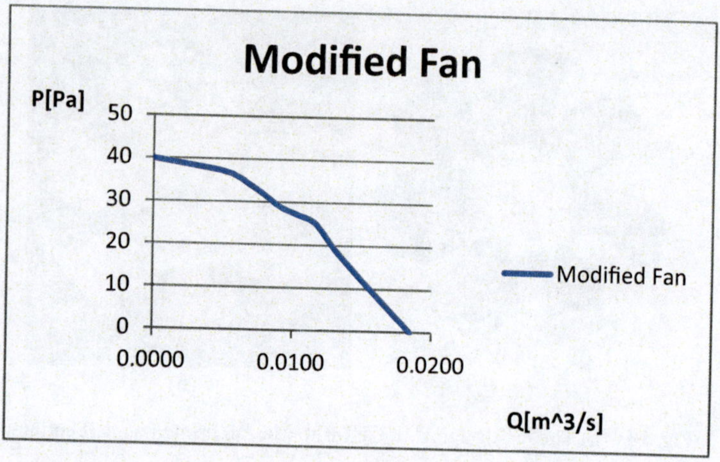

Fig. 12. The characteristic of the axial fan with blades facing forward

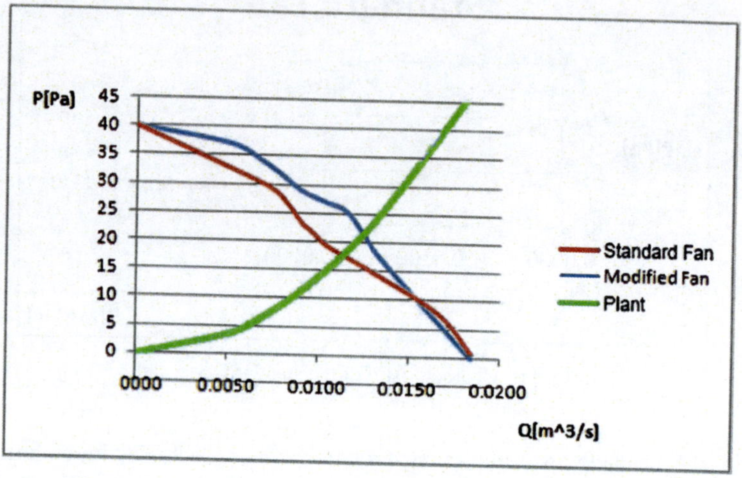

Fig. 13. Overlapping characteristics for both types of ventilators

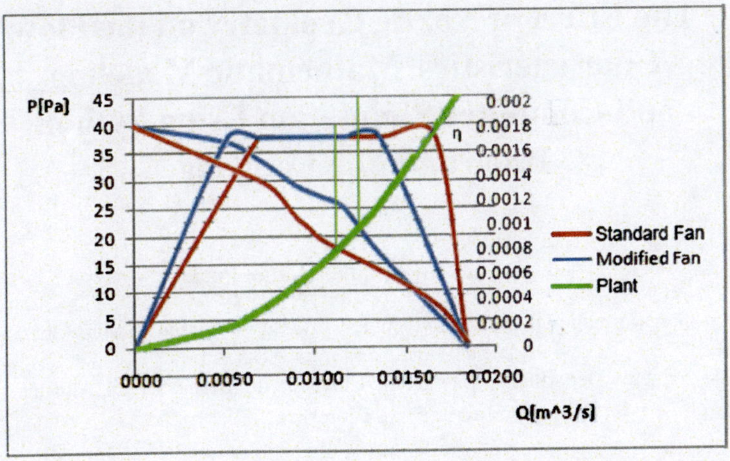

Fig. 14. Overlapping characteristics for both types of ventilators and specific efficency

References

1. Carolus, T.H.: FANS, 4th ed. Springer (2022)
2. Londo, A.: Fluid machines. SHBLU (2010)
3. Londo, A.: Fluidynamcis
4. CFturb: Users Manual. CFturbo GmbH (2023)
5. ISO 5801:2017(E): Fans—Performance Testing Using Standardized Airways, 3rd ed. (2017)
6. BS ISO 580:2007 standard (2007–2012)
7. Çengel, Y.A., Cimbala, J.M.: Fluid Mechanics, 3rd ed. (2014)
8. Mode, F.: Ventilatoranlagen. 4. Auflage, Verlag Walter de Gryter (1972)

The Effect of Nozzle Geometry on the Flow Characteristics. Mathematic Modeling and Computer Simulation Using Python Programming Language

Luis Lamani[1(✉)] and Spartak Pocari[2]

[1] Geoinformatic Division, Polytechnic University of Tirana, Tirana, Albania
luis.lamani@fgjm.edu.al

[2] Energy Department, Polytechnic University of Tirana, Tirana, Albania
spocari@fim.edu.al

Abstract. Nozzle are important elements in mechanical plants. They are used to create fluid flow with different velocity and pressures. The cross section of the nozzle varies in length. The nozzle can be convergent, divergent or combined convergent-divergent. The article deals with the influence of the geometric shape of the nozzle on the parameters of the fluid flow. A mathematical model that describes the fluid parameters during the flow in the nozzle has been modeled. The mathematical model for computer simulations in the Python languageis programmed. Meanwhile, a physical model was built with 3D printer technology. The results of the physical parameters of the fluid were measured in the physical model and were compared with the results of the mathematical simulation. For measuring parameters in the physical model the Labwiev software is used. The results of this comparison will give us the opportunity to simulate with the help of a mathematical model the performance of the nozzle for different geometric shapes of the fluid passage section. Meanwhile, we switched from the conical profile of the diffuser to the bell profile and studied the effects on the flow parameters.

Keywords: Mathematic modeling · Fluid dynamics · Computer simulation

1 Introduction

The main purpose of the use of the nozzle is to convert the potential energy at the inlet into the kinetic energy at the outlet, which determines the thrust force in rockets, airplanes, etc.

The nozzle Laval has a wide use in the technique of obtaining the thrust force.This nozzle has two parts, the one convergente to $A_{min} = A_{critical}$ and the expanded part in the form of a diffuser at the exit of which the velocity $V > V_{crit=a}$ and the pressure $p < p_{crit}$ are reached.In these nozzle, the fluid speed, temperature and pressure change during the flow as shown in Fig. 1 [2].

G. Guxho et al. (Eds.): ITC-ICEE 2023, LNMIE, pp. 220–233, 2024.
https://doi.org/10.1007/978-3-031-48933-4_21

The Laval nozzle realizes the acceleration of the fluid at supersonic speed (Ma > 1). This combination in the Laval design achieves a very efficient conversion of thermal energy into kinetic energy.

The condition of the fluid in the nozzle is generally determined by the ratio of the ambient pressure in the outlet plane to the inlet pressure p_b/p_0. Therefore, for given parameters at the inlet area, the flow in the convergent-divergent nozzle depends on the pressure in the exit plane or as we can call it the back pressure (p_b) [3].

Since the geometry of the nozzles has a determining role in the energy conversion and consequently in the performance of the thermal machines, this paper aims to study the influence of the geometric shape of the longitudinal profile of the nozzle on the behavior of flow in the Laval nozzle, using the mathematical model with the PYTHON language programming [4].

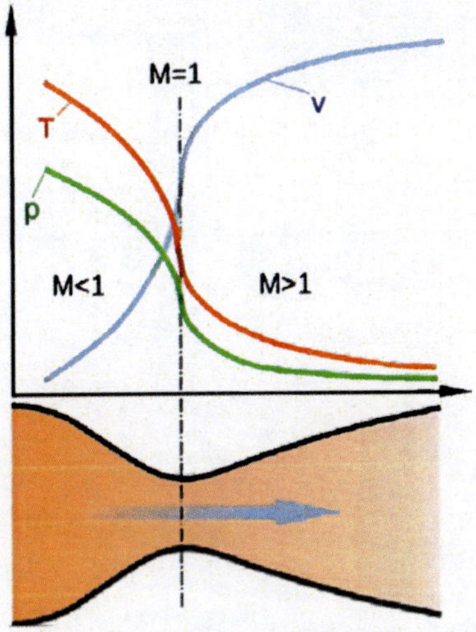

Fig. 1. Dependence of fluid velocity (v), temperature (T) and pressure (p) in the laval nozzle [3]

1.1 The Geometric Parameters of the Nozzle Taken in the Study

We have referred to the study of the parameters of the fluid in its flow in a Laval nozzle.

A **Laval nozzle** (or **convergent-divergent nozzle**, **CD nozzle** or **con-di nozzle**) is a tube which is pinched in the middle, making a carefully balanced, asymmetric hourglass shape. It is used to accelerate a compressible fluid to supersonic speeds in the axial (thrust) direction, by converting the thermal energy of the flow into kinetic energy. De Laval nozzles are widely used in some types of steam turbines and rocket engine. It also sees use in supersonic jet engines.

Figure 2 presents the geometric shape and construction parameters of a laval nozzle built in the thermofluidic machine laboratory in the energy department of P.U.T.

Referring to the Laval nozzle data shown in the Table 1, the 1:1 scale model was built in AutoCad. The model was also printed on a 1:5 scale in a 3D printer (Fig. 3).

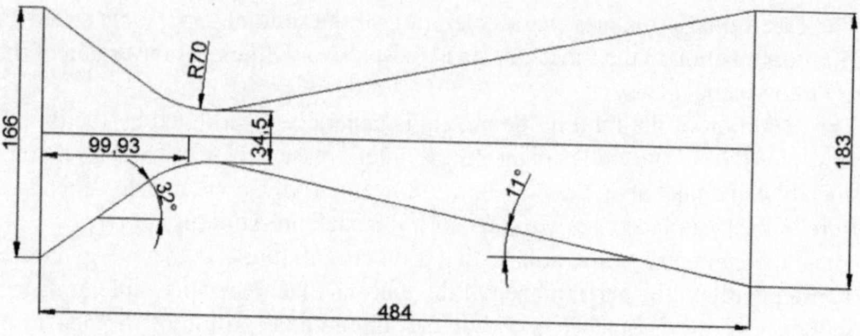

Fig. 2. Geometric shape and construction parameters of a laval nozzle [5]

Table 1. Constructive dimensions of the laval nozzle [5].

The constructive dimension	Size
Total length of nozzle	484 (mm)
Inlet diameter	166.6 (mm)
Pinched diameter	34.5 (mm)
Outlet diemater	183 (mm)
Length of the inlet area	99.93 (mm)
Angle of the convergernce	32°
Angle of the divergente	11.31°
The transition radius between the convergent and divergent section	70 (mm)
The length of the arc	40 (mm)

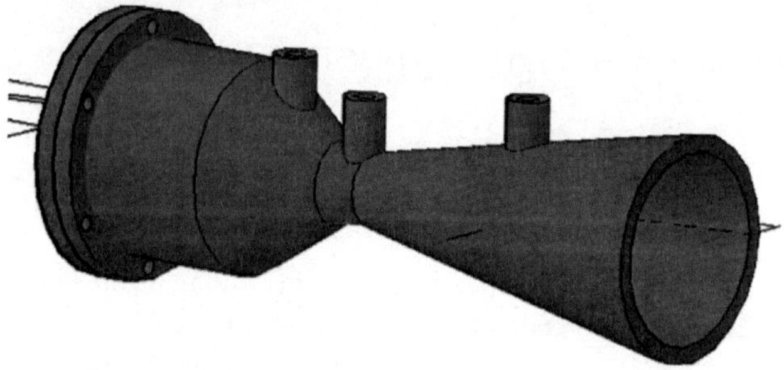

Fig. 3. 3d view of the laval nozzle modeled in AutoCAD

2 Mathematical Model

The mathematical model of the fluid flow in the Laval nozzle is based on the following relationships.

a. **Conservation mass flow equation**

$$\dot{m} = \rho \cdot v \cdot A = costan.$$

or by deriving this equation is written

$$v \cdot A \cdot d\rho + \rho \cdot A \cdot dv + \rho \cdot v \cdot dA = 0$$

after dividing by $\rho \cdot v \cdot A$, the equation takes the form

$$\frac{d\rho}{\rho} + \frac{dv}{v} + \frac{dA}{A} = 0 \qquad (1)$$

where:

- $\dot{m}$—mass flow
- v—velocity of the flow
- ρ—density of the fluid
- A—cross section of the nozzle

b. **Conservation of momentum equation**

$$\rho \cdot VdV = -dp \qquad (2)$$

Isoentropic flow equation.

$$\frac{dp}{p} = \gamma \frac{d\rho}{\rho} dp = a^2 \cdot d\rho \qquad (3)$$

By substituting dp in equation 2, we get

$$\left(1 - M_a^2\right)\frac{dV}{V} = \frac{d\rho}{\rho} \qquad (4)$$

Another important relation for the isoentropic flow of a fluid is obteined by substituting

$$\rho \cdot v = -\frac{dp}{dv}$$

$$\left(1 - Ma^2\right)\frac{dV}{V} = -\frac{dA}{A} \qquad (5)$$

The equation (5) governs the shape of a nozzle or a diffuzer in subsonic or supersonic isoentropic flow.Noting thta A anv V are positive quantities, we conclude the following:

For subsonic flow (Ma < 1), $\frac{dA}{dV} < 0$.

For supersonic flow (Ma > 1), $\frac{dA}{dV} > 0$.

For sonic flow (Ma = 1), $\frac{dA}{dV} = 0$.

For the pressure, temperature and velocity in the nozzle we use the follow equations:

$$\frac{p_t}{p_0} = \left[\frac{2}{(\gamma + 1)}\right]^{\frac{\gamma}{\gamma-1}} \tag{6}$$

$$\frac{T_t}{T_0} = \left[\frac{2}{(\gamma + 1)}\right] \tag{7}$$

$$v_t = \sqrt{\frac{2 \cdot \gamma}{\gamma + 1} R \cdot T_0} = \sqrt{\gamma \cdot R \cdot T_t} \tag{8}$$

The energy equation for steady flow is as follows (Fig. 4):

$$\frac{Q - W}{m} = \left(h + \frac{v^2}{2} + g \cdot z\right)_t \cdot \left(h + \frac{v^2}{2} + g \cdot z\right)_x \tag{9}$$

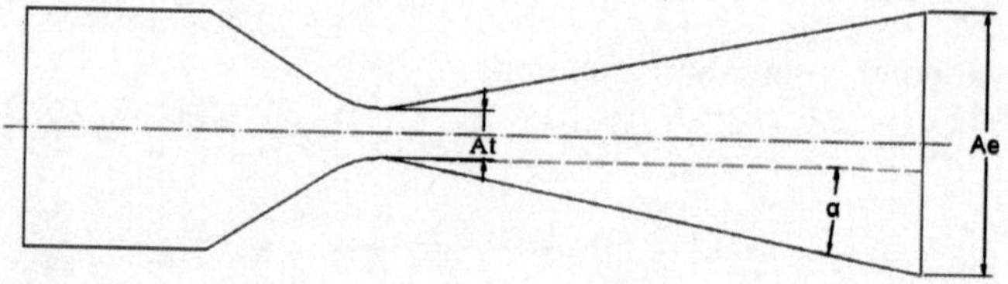

Fig. 4. Schematic view of the laval nozzle with the main parameters

The thrust force in the nozzle is calculated from the equation:

$$T = \dot{m} \cdot u_e \frac{1 + \cos(\alpha)}{2} + (p_e - p_a) \cdot A_e \tag{10}$$

where:

- α—angle of the conical geometry of the diffuser.
- $\dot{m}$—fluid flow in kg
- u_e—outlet velocity of the fluid flow
- p_e—outlet pressure of the nozzle
- p_a—atmospheric pressure pressure

The length of the diffuser we can write the relation:

$$L_{con} = \frac{r_e - r_t}{tan\theta} \tag{11}$$

In the above equations the parametras are as follow:

- γ—specific heats ratio of the fluid.
- a—speed of sound
- p—pressure of the fluid
- Ma—Mach number

3 Programming in Python language

Based on the mathematical model described above, using the Python programming language, we have built a simulation program that allows us to evaluate the impact of geometry on nozzle performance. From the simulations performed with the program, we obtained the dependence of the ratio of the exit section to the pinched section, where its influence on the thrust force is clearly visible, as shown in Fig. 5. We also obtained the dependence of the thrustforce on the influence of the conical angle, Fig. 6 [4].

import math
import numpy as np
import matplotlib.pyplot as plt
from matplotlib.patches import Arc
from bisect import bisect_left
$Re = \sqrt{\varepsilon} * Rt$
$LN = 0.8\,((\sqrt{\in} - 1) * Rt)/\tan(15)$
$x = 1.5\,Rt\cos\theta$
$y = 1.5\,Rt\sin\theta + 1.5\,Rt + Rt$
where: $-135 \leq \theta \leq -90$
$x = 0.382\,Rt\cos\theta$
$y = 0.382\,Rt\sin\theta + 0.382\,Rt + Rt$
where: $-90 \leq \theta \leq (\theta n - 90)$
$x(t) = (1-t)^{\wedge}2 * Nx + 2(1-t)t * Qx + t^{\wedge}2 * Ex,\ 0 \leq t \leq 1$
$y(t) = (1-t)^{\wedge}2 * Ny + 2(1-t)t * Qy + t^{\wedge}2 * Ey,\ 0 \leq t \leq 1$
$Nx = 0.382\,Rt\cos(\theta n - 90)$
$Ny = 0.382\,Rt\sin(\theta n - 90) + 0.382\,Rt + Rt$
$Ex = 0.8*(((\sqrt{\varepsilon} - 1) - 1)*Rt)/(\tan(15))$
$Ey = \sqrt{\varepsilon} * Rt$
$NQ = m1\,x + C1$ *and:*
$QE = m2\,x + C2$
$m1 = \tan(\theta n),\ m2 = \tan(\theta e)$ *[Eq. 8]*
and: intercept
$C1 = Ny - m1\,Nx$
$C2 = Ey - m2\,Ex$ *[Eq. 9]*
$Qx = (C2 - C1)/(m1 - m2)$
$Qy = (m1\,C2 - m2\,C1)/(m1 - m2)$ *[Eq. 10]*

4 Simulation Results

In the Fig. 7 the influence of the velocity of the fluid to the pressure in nozzle.

For the same characteristics of the nozzle under consideration, which has a conical output geometry, through the program in Python language we obtained the equivalent section for the geometry in the shape of a bell in order to study the influence of this shape on the performance of the nozzle (Fig. 8).

For a more detailed analysis of this geometry, we built a 3d model of the boiler in SolidWorks and ran some CFD simulations to analyse the performance changes for both types of profiles (Figs. 9, 10, 11, 12, 13, 14, 15, and 16).

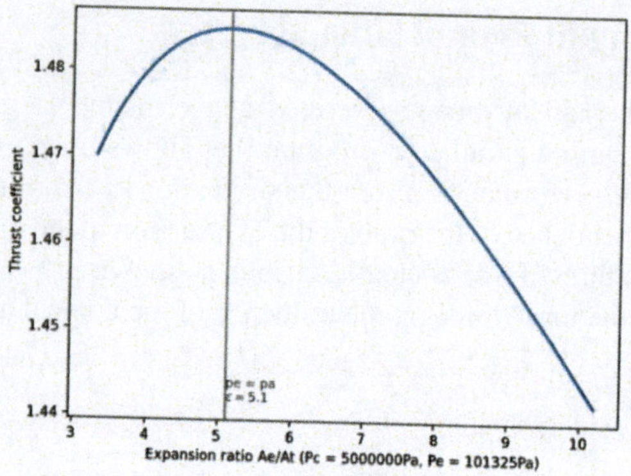

Fig. 5. The dependence of the thrust force from ratio of the exit section to the pinched section

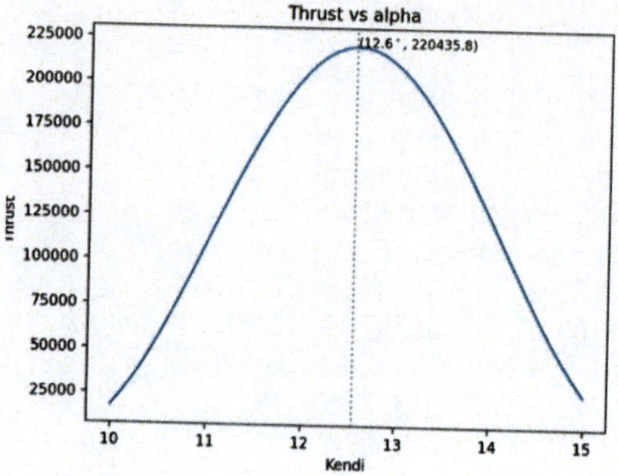

Fig. 6. The influence of the angle of the conical geometry of the diffuser to the thrust force.

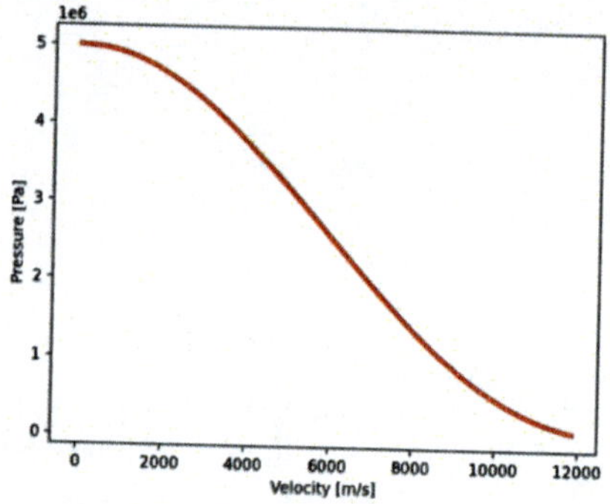

Fig. 7. The dependence of the pressure in the nozzle from velocity

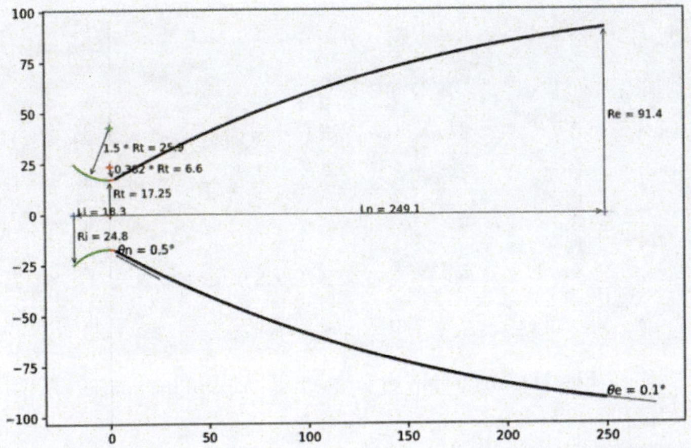

Fig. 8. The equivalent geometry in the shape of a bell confirmed by the simulation in Python referred to the conical nozzle.

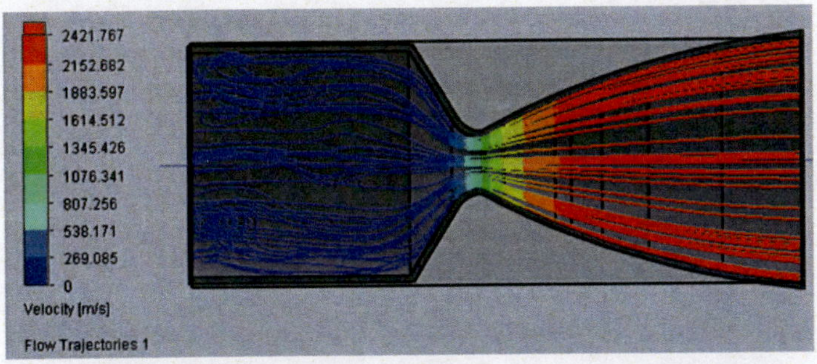

Fig. 9. Velocity gradient for the bell shape of the nozzle.

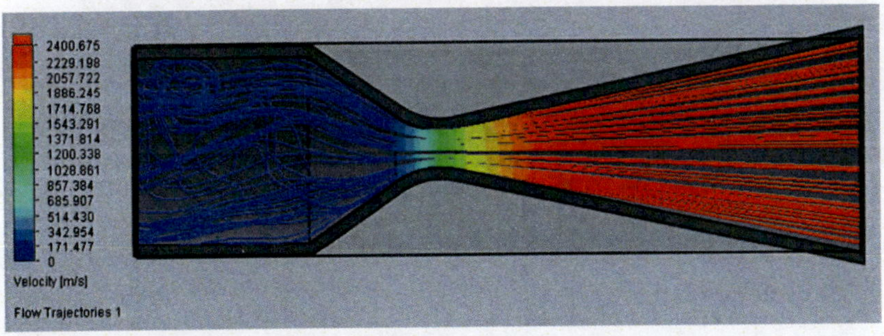

Fig. 10. Velocity gradient for the conical shape of the nozzle.

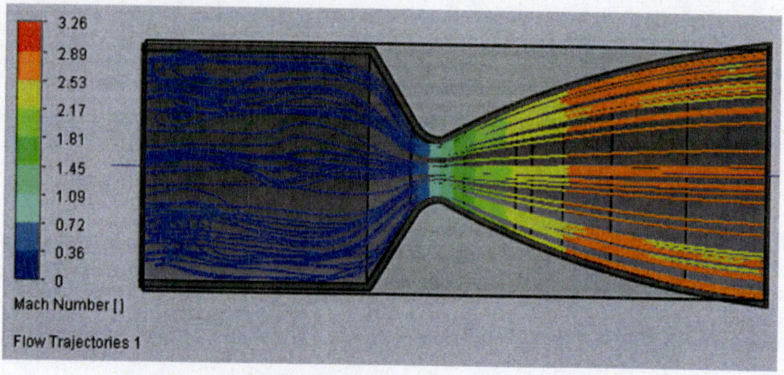

Fig. 11. Mach number for the bell shape of the nozzle

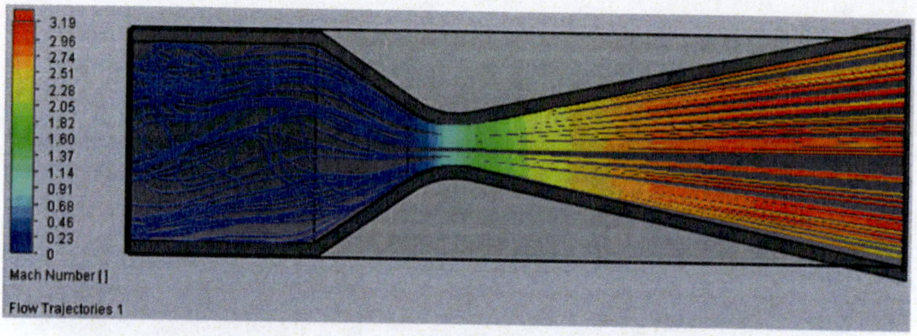

Fig. 12. Mach number for the conical shape of the nozzle.

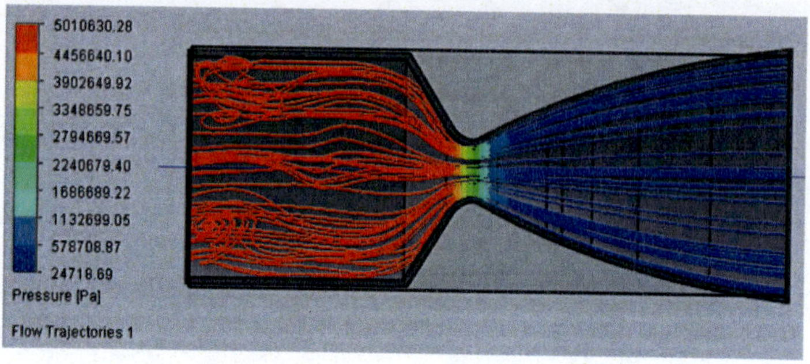

Fig. 13. Pressure for the bell shape of the nozzle

5 Experimental Results

In order to evaluate the truth of the built simulation program, we performed some measurements on the physical model built on a scale of 1:5 in the laboratory of the Department of Energy, which refers to the real model of the conical nozzle taken in the study (Figs. 17 and 18).

Input parameters for the Labwiev simulations are:

Inlet pressure of the nozzle 10 bar (1000000 Pa) and temperature 300° K.

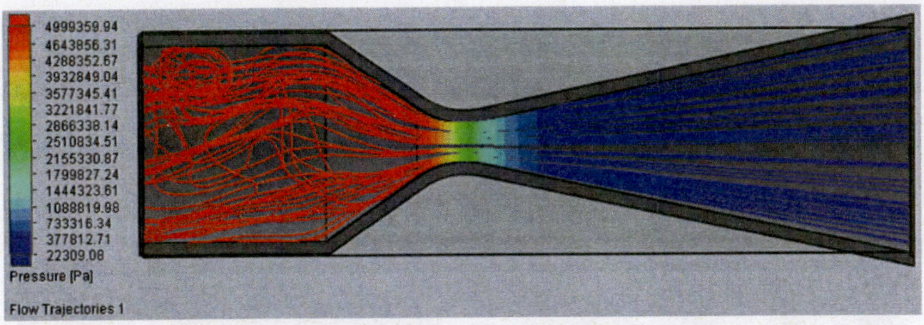

Fig. 14. Pressure for the conical shape of the nozzle

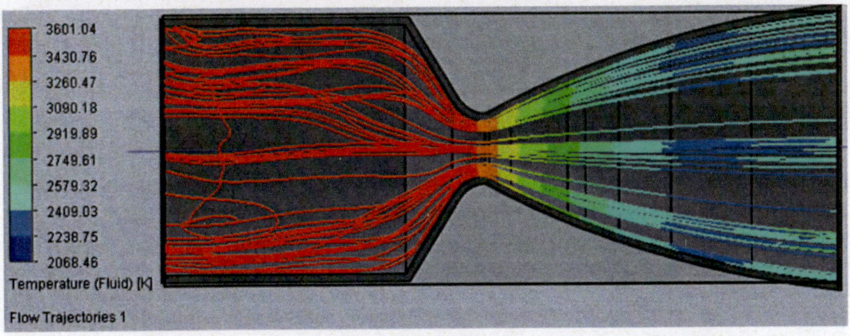

Fig. 15. Temperature gradient for the bell shape of the nozzle

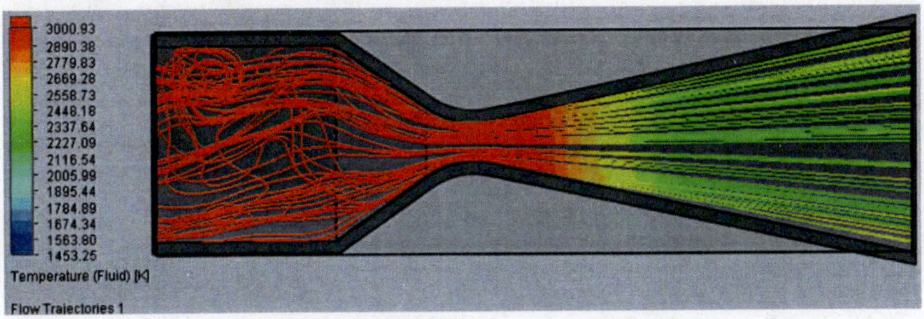

Fig. 16. Temperature gradient for the conical shape of the nozzle

The results are presented as follow (Figs. 19 and 20):

From the simulation with the Python program, for the inlet pressure of the nozzle 10 bar and temperature 300 K, the following characteristics were obtained (Figs. 21, 22, and 23):

As can be seen from the graphs above, the values corresponding to the simulation measurement points and the measured values according Labwiev model agree satisfactorily.

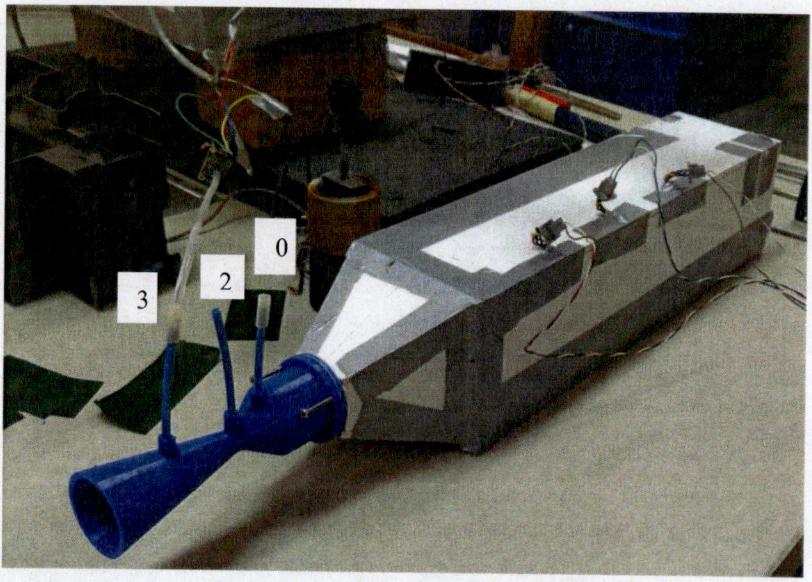

Fig. 17. Physical model built on a scale of 1:5 for the experimental measurments; 0, 2, 3 the measuring points

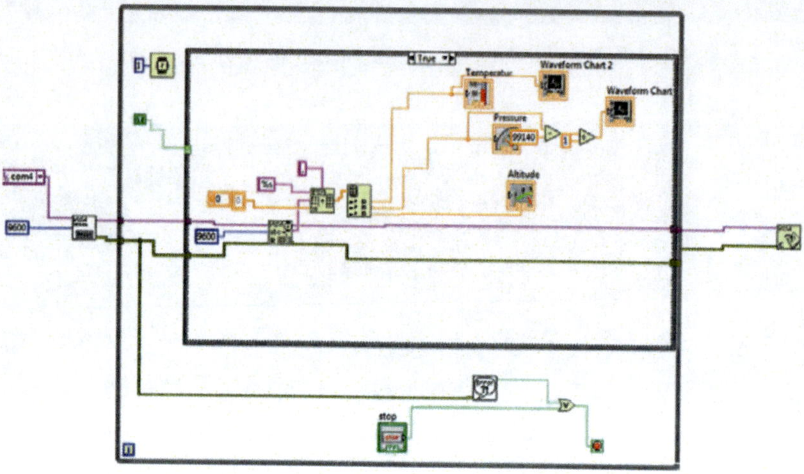

Fig. 18. LabView simulation for the experimental measurments.

6 Conclusions

1. From the simulations with the programming language Python (Fig. 6) for the conical section of the constructed nozzle diffuser, we come to the conclusion that the angle that gives the maximum driving force is 12.6 degrees.

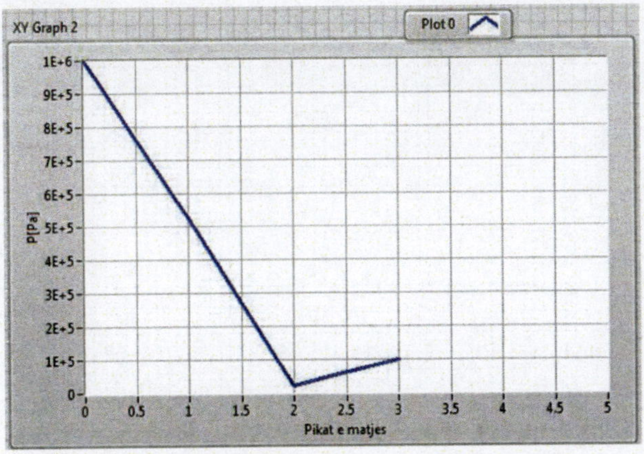

Fig. 19. Pressure measured in the physical model

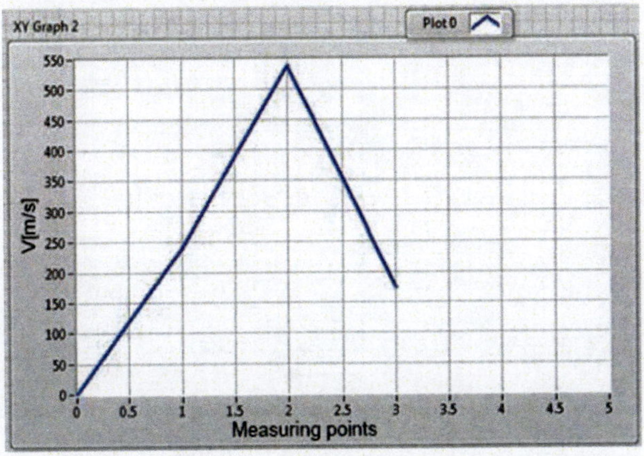

Fig. 20. Velocity measured in the physical model

2. The optimal ratio of the exit section to the section in the pinched area is $\varepsilon = \frac{A_e}{A_t} = 5.1$. This is also confirmed by the dependence of the speed on the pressure (Fig. 6) where it is clear that the maximum conversion of the pressure values into the speed of the fluid at the exit.
3. Referring to the fluid speed gradients along the length of the nozzle (Fig. 9), for the same input parameters, the fluid gains the highest speed at the exit of the bell profile. This fact is also confirmed by checking the Mach number (Fig. 11).
4. Referring to the pressure gradients, we see that the inlet pressure is converted to a higher speed in the bell-shaped nozzle (Figs. 13 and 14), this can be seen from the low values of the outlet pressure in the bell-shaped section compared to that in the conical section.
5. Referring to the temperature gradient, it is also worth noting that along the bell section we have a more complete change of temperature in the velocity of the fluid at the exit.

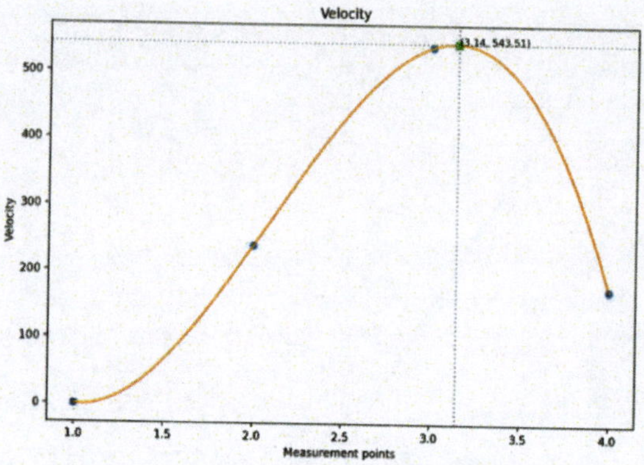

Fig. 21. Velocity from the Python simulation

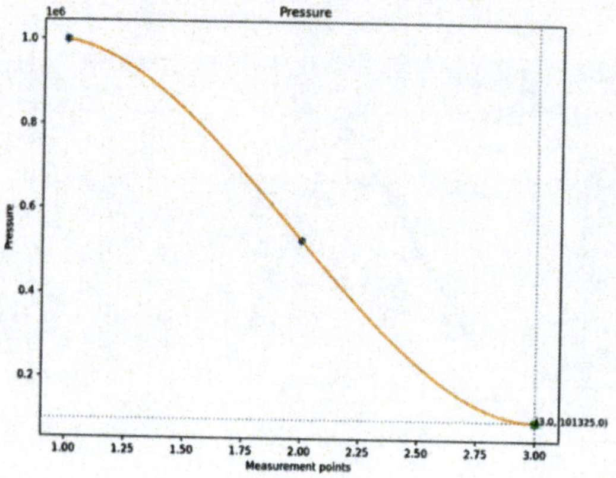

Fig.22. Pressure from the Python simulation

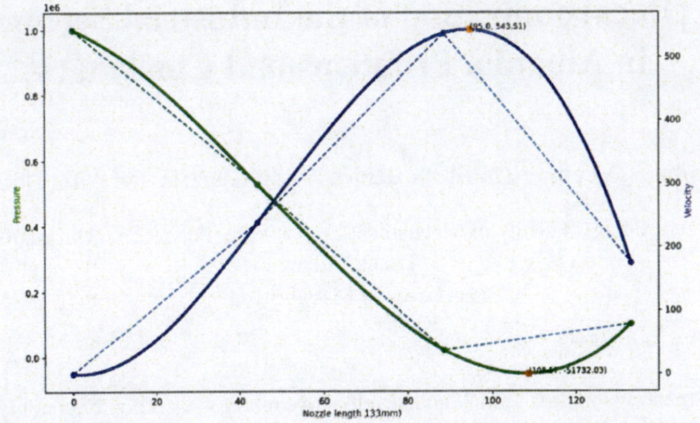

Fig. 23. Superimposition of the graphics obtained from the simulation in Python (continuous line) and the experimental measurements on the Labwiev model (dashed line) made at 3 measurement points.

References

1. Londo, A.: Mekanika e Fluideve. SH.B.L.U., Tirana (2007)
2. Shtjefni, A.: "Fizika Teknike", Rexhep Karapici. Tirane (2018)
3. Çengel, Y.A., Cimbala, J.M.: "Fluid Mechanics", Fondamentals and Applications, 3rd ed. Mc Graw Hill (2014). ISBN 978-0-07-338032-2
4. Lamani, L.: Gjuha e programimit Python. leksione te shkruara
5. International J. Mech. Product. Eng. ISSN: 2320-2092

Decarbonization of the Industrial Sector in Albania. Problems and Challenges

Majlinda Alcani(✉), Elena Bebi, Ardit Gjeta, and Altin Dorri

Department of Energy, Faculty of Mechanical Engineering, Polytechnic University of Tirana, Tirana, Albania
malcani@fim.edu.al

Abstract. Albania is a small and developing country with a fragile economy that relies mainly on the service sector, and industry and agriculture, which is facing many challenges in addressing the risks posed by climate change. In 2019 the service sector (represented by the subsectors of trade, transport, commercial activities, and telecommunication services) accounted for about 50% of the country's GDP; Industry and construction accounted for about 20% of GDP and the agriculture sector contributed to about 19% of the GDP. The structure of the industry sector in Albania shows that there are three main industry sectors that consume the highest percentage of energy: (1) food industry, (2) metal production and processing, (3) construction material sector. The National Strategy for Climate Change and Action Plan (NSCCAC) 2020–2030 is the main strategic document that addresses climate changes in Albania with objectives for 2050, approved in 2019. The focus of that strategy is climate change mitigation and adaptation in Albania. While progress has been made in the decarbonization of other sectors of the economy, the industrial sector, even at the European level, is considered the next front and the most expensive sector to decarbonize. Among the most effective measures for the decarbonization of this sector, mention: replacement of fossil fuels with non-hazardous waste in the Albanian cement industry, "electrification" of industrial processes, use of biomass and waste as a fuel, improvement of energy efficiency, application of the emissions trading system, the use of hydrogen as fuel, carbon capture and storage.

Keywords: Decarbonization · Industry sector · Effective measures

1 Introduction

The industrial sector is one of the most important ones of production and of social value on a global scale. Industrial companies produce about 1/4 of GDP and global employment and produce goods that are an integral part of our daily lives.

Industry also emits about 28% of global greenhouse gas (GHG) emissions, of which 90 percent are carbon dioxide (CO_2) emissions. Between 1990 and 2014, GHG emissions from major sectors such as residential, power, and transport, increased by 23 percent (0.9 percent per year), while emissions from the industrial sector increased by 69 percent (2.2 percent per year). Over the last decades, the outlines of energy transition pathways

G. Guxho et al. (Eds.): ITC-ICEE 2023, LNMIE, pp. 234–243, 2024.
https://doi.org/10.1007/978-3-031-48933-4_22

have emerged in the buildings, power and transport sectors. These have been driven by technological breakthroughs and cost reduction. For industrial processes, such pathways are less well-defined. [1].

The biggest efforts for the energy transition on a global scale have been directed to the buildings, energy and transport sectors, as the sectors with the highest potential to achieve satisfactory results. While the industrial sector presents more difficulties for this issue because the energy transition in industry must be seen in the context of global trends related to the demand for goods and the preferred routes for their production.

1.1 EU Climate and Energy Framework

Albania, as a country of the European continent and of the EU membership processes, has oriented its political will and efforts for the climate in the spirit of commitments towards the European framework, which represents the highest legal and regulatory standard at the global level for the environment and the climate.

In June 2021, the EU adopted the European Climate Law, setting the target of achieving net zero GHG emissions (GHG) by 2050. The law sets an intermediate target to reduce GHGs at least 55% by 2030 compared to the 1990 level.

The package includes several changes to existing policies as well as new measures to reduce GHG emissions. Some of the proposals are:

- strengthening the emission reduction targets for each Member State,
- the carbon border mechanism, placing a carbon price on imports of iron and steel, cement, aluminium, waste, electricity, fertilizers, hydrogen. Importers of those goods to the European Union will be charged a carbon border tax from 2026 for the GHGs that were created during production,
- increasing the target for renewable energy production up to 40% by 2030;
- update of the energy efficiency targets for each Member State to 36–39% by 2030.

1.2 Legal Basis for Climate Change in Albania

Albanian Ministry of Infrastructure and Energy has drawn up the National Energy and Climate Plan for implementation of the National Energy Strategy (2018–2030) and Recommendation 2018/01/MC-EC of the Council of Ministers of the Energy Community.

Among other issues, this plan defines the national objective of reducing Greenhouse Gases (GHG) until 2030 in accordance with the National Determined Contributions (NDC).

This plan has defined national objectives for each of the five dimensions of Union of Energy as well as the relevant policies and measures to fulfill these objectives, where for our case we can highlight:

- Dimension 3: Energy efficiency, which addresses existing and planned policies and measures related to achieving future energy efficiency targets to reduce energy consumption in different sectors.
- Dimension 4: Decarbonization of the economy, aimed at developing policies and measures to reduce GHG emissions in all key economic sectors to meet NDC targets of the Paris Agreement.

- Dimension 5: Research, innovation, and competition envisages policies, measures and plans to accelerate the transformation of the energy sector, including funding programs for research and development.

Despite the efforts of the Albanian government to keep pace with the European Union on this issue, the climate legislation activity in Albania appears to be slow and incomplete: the climate acts constitute a very small part of the approved legislation in relation to the legal acts approved by the Assembly and the Government on a yearly basis. Expressed as a percentage, the share of climate legal acts in relation to the totality of laws and acts adopted goes from 0.001% in 2019 to 0.003% in 2021. [2].

Among the most important documents is the Climate Change Strategy and Action Plan (2019), which focuses on the following four reporting sectors of Albania's GHGs inventory: Energy (with a special focus on Transport), Agriculture and Change sectors of Land Use and Forestry, thus covering 80% of Albania's emissions.

Considering the proposed approach, the following strategies, plans and laws have been considered and analyzed:

(1) For the energy sector:
    Replacement of fossil fuels with non-hazardous waste in the Albanian cement industry (2015–2020).
(2) Financing the mechanism for energy efficiency in buildings (2015–2020)
(3) National Action Plan for Renewable Energy Sources (2018–2020)
(4) Law 124/2015 on Energy Efficiency
(5) Draft National Plan of the Republic of Albania for Energy and Climate (July 2021).

2 The Situation of Energy Supply in Albania

In 2018, Albania's total primary energy supply (TPES) reached 2,131 kilotons of oil equivalent (ktoe) (Fig. 1). [3]. Fossil-based fuels dominate (68%, consisting of natural gas, oil and lignite), followed by electricity (24%), which is provided by hydropower plants in the country and imports. Biomass and other energy sources (such as solar thermal energy) account for 8% of TPES.

3 Energy Consumption by Sectors in Albania

The largest energy consumer in Albania is the transport sector, whose share has almost quadrupled since 1990 and reached 40% of final energy consumption in 2018.

The residential sector was the second largest (24%, 490 ktoe). It is the largest electricity consuming sector, consuming 57% of total electricity consumption (excluding losses) in 2018 [4], followed by the industrial sector (20%, 418 ktoe) (Figs. 2, 3, and 4) [5].

4 Progress Made on the Climate Monitoring System in Albania (2021)

The progress made in terms of creating the GHG inventory is in the very early stages in Albania.

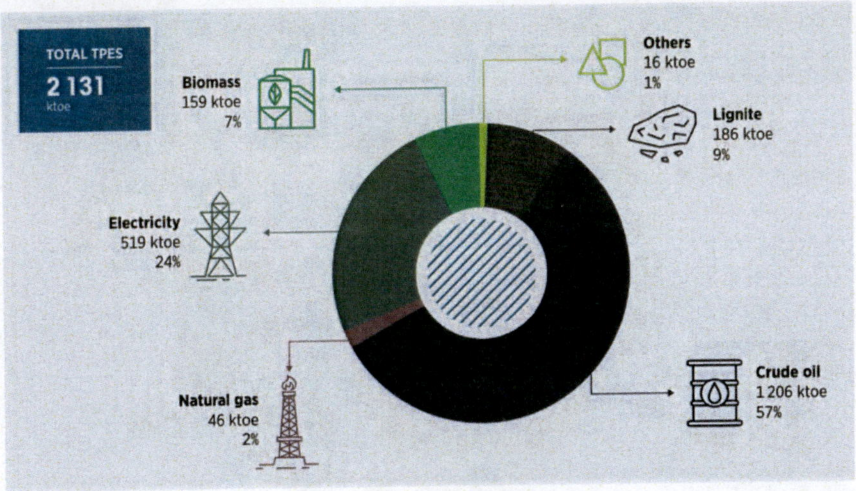

Fig. 1. Primary energy supply by sector in Albania in 2018 (ktoe). *Note* Total primary energy supply (TPES) is the sum of domestic production and energy imports minus exports and changes in its storage

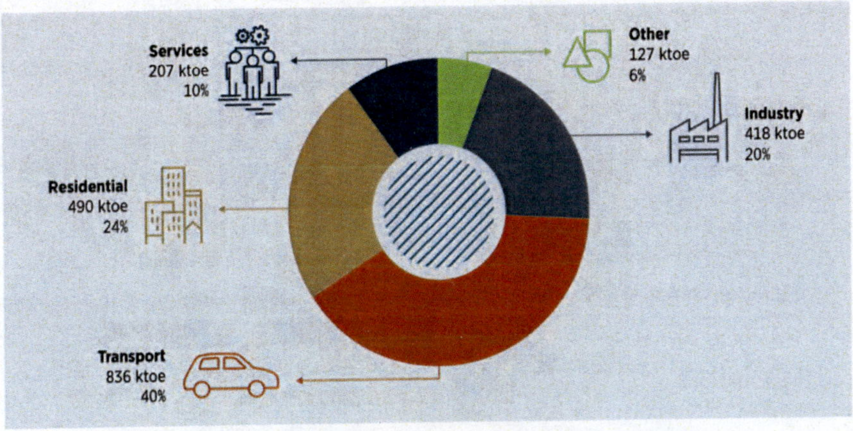

Fig. 2. Energy consumption by sector in Albania in 2019 [5, 6].

The Albania Climate Change Law specifies the country's obligation to establish a GHGs database, although its by-laws have not yet been adopted. Consequently, GHG emissions continue to be calculated indirectly. Meanwhile, there is some kind of legal and institutional progress regarding the creation of other pollution 'databases'.

The Albanian Register of Pollution Discharges and Transfers (ARPDT) is a public electronic database that helps public participation in decision-making related to the environment [7].

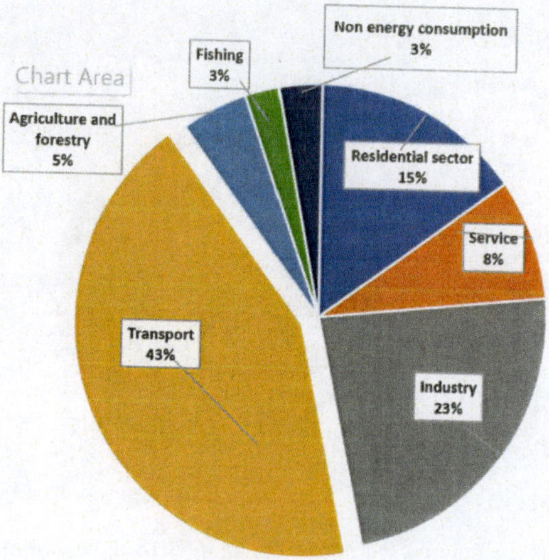

Fig. 3. The projected energy consumption by sectors in Albania in 2030 (in %) [3]

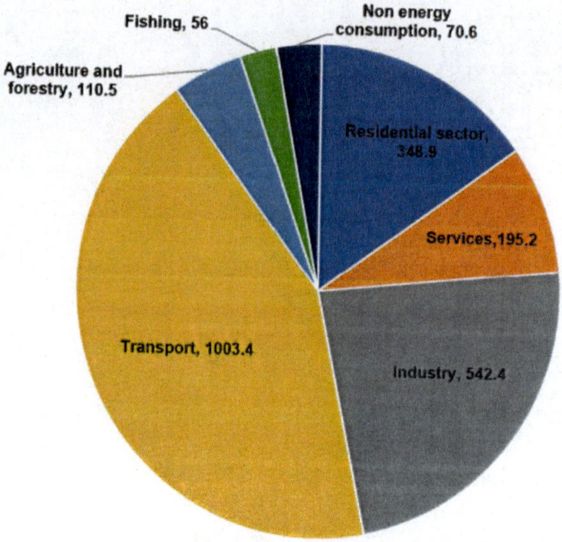

Fig. 4. The share of projected energy consumption by sector in Albania in 2030 (ktoe) (Total consumption 2327 ktoe) [3]

Specifically, ARPDT enables the public to have access to environmental information related to annual pollutant discharges into air, water, land, as well as pollutant transfers and urban wastewater treatment.

The current situation for GHG in Albania is:

- Historical and current levels of GHG emissions in Albania are low, globally and compared to other European countries.
  National GHG emissions of Albania accounted for only 0.02% of global emissions in 2016.
- While the average emissions were 8.7 tons of CO_2eq. Per capita in the 27 EU member states in 2018,
- The average emissions were 3.5 tons of CO_2eq. Per capita in Albania in 2016.

Albania's Intended National Determined Contribution (INDC) presented in September 2015 (and presented as NDC in September 2016), provides a first step towards a low carbon pathway, aiming to reach GHG emissions ap to 2 tons of CO_2eq per capita by 2050, compared to the EU's political objective for 2050 [2].

4.1 GHG Emissions from the Industrial Sector in Albania

Referring to the GHG emissions from the responsible sectors in Albania, it can be seen that:

- GHG emissions of industrial processes and product use (PIPP) reached 1356 kt of CO_2eq in 2009 and 1020 kt of CO_2eq in 2016 (-24.8%) or about 10% of the total. This decrease is mainly a result of technological improvements of the iron and steel production sector between 2009 (reduced iron production) and 2010 (use of electric arc furnaces) (Fig. 5).

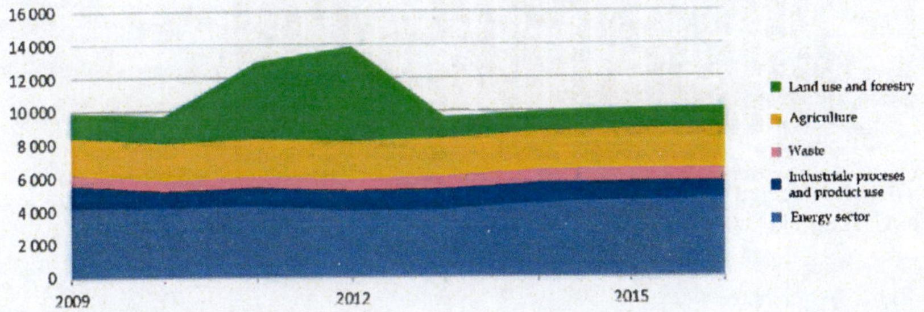

Fig. 5. Average annual emissions of GHGs in Albania (kt CO_2eq) by sector (2009–2016) [2]

Albania's National Strategy for Climate Change (2019), through mitigation measures, has the long-term objective of reducing GHGs emissions and increasing the capture of those emissions in the sectors involved, in accordance with the EU objective, to reduce CO_2 emissions by 80% to 95% in 2050, compared to 1990 emissions levels.

The strategic document covers four GHGs inventory reporting sectors:

(1) Energy with special focus on Transport
(2) Change of Land and Forest Use, Waste
(3) Buildings and Industry, which are responsible for 80% of CO_2 emissions in Albania.

a. The projected GHG emissions in Albania (Figs. 6 and 7).

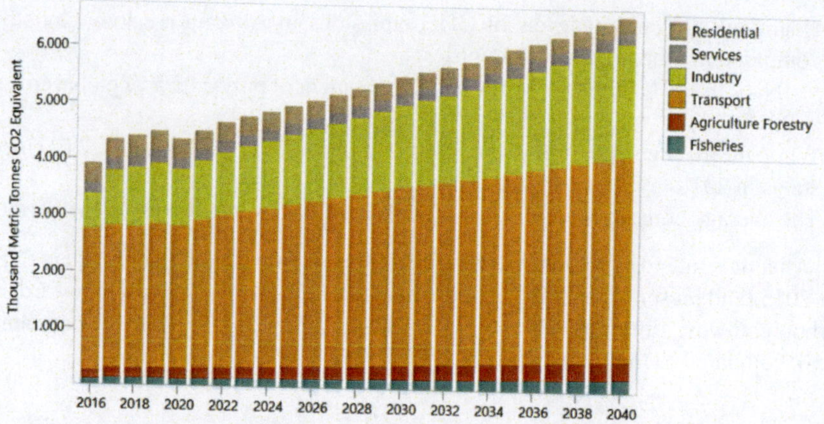

Fig. 6. The projected GHG emissions in Albania by sector (Thousand metric tonnes CO_2 eq.) [8]

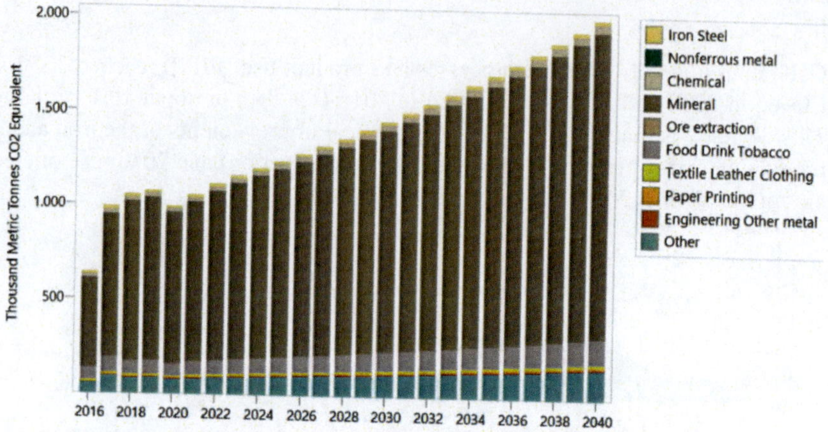

Fig. 7. The projected GHG emissions in Albanian industrial sector by sector (Thousand metric tonnes CO_2 eq.) [8]

Achieving the National Determined Contribution (NDC) objectives for reducing greenhouse gas (GHG) emissions is based on:

(1) Energy efficiency (EE): The EE target is set at 15% by 2030.
(2) Renewable energy sources (RE): The RES target is set at 42.5% by 2030.
(3) Promotion of natural gas: The objective of natural gas penetration is set at 8–10% by 2030. Natural gas in Albania will be used mainly to guarantee the security of electricity supply, since the electricity sector in Albania is entirely based on hydro resources and the import of electricity.
(4) Combined scenario using all options above.

5 Policies and Measures in the Industry Sector to Reduce GHG Emissions in Albania

Among the main measures to reduce GHG emissions in industry sector, the Albanian government has determined:

1. Implementation of the European Trade System of GHGs (2022 – 2030), through which will be achieved:
    - the reduction of GHGs emissions from the industry sector, as well as the goal of decarbonizing the economy.
    - Targeted 50% reduction of GHG emissions from the industry sector by 2030.
    - The implementation of carbon trading system at the national level of free quotas by 2025, with the participation of the cement, fertilizer, iron and steel industry.
    - The implementation of carbon trading system at the national level of free quotas by 2025, with the participation of the cement, fertilizer, iron and steel industry.
    - Setting up an appropriate monitoring, reporting and verification system for emissions monitoring, as provided in the recently approved "Climate Change" law (2020 and ongoing).
    - Efficiency improvement of the use of different energy sources.
    - Penetration and use of natural gas as fuel.
    - Increasing the use of Renewable Energy Sources.
    - Use of waste for energy purposes.
    - The possibility of reusing and recycling waste within the unit and disposal of waste (solid, liquid and gaseous) in accordance with the requirements for environmental protection.
    - Promotion of combined heat and power cogeneration in industries such as cement, metallurgy, etc.
2. Establishing and implementation of the regulation on mechanisms of monitoring.

The results to be achieved in the industry sector:

- The use of the most efficient boilers will lead to 225 kilotons of CO_2 reduction (or 32% reduction of the country's total GHGs emissions of 708 kilotons) which will be achieved by 2030;
- The change of fuels used will lead to 52 kilotons of CO_2 reduction (or 7% reduction of the country's total GHG emissions of 708 kilotons) which will be achieved by 2030.

In Albania, almost 90% of GHG emissions from industrial processes are related to cement production.

Emissions for the base scenario for this sector are expected to increase from 1020 kilotons of CO_2e by 2016 to 1854 kt CO_2e by 2030, which represents an increase of + 81.9% (Fig. 8).

The assessment of the European Commission on the issue of climate change for Albania is (2021):

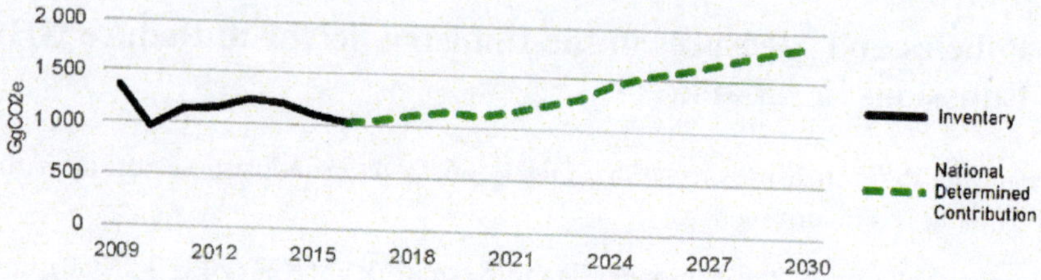

Fig. 8. Projections for industrial processes and product use sector in Albania [2]

Albania has achieved some progress in this field. There is little progress in terms of further alignment of EU framework policies and legislation, but further implementation and enforcement efforts are needed, particularly in relation to water management, air quality and climate change.

6 Conclusions

Conducting studies on issues such as the decarbonization of the economy first requires a complete database of systematic data to perform accurate and complete analysis and forecasts for the future. Currently in Albania, there is a lack of complete and accurate data, especially for the industrial sector, despite the improvements made in recent years.

Among the sectors responsible for greenhouse gas emissions in Albania and globally, it seems that very good results have been achieved in the residential, transport, services, energy sectors, but the industrial sector presents more difficulties, something for which it is also considered as "the next front" globally.

The industry sector is considered in the EU as the most expensive sector to decarbonize, because the industry would need a technology that is still under development. Even with the achievement of the net zero target by 2050 this sector will again continue to generate some emissions.

Among the ways to decarbonize the industrial sector in our country is the implementation of the emissions trading system. The EU ETS refers to CO_2 emissions from power plants, energy-intensive industries (e.g. oil refineries, steel mills and iron, aluminium, cement, paper and glass producers, etc.).

The EU ETS, a cap-and-trade system for GHGs from energy and industry, was implemented to ensure that the EU would meet its GHG emission targets in a cost-effective manner. The system offers flexibility to businesses covered by the scheme as they have the choice between reducing emissions and buying emissions from other companies depending on the carbon price.

But the unsatisfactory results in the past years in EU have led to an increase of the carbon price. Consequently, energy production from burning coal and other fossil fuels will become more expensive and clean energy sources will become more attractive.

At the same time, the companies are incentivized to become more energy efficient because they can then sell their emission allowances on the market.

The ways to decarbonize the industry in Albania as a developing country are:

(1) Reducing the demand for carbon-intensive products,

(2) Improving energy efficiency and energy management.
(3) Heat production with low and medium temperature using solar and geothermal energy, etc.
(4) Carbon capture, combined with use or storage.

References

1. https://www.mckinsey.com/~/media/mckinsey/business%20functions/sustainability/our%20insights/how%20industry%20can%20move%20toward%20a%20low%20carbon%20future/decarbonization-of-industrial-sectors-the-next-frontier.pdf. Last accessed 2023/02/10
2. Climate change strategy in Albania 2020–2030 2019–2021 Action Plan Monitoring, https://www.wfd.org/sites/default/files/2022-05/EN_Report_Monitoring%20of%20National%20Action%20Plan%20%281%29.pdf, Last accessed 2022/10/12
3. https://www.irena.org/publications/2021/March/Renewables-Readiness-Assessment-Albania, Last accessed 2022/08/27
4. Ministry of Infrastructure and Energy: https://www.infrastruktura.gov.al/, Last accessed 2022/09/23
5. https://ec.europa.eu/eurostat EUROSTAT, 2019a, Last accessed 2022/09/08
6. https://www.instat.gov.al/ INSTAT, 2020a, Last accessed 2022/05.28
7. Strategjia kombëtare e energjisë 2018–2030 Mars 2018, https://www.kesh.al/wp-content/uploads/2022/02/6-STRATEGJIA-KOMBETARE-E-ENERGJISE-PER-PERIUDHEN-2018-2030.pdf, Last accessed 2022/09/23
8. Londo, A.: Krijmi i P&M qe do te rekomandohen ne me synim dekarbonizimin e ekonomise ne Shqiperi per arritjen e objektivave te vitit 2030", Workshop, Universiteti Politeknik i Tiranes, 2023/03/16

Analysis of Energy Audit in the Architectural Design Office Located in Tirana

Klodian Dhoska[1,2,3](✉), Elena Bebi[2,4], Irida Markja[1,2], and Giorgio Mustafaraj[1]

[1] Department of Production and Management, Faculty of Mechanical Engineering, Polytechnic University of Tirana, Mother Theresa Square No.1, 1019 Tirana, Albania
kdhoska@fim.edu.al

[2] Albanian Mechanical Engineering Association (AMEA), Mother Theresa Square No.1, 38954 Tirana, Albania

[3] Association of Talent Under Liberty in Technology (TULTECH), Sopruse Pst 3, 10615 Tallinn, Estonia

[4] Energy Department, Faculty of Mechanical Engineering, Polytechnic University of Tirana, Mother Theresa Square No.1, 1019 Tirana, Albania

Abstract. Nowadays one of the important tasks in Albania is focusing on performing energy audit in residential buildings in Tirana. This can help to identify energy-saving opportunities and improve overall energy efficiency in residential buildings. Based on it, we have selected an architectural design office which is located close to the lake of Tirana. It is a new building and is constructed in 2009 where the Albanian law doesn't exist for energy efficiency during this time. In this paper it will be analyzed all the collected data and identify areas where energy-saving measures can be implemented. Afterward the recommendations are based on potential energy savings, cost-effectiveness, and the building owner's goals which includes improving insulation, upgrading windows, optimizing Heating Ventilation and Air Conditioning (HVAC) systems, adopting energy-efficient lighting, and implementing smart home technologies.

Keywords: Energy audit · Residential building · HVAC · Cost-effectiveness · Energy efficiency

G. Guxho et al. (Eds.): ITC-ICEE 2023, LNMIE, pp. 244–252, 2024.
https://doi.org/10.1007/978-3-031-48933-4_23

1 Introduction

During the last decade, most of the Western Balkans countries have been focused on the improvement of the energy efficiency in residential buildings [1–3]. This is a crucial step towards reducing energy consumption and lowering greenhouse gas emissions. The whole countries have used some potential strategies and initiatives that could be implemented to enhance energy efficiency in residential buildings [4–6]. Some important strategies and initiatives are retrofitting existing buildings, energy audits, building codes and standards, energy-efficient appliances, smart technology, renewable energy integration, green building certification, financial incentives, monitoring, reporting, public awareness and education by including training [5–7].

Nowadays the crucial task in Albania is focusing on performing energy audit in residential buildings in Tirana. This can help to identify energy-saving opportunities and improve overall energy efficiency in residential buildings. Based on it, we have selected an architectural design office which is located close to the lake of Tirana, see Fig. 1. It is a new building and is constructed in 2009 where the Albanian law doesn't exist for energy efficiency during this time. The building has a longitudinal extension in the east-west direction, and there are no natural or artificial obstacles for sunlight or wind.

Fig. 1. Building of architectural design office, located close to the lake of Tirana.

In this paper it will be analyzed all the collected data and identify areas where energy-saving measures can be implemented. Afterward the recommendations are based on potential energy savings, cost-effectiveness, and the building owner's goals which includes improving insulation, upgrading windows, optimizing HVAC systems, adopting energy-efficient lighting, and implementing smart home technologies.

2 Materials and Methods

The architectural composition of the building's facade is based on the use of large windows and the minimization of the surfaces of the external walls. At the same time, with the aim of reducing the amount of solar energy transmitted by radiation and thermal conductivity, the facade of the building has been treated with decorative low vertical concrete elements, in a wavy form, which enable the reduction of part of the radiation in the interiors at the offices. Also, the positioning of the offices on the second floor of the building and in the eastern part of the building, that is, with a view of the lake park, makes it possible to have good natural shading from the crown of trees.

Its supporting structure is made with the frame (beam-column) system, with reinforced concrete material. The soles are of the beam type with a light filling of ceramic bricks with holes. The external walls are made of bricks with holes, 20 cm thick, while in some places, the perimeter coincides with reinforced concrete supporting columns, 80 × 80 cm thick. The external windows are made of plastic material, with double transparent glass and insulating filling with gas. The cracks in it (windows and balcony doors) are closed with elements made of boxes and plastic elements with a thickness of 76 mm and 2 sheets of transparent, simple glass, each with a thickness of 4 mm. Figure 2 depict the current floor plan of the office space which is in the eastern wing of the building.

Tables 1, 2, and 3 shows respectively the space data of the design offices, office data and office devices by including for both of them the usage profiles.

Over the last 3 years 2020, 2021, 2022, the operation of all office equipment has been enabled with energy supplied from the power line. The heating and cooling of the interior is currently done through a central air conditioning system, while a small boiler with a capacity of 10 L is used for hot water, located under the kitchen sink. The work process is based on a system of 13 computers, which operate around the clock. After the end of the schedule, all computers, air conditioning and lighting system are turned off. Only the 10-L boiler and the refrigerator remain in working condition.

3 Analysis of Energy Audit

Analysis of the energy audit will be start by evaluation of the energy consumption during the last 3 years that are coming from heating/cooling system area, electronics devices and sanitary water. All the data are coming from electrical meter reading and electricity invoices as can be seen in Figs. 3, 4, and 5.

In the above data we have taken into account some specific key points which are as follows:

- Bowl with a capacity of 10 L, positioned below the kitchen sink unit works with a very reduced capacity, because the hot water in the office environment is used very rarely. For this reason, the working time for this device is calculated as minimal, only one hour a day (24 h).
- The number of computers working in the office has gradually increased throughout the three years analyzed in this report. Specifically, in 2020, 10 computers were in operation, in 2021 there were 11 computers, while in 2022 two more computers were added, so in total their number is currently 13.

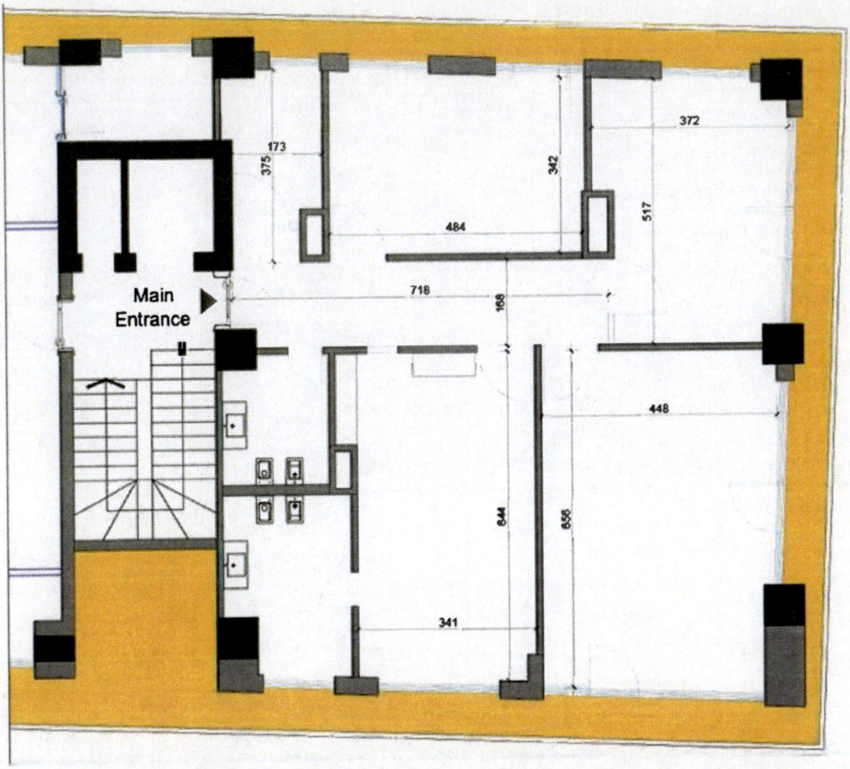

Fig. 2. The current floor plan of the office space.

Table 1. Office design data and usage profile.

Year	2009		
Area	152 m^2		
Usage profiles Standard DIN 18599	Usage type	Monday–Friday	Saturday
	Office use time and professional equipment	09:00–18:00	08:00–13:00
	Operation time of the ventilation system	09:00–18:00	08:00–13:00
	Operating time of the heating/cooling system	09:00–18:00	08:00–13:00

- The television was added in 2022, so in 2020 and 2021 it is not included in the account. Meanwhile, the use of television is very limited, and is calculated at about one hour per day per working day.

Table 2. Office data and usage profile.

Corresponding structure	Area (gross) (m^2)	Area (net) (m^2)
Eastern outer wall	40.3	19.8
East facing windows		20.8
The southern outer wall	44.8	24.3
South facing window		20.5
The northern outer wall	39.2	26.3
North facing window		12.9

Table 3. Office devices and usage profile.

Electrical and electronics devices	Power (kW)	Usage
Kitchen boiler (10 L)	1.5	1 h/day/1 year
Desktop computers (13 pieces)	2.6	9 h/5 days/1 year
TV LCD	0.071	4 h/5 days/1 year
Fridge	0.25	24 h/1 year
LED lamps	0.3	9 h/5 days/1 year
HVAC	3.79	4 h/5 days/1 year

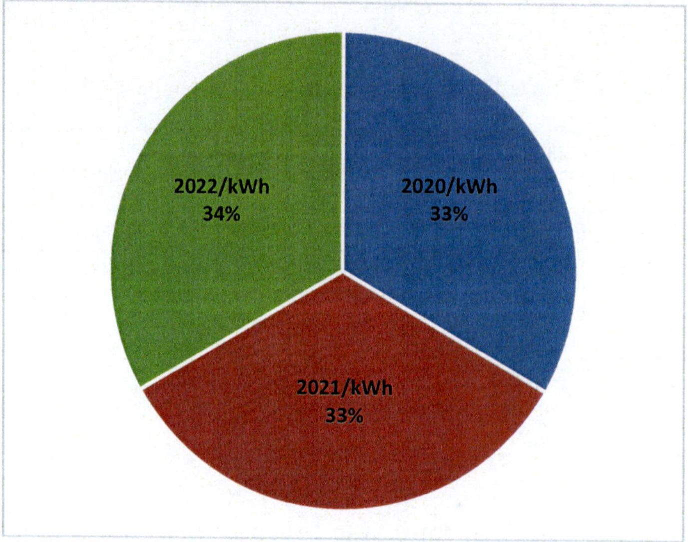

Fig. 3. Electricity invoices data, energy consumption in kWh for the last 3 years 2020–2022.

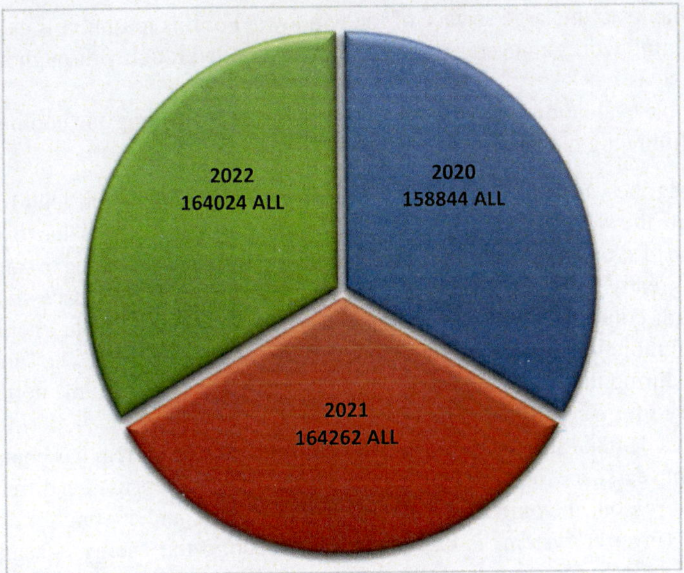

Fig. 4. Electricity invoices data, energy consumption in ALL for the last 3 years 2020–2022.

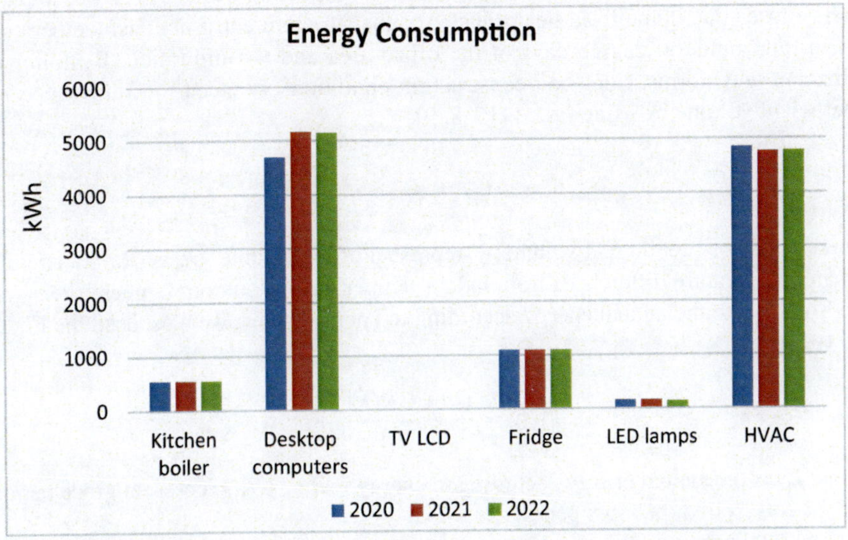

Fig. 5. Overall energy consumption in kWh for the last 3 years 2020–2022.

- The refrigerator works with limited capacity, due to the type of office function. In total, it was found that the refrigerator works with 50% of the capacity, that is, about 12 h a day. For the rest of the time, the refrigerator is on stand-by.
- The consumption of lighting in the office is very limited, since the working hours in the office are during the day, while the office spaces are very well lit with natural

light thanks to the large surface of the windows. For this reason, it is calculated that the lighting works an average of 2 h a day, distributed roughly throughout the year.

In case of the evaluation of the amount of energy used for heating and cooling is difficult to do by following reasons:

- By using the variable space inverter system in the office, it is difficult to accurately calculate the work yield and the energy consumed on a daily basis due to the different position of some office premises, the demand for heating and cooling in some premises varies greatly from one to another. In the winter months, the spaces positioned to the south consume very little energy for heating, due to the solar energy obtained through radiation and the thermal conductivity of the windows. The same situation occurs during the summer period, where rooms positioned from the north have very little need for a cooling system.
- Also, the demand for energy for the heating and cooling system has changed during different years, due to the change in the number of computers used in different offices.
- For this reason, the amount of energy for the heating and cooling air conditioning system (inverter system) is calculated by deducting the energy consumed by the internal office equipment from the total energy consumed, recorded in the relevant invoices.

Furthermore, evaluation of the energy demand for heating we need to take account the energy loses that come from heat transfer of the structure constructions, temperatures indoor and outdoor, degree days of the office area and air infiltration. By following Albanian law, norms, rules and construction conditions we need to evaluate the loss coefficient volume by using Eq. (1) [5, 8, 9].

$$Gv = \frac{Qo}{V * \Delta t} \tag{1}$$

where Gv is loss coefficient volume, V represent office volume, Qo is heat transfer and Δt are temperature differences from indoor temperature to outdoor temperature.

To estimate the annual energy according to energy losses, we have used the Eq. (2) [5, 10].

$$Ey = \frac{Qo * GRD * 24}{\Delta t} \tag{2}$$

where Ey is the annual energy according to energy losses, GRD represent grade per day and 24 correspond on hours per day.

The total consumed electrical energy is calculated with the Eq. (3).

$$E = \frac{Ey}{COP} \tag{3}$$

where E is the total consumed electrical energy and COP represent coefficient performance which correspond to the value 3.8.

Based on the Eq. (1) we will depict in the Fig. 6 the results of the energy loss of office design in current situation. Furthermore, by using Eqs. (2) and (3) we have found that energy loss corresponds to the value 5990.3 W or 5.9903 kW and annual energy needs correspond to the value 10501.85 kWh/year. Most of the loses are coming from outdoor wall and windows.

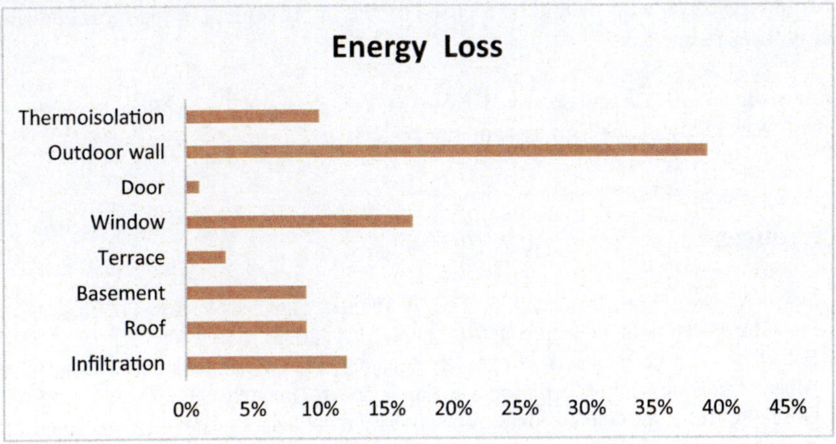

Fig. 6. Energy loss in office design located in Tirana, %.

Based on the energy consumption data, according to the relevant invoices for the three years, the office environment is classified in Class B energy, with an average value of 31.5 kWh/m^2, see Fig. 7 [11].

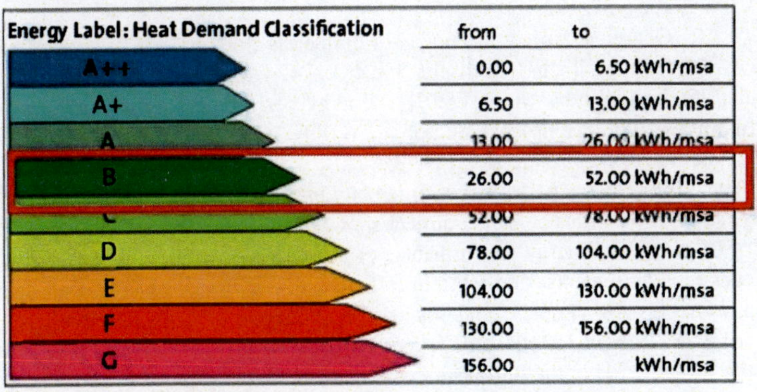

Fig. 7. Energy label: heat demand classification, class B energy.

4 Conclusion

In this paper we have analyzed an energy audit of the architectural design office that is located in Tirana. Based on our research results, it is noted that the main energy consumption in the office environment analyzed in the report is caused by the computer units as well as the heating/cooling air conditioning system with inverter. It has been investigated that most of the loses are coming from outdoor wall and windows. The results have shown that energy performance of this design office correspond to the B category which fulfil the requirements of the Albanian and European law. Furthermore, our future research work would be focused on the improvement of the most contributed energy loses parameters.

Acknowledgement. The authors would like to thank architectural design office "Archispace" for all the data and Polytechnic University of Tirana.

References

1. Dorri, A., Dhoska, K., Kodhelaj, S.: Energy efficiency analysis in Korça regional hospital. Int. J. Innov. Tech. Interdiscip. Sci. **6**(1), 1121–1129 (2023)
2. Baballëku, M., Verzivolli, A., Luka, R., Zgjanolli, R.: Fundamental basic wind speed in Albania: an adoption in accordance with eurocodes. J. Trans. Syst. Eng. **1**(2), 56–72 (2023)
3. Qosja, S., Rolle, R., Gebremedhin, A.: Solving the bottleneck issue of energy supply. Case study of a wind power plant. Int. J. Innov. Techn. Interdiscip. Sci. **5**(2), 874–891 (2022)
4. Abou EL-Hassan, A., Khairat Dawood, M.M., EL Morse, M.S., Abdel-Mageed, S.I., Sharaf, O.: The effective techniques for enhancing the turbulent flow between two parallel plates: a comprehensive review. J. Trans. Syst. Eng. **1**(1), 10–30 (2023)
5. Bebi, E., Dhoska, K., Alcani, M., Sakaj, E.,Qosja, S. Energy audit analysis in private residential apartment in Tirana. Int. Scientific J. "Industry 4" **7**(2), 79–82 (2022)
6. Spaho, E., Dhoska, K., Bylykbashi, K., Barolli, L., Koliçi, V., Takizawa, M.: Performance evaluation of energy consumption for different DTN routing protocols. Lecture Notes Data Eng. Comm. Tech. **22**, 122–131 (2019)
7. Aubell, J.P., Gebremedhin, A.: Framed- or frameless photovoltaic in snow experiencing climates. Int. J. Innov. Tech. Interdiscip. Sci. **4**(3), 742–753 (2021)
8. Albanian Legislative Law No. 116/2016: "Energy performance of the buildings", https://planifikimi.gov.al/index.php?eID=dumpFile&t=f&f=5109&token=f734e3f54cf4535ab6794ce8e5e790eb977f0113, last accessed 2023/06/17
9. EN 15603: 2008: Energy performance of the buildings—overall energy use and definition of energy ratings, https://dps.gov.al/en/project/show/dps:proj:35337, last accessed 2023/06/13
10. EN 15316-1: 2017: Energy performance of buildings—method for calculation of system energy requirements and system efficiencies—part 1: general and energy performance expression, https://standards.iteh.ai/catalog/standards/cen/1cd60602-fb14-4604-a0c3-ce15975a2e1b/en-15316-1-2017, last accessed 2023/05/22
11. Directive 2010/31/EU of the European Parliament and of the Council of 19 May 2010 on the energy performance of buildings. https://eur-lex.europa.eu/LexUriServ/LexUriServ.do?uri=OJ:L:2010:153:0013:0035:en:PDF, last accessed 2023/05/22

Energy Performance in the Industry Sector. Case Study of a Cast Iron Facility

Edmond Zeneli(✉), Albert Shira, Amanda Shameti, Flamur Bidaj, and Artan Hoxha

Department of Energy, Mechanical Engineering Faculty, Polytechnic University of Tirana, Tirana, Albania
edmondzeneli@fim.eu.al

Abstract. The industrial sector plays a crucial role in energy consumption and greenhouse gas emissions. In this study, we investigate the energy performance of the cast iron industry, focusing on a comprehensive case study. The objective is to identify energy efficiency opportunities and propose strategies for optimizing energy consumption in this specific sector. Through extensive data collection and analysis, including energy audits and process evaluations, we assess the current energy usage patterns and identify areas of inefficiency within the cast iron industry. Key parameters such as energy intensity, specific energy consumption, and energy utilization efficiency are evaluated to provide a comprehensive overview. Our findings reveal significant energy conservation potential in various stages of the cast iron production process. We identify specific areas for improvement, such as the optimization of heating systems, utilization of waste heat recovery technologies, and implementation of energy management systems. Furthermore, we explore the economic feasibility and potential environmental benefits associated with these energy efficiency measures. The outcomes of this study provide valuable insights for policymakers, industry professionals, and researchers aiming to enhance energy efficiency in the industrial sector. The findings contribute to the development of targeted energy management strategies and facilitate the transition towards a more sustainable and energy-conscious cast iron industry.

Keywords: Industry · Photovoltaic system · Energy performance · Cost

1 Introduction

This study focuses on the utilization of renewable energy sources and in particular on the application of photovoltaic systems in industry. The production of electricity from renewable sources is one of the commitments of our country to meet the objectives of the year 2030, i.e. the target of 42% of primary energy sources from the renewable part [1]. In this context, due to the favorable geographical position in which our country is located, the development of photovoltaic (PV) systems has taken on a special importance. In this prism, the significant reduction in the cost of these systems has had a positive effect, making them very attractive for various businesses in our country. Thus, one of their developers is the industry sector where the demand for electricity is constantly high.

G. Guxho et al. (Eds.): ITC-ICEE 2023, LNMIE, pp. 253–263, 2024.
https://doi.org/10.1007/978-3-031-48933-4_24

For this reason, in this thesis we have considered a foundry for melting mainly cast iron but also other products. In this context, the application of the photovoltaic system in industry has been treated as a case study. Using solar energy as a source, PV panels to generate electricity which will serve to supply the furnace for melting cast iron or for other consumption, in the industrial sector. The paper is equipped with an energy analysis of the PV plant as well as a financial analysis has also been developed to evaluate the feasibility of such projects in the industry. This study examines the implementation of a photovoltaic (PV) system in a foundry, primarily aimed at electricity production. Harnessing electricity from PV sources would lead to a substantial reduction in electricity expenses, which currently constitute the primary expenditure. This study originates from fieldwork conducted in individual foundry, during which the authors conducted many visits and gathered authentic data on both production output and energy consumption across various branches of the facility. The primary focus of our analysis will be the direct generation of electricity for internal consumption, with a particular emphasis on reduction the energy cost of the foundry. Additionally, we will provide an energy and economic assessment, taking into consideration the presence of PV plant in two of the foundry building roofs.

2 Electricity Situation in Albania

2.1 Electricity Consumption

During 2022, the available electricity decreased by 5.8%. The net domestic production of electricity reached the value of 7,003 GWh from 8,963 GWh of energy produced in 2021, marking a decrease in production by 21.9% [2]. This production was realized by public hydropower plants to the extent of 55.1%, and by private and concessionary hydropower plants to the extent of 44.2% and by other producers (Photovoltaic) to the extent of 0.7% of the net domestic production of electricity. The gross import of electricity (energy in receipt) reached the value of 3,044 GWh from 2,253 GWh that was a year ago, marking an increase of 35.1%. The gross export of electricity (energy in delivery) reached the value of 2,123 GWh from 2,800 GWh, marking a decrease of 24.2%, according to Instat data. In accordance with the categorization, the total energy consumption encompasses the utilization by residential consumers, commercial consumers, and the sum of losses within the network. In the year 2022, residential consumers accounted for a consumption of 3,081 GWh, marking a 4.2% decline in comparison to the preceding year. In contrast, commercial consumers recorded a consumption of 3,548 GWh, representing an increase of 18.2% compared to the year 2021. When compared to the region, Albania stands out as the nation with the lowest per capita electricity consumption, estimated to be approximately 2.68 MWh in the year 2022. In contrast, Serbia leads the region with the highest per capita electricity consumption, amounting to roughly 5.38 MWh. Additionally, other countries in the region also have notable per capita consumption figures, including Kosovo at 4.76 MWh per capita, Montenegro at 4.6 MWh per capita, and North Macedonia at 3.69 MWh per capita [3].

In Figs. 1 and 2 are shown the electricity consumption share by sectors for Albanian and region countries. It is evident that the industrial sector consumption comprises 12%

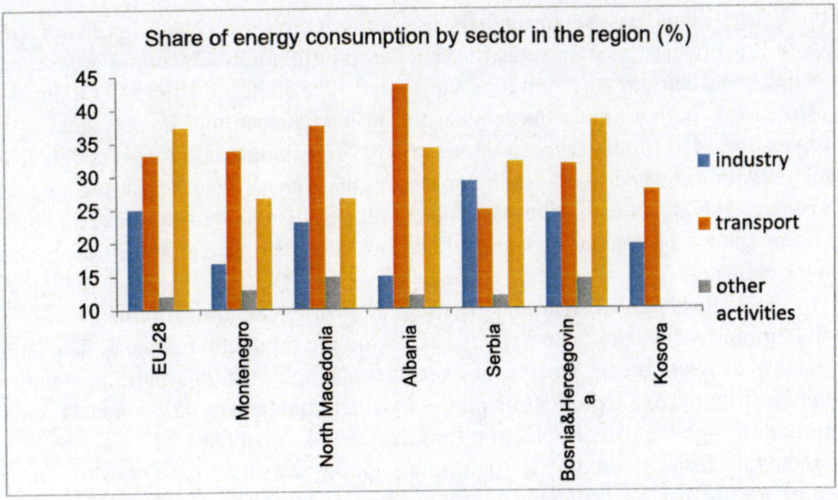

Fig. 1. Energy consumption distribution by sectors in the region countries

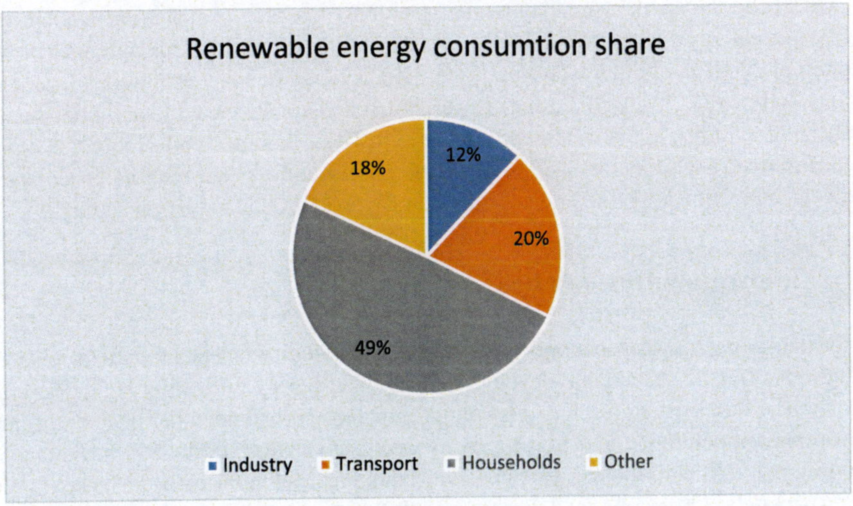

Fig. 2. Share of renewables contribution by sectors in Albania

of the total renewable energy consumption. A higher penetration of renewables it is expected considering the Albanian conditions.

2.2 Electricity Consumption in Industry

Industrial processes are known for their high energy intensity, constituting approximately one-third of the world's total energy consumption [4]. Alarmingly, approximately 70%

of this energy demand is met through the use of fossil fuels, which are significant contributors to greenhouse gas emissions. In fact, carbon dioxide (CO_2) emissions stemming from industrial activities account for a substantial 40% of the total global CO_2 emissions.

Numerous authors have undertaken the task of presenting an overview of technologies designed to mitigate emissions arising from industrial processes. Their work involves gathering and synthesizing information from diverse sources [5–8]. These efforts are crucial in identifying and implementing sustainable solutions to reduce the environmental impact of industrial activities, thus contributing to global efforts to combat climate change.

In a previously published article [9], the authors provided initial findings from energy audits, primarily focusing on energy usage within the production process. The analysis revealed that a significant portion of energy consumption in this particular industrial sector is attributed to the production process, with particular emphasis on metal melting, achieved through the use of induction furnaces and blast cupolas.

Induction furnaces emerge as the primary energy consumers, responsible for up to 80% of the factory's total energy consumption, which is predominantly in the form of electricity. In contrast, mold making accounts for approximately 10% of energy consumption, involving a mix of electricity and/or coal [10].

Blast cupolas are the second most energy-intensive aspect, contributing to about 50% of the total energy demand. Following closely are mold and core making processes, which represent 20–30% of energy consumption, with roughly equal contributions from electricity and/or coal. Minor fractions of energy demand are attributed to activities such as deburring, painting, sand treatment, and others. These findings underscore the significance of energy-efficient measures, particularly within the metal melting processes, to reduce energy consumption and environmental impact in this industrial setting.

3 Photovoltaic Development in Albania

Albania's geographical location, characterized by a Mediterranean climate, offers ideal conditions for the sustainable development of solar energy utilization [11]. The combination of factors such as high solar radiation intensity, extended periods of sunshine, favorable temperatures, and low humidity levels all contribute to this advantageous environment. Albania possesses substantial untapped potential for solar energy. With approximately 220 sunny days or about 2700 hours of sunshine per year, the country is well-suited for solar energy generation [12]. As of the latest available data, Albania has a total installed photovoltaic (PV) capacity of 23 MW, with 10 MW of solar PV connected to the grid by the end of 2022. According to ERE the total electricity generation by PV power plants at the end of 2022 was 50,092 MWh. Notably, there have been around 88 applications submitted for the construction of photovoltaic plants, contributing to the current PV plant capacity of approximately 23 MW. The Albanian government has implemented auctions to promote solar photovoltaic investments in the country. For instance, a 140 MW solar PV project in Karavasta has been contracted, with 70 MW supported through a Power Purchase Agreement (PPA) at a rate of 24.89 Euro/MWh, while the remaining capacity will be sold at market prices. Another auction was launched for the Durrës PV power plant, setting a ceiling price of 55 Euro/MWh. To maximize the generation and efficiency of photovoltaic modules, it is crucial to position the solar panels in a way that allows for optimal absorption of solar radiation throughout the day, from east to noon to west. Various factors and variables come into play when determining the ideal orientation for PV modules and multiple techniques are available in the market today to achieve this goal [13].

4 Case Study

The foundry is placed near Kashar, in Tirana. It includes a plant for smelting cast iron and other metals. This plant consists of an induction furnace with a capacity of 250 kW, a main shredder and other secondary components for the daily operation of the plant. Some important data about this plant are presented in the Fig. 1 and Table 1. Triggered by the high energy consumption in the foundry, namely in the melting furnace but also in other links of the working process that require electricity for their operation, the entity thought to increase the contribution of renewable energy, turning to self-producing energy (Fig. 3).

Fig. 3. View of the cast iron foundry a) furnace b) shredder

The electrical energy needed to heat one ton of iron to 1500 °C is about 396 kWh. Numerous losses occur in the oven, which increases the specific energy consumption to over 500 kWh. The losses are: thermal furnace losses, furnace coil losses, capacitor bank losses, converter losses and main side transformer losses. In a typical induction furnace, the energy loss in the equipment is between 100 and 130 kWh per ton. The efficiency of the furnace is about 65–75%. With new development in energy efficient coils, new refractory material, reduction of converter losses and reduction of state transformer losses Energy loss furnace equipment is reduced to 60–90 kWh per ton. New furnaces are 81 to 87% efficient. In this case study it is obvious that energy losses in the furnace indicate high values. However the company at this situation has no decision to replace the old furnace with a new one. Thus, in order to reduce high energy demand of the foundry a PV On-grid system is recommended (Table 2).

In Fig. 4 is given the installation details of PV modules on the top of the foundry. The electrical energy produced by the photovoltaic panels installed in this plant flows through electrical conductors to the vicinity of the electric meter. The inverter is located near the

Table 1. Main data of foundry processing plant

Type of furnace	Inducal 250 kW
Capacity	550 kg
Material	Steel and cast iron
Operation hours	8 h/day

Table 2. The proposed data of PV system

Location	Kashar, Tirana
Output generator power	135 kWp
PV system	On-Grid
PV area	680 m^2
PV module number	360
Inverters Fronius Symo 4.5–3-S	5
PV type	Axitec
Specific cost of PV	700 Euro/kWp
Total cost of investment	91,000 Euro

Fig. 4. Top view of the installed PV modules on the roof of facility

electric meter, which serves to convert the DC current into AC. Then, the electricity fed into this electricity meter can be used similarly to the electricity supplied by the public power distribution network. The Fronius Symo is a three-phase inverter that operates in energy ranging from 3 to 20 kW, so it is suitable for the solar parking project, which generates electricity of about 4.86 kW.

5 Results

5.1 Energy Analysis

The implementation of photovoltaic plants is suitable to be integrated with the distribution network and to reduce the expenses for electricity by facility. Operating costs and maintenance cost were estimated about 200 Euro/year. The PV system will increase the security of electricity supply and the diversification of energy sources, they also provide opportunities for energy saving. The terrace of the building is currently unused and with good characteristics for the installation of photovoltaic panels as shown in Fig. 4. The facility at current time uses electricity as the main energy source and besides it imported coal. For the calculation and extraction of the results of this study, the PV*Sol trial version software was used [14]. PV Sol is a dynamic software with 3D visualization and detailed analysis of PV systems combined with batteries, various devices or on-grid, as in our case. The aim of the calculations was to estimate the electricity generation during months and to conduct a financial analysis of the PV plant.

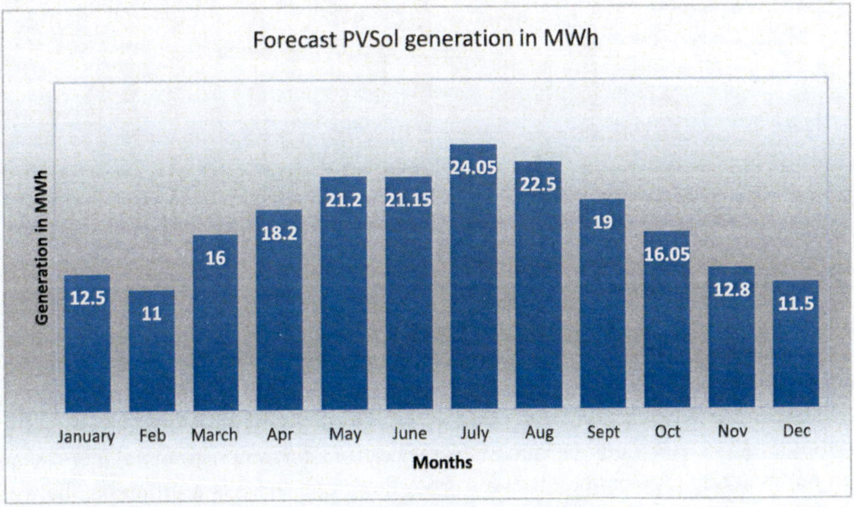

Fig. 5. Electricity generation forecast according to PV*Sol software

As can be seen from Fig. 5 the electricity generation extracted from PV*Sol calculations amount to approximately 205.95 MWh/year.

In Fig. 6 are shown the electricity generation by the PV system for the year 2022, since the operation period started in early May 2022.

In Fig. 7 is shown respectively the electricity generation for period 2023, self-consumption of the plant and electricity transmitted to the national grid. For the period

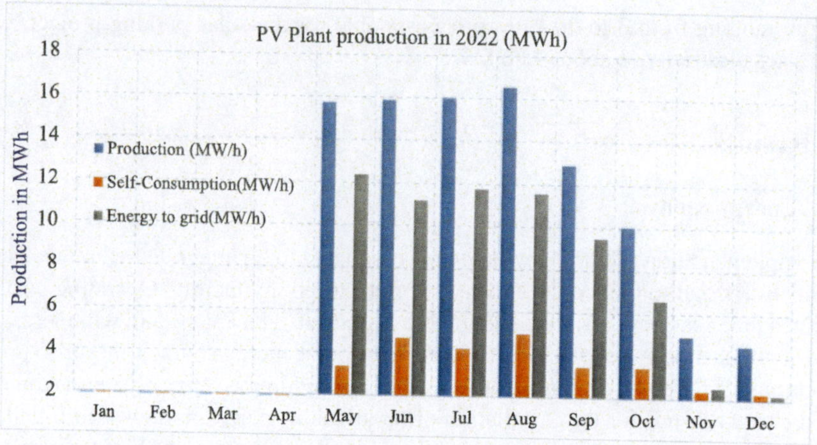

Fig. 6. Electricity production for the period May–December 2022 in MWh

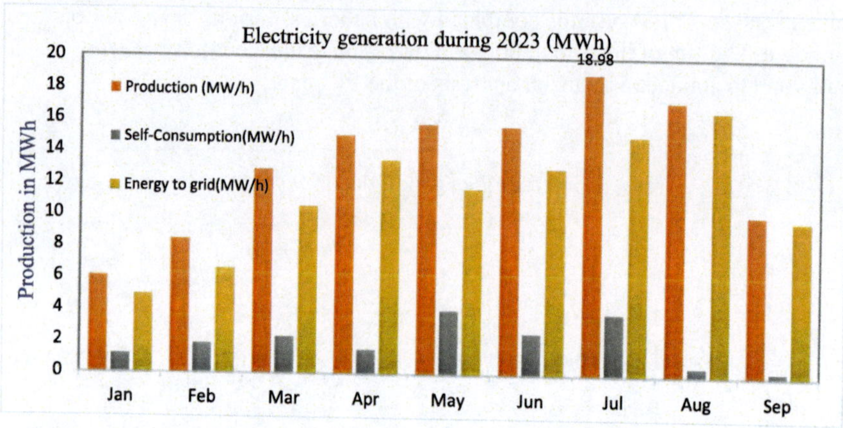

Fig. 7. Electricity generation for the period January-September 2023 in MWh

January to September 2023 the total net amount of electricity was 102 MWh and from the installation of the plant was 169.61 MWh of electrical energy. The Fig. 8 depicts the variation of gross electricity production forecast from software and real data collected from net metering. As can be observed from the graph there is a difference between two situations. According to the calculations the difference varies from 20% to 64% respectively corresponding February and July.

5.2 Cost Analysis

A simple financial cost was carried out provided by the PV*Sol software. The information required for the economic efficiency calculation such as costs, subsidies and electricity tariffs are defined. As with the surplus feed-in, the electrical energy generated is primarily used for self-sufficiency. The unused electricity is fed into the power grid via a

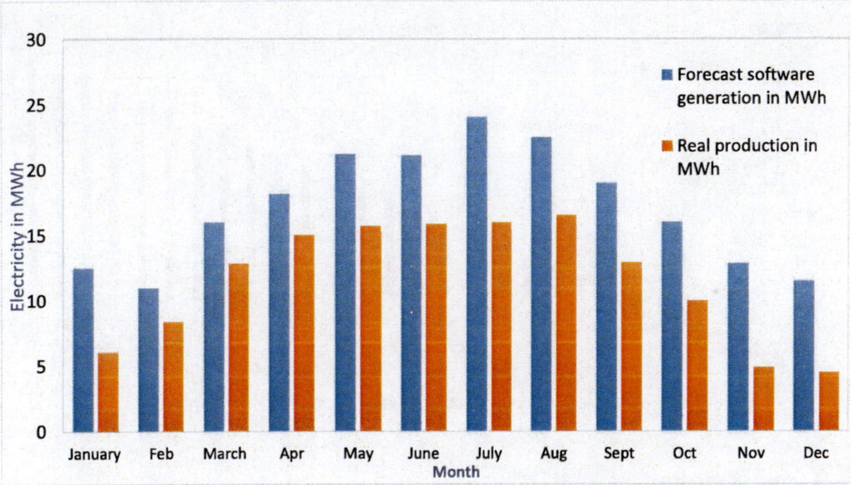

Fig. 8. Data comparison between software calculations and real generation of PV plant

double-tariff meter and offset against the electricity purchased. The tariff counter rotates backwards. If more solar power is generated than consumed in the billing period, the surplus can be carried forward as a balance to the next billing period. For the calculations no loan was predicted to the (Fig. 9, Table 3).

Table 3. Cost analysis data of the plant

Investment cost	91,000 Euro
Maintenance cost	200 Euro/year
Amortization	0.5%
Electricity price	0.14 Euro/kWh
Yearly cost of electricity	23,165 Euro
Return of investment (ROI)	3.9 Years
System lifetime	25 years

According to the data collected in the foundry facility the total expenses for electricity consumption amount to approximately 23,165 Euro/year. Electricity price calculated for the entity is 0.14 kWh. From a simple calculation for the first year of the operation the total amount of money saved was 19441.8 Euro. It is obvious that this significant amount highly contributes in expenses reduction of the foundry facility, increase the renewable contribution and CO_2 emissions avoided 98,384 kg/year. Theoretical payback period is estimated to 3.9 years. However, based on expected output electricity from PV plant the payback period results slightly higher to 5.46 years. The results show that such investments have significant contribution on decreasing the foundry electricity bills by considerable amount.

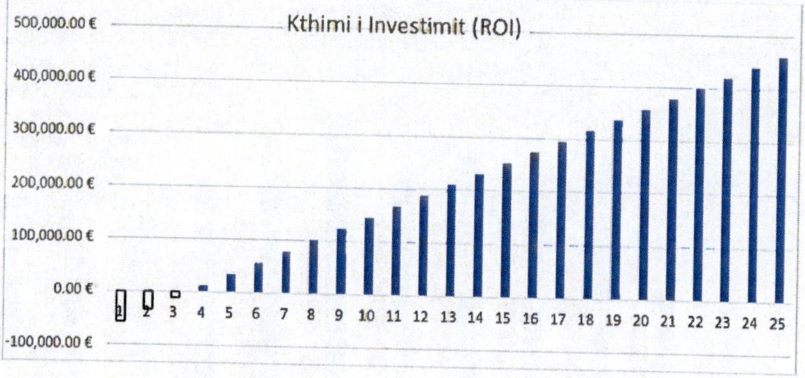

Fig. 9. Payback method evaluation of the system

6 Conclusions

This presents the application of on-grid photovoltaic system in the foundry cast-iron industry. The case study presented in this paper is located near Tirana, and the data necessary for the study were extracted from facility visits. Energy consumption in generally and application of renewable energy in the industry sector remains below the regional average. This indicates the penetration of solar energy through PV installation in small scale it is with significant importance. This study takes into consideration the application of a PV system in a foundry with the main purpose of producing electricity. The production of electricity from PV would significantly reduce electricity bills, which are the main source of expenses. In the study, we made a technical-economic analysis of the investment through PV*Sol software. Thus, on the other hand, the real generation of electricity thrown into the power grid is also excluded. An average difference of about 40% is observed between the forecast and the actual production. The financial analysis of the system results in the theoretical return of this investment being 3.9 years. However, due to the differences between the expected production and the real one, the return on investment is expected to be extended to 5.46 years, which is also a very good rate. The application of energy generation systems in industry, in addition to reducing energy dependence, increases the contribution of renewable energies to the national balance and significantly reduces CO_2 emissions in the atmosphere.

References

1. https://www.irena.org//media/Files/IRENA/Agency/Statistics/Statistical_Profiles/Europe/Albania_Europe_RE_SP.pdf?rev=ea9be058c1034a52a6339c8fdf6061e2
2. Enti Rregullator i Energjise, Raporti Vjetor (2022). www.ere.gov.al
3. Global landscape of renewable energy finance (2023). www.irena.org
4. International Energy Agency: World energy balances (2012). www.iea.org
5. Napp, T.A., Gambhir, A., Hills, T.P., et al.: A review of the technologies, economics and policy instruments for decarbonising energy-intensive manufacturing industries. Renew. Sust. Energ. Rev. **30**, 616–640 (2014)

6. Madlool, N.A., Saidur, R., Hossain, M.S., et al.: A critical review on energy use and savings in the cement industries. Renew. Sust. Energ. Rev. **15**, 2042–2060 (2011)
7. Chauhan, M.K., Varun, C.S., et al.: Life cycle assessment of sugar industry: a review. Renew. Sust. Energ. Rev. **15**, 3445–3453 (2011)
8. Ali, M.B., Saidur, R., Hossain, M.S.: A review on emission analysis in cement industries. Renew. Sust. Energ. Rev. **15**, 2252–2261 (2011)
9. Noro, M., Lazzarin, R.: Energy audit experiences in foundries. Int. J. Energ. Environ. Eng. (2014). https://doi.org/10.1007/s40095-014-0152-y
10. Lazzarin, R., Noro, M.: Energy efficiency opportunities in the production process of cast iron foundries: an experience in Italy. Appl. Therm. Eng. **90**, 509–520 (2015)
11. REN21: Renewables 2022: global status report. Renewable Energy Policy Network for the 21st Century, Paris (2022). www.ren21.net/wp-content/uploads/2019/05/GSR2022_Full_Report.pdf
12. AKBN, www.akbn.gov.al, Agjencia Kombetare e Burimeve Natyrore
13. Ministria e Energjitikes, www.infrastruktura.gov.al
14. Valentine software, https://valentin-software.com/en/

A Simplified Hydraulic Capacity-Sensitive Fluid Dynamics Numerical Model for Monitoring Aerospace Electro-Hydraulic Actuators

Matteo D. L. Dalla Vedova[1](✉), Parid Alimhillaj[2], Edmondo Minisci[3], and Paolo Maggiore[1]

[1] Politecnico Di Torino, 10129 Turin, Italy
{matteo.dallavedova,paolo.maggiore}@polito.it
[2] Polytechnic University of Tirana, 1000 Tirana, Albania
parid.alimhillaj@fim.edu.al
[3] University of Strathclyde, Glasgow G11XJ, UK
edmondo.minisci@strath.ac.uk

Abstract. Detailed models are necessary to analyze individual components or subsystems when designing modern flight control systems. However, simpler models with sufficient accuracy are needed for preliminary layout, monitoring, diagnostics, or prognostic issues. Various simplified numerical solutions are available in the literature to simulate the fluid dynamic behaviors of a given valve geometry. These models typically calculate the differential pressure the valve regulates based on its spool opening and flow rate. In some specific applications, these models are unsuitable, requiring new simplified fluid dynamic models that calculate the flow rate delivered by the valve based on the spool displacement and differential pressure. This paper introduces a new synthetic fluid-dynamic valve model that considers the effects of spool position, hydraulic capacity, variable supply pressure, and leakage between the output ports that connect the valve to the motor element. Its advantages and disadvantages are evaluated by comparing it with other simplified numerical algorithms available in the literature, analyzing the corresponding fluid-dynamic characteristics, and comparing the behaviors simulating a typical flight control servomechanism.

Keywords: Aerospace systems · EHA · Digital twin · Hydraulic capacity · Lumped parameters · Monitoring · Non-linear modelling · Servovalve · Simplified fluid dynamic numerical model

1 Introduction

Most commercial and military aircraft use electrohydraulic actuation systems (EHAs) in their powered flight controls. They represent a well-established technology and have a very high power density, allowing designers to create lightweight designs that can fit into the small places inside the aerodynamic surfaces of the aircraft. To meet the stringent safety standards of aviation regulations, proper redundancies and real-time health

G. Guxho et al. (Eds.): ITC-ICEE 2023, LNMIE, pp. 264–274, 2024.
https://doi.org/10.1007/978-3-031-48933-4_25

monitoring strategies [1–3] are required since flight controls are one of the most safety-critical systems of aircraft [4]. In particular, considering its non-negligible failure rates and the potentially catastrophic criticalities deriving from these failures [5], the electro-hydraulic servovalves (EHSVs) equipping these actuators typically require very effective and reliable monitoring algorithms. The level of detail provided by the various simulation models is inevitably correlated to multiple factors, among which the field of application of a given algorithm and the related performance requirements stand out. Therefore, the literature proposes several numerical models able to simulate the fluid dynamic behaviors of a given valve geometry with different levels of accuracy and fidelity. The design and development phases of EHA often requires accurate and high-fidelity simulations (especially regarding fluid-dynamics behaviors) capable of evaluating the system performance within its operating range [5]. In this regard, Urata has conducted extensive research on various fault modes, including internal valve leakages [6], eddy currents [7–9], asymmetric torque motor airgaps [10–12], and fringing [13]. In-depth simulations of the hydraulic and electrical performance of electrohydraulic servovalves (EHSVs) using finite electromagnetic elements and computational fluid-dynamics are proposed by Chen et al. [14], Henninger et al. [15] and Yang et al. [16]. These approaches, nevertheless, are typically expensive in terms of processor workload and computing time, making them inappropriate for some activities whose execution is time- or resource-constrained. Indeed, in activities such as the preliminary design of the actuation systems or the conception of real-time algorithms (e.g. monitoring and diagnostic routines), the main difficulty is maintaining a reasonable accuracy level while minimizing the corresponding computing load. In these cases, it is necessary to develop EHSVs simplified models designed and suitably set up for specified operations, e.g. preliminary design optimization or diagnostic/prognostic techniques, to be used as a digital mirror of the corresponding physical components. In general, these simplified models calculate the differential pressure regulated by the valve as a function of its spool opening and the flow rate disposed of by the valve itself [17, 18]. However, models with differential pressure output are inadequate in some specific applications (e.g. asymmetric hydraulic jacks, regenerative actuators, or hydraulic transmissions in which the effect of fluid compressibility is not negligible). In these cases, it is necessary to adopt new simplified fluid dynamic models that, starting from the same hypotheses and data, calculate the flow rate delivered by the valve as a function of the spool displacement and the differential pressure acting downstream of the valve (i.e. on the output ports connecting it with the hydraulic actuator). Thus, this paper proposes a new synthetic fluid-dynamic valve model (a lumped parameters model with a semi-empirical formulation) accounting for the effects of spool position, hydraulic capacity, variable supply pressure and leakage between the output ports that connect the valve to the motor element.

2 Considered Electro-Hydraulic Actuator

The numerical models proposed in this work have been developed referring to a typical electrohydraulic servoactuator architecture typically adopted in aerospace systems: i.e. a four-way control valve (supply port S, return port R, control port 1, control port 2) coupled with a symmetrical linear jack, as shown in Fig. 1.

More specifically, we aim to create a computationally light model for prognostic applications [2], focusing on the second-stage sliding spool valve. This is because its highly nonlinear behavior is responsible for most of the computational burden in high-fidelity, CFD-based models [15, 19, 20]. In addition, it has been shown that strong linearity assumptions on the spool operation may noticeably degrade the accuracy of the whole servoactuator model in some operating conditions [21].

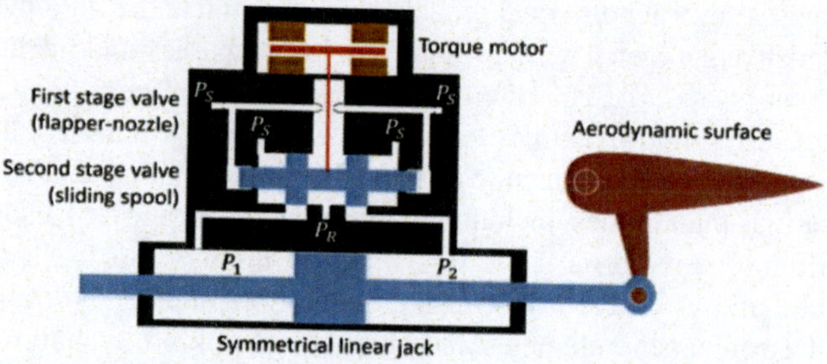

Fig. 1. Schematic of the considered electrohydraulic actuator

As shown in the schematic of Fig. 1, the valve spool displacement x_S controls the areas of the four passageways, each characterized by its overlaps or underlap, to connect each control port either to the supply or return port. This allows modulation of the hydraulic power regulated by the aforementioned piloting edges, expressed in terms of flow and absolute pressure for each control port (P1 and P2), for specific characteristics of the oil and operational conditions [22, 23]. The corresponding differential pressure, regulated between the two control ports is $P_{12} = P_1 – P_2$.

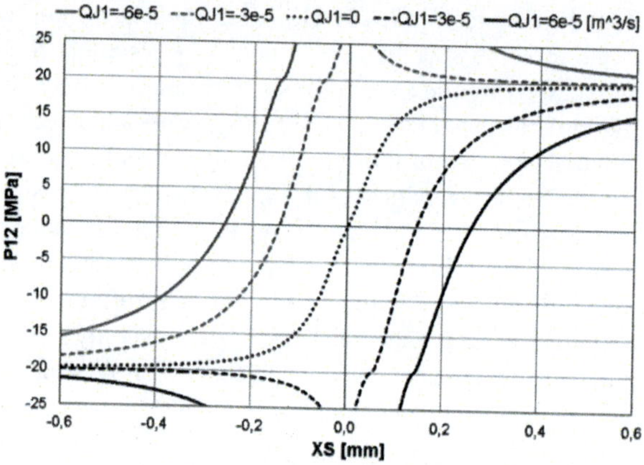

Fig. 2. Differential pressure—spool position characteristic of EHSV (HF model)

In zero-flow conditions, each control port absolute pressure is close to the supply or return pressure when the corresponding passageway is fully opened. When the spool is in an intermediate position, the control port pressures have a progressive evolution between return pressure (P_R) and supply pressure (P_S) values, as it can be seen in the valve characteristic P_{12}–x_S of Fig. 2.

3 EHSV Fluid-Dynamic HF Model

As reported in previous chapter, in this work, a high-fidelity fluid dynamic model is being used as a reference to develop lower-fidelity emulators. The high-fidelity (HF) model calculates the pressure drops across each valve passageway to determine the flow rates and pressures crossing the valve for a given spool position. This information is essential in understanding the behavior of the system. The model computes the map shown in Fig. 2: for a null flow rate (i.e. $Q_J = 0$) and small spool displacements, the differential pressure $P_{12} = P_1 - P_2$ varies approximately linearly with the spool position x_S. Large values of x_S result instead in the saturation of the differential pressure to $P_{SR} = P_S - P_R$, where P_S is the supply pressure and P_R is the return line pressure. The presence of a flow rate $Q_J \neq 0$ offsets the response of the valve as an additional pressure drop is caused by the restricted flow through the small passageways of the valve. The combination of large flow rates and small spool displacements can result in a differential pressure higher than P_{SR}. The water hammer effect is a phenomenon that significantly impacts the behaviour of servovalves.

However, the HF model is too expensive to use in real-time dynamical simulations because it requires an iterative process to solve pressure drops at each time-step. Moreover, it heavily relies on various parameters, such as the geometry of the SV and the physical properties of the hydraulic fluid. All those variables, needed to correctly set the model for the simulation of a physical system, often are not directly available cannot be measured with sufficient accuracy. For example, the model is highly sensitive to the exact value of the internal clearances between the spool and sleeve, which can be affected by manufacturing tolerances and even by elastic deformations due to pressurization of the system. A full description of the HF model is available in [5, 24].

4 EHSV Fluid-Dynamic LF Models Available in Literature

As previously reported, the HF model is generally computationally heavy and time-consuming, and is sensitive to several parameters, related to the SV geometry and the physical characteristics of the hydraulic fluid, that often are not directly available or cannot be measured with proper accuracy. Therefore, in a simpler, lighter and quicker approach, only the controlled differential pressure between the two control ports P_{12} and a single flow value Q_J (common to both control ports) are computed with linearized models [25]. When representing the hydraulic characteristic of servovalves in Low Fidelity (LF), it's common to use a local linearization of the flow-pressure-displacement map. This is usually done near the zero flow and closed valve condition.

By doing this, the operation of an electrohydraulic actuator can be described in the regulating regime, far from the effects of saturations and other nonlinear phenomena. To achieve this, a two-gains linear form is often employed:

$$\frac{P_{12}}{G_P} + \frac{Q_J}{G_Q} - x_S = 0 \tag{1}$$

where G_P is the pressure gain and G_Q is the flow gain. Fundamentally, two alternative formulations can be considered: one solves for differential pressure P_{12} with flow rate Q_J as feedback (pressure formulation), while the other solves for flow rate Q_J with differential pressure P_{12} as feedback (flow rate formulation).

Often, this former simplified model is expressed as:

$$P_{12} = G_P \cdot \left(x_S - \frac{Q_J}{G_Q} \right) \tag{2}$$

The spool displacement produces a proportional differential pressure value, which acts on the motor element (i.e. the linear jack). This displacement is reduced by the pressure loss caused by the controlled flow passing through the control passageways; this effect is accounted for with the flow gain.

It should be noted that, in general, the linear models used to describe hydraulic actuators, cannot capture several phenomena that can significantly affect their operation, such as leakages, limited supply pressure, and water hammer. For this reason, several simplified models have been proposed in the literature to extend the linear formulation and overcome its limitations.

The following is a brief overview of some simplified models previously developed by the authors, while Fig. 3 [22] summarizes the block diagram representations of these models.

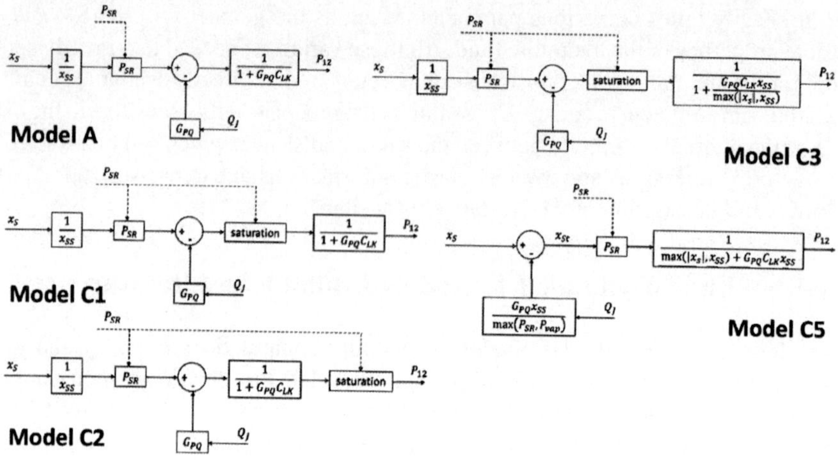

Fig. 3. Block diagrams of models A, C1, C2, C3 and C5

Model A [25] accounts for variable differential supply pressure P_{SR} and leakage flow:

$$P_{12} = \left(\frac{x_S}{x_{SS}} P_{SR} - Q_J G_{PQ} \right) \frac{1}{1 + G_{PQ} C_{lk}} \tag{3}$$

where $G_{PQ} = G_P / G_Q$. The formulation is still fully linear, although the coefficients are not constant and are corrected with respect to Eq. (2).

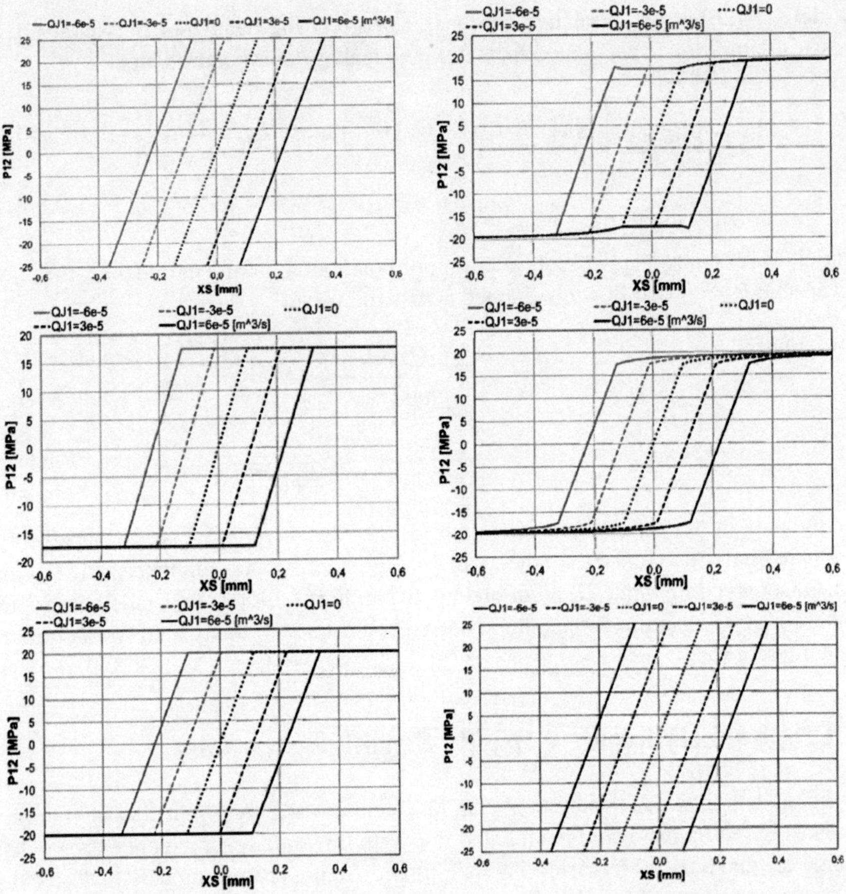

Fig. 4. Fluid-dynamic characteristic of Model A (top left), Model C1 (middle left), Model C2 (bottom left), Model C3 (top right), Model C5 (middle right) and new LF model (bottom right)

Model C1 [17, 24] extends the applicability of Model A to large spool displacements, by introducing a saturation of the differential pressure to the supply. The saturation is placed upstream the leakage estimation:

$$P_{12} = \frac{1}{1+G_{PQ}C_{lk}}\left[sgn\left(\frac{x_S}{x_{SS}}P_{SR} - Q_J G_{PQ}\right)\max\left(P_{SR}, \left|\frac{x_S}{x_{SS}}P_{SR} - Q_J G_{PQ}\right|\right)\right] \tag{4}$$

In **Model C2** [25] the pressure saturation is moved downstream the leakage flow estimation: this is to correct the underestimation of maximum differential pressure observed in Model C1. The formulation of Model C2 is modified as follows:

$$P_{12} = \mathrm{sgn}\left[\frac{1}{1+G_{PQ}C_{lk}}\left(\frac{x_S}{x_{SS}}P_{SR} - Q_J G_{PQ}\right)\right]\max\left(P_{SR}, \left|\frac{1}{1+G_{PQ}C_{lk}}\left(\frac{x_S}{x_{SS}}P_{SR} - Q_J G_{PQ}\right)\right|\right) \tag{5}$$

Model C3 [26] is an alternative approach to correct the behaviour of Model C1: the pressure saturation is accounted for within the leakage computation block:

$$P_{12} = \frac{1}{1 + \frac{G_{PQ} C_{lk} x_{SS}}{\max(|x_s|, x_{SS})}} \left[\operatorname{sgn}\left(\frac{x_S}{x_{SS}} P_{SR} - Q_J G_{PQ} \right) \max\left(P_{SR}, \left| \frac{x_S}{x_{SS}} P_{SR} - Q_J G_{PQ} \right| \right) \right] \tag{6}$$

Model C5 [27] introduces an equivalent spool-position x_{St} taking into account the effects of variable P_{SR} and oil flow Q_J drained across the valve:

$$x_{St} = x_S - \frac{Q_J (G_{PQ} x_{SS})}{\max(P_{SR}, P_{vap})} \tag{7}$$

$$P_{12} = x_{St} P_{SR} \frac{1}{\max(|x_s|, x_{SS}) + G_{PQ} C_{lk} x_{SS}} \tag{8}$$

Figure 4 shows the characteristic maps of the models reported above (representing the P_{12} regulated as a function of the spool displacement x_{St} and parameterized according to the discharged flow rate Q_J). For brevity, the authors omitted the detailed description of these maps. For more details, the readers should refer to the related works reported in the literature.

5 Proposed Hydraulic Capacity-Sensitive LF Model

All the models shown in the previous chapter calculate the differential pressure the valve regulates based on its spool opening and flow rate. In some specific applications, these models are unsuitable, requiring new simplified fluid dynamic models that calculate the flow rate delivered by the valve based on the spool displacement and differential pressure. Therefore, this section introduces a new synthetic fluid-dynamic valve model that considers the effects of spool position, hydraulic capacity, variable supply pressure, and leakage between the output ports that connect the valve to the motor element. As understandable by the schematic shown in Fig. 5, this model derives directly from Eq. (1), expressed as a function of the instantaneous flow rate Q_J.

Similarly to what has already been seen regarding Model A, a linear formulation is adopted (i.e. no saturations on the maximum value of the P_{12} pressure regulated downstream of the spool) that is based on the already known valve gains (in particular $GPQ = GP/GQ$) and it is sensitive to the hydraulic line differential pressure P_{SR}.

It should be noted that unlike models that have differential pressure as the output variable (see Fig. 4), the fluid control valve followed by a hydraulic capacity can be modeled without problems when a leak acts between the control ports, connecting the valve the same as the driving element, is present.

The leakage feedback loop cannot be instantaneous because the hydraulic capacity introduces a first-order lock within the loop, as in the right part of Fig. 5, to prevent numerical instabilities. The proposed computational algorithm considers all these factors and simulates effects due to variations in flow rates, pressure gains, and supply

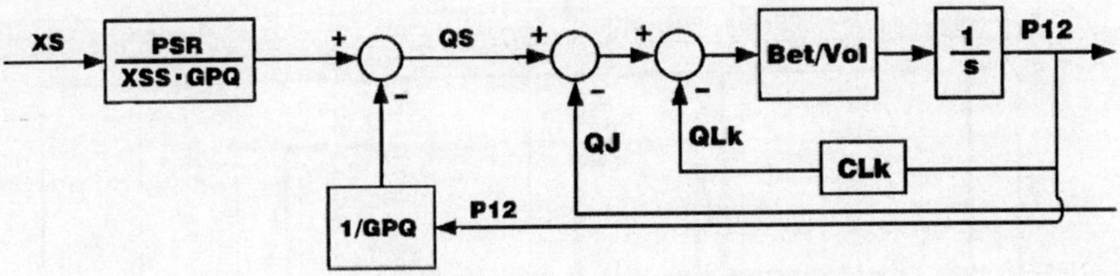

Fig. 5. Block diagram of the proposed hydraulic capacity-sensitive LF model (Model Aq)

pressure through a simplified approach. It is used to compute the fluid-dynamic characteristic of the EHSV and simulate the dynamic behavior of a typical onboard EHA. The algorithm's abilities have been verified by comparing the results with those obtained from the previously published HF model by the authors.

6 Numerical Results of the EHA Test Bench

The proposed simplified model has been integrated into a virtual test bench, which simulates a full position control electro-hydraulic actuator (EHA) consistent with this described in Sect. 2. Its behaviors have been compared with a corresponding numerical model equipped with the HF fluid dynamic model introduced in Sect. 3.

This detailed model considers the key electrical, hydraulic, and mechanical features of all relevant system components, which include inertia, the hydraulic piston dry and viscous friction, and a third-order electromechanical model of the first and second-stage dynamics of the EHSV. Figures 6 and 7 show the simulated dynamic responses of the EHA models under a proper combination of position commands (Com), external loads (FR), and changes in the hydraulic supply pressure (PSR).

The input sequence has been defined to highlight the performance of the fluid-dynamic models and their impact on the behavior of the simulated EHA [18].

Figure 6 shows the dynamic response of the full actuator implementing the HF fluid dynamic model of the valve. It can be used as a reference for comparison with the corresponding response of the proposed simplified model.

On the other hand, Fig. 7 illustrates the response of the proposed model (Aq): it can simulate with acceptable accuracy the behaviours of the reference model for the conditions of unloaded actuation (from 0 to 0.2 s) and the aiding-load condition (at 0.75 s) despite a slightly higher starting acceleration and a lower stopping deceleration (related to the model inability to properly account for the water hammer effect).

In the event of opposing load actuation at 0.3 s, the Aq model underestimates the impact of the external force FR. Therefore, when the supply pressure PSR decreases (at 0.35 s), the back movement of the jack is also underestimated.

7 Conclusions

In this paper, a new simplified fluid dynamic model of EHSV sensitive to the compressibility of the hydraulic fluid was developed and tested. The model could overcome most of the problems associated with previous low-fidelity emulators and approximate the

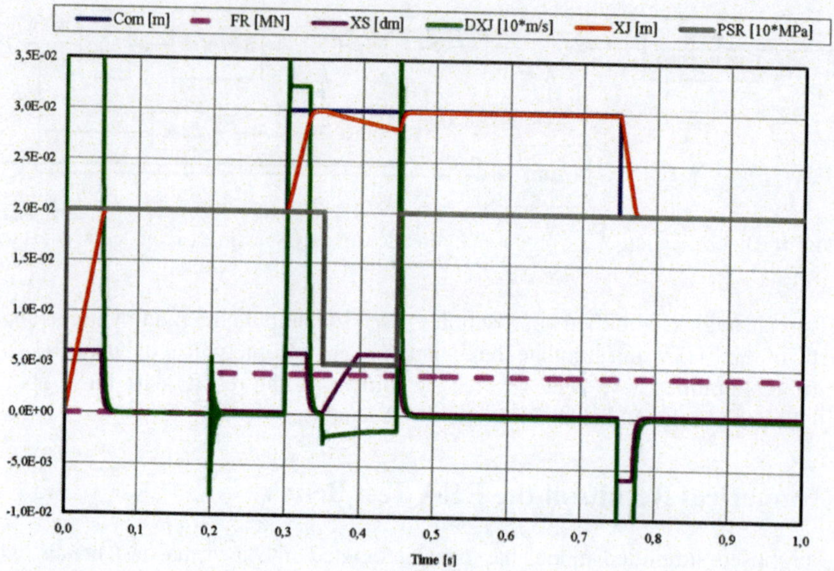

Fig. 6. Test bench simulation with the HF model; Com—position command, FR—external load, XS—spool position, DXJ—jack speed, XJ—jack position, PSR—supply pressure

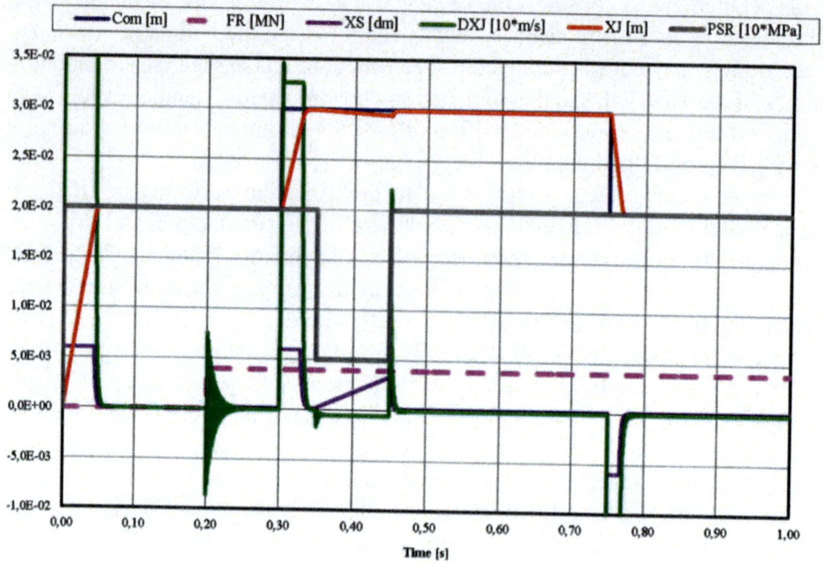

Fig. 7. Test bench simulation with the Aq model; Com—position command, FR—external load, XS—spool position, DXJ—jack speed, XJ—jack position, PSR—supply pressure

behavior of an HF simulation with reasonable accuracy (including pressure oscillations, fluid compliance, and the effects of leakage losses). At the same time, compared to the HF reference or well-known CFD codes, the computational time/effort is relatively reduced and, therefore, promising for real-time applications (especially preliminary design and diagnostic monitoring) since the formulation only requires the calculation of simple algebraic and trigonometric functions, without the need for iterative solvers. However, given that the starting model (Model A) is highly simplified, the accuracy and reliability of the proposed model are not yet compatible with applications such as model-based prognostics of hydraulic systems. Therefore, the authors aim to continue this research by implementing new simplified fluid dynamic models based on the much more performing (but complex) type C models.

References

1. Vedova, M.D.L.D., Maggiore, P., Pace, L.: Proposal of prognostic parametric method applied to an electrohydraulic servomechanism affected by multiple failures. WSEAS Trans. Environ. Develop. **10**, 478–490 (2014)
2. De Martin, A., Chiavaroli, P., Evangelista, G., Jacazio, G., Sorli M.: Real time loading test rig for flight control actuators under PHM experimentation. In: Proceedings of the ASME 2018 International Mechanical Engineering Congress and Exposition (2018)
3. De Martin, A., Dellacasa, A., Jacazio, G., Sorli, M.: High-fidelity model of electro-hydraulic actuators for primary flight control systems. In: Proceedings BATH/ASME Symposium on Fluid Power and Motion Control, Baht (UK) (2018)
4. Byington, C.S., Watson, M., Edwards, D.,Stoelting P.: A model-based approach to prognostics and health management for flight control actuators. In: 2004 IEEE Aerospace Conf. Proceedings, 6, pp. 3551–3562 (2004)
5. Borello, L., Vedova, M.D.L.D., Jacazio, G., Sorli, M.: A prognostic model for electrohydraulic servovalves. In: Proceedings of the Annual Conference of the Prognostics and Health Management Society (PHM 2009) (2009)
6. Urata, E.: Study on leakage flux from Servovalve torque motors. In: The Seventh Scandinavian International Conference on Fluid Power, 1, pp. 51–66 (2001)
7. Urata, E.: Study of magnetic circuits for servovalve torque motors. In: Bath workshop on power transmission and motion control, Bath (UK), pp. 269–282 (2000)
8. Urata, E.: Static stability of toque motors. In: The Eighth Scandinavian International Conference on Fluid Power, pp. 871–885 (2003)
9. Urata, E.: Influence of eddy current on torque-motor dynamics. In: 4th IFK Workshop, pp. 71–82 (2004)
10. Urata, E.: One degree of freedom model for torque motor dynamics. Int. J. Fluid Power **5**(2), 35–42 (2004)
11. Urata, E.: Influence of asymmetry of air-gap in servovalve torque motor. In: 9th Scandinavian International Conference on Fluid Power (2005)
12. Urata, E.: Influence of unequal air-gap thickness in servo valve torque motors. Proceedings of the Institution of Mechanical Engineers, Part C: J. Mech. Eng. Sci. **221**, 1287–1297 (2007)
13. Urata, E.: Influence of fringing on servovalve torque-motor characteristic. In: Proceedings of the Fifth JFPS International Symposium on Fluid Power, 3, pp. 769–774 (2002)
14. Henninger, D., Zopey, A., Ihde, T., Mehring, C.: High-fidelity 1D dynamic model of a hydraulic servo valve using 3D computational fluid dynamics and electromagnetic finite element analysis. Int. J. Mech. Mechat. Eng. **11**(8), 1476–1482 (2017)
15. Chen, Q., Stoffel, B.: CFD simulation of a hydraulic servo valve with turbulent flow and cavitation. In: ASME/JSME 2004 Pressure Vessels and Piping Conference (2004)
16. Yang, H., Wang, W., Lu, K.: Cavitation and flow forces in the flapper-nozzle stage of a hydraulic servo-valve manipulated by continuous minijets. Adv. Mech. Eng. **11**(5), 1–9 (2019)

17. Vedova, M.D.L.D., Berri, P.C., Corsi, C., Alimhillaj, P.: New synthetic fluid dynamic model for aerospace four-ways servovalve. Int. J. Mech. Cont. **20**(2), 105–112 (2019)
18. Vedova, M.D.L.D., Berri, P.C.: A new simplified fluid dynamic model for digital twins of electrohydraulic servovalves. Aircraft Eng. Aerospace Tech. **94**(1), 79–88 (2022). https://doi.org/10.1108/AEAT-12-2020-0321
19. Henninger, D., Zopey, A., Ihde, T., Mehring, C.: Int. J. Mech. Mecham. Eng. **11**(8) (2017)
20. Kim, D.H., Tsao, T.C.: A linearized electrohydraulic servovalve model for valve dynamics sensitivity analysis and control system design. J. Dyn. Syst. Measure. Cont. **122**(1) (2000)
21. Berri, P.C., Vedova, M.D.L.D., Santaera, S.: Digital twins for prognostics of electro-hydraulic actuators: novel simplified fluid dynamic models for aerospace valves. In: IOP Conference Series: Materials Science and Engineering, 1226 (012106) (2022)
22. Merritt, H.E.: Hydraulic Control Systems. Wiley (1967)
23. Viersma, T.J.: Analysis, Synthesis and Design of Hydraulic Servosystems and Pipelines. Elsevier (1980)
24. Alimhillaj, P., Borello, L., Dalla Vedova, M.D.L.: Proposal of innovative fluid dynamic nonlinear servovalve synthetic models. Int. J. Mech. Cont. **14**(2), 39–49 (2013)
25. Borello, L., Villero, G.: Confronto fra modelli semplificati di valvole di comando. In: XI A.I.D.A.A. National Congress (Paper in Italian language) (1991)
26. Alimhillaj, P., Vedova, M.D.L.D.: Study of new fluid dynamic nonlinear servovalve numerical models for aerospace applications. In: 2nd European Conference on Electrical Engineering and Computer Science (2018)
27. Berri, P.C., Vedova, M.D.L.D.: A review of simplified servovalve models for digital twins of electrohydraulic actuators. J. Phys.: Conf. Seri. **1603**(012016) (2020)

Flow Induced Vibration Assessment of a Gas Piping System

Deborah Ibrahimi[1], George E. Varelis[1(✉)], Deng Peng[2], Graeme Campsie[1], and Jeroen Hoefakker[2]

[1] Worley, Brentford, UK
Deborah.Ibrahimi@Worley.com, Georgios.Varelis@intecsea.com
[2] INTECSEA, The Hague, The Netherlands

Abstract. Internal fluid flow within a piping system is known to create structural vibrations when forced to pass through directional changes, T-junctions, and piping systems with a varying pipe diameter. This paper presents a numerical application example where the flow induced vibrations (FIV) of a gas transmission system containing a flow control valve (FCV) are analysed using ANSYS CFD and FEA tools, considering a one-way fluid-structure interaction (FSI). To establish the variation of the flow behaviour due to the FCV presence, three specific base cases are considered, corresponding to the maximum, average and minimum pressure drops across the FCV, determining the effects of the corresponding fluid velocities on the vibrations created. The assessment is conducted in line with the provisions included in the Energy Institute guidelines to determine the likelihood of failure of the pipework as a result of the FIV.

Keywords: FIV · FSI · Piping system · FEA · CFD

1 Introduction

ANSYS computational fluid dynamics and structural finite element analyses were used to investigate the response of a 24-inch diameter mainline section of a Gas Metering Train and the associated Blow Dow Valve (BDV) pipework, located downstream of a Flow Control Valve (FCV), to structural vibrations. It should be noted that the reduced diameter BDV pipework is effectively a 'dead leg' branch, at which the structural vibration has been observed during the normal operations. This study aims at investigating the cause of the vibration occurred in the main 24-inch pipeline as well as the BDV dead leg pipework.

The structural vibration is believed to be caused by the fluid behaviour in the piping system. Computational Fluid Dynamics (CFD) analysis was therefore implemented to establish the variation of the flow behaviour in the mainline and smaller diameter branch. Analysis of supplied process data was undertaken, resulting in the definition of three analysis cases that correspond to maximum, average and minimum pressure drops across the FCV. To simulate those conditions, three simplified local analysis models were developed, allowing for the representation of the pressure drop conditions upstream

G. Guxho et al. (Eds.): ITC-ICEE 2023, LNMIE, pp. 275–284, 2024.
https://doi.org/10.1007/978-3-031-48933-4_26

of the BDV pipework and establishing fully developed flow downstream of the FCV, without the need for resource-expensive fully detailed 3D representation of the FCV. The results from these numerical models correlated well with the supplied process data. As such, a further FIV assessment was carried out.

The three local models were integrated into a global model of the Gas Metering Train and computationally intensive steady state and transient CFD simulations were then employed to capture the variation of the internal pressure profile over 10 s. The CFD-predicted pressure profile was then mapped to a structural model of the piping system. The assessment then focused on the predicted vibration characteristics (amplitude, frequency) and associated Likelihood of Failure (LOF) index and the effects of vibration on the fatigue life of the piping system. The vibration assessment was conducted in line with the provisions included in the Energy Institute Guidelines.

2 Local Model

An initial local model was developed in ANSYS Spaceclaim [1] to represent the Flow Control Valve (FCV), which is used to adjust and control the volumetric flow of the gas within the system, and the adjacent piping components. The extent of the local model is as shown in Fig. 1.

The FCV was represented by two 24-inch diameter pipe segments with a reduced diameter central pipe segment in between. The diameter of the central pipe segment, 'd', was varied as required to achieve the pressure drop across the FCV for the given process conditions. As shown by Fig. 1, the larger diameter pipe segments were modelled with lengths equal to 15D, where 'D' is the diameter of the 24- inch pipe segments. These lengths were selected so that a fully developed flow within the system could develop, and the desired pressure for each case is matched.

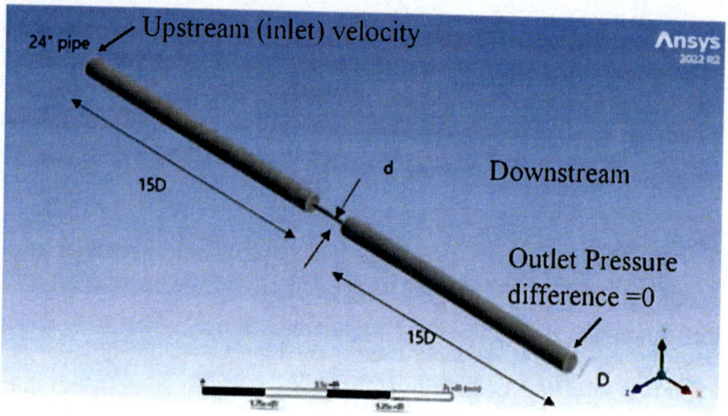

Fig. 1. Extent of local model.

The CFD analyses were conducted using the ANSYS CFX software, within ANSYS Workbench version 2022 R2 [1]. Three analyses were undertaken in order to calibrate

Table 1. Field process data for validation of FCV local model

Case	ΔP across FCV [bar]	Downstream pressure [bar]	Upstream velocity [m/s]	Gas density [kg/m^3]
1	14.40	57.5	3.3	51.7
2	5.10	66.9	6.4	53.3
3	0.1	71.9	6.0	52.0

the local model. These analyses considered a maximum, average and minimum pressure drop (ΔP) cases across the FCV, as seen in Table 1.

The applied boundary conditions for each analysis are summarized below:

- Upstream velocity applied at inlet
- Outlet pressure difference set to zero
- Downstream pressure used as the domain pressure
- Wall condition defined as no slip, smooth

The pressure drop across the FCV was taken to be the pressure at the inlet minus the pressure at the outlet, such that:

$$\Delta P = P_{inlet} - P_{outlet}$$

An iterative process was then undertaken whereby the diameter of the central pipe segment was modified in the range 30 mm to 160 mm to develop the appropriate pressure drop, ΔP, as given in Table 1. The results are summarized in Table 2 and agree with the target pressure drop across the FCV seen in Table 1.

Table 2. Central pipe segment diameter and corresponding pressure drop.

Case	Diameter, d [mm]	ΔP across FCV [bar]
1	39.0	14.6
2	68.3	5.2
3	157.0	0.1

From the values reported in in Tables 1 and 2 it can be seen that the CFD predicted pressure drop values and field process data are well matched. The relationship of equivalent length and ID variation was established based on this benchmarking exercise, which allowed for the simplified equivalent approach to be adopted for the representation of the FCV.

3 Global Model

Three separate analysis models were subsequently generated by integrating the three local models, representing the FCV for either the maximum, average or minimum pressure drop condition, into the global model at the appropriate location, as shown by Fig. 2.

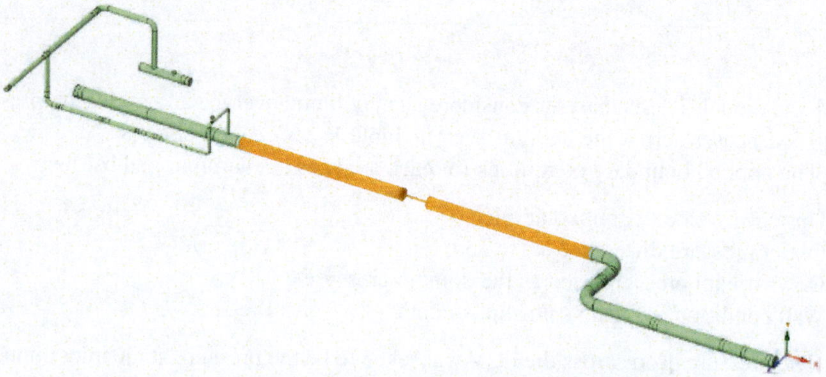

Fig. 2. Extent of global model with integrated local model—highlighted in orange.

The pipework is fabricated from EN 3183 L450Q PSL2 carbon steel line pipe. The material yield strength and ultimate tensile strength are in accordance with ISO 3183:2019 [2] and are reproduced in Table 3.

Table 3. Summary of material properties

Material	Yield strength [MPa]	Ultimate tensile strength [MPa]	Young's modulus [MPa]	Poisson's ratio
EN 3183 L450Q PSL2	450	535	2.10×10^5	0.3

Note that to account for the stiffness of the FCV relative to the pipework, the Young's modulus has been increased to 2.10×10^8 MPa to the extent of the model representing the FCV. Distributed masses (kg) were also applied in accordance with provided details to represent the flange connections at the ends of pipe spools and various equipment weights.

The support boundary conditions were applied over short pipe lengths by constraining displacements in certain DOFs and allowing free displacement in others, depending on the support type at each location. A summary of the boundary conditions applied is shown in Fig. 3. For the fully constrained DOFs, at either end of the model, rigid (infinite stiffness) supports are assumed, while for the free DOFs no restrictions in displacement (e.g., friction, gaps/stops) were considered for the base cases. The two types of conditions were as follows:

- Type A—DOF: x = free, y = 0, z = free,
- Type B—DOF: x = 0, y = 0, z = 0

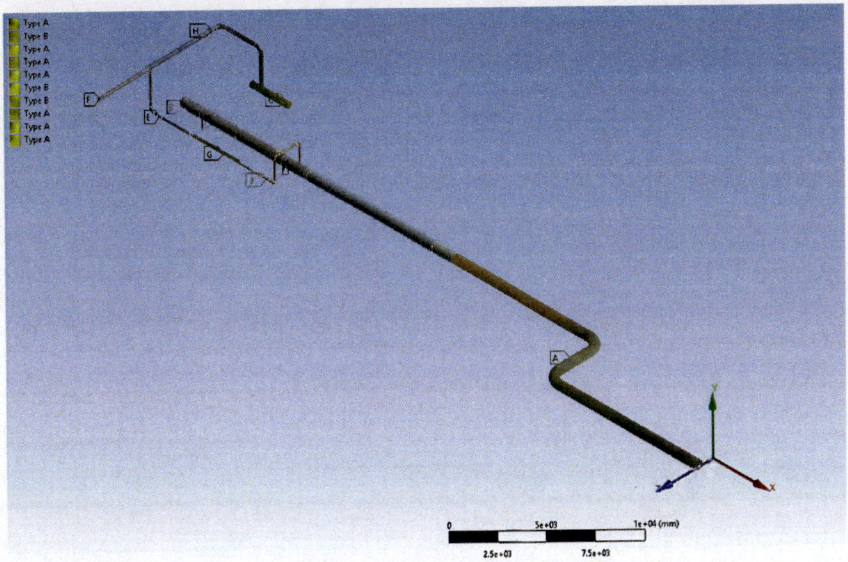

Fig. 3. Definition of boundary conditions.

4 Results

An analysis time period of 10 s has been selected for each transient structural analysis. The predicted maximum deflections for Case 1 to Case 3 conditions are summarized in Table 4, with a total deformation plot for Case 3 shown in Fig. 4.

Table 4. Maximum deflections in pipework

Location	Case 1			Case 2			Case 3		
	X [mm]	Y [mm]	Z [mm]	X [mm]	Y [mm]	Z [mm]	X [mm]	Y [mm]	Z [mm]
Mainline	0.37	1.74	1.07	0.29	1.73	0.63	0.13	1.74	0.17
BDV	0.61	0.57	0.51	0.69	0.57	0.64	0.59	0.57	0.47

The vibration response of the pipe work was also assessed in accordance with the Energy Institute (EI) [3] guidelines to categorise the risk of fatigue. The assessment is based on the calculated frequency, the vibration velocity (RMS value) and the zones defined in Fig. 5.

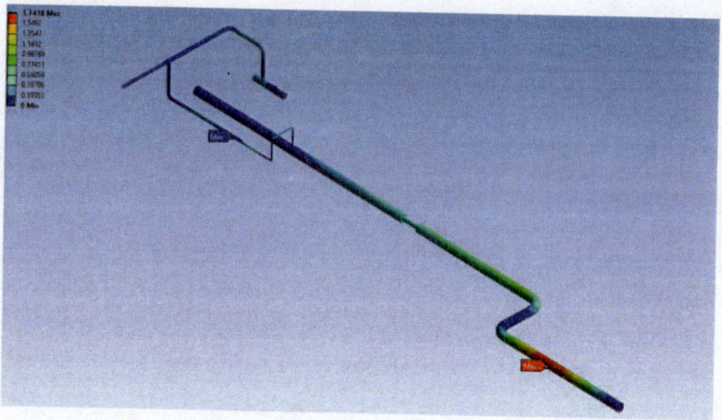

Fig. 4. Case 3: total deformation.

The frequency of vibration and RMS values are obtained by interpreting the velocity time history from the FEA. The following equation if applied for the RMS evaluation:

$$x_{RMS} = \sqrt{\frac{1}{n}\left(x_1^2 + x_2^2 + \ldots + x_n^2\right)}$$

where x_i corresponds to the quantity of which the RMS value is to be evaluated at the i-th time increment.

Based on the observed maximum deflections, one location on the mainline and one location on the blow down valve (BDV) pipework were selected for further consideration, as detailed in Fig. 5.

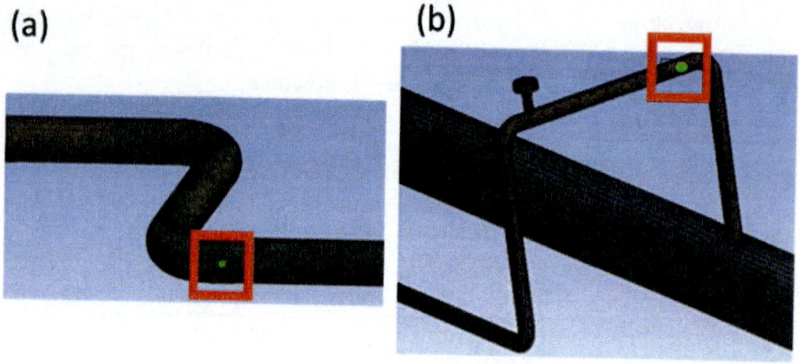

Fig. 5. Locations for further assessment: (a) Mainline Elbow (b) BDV Elbow

The extracted velocity time histories for Location 1 and 2 are therefore as given in Figs. 6 and 7, respectively, for all cases. For Location 1, Case 3 conditions results in higher velocity amplitude compared to the other two cases. For Location 2, Case 2 conditions result in higher response in terms of vibration velocities. In both locations, the higher flow velocity appears to affect the vibration response more than the pressure drop across the FCV.

The calculated frequencies and RMS values are summarized in Table 5. These values were superimposed onto the EI Guidelines Pipework vibration criteria for the mainline and BDV pipework, as shown in Fig. 8.

Table 5. Calculated Frequencies and RMS values

Location		Case 1	Case 2	Case 3
1	Frequency [Hz]	16.67	16.67	16.67
	RMS [mm/s]	5.57	5.95	6.89
2	Frequency [Hz]	14.29	14.29	14.29
	RMS [mm/s]	3.09	3.89	2.30

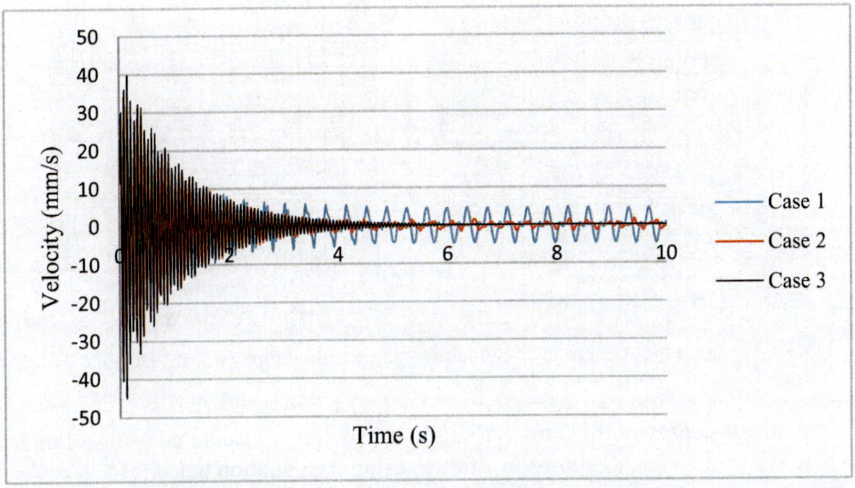

Fig. 6. Location 1, velocity time histories for mainline elbow.

As shown in Fig. 8, Case 1 and 2 mainline vibrations fall in the "acceptable" range but are very close to the "concern" zone. Case 3 mainline vibrations fall in the "concern area". For the BDV pipework vibrations all cases fall in the "acceptable" range.

5 LOF Assessment

The Likelihood of Failure (LOF) calculation is a screening process in line with the Energy Institute (EI) [3] guidelines, which is used to determine the fluid velocity up to which the system can operate safely. If LOF takes a value of 0.5 or higher, a detailed analysis should be undertaken. The LOF value for each case was evaluated and the details of the calculations are presented in the following sections.

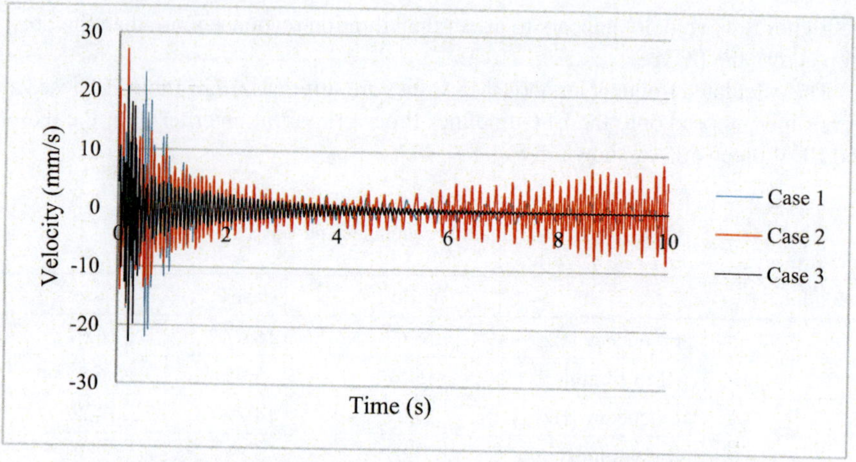

Fig. 7. Location 2, velocity time histories for BDV elbow.

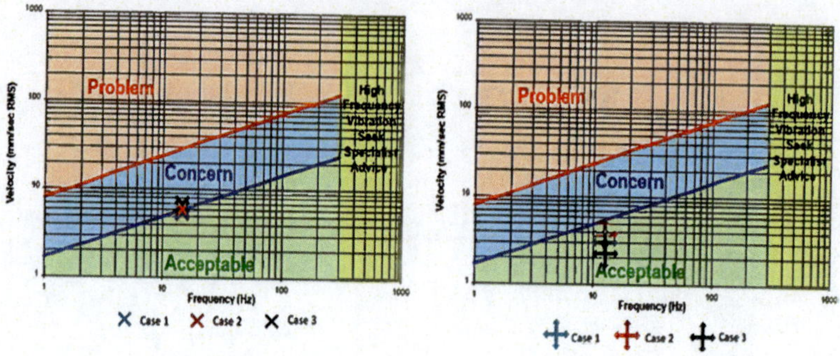

Fig. 8. Vibration assessment to EI guidelines: (a) mainline pipework (b) BDV pipework

The kinetic energy of the fluid as a single-phase flow is calculated using all three of the base case velocities, as shown in Table 6, using the equation below:

$$Kinetic\ energy : \rho v^2 = (actual\ Density) \times (actual\ Velocity)^2$$

Table 6. Kinetic energy at three base case flow velocities

	Fluid velocity [m/s]		
	3.3	6.4	6.0
ρv^2 [kg/ms^2]	563.01	2.118×10^3	1.861×10^3

The amount of turbulent energy partially depends upon the fluid viscosity, which is taken into account by the Fluid Viscosity Factor (FVF) [2]. To determine this value, the dynamic viscosity (μ_{gas}) is needed.

Determining Fluid Viscosity Factor, FVF:

$$\mu_{gas} = 1.08 \times 10^{-5} Pas$$

$$FVF = \frac{\sqrt{\mu_{gas}}}{\sqrt{1 \times 10^{-3}}} = 0.104$$

The FVF for the system is 0.104 and this value was used in the LOF calculation.

For each vibrational analysis, the system was split into three sections of straight pipe to be used in the calculation of LOF. The support arrangements for each length of pipe and the length between supports L_{Span} were used to determine the Flow Induced Vibration factors (Fv), both shown in Table 7.

Table 7. Support arrangements for each pipe length

	L_{Span} [m]		
	10.9	3.019	7.182
Support arrangement	Medium stiff	Stiff	Stiff
Fv	3.061×10^4	7.769×10^4	7.769×10^4

The likelihood of failure for flow induced turbulence was then determined using the equation below:

$$LOF = \frac{\rho v^2}{Fv} FVF$$

where ρv^2 is shown in Table 6, FVF calculated as 0.104 and Fv shown in Table 7. The results for L_{Span} for each case are shown in Table 8.

Table 8. LOF results

		L_{Span} [m]		
		10.9	3.019	7.182
LOF	Case 1	0.002	7.532×10^{-4}	7.532×10^{-4}
	Case 2	0.007	0.003	0.003
	Case 3	0.006	0.002	0.002

In all cases, the LOF value is less than 0.3. According to the EI guidelines, 'A visual survey should be undertaken to check for poor construction and/or geometry and/or support for the main line and/or potential vibration transmission from other sources'.

6 Conclusions

ANSYS computational fluid dynamics and structural finite element analyses were used to investigate the response of a 24″ diameter mainline section of a Gas Metering Train and the associated BDV pipework, located downstream of the FCV, to structural vibrations.

Three local CFD models, simulating the desired maximum, average and minimum pressure drops upstream of the BDV pipework, were developed, and integrated into a global CFD model of the piping system. Steady state and transient coupled CFD and FEA simulations were then used to predict the system vibrations. The associated LOF values were also evaluated in line with the provisions included in Energy Institute Guidelines. Based on the analysis results presented, it can be concluded that the effect of higher flow velocities can be more significant in terms of the predicted line vibrations compared to the pressure drop conditions. In cases 2 and 3 where the flow velocity was above 6.0 m/s, the predicted vibration levels for the mainline section fall in and around the "concern" zone. However, the LOF values calculated for each case are lower than 0.3 meaning that the pipe system demonstrated no immediate fatigue failure concerns and requires only a visual survey to assess the pipeline.

References

1. Ansys Workbench, 2022 R2, ANSYS, Inc.
2. ISO 3181: Petroleum and Natural Gas Industries—Steel Pipe for Pipeline Transportation Systems, 4th ed. (2019)
3. Energy Institute: Guidelines for the Avoidance of Vibration Induced Fatigue Failure in Process Pipework, 2nd ed. (2008)

Flow Induced Forces on Pipe Elbows

Lydia Lymperopoulou[1] and George E. Varelis[2](✉)

[1] University of Thessaly, 38221 Volos, Greece
[2] Worley, Great West Road 27, Brentford TW89BW, UK
Georgios.Varelis@intecsea.com

Abstract. Internal flow in piping systems can be the source of significant forces of dynamic nature caused by variations of the flow characteristics along the length of the piping system and over time. The resulting Flow Induced Vibrations (FIVs) are often associated with fatigue failures. This paper attempts to establish a simple but robust approach to address FIV problems in piping systems resulting from complex flow-related phenomena, focusing on pipe elbows and single-phase flow conditions. The analysis approach presented in this study is numerical and it is based on a combination of advanced computational fluid dynamics (CFD) models with detailed finite element (FE) models of the geometries considered. The results obtained are used to establish analytical closed-form expressions to correlate the piping system's primary geometrical characteristics with the maximum resulting flow-induced forces.

Keywords: FIV · Pipe elbows · FEA · CFD

1 Introduction

Fluid flow through piping systems is a known source of structural force generation that can cause deformations and vibrations of the system. These forces act on top of the pressure forces on the piping system walls and are generated due to changes in the fluid inertia, e.g., due direction change in pipe bends, and due to the development of turbulence as for example downstream of elbows or at small bore connection areas. Turbulence is the source of pressure fluctuation, which in turn causes structural vibrations often associated with fatigue failures. Understanding the impact of all these causes is important for the safe design of piping systems and the avoidance of catastrophic failures.

Elbows are important and widely used components of piping systems. As fluid flows through a system containing an elbow, the fluid particles will alter their principal direction of motion due to the elbow presence. A pressure gradient is caused by the elbow curvature, resulting in drop of the fluid velocity near the elbow extrados. The opposite will occur towards the elbow intrados, where the fluid flow accelerates. As a result, an imbalanced force occurs in the fluid volume because of the pressure gradient, while the considerable pressure differential causes vortex pair and turbulence which, in turn, result in structural vibration.

G. Guxho et al. (Eds.): ITC-ICEE 2023, LNMIE, pp. 285–296, 2024.
https://doi.org/10.1007/978-3-031-48933-4_27

The study of Flow Induced Vibrations (FIVs) in piping systems is a multidisciplinary problem, quite challenging and highly nonlinear in nature. Over the last decades, FIVs were investigated both experimentally as well as using numerical tools. For the specific case of piping systems containing elbows, a substantial amount of scientific research has been conducted to examine the response of elbows in unstable flow conditions. Ramanathan [1] presented a thorough numerical investigation on how elbows affect the fluid flow and how they should be arranged in piping systems. Homicz [2] collected numerical data on the erosion and corrosion of elbows by performing analysis on three different turbulent models. Kalpakli [3] used experimental techniques to study the form of the secondary flow passing through an elbow and Kim et al. [4] used computational methods to study the intensity of the vortices created by the presence of elbows and its dependence on the radius of curvature and Reynolds number. Dutta et al. [5, 6] numerically investigated the separation of the turbulent flow and the complex secondary flow, alongside with the effect of Reynolds number and curvature ratio. Experimental investigation was performed by Jain [7] to study the two oscillating eddies using Particle Image Velocimetry and by Shwin et al. [8] using Ultra-sound Technique. Apalowo [10] compared different CFD models with experimental data to conclude which one presents the most accurate results and Zhang et al. [11] investigated the effect of different bending angles of pipe elbows. Finally, Abuhadira [12] developed a strategy on how the phenomenon of Flow Induced Vibrations, occurring due to the presence of elbows, can detect whether pipe leakage is present in a pipeline system. At the same time, attention was also paid to the case of multiphase flow. An attempt to understand the phenomenon of Flow-Induced Vibration in multiphase flow was made by Shuichiro et al. [9] among others. Finally, Riverin et al. [13] analysed the vibrational forces resulting from an air–water mixture flowing through a bend and a tee and introduced a dimensionless formula for the early design stage.

For everyday design practice purposes, to screen for and to avoid FIV phenomena, the Energy Institute "Guidelines for the avoidance of vibration induced fatigue failure in process pipework" [14] are widely used. A full description of the mechanisms that cause FIV is provided, as well as recommendations for avoiding known sources of turbulence generation. However, it is important to note that these guidelines are only used for screening and do not provide a quantitative assessment methodology to define stresses or vibrations for flow-induced turbulence cases. In the case of concern, advanced analysis is recommended if redesigning the system is not possible or desirable.

The present study is an effort to suggest an analytical tool which can be used in the early design stage to quantify the flow induced forces on 90-degrees pipe elbows for single-phase flow conditions. The research consists of numerical simulations performed using ANSYS Workbench commercial software package. Coupled computational fluid dynamic (CFD) and finite element analyses (FEA) were conducted. The CFD models were able to accurately represent the fluid flow through a piping system containing a 90-degrees elbow and its specific characteristics, including the two counter-rotating vortex pair and the resulting pressure pulsation. At the same time, the Fluid-Structure Interaction (FSI) effects were investigated by mapping the CFD predicted pressure profile on the piping system walls to evaluate its effects in terms of resulting stresses and system vibrations. Five different flow conditions corresponding to different Reynolds numbers were investigated, combined with three different pipe support locations. In each analysis case, the maximum force reaction at the pipe support location due to the fluid flow and associated turbulence phenomena was monitored.

2 Numerical Analysis

The CAE/Multiphysics engineering simulation software Ansys Workbench was employed for the parametric study. The analysis involved both CFD and FEA models, combined assuming one-way coupling to investigate FSI effects. One-way coupling was considered adequate for the purposes of the present study since no substantial deformations of the pipe geometry were expected that could significantly affect the fluid flow.

A 4-inch (nominal diameter) SCH40 elbow was analyzed. The pipe outer diameter D was 114.3 mm, its bend radius R was 152 mm, and the pipe thickness t was 6.02 mm. Each straight part of the piping specimen L was 1714.5 mm, equal to 15D. The pipe material was assumed to be P355 steel grade with a nominal yield strength σ_y equal to 355 MPa, a Young's modulus E equal to 200 GPa and Poisson's ratio set to 0.3. The fluid running through the pipe was assumed to be water with density equal to 997 kg/m^3. It was further assumed that the piping assembly is horizontal and therefore gravity effects were not considered. A schematic representation of the overall geometry is shown (see Fig. 1).

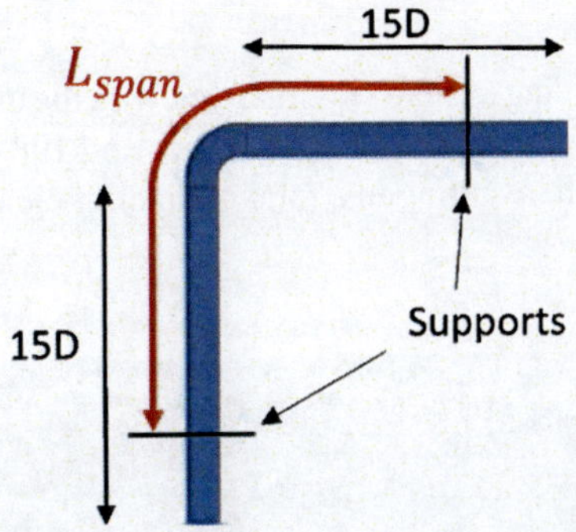

Fig. 1. Piping system schematic representation

The CFD analysis was conducted first using ANSYS CFX software. At the inlet of the pipe, five different velocities were set, namely 1, 5, 10, 15, 20 m/s. At the pipe outlet the relative pressure was fixed at 1 atm, while the domain pressure was also set to 1 atm. No slip conditions and smooth wall setup was chosen for the boundary condition at the pipe wall (see Fig. 2). The fluid domain contained within the piping system was modeled employing 820,500 solid elements (see Fig. 2). The "mesh inflation" method was used to generate thinner element layers near the pipe walls to capture the velocity gradient within the boundary layer of the flow. The straight part of the piping specimen L = 15D was selected long enough to allow for the full development of the flow before reaching the elbow area.

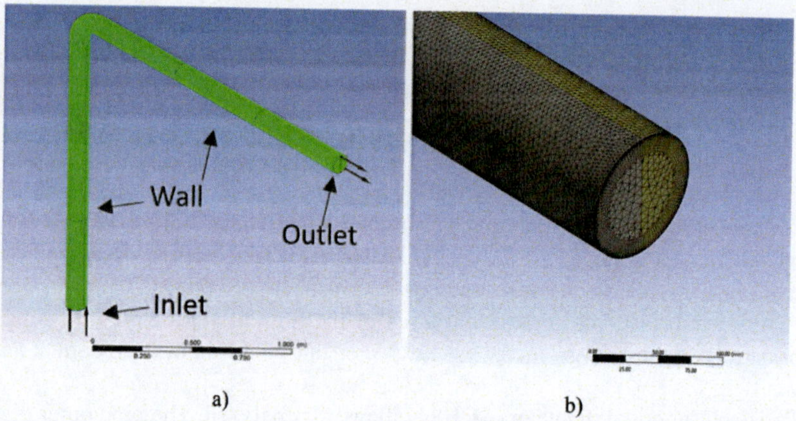

Fig. 2. a) CFD domain and boundary conditions, b) CFD mesh detail The FE analysis considered geometrical nonlinearities through a large-deformations formulation. Three different support locations were considered on each straight pipe segment of the piping system assembly, namely at 5D, 10D and 15D away from the start/end of the elbow component. The boundary conditions were enforced using "remote displacement" setups. At the boundary condition locations all displacement and rotational degrees of freedom were set to zero apart from the displacements along the axes of the straight pipe segments, which were allowed. Typical boundary conditions application is shown in Fig. 3. In each case, the length of the piping system between supports (L_{span}) was defined as shown in Fig. 1. L_{span} is equal to 3.67m, 2.23m and 1.38m for support at locations 15D, 10D and 5D, respectively.

The steel part of the piping system was modeled with the use of linear shell elements; 32 elements were used along the pipe circumference and 198 elements were used along the piping assembly length, resulting in a total number of elements equal to 11,700 (see Fig. 4).

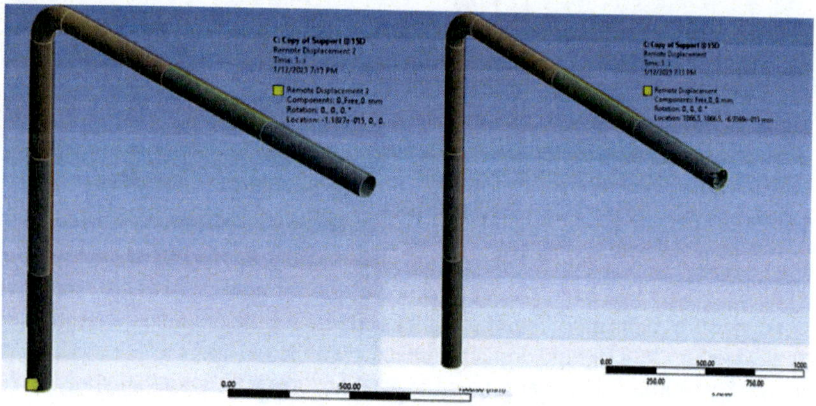

Fig. 3. FEA model boundary conditions

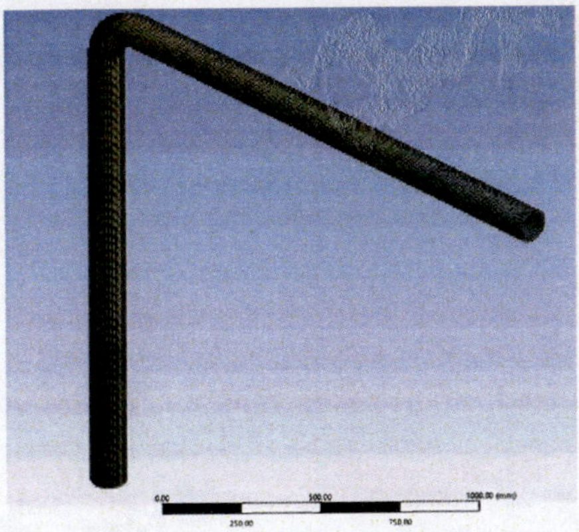

Fig. 4. FEA mesh

3 Analysis Results

3.1 CFD Analysis Results

The analysis initially focused on understanding the internal flow characteristics and how these are affected by the presence of the elbow. A steady state CFD analysis was performed first for the preset flow conditions. Subsequently, the steady-state analysis results were set as initial conditions of each transient analysis that followed. A suitably small time increment of 0.001 s and the k-epsilon turbulence model were employed in the transient analyses to adequately capture turbulence phenomena and pipe wall pressure fluctuations throughout the analysis time. Figure 5(a) illustrates the streamlines predicted along the piping system, while Fig. 5(b) depicts the two counter-rotating vortex pair that appear downstream of the elbow. Finally, Fig. 6 shows the pressure profile variation along the piping system.

For each analysis conducted, the pressure profile on the pipe wall was transferred to a coupled structural transient analysis module. The analysis settings were adjusted to be compatible between the CFD and structural transient FEA modules in terms of total analysis time and time stepping. The pressure profile evaluated from the CFD analysis was automatically imported to the structural analysis module and applied at the piping wall, which was the interface between the fluid and structural domains.

3.2 Structural Analysis Results

A modal analysis was performed first for each one of the three different configurations of support locations. The following Table 1 lists the first six natural frequencies for each support configuration.

The sinusoidal form of the force reaction at the inlet of the pipe for the case of Reynolds number Re= 101953 and support at 15D is presented as an indicative result (see Fig. 7). From each response diagram, the response frequency was calculated. It was observed that for each support condition, a different response frequency developed which was independent of Reynolds number. This frequency is also listed in Table 2.

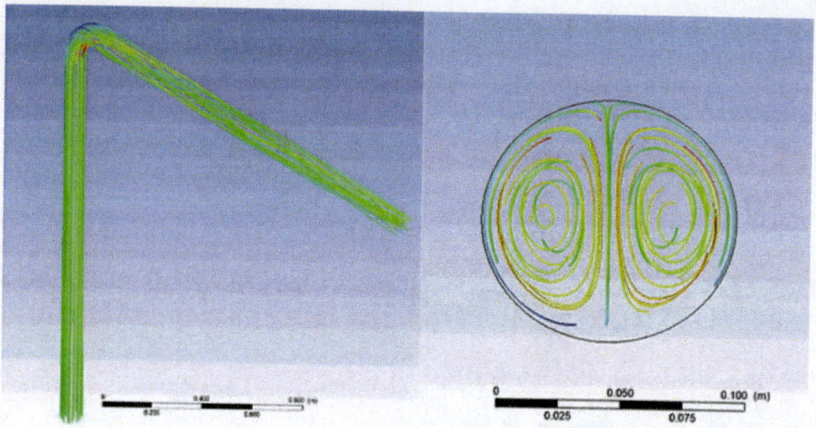

a) b)

Fig. 5. CFD results: (a) streamlines along the piping system, (b) two counter-rotating vortex pair.

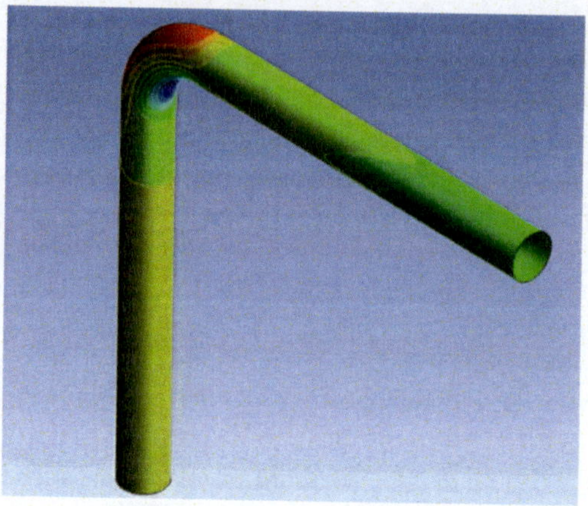

Fig. 6. CFD predicted pressure profile along the piping system.

However, it was evident that these frequencies were close to the natural frequencies evaluated from the modal analysis. For the cases of support locations at 10D and 15D the second vibration mode was excited, while for the case of support locations at 5D the sixth vibration mode was excited.

A summary of the analysis results in terms of the maximum force reaction F_{max} and the corresponding RMS force F_{rms} for each case are presented in Table 3. It is also noteworthy to compare the results of the current study to those of Reverin et al. [13], which refer to similar two-phase flow experiments. In this work it was conducted that for two-phase flow conditions $F_{max} / F_{rms} \approx 5$. The same ratio was calculated for the results of this study to be between 1.66–1.94, indicating that two-phase flow conditions can be more severe in terms of resulting flow induced forces.

Table 1. Natural frequency characteristics for different support locations

Support location	@15D	@10D	@5D
Modes	Natural frequencies (Hz)		
1	13.36	24.17	57.74
2	21.88	38.95	78.74
3	34.31	73.00	79.04
4	130.09	264.14	79.18
5	131.86	264.99	79.32
6	172.95	293.58	93.05
System response frequency	22	39	90

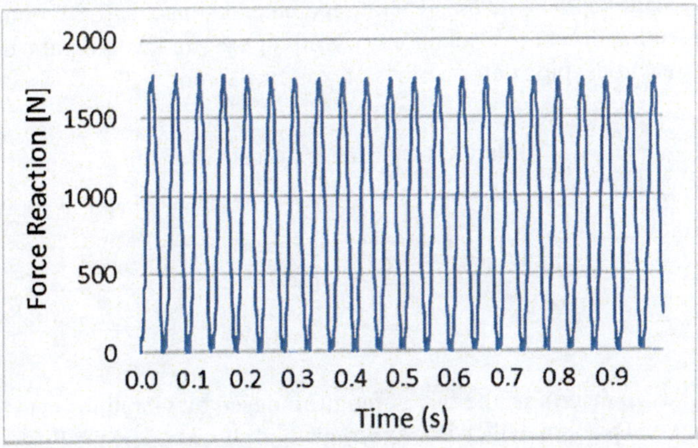

Fig. 7. Force reaction time history, Re = 101953, Support location @15D

3.3 Energy Institute Guidelines

According to the Energy Institute Guidelines [15], proactive assessments should be conducted on a central pipeline to check for failure. Potential excitation mechanisms should be identified, and their risk shall be considered. For each case of external excitation, a likelihood of failure (LOF) number should be calculated. This number is calculated for the preventive assessment of an installation and predicts a possible failure. It is not an absolute probability, nor an absolute measure, but a conservative measure for designing purposes. In the specific case where the excitation is caused by Flow Induced Turbulence, the fluid flow produces the turbulent energy and the LOF is given by the following formula:

$$LOF = \rho v^2 FVF/Fv \tag{1}$$

where ρ is the density of the fluid, v the velocity of the fluid, FVF is the Fluid Viscosity Factor, which is equal to 1 for liquids, and Fv is the Flow Induced Vibration factor. This factor depends on the type of support and the geometrical characteristics of the pipe. For Stiff pipes with an outside diameter of 60–762 mm the formula given for Fv is as follows:

$$Fv = exp[\alpha(\mathrm{D}/t)^{\beta} \tag{2}$$

D is the external diameter, t is the pipe thickness and α, β are defined as:

$$\alpha = 446187 + 6,46\mathrm{D} + 9,17 * 10 - 4D^3 \tag{3}$$

$$\beta = 0,1\ ln(D){-}1,3739$$

The results for the LOF number for each case analyzed are listed in Table 2. Apart from the cases with inlet velocity of 1 m/s, all other cases analyzed result in a LOF > 0.5, which means that the line shall be redesigned, reanalyzed, and monitored for vibrations. Also, corrective measures shall be taken, and the bore connections must be assessed. It should be stated that despite the general recommendations, the LOF number cannot be directly used for structural evaluations of a piping system involving the definition of pipe support reactions, pipe deformations or resulting forces.

Table 2. LOF number calculations.

Inlet velocity (m/s)	1	5	10	15	20
Re	101953	509766	1019532	1529298	2039064
LOF	0.02	0.54	2.15	4.83	8.59

Another way to determine the risk of fatigue induced by vibrations is to employ the Vibration Criteria Diagram, which is a survey method suggested by the Energy Institute. The measured vibration velocity (RMS) and the peak frequency of the response are used to characterize a specific vibration into three zones, namely "Acceptable", "Concern" and "Problem", as shown in Fig. 8. If the vibration level falls in the "Problem" area, then control measures should be taken immediately to define the likelihood of failure. If it falls in the "Concern" area, then there is a possibility for fatigue damage occurrence and control measures should also be taken. For both cases checks on the welds should be performed to ensure a fatigue crack has not been initiated. Vibrations levels that fall into the "Acceptable" area need not to be further investigated.

This methodology was used to categorize all the vibration levels predicted numerically, based on the different input velocities and support configurations analyzed. It can be seen in Fig. 8 that most of the cases fall within the "Problem" region and only the cases for inlet velocity 1,5 and 10 m/s for support location at 5D fall within the "Concern" region.

3.4 Maximum Flow Induced Force Evaluation

The main aim of this study was to establish an analytical equation to predict the maximum flow induced force for a piping system containing an elbow, which can be used to

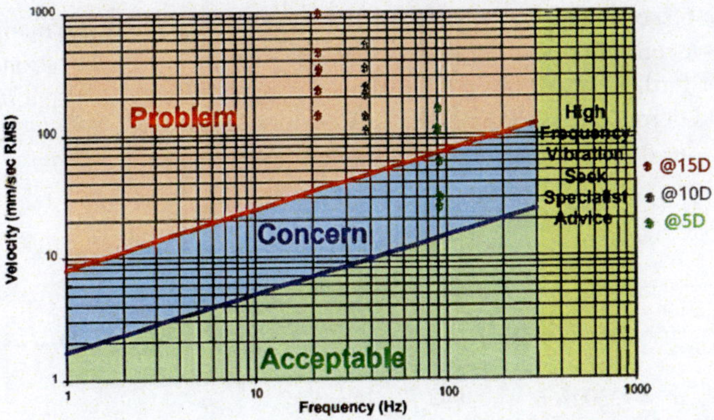

Fig. 8. Summary of analysed cases shown in the vibration assessment diagram

assist engineers at the initial stages of the piping system design. For this purpose, the dimensionless force ratio c_f is defined as:

$$c_f = \frac{F_{max}}{F_{pressure}}, \quad F_{max} = F_{flow} + F_{pressure}$$

F_{max} represents the maximum force occurring at the support of the piping system due to internal fluid flow (F_{flow}) and pressure ($F_{pressure}$). For the specific geometry and pressure conditions, $F_{pressure} = P * A$, where P is the line pressure and $A = \pi D_{in}^2/4$. It should be noted that if the line is pressurized but no flow is allowed, then $F_{max} = F_{pressure}$ and $c_f = 1$.

Furthermore, the dimensionless parameter λ is introduced, defined as:

$$\lambda = \frac{I}{L_{span}^4} \tag{5}$$

where I is the pipe section moment of inertia and L_{span} is the length of the pipe between supports. The parameter λ can be interpreted as a measure of the systems stiffness. A summary of the analysis results presented using the parameters c_f and λ is shown in Table 3 and in Figs. 9 and 10

From Fig. 9 it can be observed that the correlation between the c_f parameter and the Re number follows a second order polynomial. It can be also observed from Fig. 10 that there is a very small dependency of the c_f value to the range of λ parameter considered for given Re number. Based on the above, the following equation is proposed:

$$c_f(Re) = aRe^2 + bRe + c \tag{6}$$

$$a = 8.4517 * 10^{-09}, b = 1.2115 * 10^{-05}, c = 1$$

Equation (6) is also plotted in Figs. 9 for comparison with the numerical results. The proposed analytical expression allows for the definition of the maximum force at an elbow with respect to the force due to pressure solely based on the Re number. Hence it constitutes a convenient tool for ever day engineering practice applications, which can be used for further evaluation of the anticipated reaction forces at supports, as well as stresses and deformations of the piping system by using simple static FEA models and without the need for complicated and resource-expensive coupled CFD/FEA analyses.

Table 3. Summary of analysis results

Support location	Re	F_{max}	F_{rms}	F_{max}/F_{rms}	c_f	λ
@5D	101953	1684.3	923.8	1.82	1.01	8.26E-07
	509766	1994.1	1142.2	1.75	1.20	
	1019532	3846.3	2000.3	1.92	2.31	
	1529298	6478.2	3337.4	1.94	3.89	
	2039064	9556.6	5090.6	1.88	5.74	
@10D	101953	1727.0	923.8	1.66	1.04	7.41E-08
	509766	2220.3	1142.2	1.66	1.33	
	1019532	3679.4	2000.3	1.67	2.21	
	1529298	6105.8	3337.4	1.67	3.67	
	2039064	9457.5	5090.6	1.67	5.68	
@15D	101953	1784.9	1061.9	1.68	1.07	1.66E-08
	509766	2275.9	1354.6	1.68	1.37	
	1019532	3766.8	2244.2	1.68	2.26	
	1529298	6246.2	3710.0	1.68	3.75	
	2039064	9696.7	5739.4	1.69	5.83	

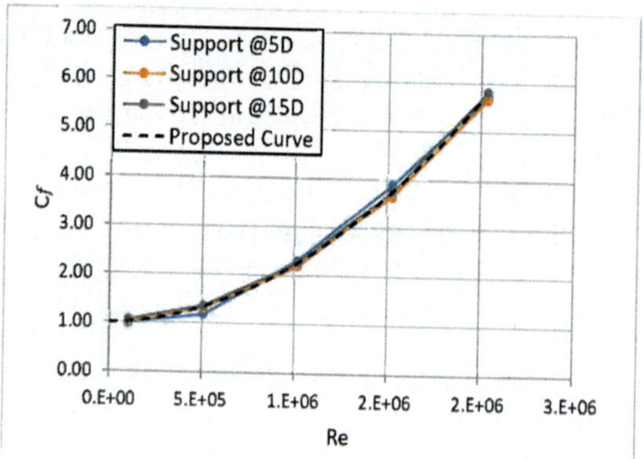

Fig. 9. Correlations between Reynolds number and C_f ratio

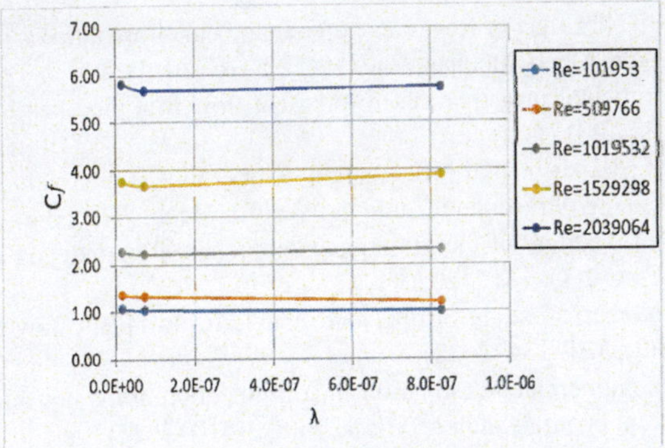

Fig. 10. Correlations between λ ratio and C_f ratio

4 Conclusions

Flow induced vibration phenomena were investigated using coupled CFD and FEA models of a piping system containing a 90-degrees elbow and single-phase liquid (water) flow conditions, considering different flow conditions and support arrangements. It was demonstrated that the turbulence generated in the fluid flow at the elbow area results in fluctuations of the pressure profile acting on the pipe walls, which is the source of structural vibrations and potential fatigue failure of the piping system.

The vibration characteristics of the piping assembly were investigated using the traditional approach proposed by the Energy Institute guidelines, as well as based on the numerical results. A good agreement between the two approaches in terms of the severity classification of each case analyzed was found. Furthermore, based on the numerical results, an analytical equation was proposed, correlating the internal flow conditions expressed through the Reynolds number with the maximum expected structural force generated by the internal flow. This analytical expression can be used in early design stages alongside with the Energy Institute LOF classification for quantification of the FIV phenomena of piping systems containing elbows, without the need for complicated and computationally expensive numerical analyses.

References

1. Ramanathan, K.: Computational Simulation of Flow Through Elbows in the Approach Piping Systems to a Headbox. Western Michigan University (1992)
2. Homicz, G.F.: Computational Fluid Dynamic Simulations of Pipe Elbow Flow. No. SAND2004-3467. Sandia National Laboratories (SNL), Albuquerque, NM, and Livermore, CA (United States) (2004)
3. Kalpakli, A.: Experimental study of turbulent flows through pipe bends. Diss. KTH Royal Institute of Technology (2012)
4. Kim, J., Yadav, M., Kim, S.: Characteristics of secondary flow induced by 90-degree elbow in turbulent pipe flow. Eng. Appl. Comp. Fluid Mech. **8**(2), 229–239 (2014)
5. Dutta, P., et al.: Numerical study on flow separation in 90° pipe bend under high Reynolds number by k-ε modelling. Eng. Sci. Tech., an Int. J. **19**(2), 904–910 (2016)

6. Dutta, P., Nandi, N.: Effect of Reynolds number and curvature ratio on single phase turbulent flow in pipe bends. Mech. Mechan. Eng. **19**(1), 5–16 (2015)
7. Jain, A.: Experimental investigation of turbulent flow in a pipe bend using particle image velocimetry. Diss (2017)
8. Shwin, S., et al.: Experimental investigation of two-dimensional velocity on the 90 double bend pipe flow using ultrasound technique. World J. Mech. **7**(12), 340–359 (2017)
9. Miwa, S., Hibiki, T.: State-of-the-art in plant component flow-induced vibration (FIV). Exp. Comp. Multi. Flow **2**(1), 1–12 (2020)
10. Apalowo, R.: Numerical study of different models for turbulent flow in 90° pipe bend. U. Porto J. Eng. **8**(2), 121–134 (2022)
11. Zhang, J., et al.: Numerical investigation and optimization of the flow characteristics of bend pipe with different bending angles. Proces. **10**(8), 1510 (2022)
12. Abuhatira, A., Salim, S.M., Vorstius, J.: Computational study of vibration-based leak detection approach in a 90-degree pipe elbow. In: UK Fluids Conference 2021. The Interdisciplinary Teaching and Learning Group, University of Southampton (2021)
13. Riverin, J.L., De Langre, E., Pettigrew, M.J.: Fluctuating forces caused by internal two-phase flow on bends and tees. J. Sound Vib. **298**(4–5), 1088–1098 (2006)
14. Energy Institute: Guidelines for the Avoidance of Vibration Induced Fatigue in Process Pipework, 2nd ed. (2008)

The Influence of Thermodynamic Parameters on Internal Combustion Engines and the Graphical Presentation of Cycles Diagrams Using Matlab

Ardit Gjeta(✉), Majlinda Alcani, and Altin Dorri

Department of Energy, Faculty of Mechanical Engineering, Polytechnic University of Tirana, Tirana, Albania
agjeta@fim.edu.al

Abstract. In this paper, a software package written in Matlab that allows setting up the engine thermodynamic parameters of internal combustion engine cycles has been developed. A Matlab-based graphical user interface is used to assist the lecturer in reducing the efforts in teaching applications for the variation of engine parameters; it is also a perfect tool for graphical presentation. Students can also benefit from the program code, where they can solve examples and compare the results they will get manually while solving their project. To derive the engine cycle data, the engineering model is used with simplifications such as: a constant specific heat is assumed for the air in the model; the piston-cylinder coupling is accepted as a closed system; every process is internally reversible, including the compression and expansion processes; and kinetic and potential energy effects are negligible. An application has been made by using the software to demonstrate the effect of thermodynamic parameters on internal combustion engines, such as engine work output, cycle thermal efficiency, Carnot efficiency, etc., and the graphical presentations of cycle diagrams.

Keywords: Thermodynamics cycle · Internal Combustion Engines (ICE) · Otto · Diesel · Sabathe · Matlab GUI

1 Introduction

Designing an internal combustion engine requires careful consideration of a wide range of parameters. A new internal combustion engine project begins in the industrial world with the creation of a clearly defined solution and the expertise of the designer [1]. In the academic field, students studying engine cycle design possess limited resources to comprehend the design process and the methods for determining component dimensions. For that reason, in this paper, a series of procedures are implemented using graphical user interfaces (GUIs) in the Matlab programming language. The purpose of this GUI is to provide engine designers and students with an easy tool that enables them to assess various engine parameters, compare engine types, and provide a graphical presentation

G. Guxho et al. (Eds.): ITC-ICEE 2023, LNMIE, pp. 297–303, 2024.
https://doi.org/10.1007/978-3-031-48933-4_28

of cycle diagrams. The choice of Matlab was made due to the possibility of creating an independent interface through which each designed parameter can be executed after entering some thermodynamic parameters of the remodeled internal combustion engine [2]. We consider three air-standard cycles: Otto, Diesel, and Sabathe. These cycles differ from each other only in the way the heat transfer process is modeled, which replaces the combustion process in the current cycle.

1.1 Otto Cycle

An ideal cycle known as the air-standard Otto cycle posits that heat addition takes place instantly when the piston is at top dead center [3–5]. The Otto cycle is shown on the pressure-specific volume and temperature-entropy diagrams in Fig. 1. Four internally reversible processes are sequentially performed during the cycle. Process 1–2: an adiabatic compression; Process 2–3: a constant-volume heat transfer to the air; Process 3–4: an adiabatic expansion; Process 4–1: a constant-volume process that completes the cycle.

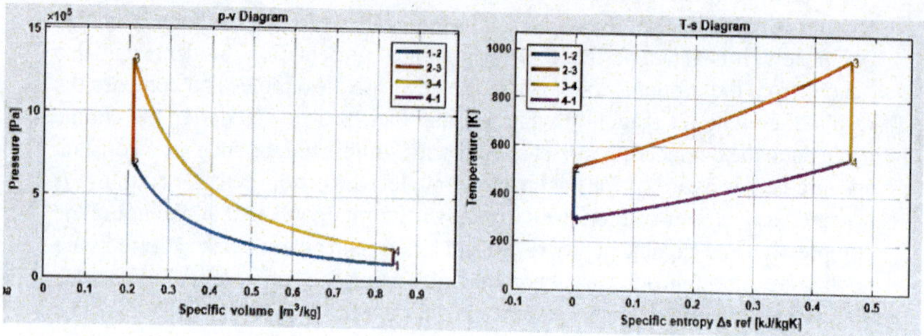

Fig. 1. Pressure–specific volume and temperature–entropy diagram for the ideal OTTO cycle

1.2 Diesel Cycle

The air-standard Diesel cycle is a hypothetical cycle that starts with the piston at the top dead center and implies heat addition takes place throughout a constant-pressure process [3–5]. The Diesel cycle is shown on the pressure-specific volume and temperature-entropy diagrams in Fig. 2. Four internally reversible processes are sequentially performed during the cycle. Process 1–2: an adiabatic compression; Process 2–3: a constant-pressure heat transfer to the air; Process 3–4: an adiabatic expansion; Process 4–1: a constant-volume process that completes the cycle.

1.3 Sabathe Cycle

Thermal cycles like the Otto cycle and the Diesel cycle are combined to create the Sabathe cycle, otherwise known as the mixed/dual combustion cycle. The pressure-specific volume and temperature-entropy diagram of the Sabathe cycle is shown in Fig. 3.

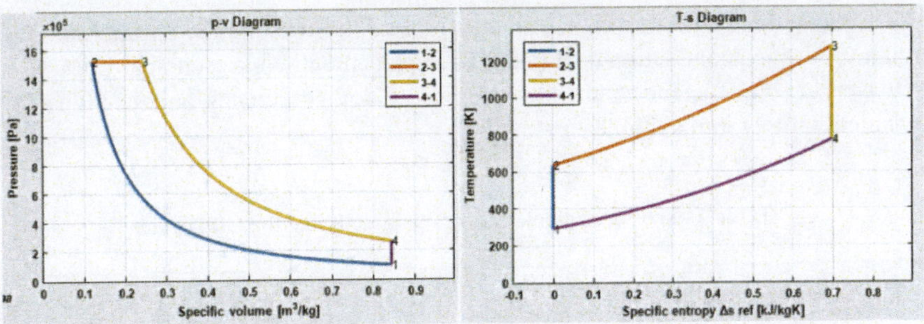

Fig. 2. Pressure–specific volume and temperature–entropy diagram for the ideal Diesel cycle

Unlike the previous two cycles, heat addition takes place in two steps: the addition of heat at a constant volume during process 2–3, and the addition of heat at a constant pressure during process 3–4. As in the situations of the Otto and Diesel cycles, regions on the temperature-entropy and pressure-specific volume diagrams can be understood as heat and work, respectively.

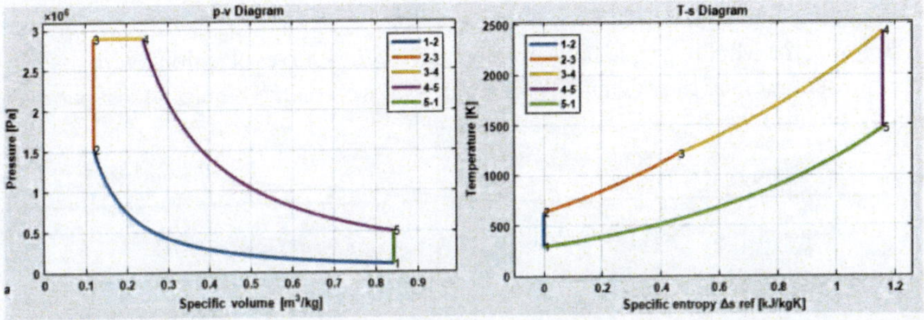

Fig. 3. Pressure–specific volume and temperature–entropy diagram for the ideal Sabathe cycle

2 Matlab GUI for Internal Combustion Engine Design

In this paper, a graphical user interface (GUIs) is used to implement several operations in the Matlab programming language to develop and validate each of the engine's key parts. Each of the initial thermodynamic design parameters can be changed according to the desired value imposed by the user, such as the mass of air, ambient pressure, inlet temperature, polytropic index n1, n2 (1.4 is used as an isentropic exponent in this case), compression ratio, pressure ratio, cutoff ratio, air gas constant, and ambient temperature.

After all the desired parameters have been set, click on the button with the name of the cycle and immediately get the graphic representation in the pressure-specific volume and temperature-entropy diagrams of the required cycle, as well as table data with all the parameters of the cycle, such as state parameters (pressure, temperature in Kelvin and Celsius, specific volume).

The next table contains all the parameters of the cycle, such as internal energy, enthalpy, entropy, conduction heat transfer, displacement work, technical work, exergy, cycle work, cycle efficiency, indicators pressure, Carnot efficiency, and the high and low equivalent temperatures (Table 1).

Table 1 Thermodynamic process parameters of the Otto cycle.

State	Pressure [Pa]	Temperature [K]	Temperature [C]	Specific Volume [m3/kg]	Process	Cpm [kJ/kgK]	Cvm [kJ/kgK]	k
1	40000	280	7	2.0090	1-2	1.0045	0.7175	1.4000
2	8.6696e+05	674.3029	401.3029	0.2232	2-3	1.0045	0.7175	1.4000
3	2.4275e+06	1.8880e+03	1.6150e+03	0.2232	3-4	1.0045	0.7175	1.4000
4	1.1200e+05	784.0000	511.0000	2.0090	4-1	1.0045	0.7175	1.4000

Process	deltU [kJ]	deltH [kJ]	deltS [kJ]	Q [kJ]	L [kJ]	Lt [kJ]	EXQ [kJ]
1-2	282.9123	396.0773	0	0	-282.9123	-396.0773	0
2-3	870.8622	1.2192e+03	0.7388	870.8622	0	-348.3449	669.0721
3-4	-792.1545	-1.1090e+03	0	0	792.1545	1.1090e+03	0
4-1	-361.6200	-506.2680	-0.7388	-361.6200	0	144.6480	-159.8299
Closed Cycle	2.8422e-13	1.7053e-13	0	509.2422	509.2422	509.2422	509.2422

Cycle work [kJ]	Cycle efficiency [%]	Indicatorial Pmi [bar]	Carnot Efficiency [%]	TeqH [K]	TeqL [K]
509.2422	58.4756	0.9094	85.1699	1.1788e+03	489.5013

Below is the Matlab graphical user interface for the Otto cycle, along with the graphical representation of the p-v and T-s diagrams as well as the results of the parameters (Fig. 4; Table 2).

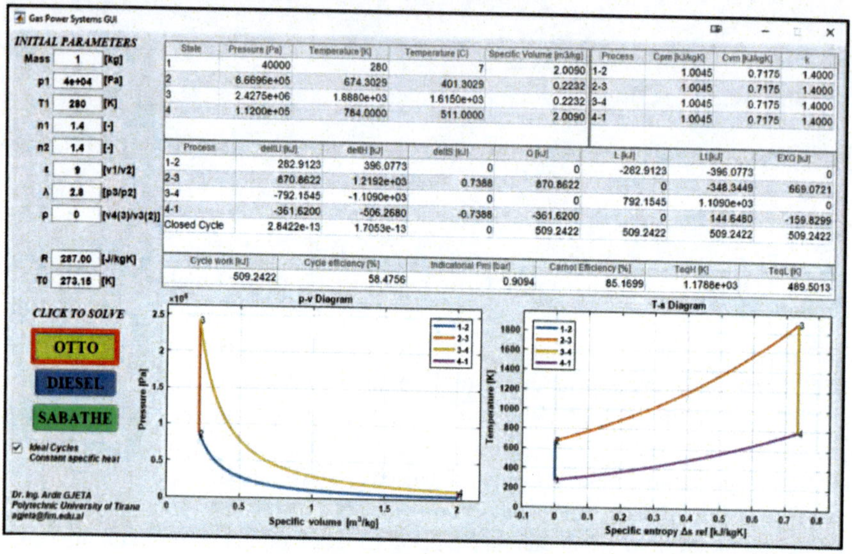

Fig. 4. Matlab GUI for Otto cycle

Table 2 Thermodynamic process parameters of the diesel cycle.

State	Pressure [Pa]	Temperature [K]	Temperature [C]	Specific Volume [m3/kg]	Process	Cpm [kJ/kgK]	Cvm [kJ/kgK]	k
1	40000	280	7	2.0090	1-2	1.0045	0.7175	1.4000
2	2.6516e+06	928.0471	655.0471	0.1005	2-3	1.0045	0.7175	1.4000
3	2.6516e+06	1.6705e+03	1.3975e+03	0.1808	3-4	1.0045	0.7175	1.4000
4	9.1084e+04	637.5871	364.5871	2.0090	4-1	1.0045	0.7175	1.4000

Process	deltU [kJ]	deltH [kJ]	deltS [kJ]	Q [kJ]	L [kJ]	Lt [kJ]	EXQ [kJ]
1-2	464.9738	650.9633	0	0	-464.9738	-650.9633	0
2-3	532.6990	745.7787	0.5904	745.7787	213.0796	0	584.5022
3-4	-741.1041	-1.0375e+03	0	0	741.1041	1.0375e+03	0
4-1	-256.5688	-359.1963	-0.5904	-256.5688	0	102.6275	-95.2923
Closed Cycle	0	1.1369e-13	-3.3307e-16	489.2099	489.2099	489.2099	489.2099

Cycle work [kJ]	Cycle efficiency [%]	Indicatorial Pmi [bar]	Carnot Efficiency [%]	TeqH [K]	TeqL [K]
489.2099	65.5972	0.9337	83.2384	1.2631e+03	434.5444

Below is the Matlab graphical user interface for the Diesel cycle, along with the graphical representation of the pressure-specific volume and temperature-entropy diagrams as well as the results of the parameters (Fig. 5).

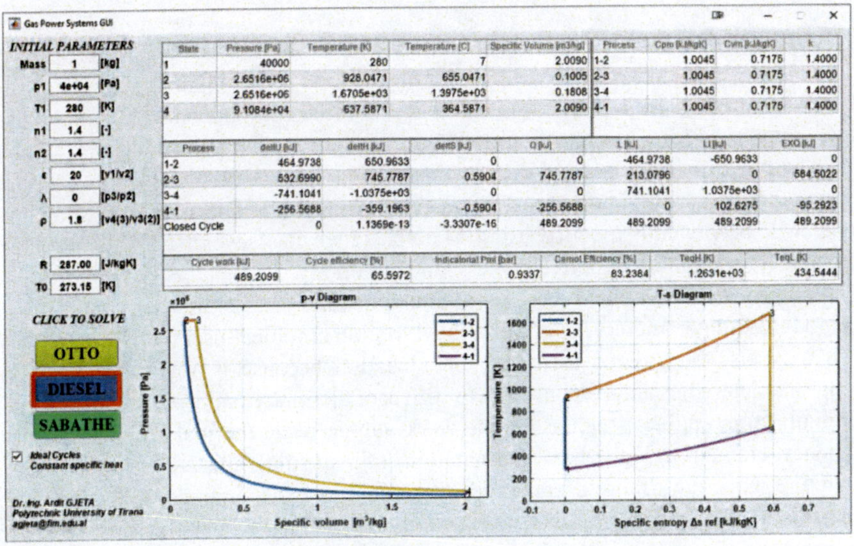

Fig. 5. Matlab GUI for diesel cycle

Below is the Matlab graphical user interface for the Sabathe cycle, along with the graphical representation of the pressure-specific volume and temperature-entropy diagrams as well as the results of the parameters (Fig. 6).

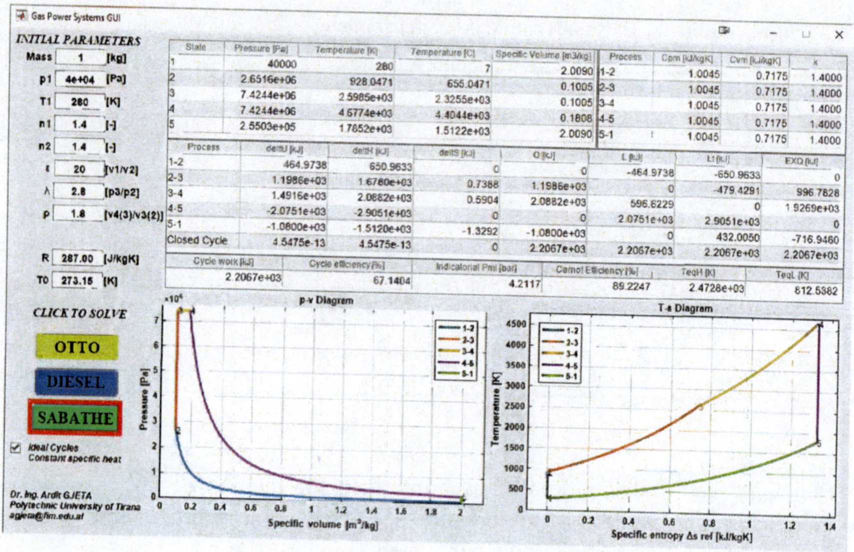

Fig. 6. Matlab GUI for dual cycle engine

3 Conclusion

Due to their high energy density and efficiency, internal combustion engines have been widely employed for a variety of purposes, including transportation and power production. The paper presents a graphical user interface (GUI) to facilitate finding thermodynamic parameters and comparing between engine cycles. The suggested simulation model will be put into practice using Matlab, which has several benefits, including adaptability, computational effectiveness, and simplicity.

To study various engine types and alternative solutions, the user may alter several thermodynamic parameters using the created Matlab graphical user interface.

It may also be utilized as a tool for education because it offers a summary of engine characteristics and demonstrates the effect of thermodynamic parameters on internal combustion engines, such as engine work output, cycle thermal efficiency, Carnot efficiency, etc., through graphical presentations of cycle diagrams.

Future work for a more accurate design of engine cycles includes the study of the effect of non-constant specific heat, non-ideal cycles (polytropic compression and expansion processes), and further development of the code for application to gas turbine cycles.

References

1. Delprete, C., Pregno, F., Rosso, C.: Internal combustion engine design: a practical computational methodology. SAE Int. J. Engines **2**(1), 263–270 (2009)
2. Bhowmick, S., Singh, B., Ansari, M.M., Sharma, M.: Simulation of dual combustion cycle in internal combustion engines using MATLAB. Int. J. Innov. Res. Sci. Eng. Technol. (IJIRSET), pp. 9308–9311 (2023)

3. Borgnakke, C., Sonntag, R.E.: Fundamentals of Thermodynamics, 7th edn. Wiley & Sons, Inc. University of Michigan (2009)
4. Çengel, Y.A., Boles, M.A.: Thermodynamics: An Engineering Approach, 5th edn. McGraw-Hill (2006)
5. Moran, M.J., Shapiro, H.N., Boettner, D.D., Bailey, M.B.: Fundamentals of Engineering Thermodynamics, 8th edn. Wiley (2014)

A New Approach for Drought Evaluation by Using the SPI Methodology

Petrit Zorba[1](✉) and Elsuida Hoxha[2]

[1] Department of Meteorology, IGEO, UPT, Tirana, Albania
p.zorba@geo.edu.al, petrit.zorba@gmail.com
[2] Grove School of Engineering CCNY, New York, NY, USA
ehoxha1@ccny.cuny.edu

Abstract. Drought management and its evaluation are directly related to the appropriate methodology used for such purposes. In this context, this paper aims to present a model recently developed that provides reliable outputs of the Standard Precipitation Index (SPI). The products obtained by applying this platform are computed based on monthly precipitation data from various meteorological stations, using at least a 30-year series. Additionally, the SPI values obtained are compared with other known software used in this field and analyzing as well as the errors produced by some factors to the final SPI model outputs. Spearman's rank correlation coefficient [ρ] between paired SPI series computed for the selected stations brings up the strength of the association between two SPI series produced by different models. It was computed also referring to different lengths of SPI data series, coming from each software, but for the same time frame of data. Also, to see and compare the product of 4 analyzed software has been calculated the Agreement Index [d], which is sensitive to differences between two variables. Furthermore, the new model provides data on the frequency of the SPI drought index and monthly precipitation, categorized and presented in graphs and various formats that are compatible and useful for other purposes.

Keywords: Standard Precipitation Index—SPI · Drought · Meteorology · Data processing · Climate

1 Meteorological Data Used for SPI Estimation

1.1 Meteorological Database and Preliminary Works

The application of the SPI methodology needs first of all to have in the right format in [(.csv) and (.xlsx)] the meteorological precipitation data. In this study, monthly precipitation data was calculated on the basis of the daily data for length series of 30, 60, and 90 years for one meteorological station of Korça, Albania, part of the Mediterranean climate, and another one of City Park of NY, USA, for the Humid Continental climate, observed in conformity with WMO standard. Referring to the input requests of 4 software's used for further analysis of SPI index, at the beginning the data are processed and adjusted in format to be appropriate and compatible for each software. After that, the provided outputs are transformed into the right format to be compatible for comparisons and further analyses between them.

G. Guxho et al. (Eds.): ITC-ICEE 2023, LNMIE, pp. 304–316, 2024.
https://doi.org/10.1007/978-3-031-48933-4_29

1.2 The Main Objective of this Study—The Evaluation of Different Error Types

The main objective of this paper is to bring up and show some types of errors that can be produced by applying different software, for the same time frame of 30 years that influence the final output of the estimated SPI index. The origin of these types of errors results as part of different software's used for data processing; but also other elements like length of data series even using the same time frame of a minimum of 30 years can provide errors; certain corrections suggested on the methodology by some authors [1]; the incorporated errors on data himself originated by rain gauge measurements [2]; etc., have their impact on the produced outputs. Of course, in that context, the *development* of our model proved to be helpful, named "SPI-AlbPZ", not only functioned as a tool for assessing the SPI index but also facilitated the comparison of data outputs among various models, thereby helping identify errors and enhancing the clarity of certain concepts.

1.3 Methodology Used and the Composition of Our Model "SPI-AlbPZ"

In recent years different methodologies and models have been produced to evaluate the drought. One of them is SPI index which has a number of users in different countries. It is based on the idea of McKee et al., 1993, which produced and presented the base methodology to compute the Standard Precipitation Index (SPI) [3]. Throughout the years different authors have made many efforts to improve such models.

The analysis of drought in this study is done using the Standard Precipitation Index (SPI) methodology that relies on software tools such as the "RDIT" software available for purchase at AgriMetSoft.com [4], the "DrinC" software developed by the National Technical University of Athens [5], and the "SPI Generator" application provided by the National Drought Mitigation Center at the University of Nebraska-Lincoln [6], which are accessible for free on their respective websites and another one developed by our team named "SPI-AlbPZ".

For the same database series with monthly data on 30, 60, and 90-year precipitation, the products obtained from different models applied for the assessment of the SPI index, were analyzed. For our model the gamma distribution for precipitation data series was also applied, assessing the respective parameters for alpha and beta,

$$\hat{\alpha} = \frac{1}{4A}\left(1 + \sqrt{1 + \frac{4A}{3}}\right) \quad \hat{\beta} = \frac{\overline{x}}{\hat{\alpha}}$$

where X—is the average value of the precipitation series of data, $A = (X_{mean} - \Sigma L_n)/n$ followed by further computing, finally enabling the calculation of the SPI values in conformity with the methodology used. This model was composed in the Excel platform.

2 Analysis of SPI Monthly Data Processed for Series with the Same Length But According to Different Models for 2 Climate Zones

2.1 The Estimated Errors, and Correlation Relationships

The “DrinC”, “SPI Generator” and “SPI AlbPZ” models produced SPI index value series that had, on average, quite small discrepancies or differences, compared to the products obtained from the “RDIT” model for the same database series with monthly precipitation used as inputs for these models.

The correlation relationships between the SPI series produced from different models were calculated, showing a correlation coefficient [r] as expected, quite high at 0.999. These results were obtained from the 30-year length series (a total of 5 analyzed series, 3 for CpNy and 2 for Korça), the 60-year series (a total of two analyzed series, one for CpNy and one for Korça), and the single 90-year series of the CpNy station, for which the same abovementioned models were applied in all cases. In meantime it has to be noted also the fact that our model performs as well as the other software’s used for processing the data and the estimation of the final SPI index [7].

But it’s very important to note that regarding the maximum and minimum absolute values provided by this type of differences between the compared series of various outputs models showed up in very few cases some values that can’t be neglected.

A preliminary conclusion from this analysis is that by the application of different models, the error of the obtained results or discrepancies (the absolute values marked between these series as Max. A. and Min. A.) from one model to another generally move very little in magnitude, averaging from −0.008 to +0.007, which can be considered negligible.

However, some limited and rare cases have been identified specific isolated data, where the absolute values of discrepancies range from −0.112 to +0.051 for the “RDIT”—“DrinC” differences comparison and from −0.109 to +0.051 for the “RDIT”—SPI Generator” comparison for the Korça station as well as from -0.087 to + 0.363 for the “RDIT” differences with the “SPI AlbPZ” model for the 30-year series of the CpNy station. For the 60-year series, no high deviations (more than ± 0.1) were noted, while for the 90-year series of CpNy, the fluctuations ranged from −0.837 to + 0.479 only for the “RDIT"- “SPI Generator” comparisons [7].

Regardless, those rare instances of SPI results that exhibit fluctuations from ± 0.1 to ± 0.8 cannot be ignored. They result in an error generated for the same series of data, but applying different models.

2.2 The Agreement Index

Additionally, certain assessments were conducted concerning the agreement index. “The index of agreement (d—index) is sensitive to differences between two variables and reflects the degree to which one variable is accurately estimated by another [4]. It is calculated according to the following formula,

$$d = 1 - \frac{\sum_{i=1}^{n} (O_i - P_i)^2}{\sum_{i=1}^{n} \left(\left|P_i - \overline{O}\right| + \left|O_i - \overline{O}\right|\right)^2}, \quad 0 \le d \le 1$$

where [O_i] is the notation for the values of the series indicator, in our case the base series with SPI values obtained from the application of the "RDIT" model, and [P_i] are the predicted values or, in our case, those of the SPI indicator series produced by one of the other models "SPI DrinC," "SPI Generator," and "SPI AlbPZ" compared to the reference series of the "RDIT" model.

In every instance involving comparisons and associations among products, pairs of series, "RDIT" with "DrinC," "RDIT" with "SPI-Generator," and "RDIT" with "SPI-AlbPZ" resulted in agreement index values [d] equal to 0.999 for the series with 30-year data for Korça for the periods 1931–1960 and 1961–1990 and for CpNy for the periods 1931–1960, 1961–1990, and 1991–2020.

It is mentioned at (https://agrimetsoft.com/calculators/Index%20of%20Agreement) that "The agreement value of 1 indicates a perfect match, and 0 indicates no agreement at all. The index of agreement can detect additive and proportional differences in the observed and simulated means and variances; however, [d] is overly sensitive to extreme values due to the squared differences" [4, 8].

In our final data outputs development by our model "SPI AlbPZ" in relation to data output of RDIT model are noted also the values of near 1, for station of Korça for the period 1961–1990 and for CpNy for the 3 periods 1931–1960, 1961–1990 and 1991–2020 analyzed. The same value of near 1 was noted also by the analysis of the pair between the model "DrinC" with "RDIT" for CpNy for only the 2 periods 1961–1990 and 1991–2020.

3 Analysis of SPI Monthly Data Processed for Series with the Same Time Frame, But Coming from Different Length Series

This section of the paper centers on comparing various data lengths series while maintaining the same time frame, but always with a minimum duration of 30 years. Hence, the calculations are executed using identical models, stations, and time frames, albeit on data series that belong to different lengths.

To evaluate these SPI discrepancies resulting solely from variations in the length of precipitation data series used, Spearman's rank correlation coefficient and other relevant indexes are employed. By examining the SPI changes, this analysis aims to shed light on the influence of data series length on drought evaluations, independent of the methodology or climate factors. A similar study tray to bring up some results in that context [8].

It is widely recognized in meteorology that longer data series yield more accurate and reliable results. Therefore, for SPI estimation, a minimum data series length of 30 years is generally accepted. However, when longer data series are used, how do the results fare?

This paper, which is part of a broader study with the main objective of identifying errors that may arise from various factors when applying the same or different software tools, etc., clearly demonstrates the need for caution.

The analysis includes assessing differences, standard deviation, maximum (Max.A) and minimum (Min.A) values, the percentage of cases above or below a certain threshold, as well as calculating the Spearman coefficient for pairs of 30-year series compared to corresponding 60 or 90-year series, and 60-year series compared to the same time frame or corresponding to 90-year series. This analysis is conducted for both CpNy and Korça locations representing respectively the Humid Continental climate and the Mediterranean climate.

3.1 The Differences Between the SPI Values Obtained from Different Time Series

Only the SPI values corresponding to the same common 30-year period were compared, belonging to the 90-year period or 60-year and the other one to the 30-year period.

Precipitation time series of different lengths, which are used as input for the models, naturally exhibit variations in average values, median, standard deviation, and other statistical indexes. Consequently, the outputs of the models, i.e., the obtained SPI series, also show differences. When comparing the SPI values, it was observed that they fluctuate within an average deviation range of −0.143 to +0.124 for the "SPI Generator" model, −0.145 to +0.117 for the "DrinC" model, and −0.146 to +0.118 for the "RDIT" model. Table 1 presents these results for the CpNy station.

Table 1. Differences produced by each one of the 3 models of SPI calculation "SPI RDIT", "SPI DrinC" and "SPI Generator" for the pair of 30 years data series between different length series 90 and 30 years used like input, for CpNy station.

Index	SPI RDIT	SPI DrinC	SPI Gener
Average	−0.146	−0.145	−0.143
Max. A	0.465	0.461	0.400
Min. A	−0.908	−0.910	−0.800
SD	0.187	0.187	0.183
≥+0.1	38	38	34
≤−0.1	235	236	220
≥+0.1 (%)	10.6%	10.6%	9.4%
≤−0.1 (%)	65.3%	65.6%	61.1%
ρ	0.989	0.989	0.990
Index	SPI RDIT	SPI DrinC	SPI Gener
Average	0.028	0.028	0.018

(*continued*)

Table 1. (*continued*)

Index	SPI RDIT	SPI DrinC	SPI Gener
Max. A	0.640	0.640	0.610
Min. A	−0.640	−0.640	−0.660
SD	0.179	0.179	0.211
≥+0.1	126	125	143
≤−0.1	75	75	113
≥+0.1 (%)	35.0%	34.7%	39.7%
≤ −0.1 (%)	20.8%	20.8%	31.4%
ρ	0.992	0.992	0.982
Index	SPI RDIT	SPI DrinC	SPI Gener
Average	0.118	0.117	0.124
Max. A	1.043	1.044	0.830
Min. A	−0.476	−0.476	−0.840
SD	0.196	0.196	0.194
≥+0.1	193	193	215
≤−0.1	40	40	29
≥+0.1 (%)	53.6%	53.6%	59.7%
≤−0.1 (%)	11.1%	11.1%	8.1%
ρ	0.987	0.987	0.987

90 years of CpNy: 1931–1990 compared with 30 years 1931–1960 (left), 1961–1990 (left down), and 1991–2020 (down right)

- The results not only indicated an average error or mismatch of approximately 0.1, as mentioned earlier but they were also accompanied by a standard deviation (SD) of around 0.2 (ranging from 0.179 to 0.211).
- Additionally, the maximum and minimum absolute anomalies observed in these series varied from −0.908 to + 1.043 for the "RDIT" model, −0.910 to + 1.044 for the "DrinC" model, and −0.840 to + 0.830 for the "SPI Generator" model, as detailed in the table.

The analysis conducted on the three 30-year series from the CpNy station matched with the corresponding parts of the SPI indicator values in the 90-year base series, revealed that approximately 10.6–53.6% of the SPI values obtained from the "RDIT" model exceeded the threshold of +0.1, while 11.1% to 65.3% of the cases recorded values lower than the threshold of −0.1. For the "DrinC" evaluation model, around 10.6% to 53.6% of SPI values were higher than the threshold of + 0.1, while 11.1% to 65.6% fell below the threshold of -0.1. In the case of the "SPI Generator" model, the results showed that approximately 9.4–59.7% of SPI values exceeded the threshold of +0.1, while 8.1–61.1% had values below the threshold of −0.1.

However, it is important to note that overall results did not demonstrate any clear and consistent trend indicating higher or lower values for the longer 90-year period compared to the shorter 30-year period, or vice versa.

Regarding the meteorological station of Korça, the results indicated an average deviation value for all models in the range of −0.088 to +0.093. These deviations were accompanied by a standard deviation ranging from 0.200 to 0.206. Furthermore, the maximum and minimum absolute anomaly values varied from −0.972 to +1.362 for the "RDIT" model, −0.972 to +1.362 for the "DrinC" model, and −0.970 to + 1.360 for the "SPI Generator" model. Detailed information regarding the different calculation methodologies can be found in Table 2.

Table 2. Differences produced by each one of the 3 models of SPI calculation "SPI RDIT", "SPI DrinC" and "SPI Generator" for the pair of 30 years data series between different length series 60 and 30 years used like input, for Korça station.

Index	SPI RDIT	SPI DrinC	SPI Gener
Average	0.089	0.093	0.093
Max. A	1.362	1.362	1.360
Min. A	−0.559	−0.559	−0.550
SD	0.204	0.206	0.206
≥+0.1	163	163	169
≤−0.1	60	54	55
≥+0.1 (%)	45.3%	45.3%	46.9%
≤−0.1 (%)	16.7%	15.0%	15.3%
P	0.989	0.989	0.989
Index	SPI RDIT	SPI DrinC	SPI Gener
Average	−0.088	−0.088	−0.088
Max. A	0.708	0.707	0.710
Min. A	−0.972	−0.972	−0.970

(*continued*)

Table 2. (*continued*)

Index	SPI RDIT	SPI DrinC	SPI Gener
SD	0.201	0.201	0.200
≥+0.1	49	49	49
≤−0.1	186	185	192
≥+0.1 (%)	13.6%	13.6%	13.6%
≤−0.1 (%)	51.7%	51.4%	53.3%
ρ	0.988	0.988	0.988

60 years of Korça: 1931–1990 compared with 30 years 1931–1960 (left) and 1961–1990 (right)

Additionally, it was found that approximately 13.6–45.3% of the SPI values obtained from the "RDIT" model, when comparing the 60-year series to the results obtained from the 30-year series (taking into account temporal alignment for the corresponding years and months), exceeded the threshold of +0.1. Furthermore, about 16.7–51.7% of the SPI values indicated values lower than the threshold of −0.1.

Regarding the "DrinC" evaluation model, approximately 13.6–45.3% of the SPI values exceeded the threshold of +0.1, while about 15.0–51.4% of the values fell below the threshold of −0.1.

In the case of the "SPI Generator" evaluation model, it was observed that around 13.6–46.9% of the SPI values were higher than the threshold of +0.1, and approximately 15.3–53.3% of the values indicated values lower than the threshold of −0.1.

As well as the data comparing the SPI indicator of the CpNY station in the USA, calculated for a 60-year period from 1931 to 1990 (serving as the reference baseline period), with those calculated for the 30-year series, specifically for the periods 1931–1960 and 1961–1990, and those referring to the base series of 90 years (1931–2020) and the 60-year series (1931–1990) were calculated, and the results similarly with above examples for CpNy are presented largely at the paper [9].

3.2 The Spearman's Rank Correlation Coefficient

Another indicator analyzed in this context was Spearman's rank correlation coefficient. According to certain authors, "The rankings of SPI values during a specific time period are an important indicator of the intensity of drought/wetness events." The Spearman's rank correlation coefficient [rs] or [ρ] is calculated between paired values of SPI, which uses rankings as a basis for measuring the strength of the relationship between two SPI values. This indicator assesses how well the relationship between two variables can be described using a monotonic function. Spearman's correlation is also described as Pearson's correlation coefficient between two ranking variables.

The Spearman's rank correlation coefficient [rs] or [ρ] was calculated using the formula:

$$\rho = 1 - \frac{6\sum d_i^2}{n(n^2 - 1)}$$

where [d] is the difference between the ranked SPI values from smallest to largest, respectively, according to the pairs created from this ranking, for the analyzed series, and [n] is the number of values in the series.

The data obtained from the analysis showed values of this coefficient [ρ] ranging from 0.982 to 0.999 for the pairs of 30, 60, and 90-year series analyzed for both meteorological stations, and are presented in more detail together with other indicators of average anomaly values, those of Max.A, Min.A, SD, the number of cases in % ≥ 1.0 and ≤1.0 in the tables mentioned above, Tables 1 and 2.

3.3 Some Specific Considerations

The preliminary assumption from this analysis is that there are discrepancies in the SPI values obtained for different lengths of rainfall series (30 years or longer) compared to longer ones. These differences result in series with SPI values that vary from each other. On average, these differences range from −0.1 to +0.1, with a standard deviation of 0.2. In some cases, these differences can be as high as 10–50% above or below the average fluctuations.

Additionally, the anomalies or discrepancies in the absolute values derived from these processed series are significant, ranging from −1.050 to +1.362. These magnitudes indicate substantial variations in the SPI values, further highlighting the differences between the series obtained from different lengths of rainfall data.

These findings emphasize the importance of considering the length of the rainfall series when analyzing SPI values and their impact on drought/wetness events. The observed discrepancies and fluctuations suggest that the choice of the time period for SPI calculation can have a notable influence on the results and should be carefully considered in drought monitoring and assessment.

For example, let's consider a specific situation where the SPI value indicates severe dryness with a value of −1.9. If this value is modified by another value of −0.1 due to the length of the series, resulting in a value of −2.0, it immediately falls into the category of extreme dryness according to the classification criteria as presented in Table 3.

This example illustrates how even a small change in the SPI value can lead to a shift in the drought severity category. It highlights the sensitivity of the SPI index and the importance of considering the accuracy and consistency of the data used for its calculation, particularly when assessing and comparing drought conditions over different time periods or lengths of rainfall series.

It is essential to carefully evaluate and interpret SPI values in the context of the specific classification thresholds and take into account any discrepancies or anomalies that may arise due to variations in data length or calculation methodologies.

Based on the described scenarios, it becomes evident that when changes in SPI values exceed ±0.1, which occur in a significant percentage of cases (up to 50%), the

Table 3. SPI index classification.

SPI values	Classification
+2.0 and more	Extreme wet
+1.5 to +1.99	Severe wet
+ 1.0 to +1.49	Moderately wet
−0.99 to +0.99	Near normal
−1.0 to −1.49	Moderately dry
−1.5 to −1.99	Severely dry
−2 and less	Extremely dry

classification of drought severity can be significantly influenced. These modifications are not due to any fault in the model output but rather arise from using precipitation data with different durations (longer or shorter series) as input.

For instance, a severe dryness state evaluated with an SPI value of −1.7, when mathematically modified due to a longer or shorter duration of the precipitation series, by adding a value of +0.5, would result in an SPI value of −1.2, indicating moderate dryness. On the other hand, if the correction is a subtraction of -0.5, the SPI value would become −2.2, indicating extreme dryness.

Similarly, a modification of +1.2 would change the assessment from SPI = −1.7 (severe dryness) to SPI = −0.5, indicating a situation close to normal. These examples highlight the sensitivity of the SPI index to changes in input data and the potential impact on the evaluation of drought severity.

Therefore, it is crucial to consider these potential modifications when analyzing dryness conditions. The results obtained from the SPI calculations should not be neglected, as they can deviate from the original assessment due to variations in the length of the precipitation series used as input data.

In summary, it is important to be aware of the potential impact of data modifications on SPI values and their corresponding drought severity classifications. Careful consideration should be given to the reliability and consistency of the input data when interpreting and comparing SPI results over different time periods or lengths of precipitation series.

3.4 SPI Frequency

In the context of this study, the frequency indicator for the SPI values provided by RDIT model has been evaluated in a second step by our model. Figure 1 shows the results of the frequency indicator calculated for the SPI values for the Korça meteorological station for the period 1931–1960. Furthermore, this indicator was also evaluated for the number of cases with monthly precipitation from 0 to 1 mm, from 1 to 2, from 3 to 4, and from 4 to 5 mm, graphically presented with the corresponding values expressed in %.

In summary, for the Korça meteorological station, about 2.5% of cases (9) are in the extreme drought category, while precipitation levels up to 5 mm are 14 cases or 4% of the total cases.

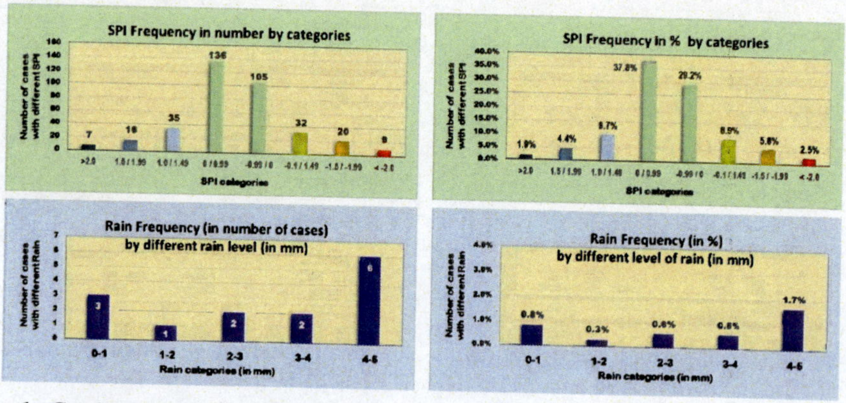

Fig. 1. Graph presentations of SPI values related to their division in groups by number of cases and by % and the precipitation according to values (from 0 mm to 1 mm, 1–2, 2–3 and 3–4 mm) for Korça station for the 30 years period 1931–1960.

Additionally, for a 40-year series, these values were calculated for 2 other meteorological locations near the Korça area in the SE part of Albania, Pogradec, and Liqenas, presented in Table 4. These indicators show that for the Mediterranean Climate, there is an increase in the frequency indicator of observing low precipitation values compared to the Humid Continental Climate.

The data for all the analyzed series in the study are presented in Table 4, where it is clearly evident that SPI $\leq$ 2.0, which belongs to and category "extreme drought", accounts for 1.9–2.7% of the total observations by processed periods for CpNy and 2.5–3.2% for Korça station.

The assessment of the frequency indicator highlights once again the important fact that situations with extreme and problematic droughts are relatively more susceptible to the errors and anomalies mentioned above, so it is very important that this issue, especially SPI for extreme drought values of SPI $\leq$ 2.0, should have a more thorough treatment and analysis, seen both in the context of the preceding or following month, in order to bring to light what is really a natural extreme drought and not the result of numbers or values that result or convey mathematical models, series with different lengths, instrument errors, etc.

4 Conclusions

- Even by using the same database and the same length of series and time frame of a minimum of 30 years period, for the same meteorological stations by applying different software in a few cases SPI output values fluctuate from ±0.1 to ±0.8 from one model to another, that is a fact that can't be neglected.

Table 4. The values of frequencies for SPI ≤ 2.0 (Extreme drought), in a number of cases and in % related to different length series, and for the analyzed meteorological stations.

No	Period (in years)	Length of series (in years)	Stations name	SPI < -2.0	
				Frequency (in number of cases)	Frequency (in %)
1	1931–1960	30	Kor^e	9	2.5
2	1961–1990	30	Kor^e	10	2.8
3	1931–1960	30	CpNy	7	1.9
4	1961–1990	30	CpNy	7	1.9
5	1991–2020	30	CpNy	9	2.5
6	1951–1990	40	Kor^e	13	2.7
7	1981–2020	40	Pogradec	10	2.1
8	1981–2020	40	Liqenas	17	3.5
9	1981–2020	40	CpNy	14	2.9
10	1931–1990	60	Kor^e	23	3.2
11	1931–1990	60	CpNy	19	2.6
12	1931–2020	90	CpNy	29	2.7

- Different lengths of monthly precipitation data series can introduce as well as errors in the final SPI index values. Therefore, it is crucial to exercise caution and remain aware while analyzing drought patterns.
- When processing drought maps using GIS or other software, it is vital to ensure that the precipitation data series not only have the same length but also correspond to the same time frame as the year's period being analyzed. Failing to do so can lead to significant errors and incorrect estimations of drought severity and its spatial distribution.
- These types of errors are produced by each of the programs when applied independently to the same database. Therefore, no preference should be given at the end of the analysis process to any software, because their comportment in data processing and final data provided like outputs don't allow to find out something specific.
- Other types of errors like the different time frames of various lengths of months, instrument errors, nature of precipitations, etc., are other elements that need to be analyzed and see their impact on the final outputs of different SPI models used for drought evaluations.

References

1. Sabău, N.C.: Comparative Study Regarding Performance of Some Software for the Calculations of the Standardized Precipitation Index (SPI). University of Oradea, Faculty of Environmental Protection, Oradea, January 2014, Romania, vol. XXIII (2014)

2. Guide to Instruments and Methods of Observation, Volume V—Quality Assurance and Management of Observing Systems, WMO-No. 8 (2018)
3. Standardized Precipitation Index User Guide, WMO-No. 1090 © World Meteorological Organization, Geneva 2, Switzerland (2012)
4. AGRIMETSOFT homepage: https://agrimetsoft.com/calculators/Index%20of%20Agreementof Agreement. Last accessed 18 Sept 2023
5. DrinC homepage: https://drought-software.co. Last accessed 18 Sept 2023
6. SPI Generator homepage: https://drought.unl.edu/monitoring/SPI/SPIProgram.aspx. Last accessed 18 Sept 2023
7. Zorba, P., Hoxha, E.: A novel model for Standardized Precipitation Index (SPI) analysis and the calculation of errors produced by various software programs using meteorological data from stations of diverse climate types. Monthly Clim. Bull./Buletini Mujor Klimatik, **7**(75), 22–31 (2023). ISSN: 2521-831X
8. Wu, H., Hayes, M.J., Wilhite, D.A., Svoboda, M.D.: The Effect of the Length of Record on the Standardized Precipitation Index Calculation. National Drought Mitigation Center, University of Nebraska, Lincoln, NE 68583-0728, USA (2014)
9. Zorba, P., Hoxha, E.: Comparative analysis of SPI errors generated by software tools using data series of different lengths for meteorological stations in Mediterranean and humid continental climates. Monthly Clim. Bull./Buletini Mujor Klimatik **7**(73), 24–30 (2023). ISSN: 2521-831X

Comparison of Different Energy Efficiency Techniques Applied to Building Envelope and Air Conditioning Plants

Giorgio Mustafaraj[1](✉) and Klodian Dhoska[1,2,3]

[1] Department of Production and Management, Faculty of Mechanical Engineering, Polytechnic University of Tirana, Mother Theresa Square No. 1, 1019 Tirana, Albania
gmustafaraj@fim.edu.al

[2] Albanian Mechanical Engineering Association (AMEA), Mother Theresa Square No. 1, 38954 Tirana, Albania

[3] Association of Talent under Liberty in Technology (TULTECH), Sopruse Pst 3, 10615 Tallinn, Estonia

Abstract. The building analyzed is a structure where the facades are made mainly by wood, while the remaining is concrete reinforced. Building is mainly naturally ventilated on perimetral parts, while the middle areas such as toilets, kitchen and stores are air conditioned. Different energy efficiency techniques were applied to building envelopes and HVAC plants to reduce energy usage by the facility. The aim of this work is to identify the appropriate techniques to be applied for reducing building electricity consumption. Furthermore, analysis will include economical comparisons of these techniques to identify the best ones in terms costs and electricity savings. Three techniques were selected such as: Heat Pump time turning ON/OFF and replacement of motors and light with high efficiency ones. In conclusion, after applying the selected most suitable energy efficiency techniques, yearly electricity savings was about 30%, which correspond to 28260 kWh.

Keywords: Electricity savings · Natural ventilation · Heat pumps

1 Introduction

To save more electricity on new and existing buildings, improvements need to be done by implementing new efficiency techniques. The European research group Ecofys for EURIMA [1], analyzed the impact of applying different retrofitting technologies on reducing the consumption of oil and gas in buildings. They have concluded that by implementing retrofitting technologies on existing building will lead to 75% decrease in energy use by 2050 (compared to 2010). Furthermore, implementation of retrofitting technologies will support the economic growth due to increasing on investment.

Numerous research analysis [2–5] were conducted in the past by applying different efficiency techniques. They have demonstrated that savings could varied between 15% up to 35%. A survey of various retrofitting technologies was investigated by Ma et al.

G. Guxho et al. (Eds.): ITC-ICEE 2023, LNMIE, pp. 317–322, 2024.
https://doi.org/10.1007/978-3-031-48933-4_30

[6]. Their objective was then to provide a methodology for the identification of the most suitable retrofit technologies to be applied.

Despite efforts of past researchers, there are still concerns that were not solved and these are addressed by the present research work. An accurate energy model was built based on calibration methodology developed previously [7–9]. Past research works place more attention on analyzing various retrofitting technologies and identifying the most suitable in terms of cost of investments and measures uncertainties costs and ignoring building model calibration [10, 11]. Obtaining accurately calibrated energy models for buildings is one of the most important steps to be considered before choosing the most suitable retrofitting technologies.

The aims of this research works include: firstly, to implement different energy retrofit technologies. Secondly, three techniques were selected such as: Heat Pump time turning ON/OFF and replacement of motors and light with high efficiency ones.

Finally, next sections include: (1) A brief overview of calibrated techniques and existing air conditioning plants, (2) Results analysis were presented together with the selected electricity efficiency techniques, and (3) Gives a summary of result together with other considerations.

2 Findings of Examination

In this research, model of the building was built utilizing Energy-Plus that is an open-source software [12]. The calibration process was developed previously by the present researcher and for more details refer to Mustafaraj et al. [13] (Fig. 1).

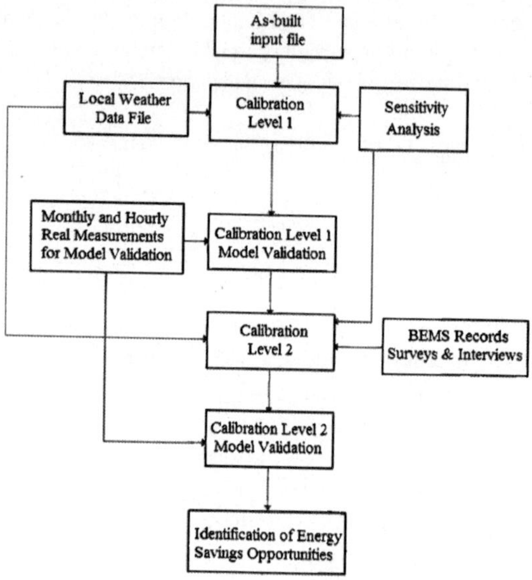

Fig. 1. Calibration methodology [13]

General overview of the edifice architecture built by Energy-Plus [12] is presented in Fig. 2, while the schematic Heat Pump used for heating is shown in Fig. 3. Detail information can be found in Mustafaraj et al. [13].

Fig. 2. General overview of simulation [12]

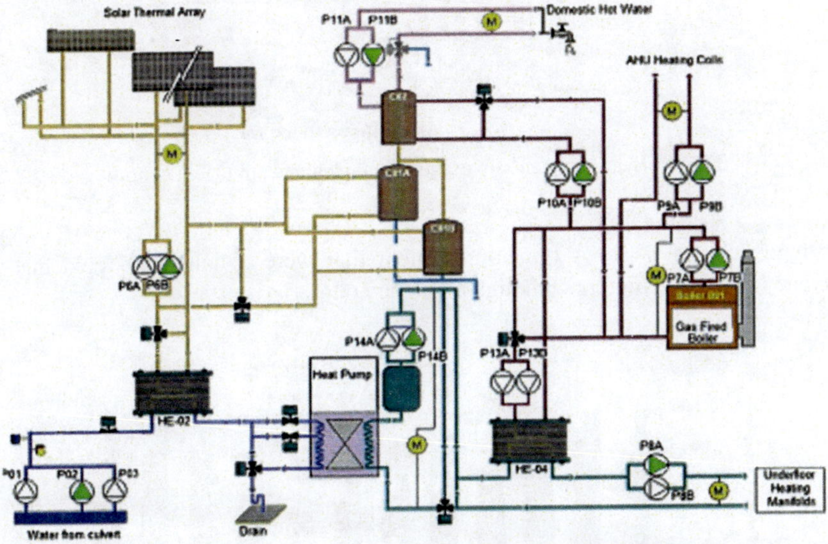

Fig. 3. Schematic representation of heating plant [13]

In this analysis different techniques were applied to reduce electricity consumption by the facility. After an economic analysis in term of budget available, three techniques were selected such as: Heat Pump time turning ON/OFF, replacement of motors and lighting with more efficiency ones. Heat pump is used to heat up the building during cooling season and its regulation ON/OFF was performed manually. In this work, simulation of the Heat Pump was performed automatically based real weather data.

3 Summary of Results

This research works presents three electricity techniques applied on calibrated model developed in previous research work [8]. Three techniques were selected such as: Heat Pump time turning ON/OFF and replacement of motors and light with high efficiency ones. Figure 4 shows a comparison between manual and simulated regulation of Heat Pump. A decrease of energy usage by 6500 kWh yearly was obtained by controlling the Heat Pump based on real weather data compare to that performed manually by the technician.

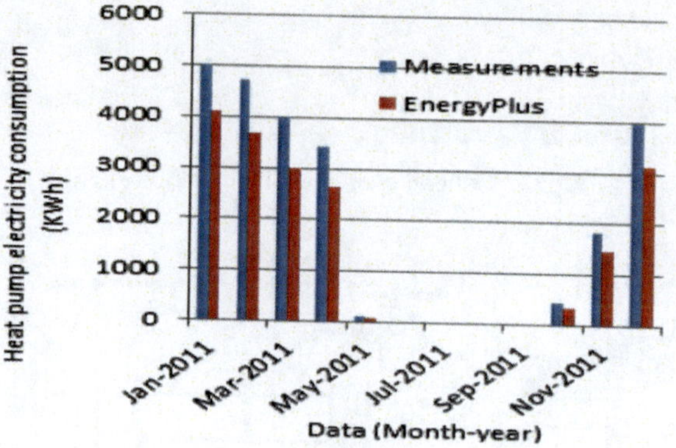

Fig. 4. Heat pump energy usage for manual and simulated control

Figures 5 and 6 shows the improvement that were obtained by replacing low efficiency lights and motors with high efficiency ones.

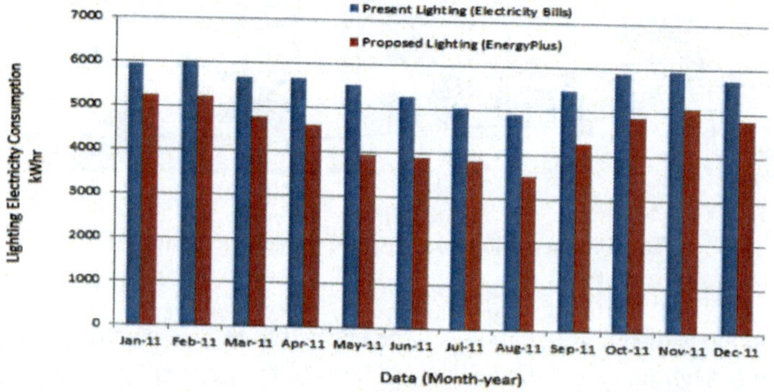

Fig. 5. Comparison between present and high efficiency lighting

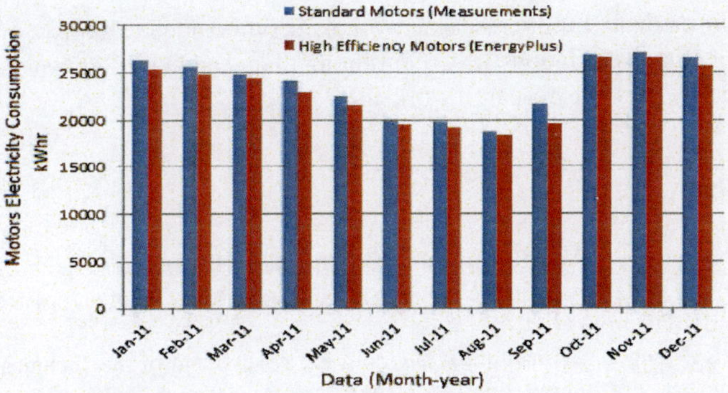

Fig. 6. Comparison between existing versus high efficiency motors

Table 1 present a summary of three techniques selected for saving electricity where the results were shown through Figs. 4, 5 and 6. From Table 1 the simple payback period by implementing the three electricity savings techniques is less than one year.

Table 1. Summary of three techniques selected for reducing energy usage.

Description of energy retrofit technologies	Potential savings (€/year)	Implementation cost (€)	Simple payback period (years)	Energy savings (kWh/year)	Demand reduction (kW)
Process improvements					
1. Time schedule heat pump modification	1500	0	Immediate	6500	0
2. Replacement with high efficiency lighting	2600	1146	0.6	11,520	0.450
3. Replacement with high efficiency motors	2400	1590	0.8	10,240	1.567

4 Conclusions

In conclusion, after applying the selected most suitable energy efficiency techniques, yearly electricity savings was about 30%, which correspond to 28,260 kWh. In particular, the most convenient technique is by controlling Heat Pump more efficiently based on real weather data, where the cost of investments is negligible, and the simple payback is immediate. Future research work is recommended to apply some other techniques on Heating Ventilation Air Conditioning plants and building envelope that were implemented by previous research works [9–13].

References

1. Bettgenhäuser, K., de Vos, R., Grözinger, J., Boermans, T.: Deep renovation of buildings: an effective way to decrease Europe's energy import dependency. Ecofys for EURIMA, Project number: BUIDE 14901 (2014)
2. Griffith, B., Long, N., Torcellini, P., Judkoff, R.: Assessment of the Technical Potential for Achieving Net Zero-Energy Buildings in the Commercial Sector. Technical Report NREL/TP-550–41957 (2007)
3. Chidiaca, S.E., Catania, E.J.C., Morofskyb, E., Foob, S.: A screening methodology for implementing cost effective energy retrofit measures in Canadian office buildings. Energy Build. **43**, 614–620 (2011)
4. Mills, E.: Building Commissioning: A Golden Opportunity for Reducing Energy Costs and Greenhouse Gas Emissions. Technical Report, Lawrence Berkeley National Laboratory (2009)
5. Aste, N., Pero, C.D.: Energy retrofit commercial buildings: case study and applied methodology. Energ. Effi. **6**, 407–423 (2011)
6. Ma, Z., Cooper, P., Daly, D., Ledo, L.: Existing building retrofits: methodology and state-of-the-art. Energy Build. **55**, 889–902 (2012)
7. Dorri, A., Dhoska, K., Kodhelaj, S.: Energy efficiency analysis in Korça regional hospital. Int. J. Innov. Technol. Interdisc. Sci. **6**(1), 1121–1129 (2023)
8. Dhoska, K., Hofer, H., Rodiek, B., López, M., Kübarsepp, T., Kück, S.: Improvement of the detection efficiency calibration and homogeneity measurement of Si-SPAD detectors. Springerplus **5**, 2065 (2016)
9. Dawood, M.M.K., Morse, M.S.E., Abdel-Mageed, S.I., Sharaf, O.A.: The effective techniques for enhancing the turbulent flow between two parallel plates: a comprehensive review. J. Trans. Syst. Eng. **1**(1), 10–30 (2023)
10. Qosja, S., Rolle, R., Gebremedhin, A.: Solving the bottleneck issue of energy supply. Case study of a wind power plant. Int. J. Innov. Technol. Interdisc. Sci. **5**(2), 874–891 (2022)
11. Dorri, A., Dhoska, K., Kodhelaj, S.: Computational simulation of heat transfer through fins of different shapes in an air-cooled internal combustion engine. Int. J. Techn. Phys. Probl. Eng. **15**(1), 248–254 (2023)
12. EnergyPlus Version 8.3. U.S. Department of Energy: Energy Efficiency and Renewable Energy. Office of Building Technologies (2017). http://www.energyplus.gov
13. Mustafaraj, G., Marini, D., Costa, A., Keane, M.: Model calibration for building energy efficiency simulation. Appl. Energy **130**, 72–85 (2014)

Management and Production

Evaluation of Surface Flatness and Roughness for Marble Products Produced with CNC Machining

Ilo Bodi[1](✉), Erald Piperi[1], Mariel Hoxha[1], and Jorgaq Kaçani[2]

[1] Polytechnic University of Tirana, "Mother Theresa" Square, Nr. 4, Tirana, Albania
ibodi@fim.edu.al

[2] Albanian Academy of Sciences, "Murat Toptani" Str., Tirana, Albania

Abstract. Industrial parts quality manufactured with CNC machining is related to the cutting parameters, the material of the part, and the part geometry. Different materials behave differently according to the machining process. The widespread use of marble products for different applications for granite surface tables for CMMs, calibration reference gauges, and in-home furniture has led to revising the quality of these products produced from CNC machining processes. Four marble cubes 600 × 600 mm are produced with CNC milling with different parameters and will serve as specimens. Geometric tolerances define the deviation of different geometric elements more sophisticatedly than dimensional tolerances. This paper presents two methods for the evaluation of surface flatness and roughness by a limited number of measured points in manual and automated measurement methods with GD&T by 3D scanning technique. The results obtained indicate that surface roughness affects surface flatness measurements, by identifying the problem, and can help to find a relationship between the manufacturing process parameters and the quality of the parts.

Keywords: CNC machining · Surface flatness · Surface roughness · Geometric tolerances · GD&T

1 Introduction

Product quality is one of today's most significant targets for all product manufacturers. Different technological developments have led to product design and production process automation. To respond to these requests, the process of measuring and evaluating the quality of the products must be at the same level.

During the machining process, several deviations are observed from the product design on a machined part. The origin of these deviations can be found in the uncertainty selection process of the parameters for the manufacturing process. The main factors are related to the material, machine tool, measuring technology, manufacturing environment, and the man, who controls and does the process [1]. The factors that create errors can be reduced by good process planning, more accurate equipment, and automated control, but they cannot be eliminated.

G. Guxho et al. (Eds.): ITC-ICEE 2023, LNMIE, pp. 325–333, 2024.
https://doi.org/10.1007/978-3-031-48933-4_31

Part tolerances define the product's allowed deviations and are estimated in thousandths of a millimeter. The process of dimensional measurements for industrial parts is an activity whose importance is increasing above in the manufacturing process. From an economic viewpoint, it can increase the parts costs from 10 to 70%. This assessment in measurement came from the request for a universal, flexible, and accurate approach to interpretation, which can provide full capabilities for the unification of the control process. From today's perspective, the use of coordinates measurement techniques and later of CMM—Coordinate Measuring Machines, in dimensional metrology can be seen as revolutionary for measurement techniques in production lines. Evaluation of parts quality for CNC machined products is very needed and is related directly to the industrial needs of the companies. A standardized procedure didn't exist until now and the CMM machines and 3D optical scanning systems are used these days for dimensional measurements. As the geometrical complexity of industrial components increases, the advantages of using these technologies increase as well.

According to the standards [2, 3] the dimensional or size tolerances can be defined in an indirect way, or the deviation can be defined by tolerance values or IT standards can be used. Roughness measuring features can be used to characterize the quality of the surface quality in the technical practice which can be 2D, they are micro geometric parameters and 3D so-called micro topographic characteristics. The surface micro geometric measuring and characterizing are mostly standardized so it is suitable to compare different types of surfaces. The surface quality can be controlled by different surface roughness parameters, like Ra, Rz, Ry, Rsk, Rku, etc. [3, 4]. A wide-ranging set of parameters is available to characterize the surface texture, but the industrial practice uses only 2–3 measuring characteristics (e.g., Ra, Rz, Rmax).

There are several ways to predict surface roughness. Felhő and Kundrák [5] define the surface roughness of a face-milled plain surface based on a 3D CAD model. Based on cutting experiment data Horvath and Drégelyi-Kiss [6] present a phenomenological model, Markopolous et al. [7] present a neural network based on the estimation method.

The more complex requirements can be defined as geometrical tolerances [2]. The standard defines four groups of tolerances. In the case of form tolerances, the accuracy of geometric elements should be defined, in that case, there is no need to use datum features. The four tolerances are straightness, flatness, roundness, cylindricity, profile any line, and profile any surface. The other three groups need datum features.

The article presents the result of a preliminary investigation, where the evaluation of surface roughness and the flatness deviation is studied. Four marble cubes were manufactured by the CNC milling process with different cutting parameters, and flatness and surface roughness parameters were measured. This work tends to find a relationship between the manufacturing parameters and the flatness deviation.

2 Materials and Methods

Four test specimens were produced of marble material with CNC machining center Model LEXTA 36 FULL of four machining axes. The dimensions of the specimens are 600 × 600 mm, as shown in Fig. 1. To prevent hand cutting by the marble cubes, every corner was de-signed and machined with a 1.5 mm chamfer. The cutting parameters are shown in Table 1. The strategy of the machining was: all the cubes were first cut in dimensions 610 mm and after every surface was machined with the milling process to realize the dimensions 600 mm. All the surfaces were polished with abrasive materials starting from 60, 120, 220, 400, 600, and 800 grit.

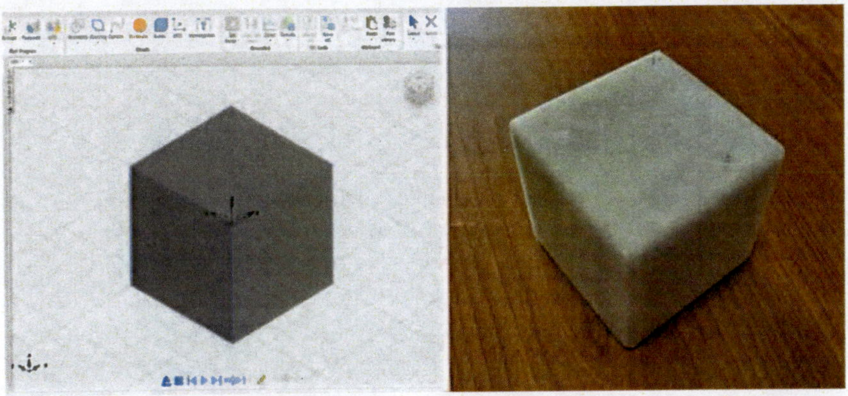

Fig. 1. Marble specimen 600 × 600 mm.

Table 1. Machining parameters.

Cube	Dimensions (mm)	Material	Machining process	Cutting parameters
Cube 1	600 × 600	White middle marble	Milling	S = 800 m/min F = 80 mm/rev P = 0.5 mm
Cube 2	600 × 600	White marble	Milling	S = 1000 m/min F = 100 mm/rev P = 0.5 mm
Cube 3	600 × 600	Black marble	Milling	S = 1200 m/min F = 100 mm/rev P = 0.8 mm
Cube 4	600 × 600	Black middle marble	Milling	S = 1400 m/min F = 100 mm/rev P = 0.8 mm

The surface roughness was measured with a KAIRDA KR220 surface tester with measuring range (μm) Ra: 0.005–16.000 and Rz: 0.02–160.00 and indication accuracy 0.001 [8]. The surface deviation was measured first with FARO® Quantum Max Scan Arms Model FARO Blu Max with accuracy [9] (Table 2).

The values of the surface flatness were elaborated by FARO software. Evaluation of surfaces was done according to ISO 4288:1996 standard: evaluation length, lm = 4 mm; the prescribed filter, lc = 0.8 mm. The surface roughness was measured in X and Y directions in 580 × 580 mm as the zone of interest.

Table 2. FARO Blu Max accuracy level.

FARO Blu Max	xR	xP	xS
2.0 m (6.6 ft)	0.03 mm 0.0012 in	0.038 mm 0.0015 in	0.046 mm 0.0018 in

3 Results

3.1 Surface Roughness

The surface roughness was measured with a KAIRDA KR220 surface tester and before starting the measurements the instrument was calibrated with its calibration plate according to the manufacturer's procedure as shown in Fig. 2.

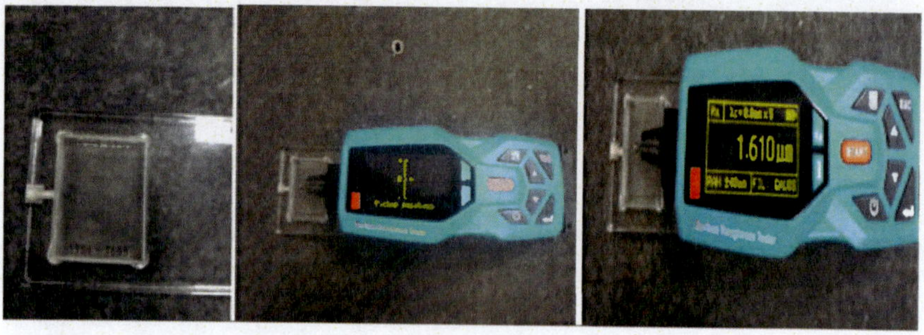

Fig. 2. KAIRDA KR220 surface tester calibration procedure.

All surfaces were measured in eight lines of each cube face to eliminate measurement errors. After that, the most characteristic measurement result was selected as a representation of that surface. Based on 24 measured surface roughness with 48 results per cube are shown in Table 3. With these values, it is possible to evaluate the average parameters that represent the surface roughness value. For all the Ra values shown in Fig. 3, is calculated the standard deviation and by observing the results, the average values for surface roughness represent the "real" surface roughness value.

By analyzing the results obtained from the measurement obviously, it is noticed for the CNC machining centers, that the increment of machining parameters will result in quality surface improvement. The machining parameters sets show that cutting speed and the feed rate increment, the forces also increase, thus the whole system has much more stress which is reflected in the change of the topography of the surface which directly affects surface roughness. The standard deviation indicates that the surface quality is not so homogenous, as it can be followed from the geometric model or the look of the surfaces. In the milling and the polishing process, the X and Y directions show different value characteristics, because of the milling lanes and the random position of the measuring probe of the surface tester in the measurement area.

Table 3. Surface roughness results per each Cube.

Cube 1	Ra 1	Ra 2	Ra 3	Ra 4	Ra 5	Ra 6	Ra 7	Ra 8	Average	Standard Deviation
Face 1	0.898	0.71	0.781	0.726	0.708	0.642	0.428	0.649	0.69275	0.125565471
Face 2	0.816	0.789	0.982	0.955	0.987	0.926	0.553	1.084	0.8865	0.154182522
Face 3	0.62	0.834	0.782	0.798	0.668	0.546	0.651	0.549	0.681	0.104724639
Face 4	0.954	0.898	0.892	0.627	0.801	0.82	0.641	0.811	0.8055	0.110268989
Face 5	0.543	0.618	0.177	0.505	0.48	0.779	0.204	0.295	0.450125	0.196248846
Face 6	1.241	0.663	0.637	0.745	1.231	0.397	0.473	0.33	0.714625	0.327973298

Cube 2	Ra 1	Ra 2	Ra 3	Ra 4	Ra 5	Ra 6	Ra 7	Ra 8	Average	Standard deviation
Face 1	0.66	0.668	0.781	0.579	0.507	0.633	0.575	0.542	0.618125	0.081028448
Face 2	0.377	0.432	0.231	0.377	0.602	0.603	0.511	0.678	0.476375	0.139461408
Face 3	0.444	0.669	0.541	0.527	0.706	0.748	0.636	0.695	0.62075	0.098693655
Face 4	0.788	0.498	0.703	0.587	0.606	0.477	0.414	0.638	0.588875	0.115510213
Face 5	0.687	0.742	0.641	0.64	0.402	0.778	0.705	0.734	0.666125	0.109489083
Face 6	0.672	0.866	0.709	0.942	1.029	1.022	0.83	0.874	0.868	0.122624427

Cube 3	Ra 1	Ra 2	Ra 3	Ra 4	Ra 5	Ra 6	Ra 7	Ra 8	Average	Standard Deviation
Face 1	0.65	0.68	0.63	0.62	0.608	0.584	0.428	0.627	0.603375	0.071363397
Face 2	0.716	0.758	0.754	0.637	0.621	0.587	0.584	0.823	0.685	0.083889809
Face 3	0.43	0.435	0.532	0.568	0.572	0.546	0.483	0.468	0.50425	0.053899791
Face 4	0.684	0.842	0.325	0.24	0.251	0.36	0.27	0.187	0.394875	0.221671174
Face 5	0.243	0.318	0.177	0.305	0.41	0.379	0.204	0.295	0.291375	0.075592555
Face 6	0.241	0.263	0.237	0.245	0.231	0.397	0.273	0.33	0.277125	0.054148955

Cube 4	Ra 1	Ra 2	Ra 3	Ra 4	Ra 5	Ra 6	Ra 7	Ra 8	Average	Standard deviation
Face 1	0.46	0.368	0.381	0.379	0.307	0.433	0.375	0.342	0.380625	0.044930328
Face 2	0.277	0.232	0.231	0.277	0.302	0.303	0.311	0.378	0.288875	0.044259003
Face 3	0.344	0.469	0.341	0.427	0.206	0.248	0.336	0.295	0.33325	0.080872971
Face 4	0.488	0.298	0.303	0.287	0.306	0.277	0.314	0.238	0.313875	0.069454009
Face 5	0.287	0.242	0.341	0.24	0.202	0.278	0.205	0.334	0.266125	0.049878446
Face 6	0.172	0.466	0.209	0.442	0.229	0.322	0.23	0.374	0.3055	0.104907102

3.2 Surface Flatness

One of the main problems for research and industry is obtaining the best surface shape for the designed models to meet the product design. According to the ISO 12781-1:2011 Geometrical product specifications (GPS)—Flatness—Part 1: Vocabulary and flatness parameters, a surface's flatness represents the distance between the two closest parallel planes within [10].

A lot of techniques are used to evaluate the flatness of a surface including height gauges, CMM machines, and 3D scanning techniques. Flatness can be measured by a height gauge and CMM (run across the surface of the part) only if the reference surface (on which the height gauge is placed) is parallel to the surface whose flatness is to be checked.

The flatness parameters that can affect the surface quality are very different, of course, but only the values cannot characterize the nature of the surfaces. For this reason, it was decided to use a 3D scanning technique to evaluate the surface's flatness [11].

The surface flatness was measured with FARO Blu Max scanner and before starting the measurements the instrument was calibrated with its calibration plate according to the manufacturer's procedure as shown in Fig. 4.

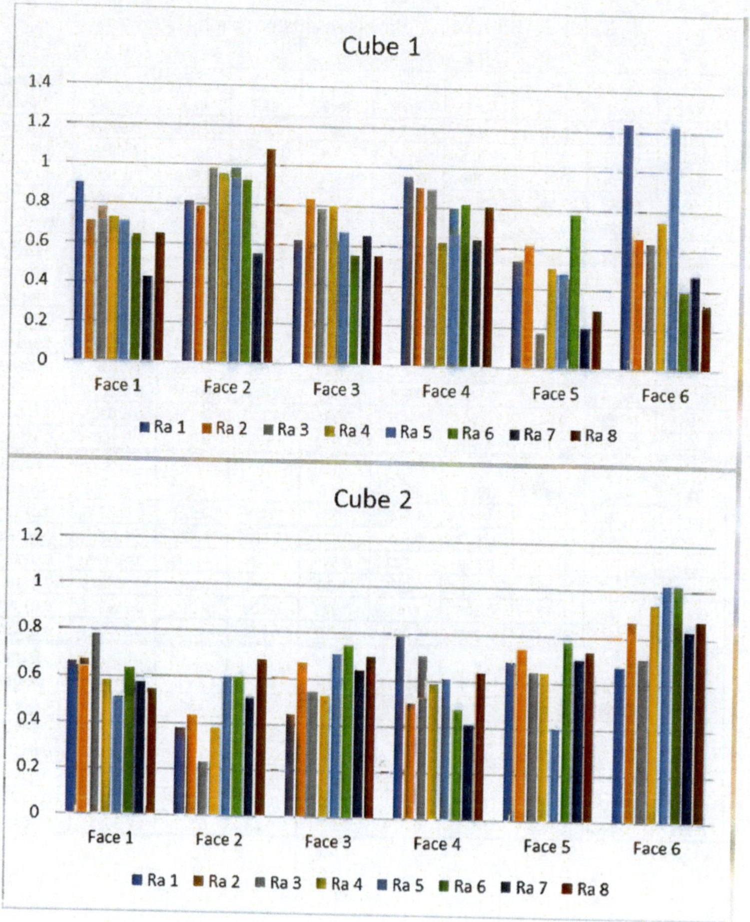

Fig. 3. Surface roughness results of Ra value

The results of the surface flatness measured with FARO Arm are shown in Fig. 5 (Table 4).

Based on the topology of the reconstructed surfaces in FARO software, the "micro scratches" of the milling and polishing instrument are seen and are evident. Also, the surface presentation is more realistic in the FARO software.

When comparing and analyzing the results of the surface flatness it is seen that for all the cubes only one of six faces resulted in a bigger deviation from the standard deviation. This problem is related to the machining process because it was the first machined surface. The increase of material removal during the cut reduces the rigidity of the component with the progress of operation and will increase workpiece deflections. For the marble material, the de-flections caused by cutting forces are evident in thin wall products less than 5 mm and with lengths bigger than 100 mm. For other dimensions, marble is very consistent compared to other metal products. It can be inferred that the

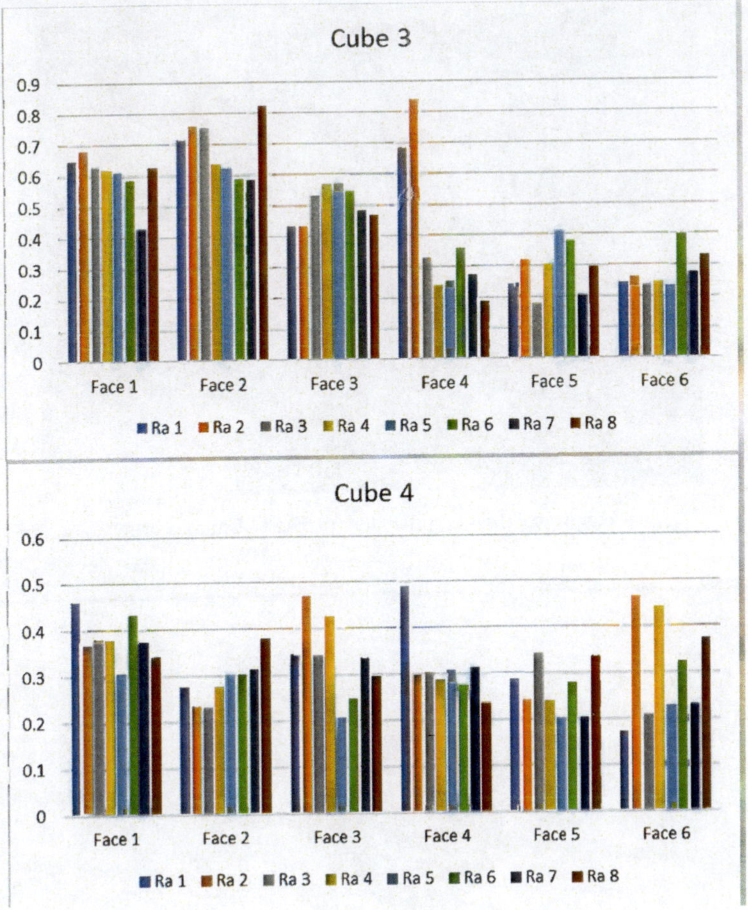

Fig. 3. (*continued*)

contribution of both tool and workpiece deflections is significant at elevated cutting conditions causing surface deviations, related to surface flatness.

4 Conclusions

The main objective of this article was to determine the relations between the machining parameters and measured values of surface roughness and surface flatness of four marble cubes. The selection of proper cutting parameters for the milling process is needed for the reduction of rework and to increase production productivity. The CNC machining process and polishing of the flat surfaces cause different topologies and have an influence on the flatness deviation, thus causing surface roughness effects. The inhomogeneous surface structure requires that the selection of measured points covers the whole surface. The increasing number of measured points will help to enhance the accuracy of flatness

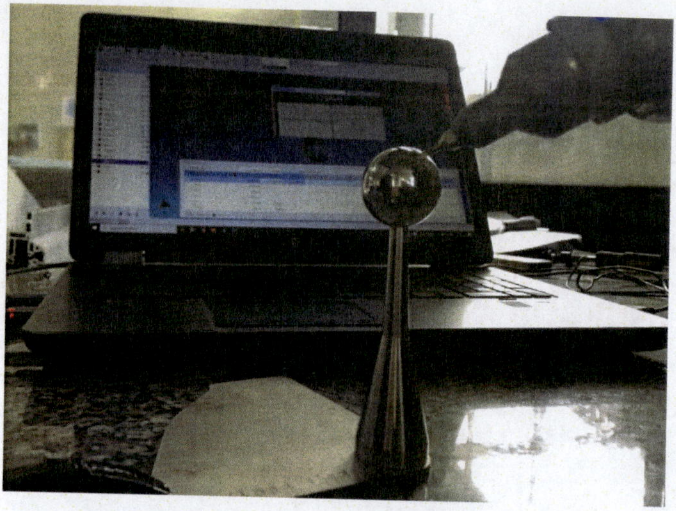

Fig. 4. Calibration procedure of FARO Blu max arm

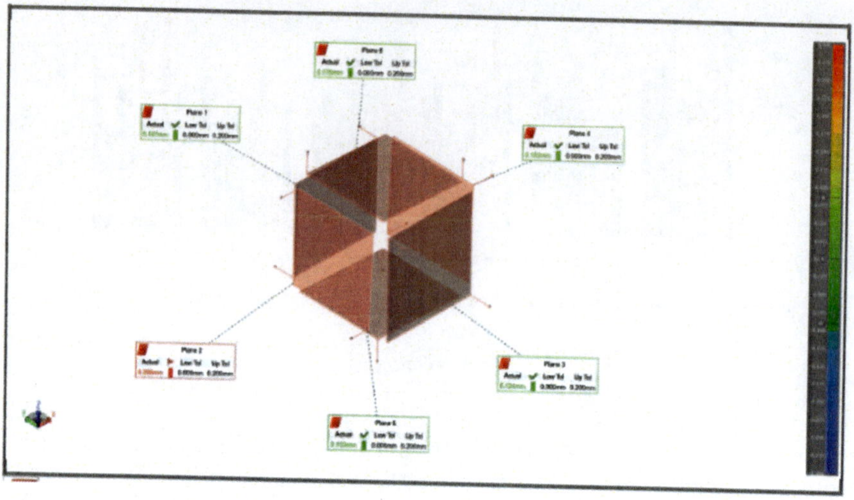

Fig. 5. Flatness measurement results with FARO arm

and surface roughness evaluation but increases the time of the measuring and tends to increase the total costs of the product.

Table 4. Surface flatness results from the scanning technique.

Cube 1 White Middle Marble 600 x 600 mm			
Face No	No of Points	STD Deviation	Flatness FARO (mm)
Face 1	20249	0.057	0.197
Face 2	31152	0.043	0.268
Face 3	25325	0.048	0.124
Face 4	24312	0.051	0.182
Face 5	27314	0.047	0.103
Face 6	24822	0.05	0.176

Cube 2 White Marble 600 x 600 mm			
Face No	No of Points	STD Deviation	Flatness FARO (mm)
Face 1	21152	0.057	0.148
Face 2	28455	0.053	0.265
Face 3	27385	0.049	0.174
Face 4	22874	0.054	0.137
Face 5	24984	0.047	0.166
Face 6	28328	0.05	0.1

Cube 3 Black Marble 600 x 600 mm			
Face No	No of Points	STD Deviation	Flatness FARO (mm)
Face 1	20249	0.057	0.065
Face 2	31152	0.043	0.067
Face 3	25325	0.048	0.072
Face 4	24312	0.051	0.069
Face 5	27314	0.047	0.051
Face 6	24822	0.05	0.076

Cube 4 Black Middle Marble 600 x 600 mm			
Face No	No of Points	STD Deviation	Flatness FARO (mm)
Face 1	21152	0.057	0.075
Face 2	28455	0.053	0.057
Face 3	27385	0.049	0.062
Face 4	22874	0.054	0.079
Face 5	24984	0.047	0.054
Face 6	28328	0.05	0.086

References

1. Mikó, B., Farkas, G.: Comparison of flatness and surface roughness parameters when face milling and turning. Dev. Mach. Technol. **7**, 18–27 (2016)
2. ISO 1011-2012: Geometrical Product Specifications (GPS)—Geometrical Tolerancing—Tolerances of Form, Orientation, Location, and Run-Out (2012)
3. ISO 286-2010: Geometrical Product Specifications (GPS)—ISO Code System for Tolerances on Linear Sizes—Part 1: Basis of Tolerances, Deviations, and Fits (2010)
4. ISO 4287-1997: Geometrical Product Specifications (GPS)—Surface Texture: Profile Method—Terms, Definitions, and Surface Texture Parameters (1997)
5. Felhő, C.S., Kundrák, J.: CAD-based modeling of surface roughness in face milling. Int. J. Mech. Aerosp. Ind. Mechatron. Eng. **8**(5), 814–818 (2014)
6. Horváth, R., Drégelyi-Kiss, A.: Analysis of surface roughness of aluminum alloys fine turned: united phenomenological models and multi-performance optimization. Measurement **65**, 181–192 (2015)
7. Markopoulos, A.P., Georgopoulos, S., Manolakos, D.E.: On the use of backpropagation and radial basis function neural networks in surface roughness prediction. J. Ind. Eng. Int. **12**(3), 389–400 (2016)
8. https://www.equipmentimes.com/product/details/KAIRDA-KR220-Surface-Roughness-Tester_8835.html. Last accessed 21 July 2023
9. https://www.faro.com/en/Products/Hardware/ScanArms. Last accessed 21 July 2023
10. ISO 12781-1:2011: Geometrical Product Specifications (GPS)—Flatness—Part 1: Vocabulary and Parameters of Flatness (2011)
11. Piperi, E., Bodi, I., Doumalin, P., Kaçani, J.: Scanning, 3D GD&T inspection for automotive plastic elements. In: 9th International Textile Conference & 3rd International Conference on Engineering and Entrepreneurship Proceedings, pp. 264–269, Tirana, Albania. (2021)

Multi-depth Camera Setup for Body Scanning

Erald Piperi(✉), Ilo Bodi, and Elidon Avrami

Mechanical Engineering Faculty, Department of Production and Management, DBS Group Albania, Kashar, Albania
epiperi@fim.edu.al

Abstract. Recently the use of depth sensors in different areas of application related to 3D scanning has become a game changer when dealing with budgets. In these paper real-time object recognition and 3D data acquisition are obtained by estimating a depth map and RGB color data by so called RGB-D sensor. Real time acquisition can anticipate corrupted data scan and permitting the operator to intervene for the best 3D acquisition during scan. These studied tend to show the creation of a low-cost system based on multi-depth sensors and a rotary table to perform a full body scan in a single posture. The approach solution can be used for different applications such as avatar creation, 3D printed figurines, etc., when accuracy and resolution delivered by professional 3D body scanning systems is not the main attribute.

Keywords: Depth sensor · 3D scanning · Multi-sensor scanning · 3D printing · Body scan

1 Introduction

Many solutions exist on the market for full-body scanning technology with different kinematics toward the increased speed of data capture, data manipulation, etc. Different strategies used for fast and accurate data capture are used for body-garment fitting [1], Sabina et al. [2] the creation of a human body model for advanced anthropometric analysis [3, 4]. Most of the commercial 3D body scan on the market are based on laser triangulation technology [5], portable handheld professional 3D scanner with real-time acquisition display [6] and photogrammetry with higher texture level of details. These systems offer different output 3D file formats in high resolution with or without color scan texture in just a few seconds. Their price ranges from medium to high, restrict their use, and leave space for the development of low to medium price range systems. With the rapid growth of 3D sensor technology, some custom open systems have been created by permitting the use of shelf hardware such as depth cameras, projectors, rotational platforms, etc. There are a lot of factors when it comes to the decision of which body scanning system is most suitable for human scanning and then 3D printing in scale replica so called "the figurine". Photogrammetry is used when photorealistic surface texture is needed most [7] compared to the accuracy of the model. Most of the photogrammetry systems are based on the combination of multiple DSLR cameras capable of shooting in

G. Guxho et al. (Eds.): ITC-ICEE 2023, LNMIE, pp. 334–338, 2024.
https://doi.org/10.1007/978-3-031-48933-4_32

faster time the object that is positioned in the center of the platform. The system should be calibrated, and all cameras should be focused on in advance by the operator. Also, for better acquisition the scene needs to be illuminated and the trigger system needs to be fast enough for data acquisition.

Instead, the new emerging technologies in the 3d sensor hardware structure and the decrease in its price range have given the researchers and producers a bigger range of possibilities in different areas of application as a final solution for different 3D applications. The areas of interest include robotic application [8], anthropometry [9], object datasets [10], etc. In this work, we use six Primesense Carmine 1.09 (short-range) depth sensors mounted in a vertical aluminum stand at a distance of 0.5 m away from the rotation axis of the turntable. This PrimeSense RGB-D camera combines both depth information and color information with a real-time structure light IR frame.

2 Experimental Work

The fast development of 3D sensors has increased the possibilities for developing fast and rapid data acquisition for indoor and outdoor environments. In this work, we have tried to build up and upgrade the body scan system which is based on six Primesense Carmine version 1.09 depth sensors. The specifics of the RGB-D sensor are shown in Fig. 1.

The 3D depth Sensor Primesense Carmine 1.09	
Field of View	54 degrees horizontal, 45 degrees vertical
Depth Image Size	VGA(640x480)
Spatial X/Y Resolution	@0.5m: 1mm
Depth Z Resolution	@0.5m: <1mm
Maximum Image Throughput	(QVGA) 60fps, (VGA) 30fps
Operation Range	0.35-3m
Color Image Size	1280x960
Data Interface	USB 2.0, USB 3.0

Fig. 1. Depth RGB-D sensor specification

The sensors are mounted in a vertical aluminum stand with a regulable vertical "Z" position and "Y" rotation angle for better field of view regulation by having overlapping areas for a best 3d acquisition of a human body (see Fig. 2a). The development of the turntable platform for carrying and rotating a considerable weight (up to 300 kg) with low rotation speed and reverse spin was designed and produced for the purpose of this work (see Fig. 2b). The operator can reverse the rotation direction with the joystick if some 3D data corruption occurs during the 3D digitalization of body capture. The standard workflow used in this work is presented in Fig. 3. Post-processing steps are defined by the software chosen for this study as Smooth, Decimate, Clean, Crop, Align, Remove edges, Fill holes, and Re-texture [11].

In the preliminary steps for multi-sensor scanning, it is necessary to follow the self-calibration marker-based procedure to minimize depth errors and distortions of the entire system [11]. The data acquired have also the colorized vertex attribution and can be used for colorized 3D printing. In our work we used a simple FDM 3D printer and printed in PLA (Polylactic acid) mono-color in a scale figurine (Fig. 4).

Fig. 2. The proposed solution for 3D scanning (**a**) and the design and production of the turntable with reverse spin for heavy weights (**b**)

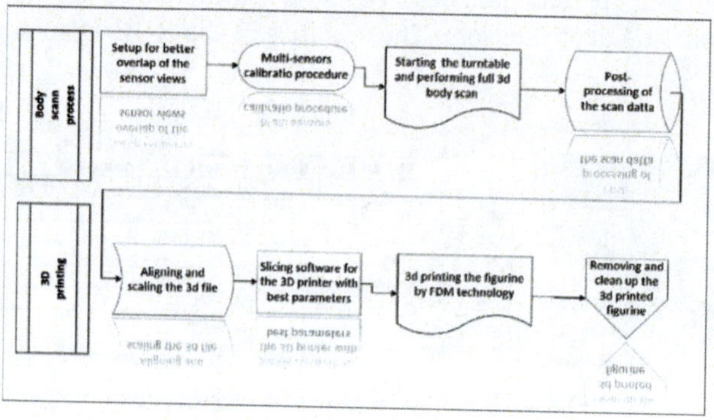

Fig. 3. General workflow from body scan to 3D printing figurine

It needs to be stated that according to different subjects (people with different anthropometric characteristics) the setup needs to be adjusted for better overlap and the calibration procedure needs to be redone. Figure 5 (bottom part) shows the 3D mesh file taken from the data acquisition process and the 3D printed figurine for each subject (upper part).

3 Discussion and Conclusion

In this paper, we have tested a full workflow from body scan to 3D printed in-scale figurine with an FDM 3D printer. The system is based on a low-cost RGB-D multi-sensor with PrimeSense technology and a custom-made turntable with inverse rotation directions. The system offers some advantages in terms of adjustments for different body sizes and can be upgraded, transportable, and is very fast for 3D data acquisition in real

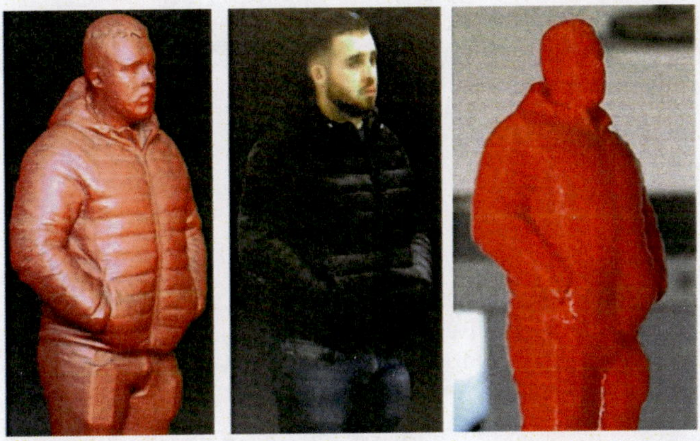

Fig. 4. From left to right: 3D mesh scan, vertex colorized attribute, and PLA 3D printed figurine.

Fig. 5. 3D mesh file (bottom part) and the 3D printed figurine (upper part)

time. The reconstructed mesh of the selected people has been 3D printed in single color PLA filament, but the system can capture also the RGB color information for an updated 3D printed color figurine if needed. The first tests showed that the system quality mesh output for that kind of application is very good, but we didn't evaluate the accuracy of the setup. In future work, this system is going to be tested for anthropometric 3D data capture for scientific research.

Acknowledgements. This work has been supported by two companies in Albania: © Artistic Jewellery Pirro and © DBS Group Albania throw a long collaboration, a positive impact in establishing new technologies and approaches as 3D printing in Albania.

References

1. Kareem, A., Najem, A.-A.A., Shaban, M.A.-A.: Human body measurements and clothing virtual for men adult consumers. Int. J. Comput. Appl. **127**(9) (2015)
2. Sabina, O., Elena, S., Emilia, F., Adrian, S.: Virtual fitting—innovative technology for customize clothing design. Procedia Eng. **69**. In: 24th DAAAM International Symposium on Intelligent Manufacturing and Automation, pp. 555–564 (2013)
3. Lovato, C., Castellani, U., Zancanaro, C., Giachetti, A.: Automatic labelling of anatomical landmarks on 3D body scans. In: Graphical Models. Elsevier Inc. (2014)
4. Stewart, A., Ledingham, R., Furnace, G., Nevill, A.: Body size and ability to pass through a restricted space: Observations from 3D scanning of 210 male UK offshore workers. Appl. Ergon. **51**, 358–362 (2015)
5. Scanner, V.V. 3. b.: "aniwa". [Online]. Available: https://www.aniwaa.com/product/3d-scanners/vitronic-vitus-3d-body-scanner/. [Accessed 24 November 2022]
6. Leo, A.: Artec 3D. Artec, [Online]. Available: https://www.artec3d.com/portable-3d-scanners/artec-leo. Accessed 24 Nov 2022
7. Dixit, I., Dunne, C., Logan, C.J., Fejtek, M., Blumer, P., Prakitpong, R., Krebs, C.: A workflow for digitizing anatomical specimens by combining photogrammetry and 3D scanning. FASEB J. **36**(S1) (2022)
8. Calderita, L., Manso, L., Bandera, J., Núñez, P.: Assessment of PrimeSense RGB-D camera for using in robotics: application to human body modeling. In: WAF 2012. XIII Workshop of Physical Agents (2012). https://doi.org/10.13140/2.1.1550.9764
9. Lovato, C., Bissolo, E., Lanza, N., Stella, A., Giachetti, A.: A Low Cost and Easy to Use Setup for Foot Scanning (2014). https://doi.org/10.15221/14.365
10. Garcia-Garcia, A., Orts-Escolano, S., Oprea, S., Garcia-Rodriguez, J., Azorin-Lopez, J., Saval-Calvo, M., Cazorla, M.: Multi-sensor 3D object dataset for object recognition with full pose estimation. Neural Comput. & Appl. (2015)
11. RecFusion: RecFusion [Online]. Available: https://www.recfusion.net/user-guide/. Accessed 30 Aug 2023

Experimental Investigation of Build Time During ABS Filament Material Extrusion Process

Aristidis Tsiolikas[1], Tatjana Spahiu[2], and John D. Kechagias[1](✉)

[1] FWSD Department, University of Thessaly, 43100 Karditsa, Greece
jkechag@uth.gr

[2] Textile and Fashion Department, Polytechnic University of Tirana, 1019 Tirana, Albania

Abstract. Filament material extrusion (f-ME) is a widespread process for thermoplastic prototypes or final parts, which is characterized as an additive manufacturing process, i.e., adding materials in layers utilizing the material extrusion mechanism. Build time is significant in this process and defines the productivity rates, energy and material consumption. In this work, is presented the study about the influence of deposition angle, layer thickness, infill ratio and infill pattern on process build time experimentally. The Taguchi experimental design was selected, and nine ABS (Acrylonitrile—Butadiene—Styrene) thermoplastics were 3D printed following the L9 experimental orthogonal array. The measured build time was analyzed by using analysis of means (ANOM) and analysis of variance (ANOVA) techniques, followed by a validation experiment in optimal parameter combination. This research serves as a valuable source and comprehensive guide to 3D printing parameters selection in the filament ME process.

Keywords: Filament · Material · Extrusion · ABS · Time · Design of experiment · Taguchi · Optimization

1 Introduction

3D printing fused deposition modeling (FDM) or filament material extrusion (f-ME) technology in manufacturing functional parts has been widely used since the end of the 80s [1–6]. Acrylonitrile-butadiene-styrene (ABS) stands out as the most practical and cost-effective alternative option among the materials used in FDM technology [7–10]. ABS has adequate short-term strength properties, making it increasingly popular in industrial applications [11, 12]. Moreover, exhibits recoverable and degradable properties, providing recyclability that enhances environmental and economic performance [13–15].

The FDM technology part quality and mechanical strength is affected by many factors, such as manufacturing parameters, part orientation, etc. [16–19], also explained in Fig. 1.

Even though the FDM 3D printing process has been investigated in the literature extensively for ABS material and the effects of the processing parameters on the mechanical properties have been examined, the simultaneous screening of four parameters such

G. Guxho et al. (Eds.): ITC-ICEE 2023, LNMIE, pp. 339–345, 2024.
https://doi.org/10.1007/978-3-031-48933-4_33

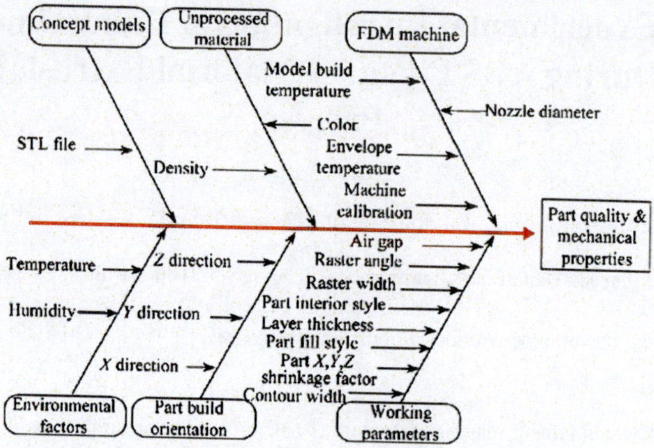

Fig. 1. Factors that affect filament material extrusion efficiency.

as the deposition angle, layer thickness, infill ratio, and infill pattern have not been researched regarding their effects on the manufacturing time. Note here that all these four parameters have considerable influence on the FDM 3D printing process efficiency and sustainability issues, and therefore, their tuning is vital for process optimization and zero-defect products.

2 Materials and Methods

The experimental setup in this research consisted of an economical FDM 3D printer, Zortrax M200, with a nozzle diameter of 0.4 mm and an ABS filament of 1.75. Bed and material deposition temperatures are maintained at 80 and 275 °C while the raster angle is kept constant at ±45°. The rest of the process parameter values are kept fixed. Printing time is measured by a digital timer and not from the CAM software, as shown in Fig. 2.

The 3D printing process variable parameters are presented in Table 1. The specimen that was 3D printed nine times had a base of 35x15mm with a height of 10mm and a cylindrical detail with a 7mm diameter on the top (see Fig. 3a). The deposition angle is the angle of the specimens in the Y-axis, as presented in Fig. 3b. Figure 3c shows the three deposition angles (0°, 15°, and 30°) and the three layers' thickness (0.09, 0.19, and 0.29 mm). The infill ratio is the part density, and it is set at 10, 20 and 30%. Last but not least, in Fig. 3d, the three different infill styles are presented (Rectangular, Honeycomb, and Octagonal).

3 Results and Discussion

The experimental design follows the Taguchi orthogonal array known as L9, which consists of four parameters with three levels and nine repetitions with different parameter values (combinations), as depicted in Table 2. The required experiments were reduced from 81 to only 9 using the Taguchi method. Total build time consists of the time needed

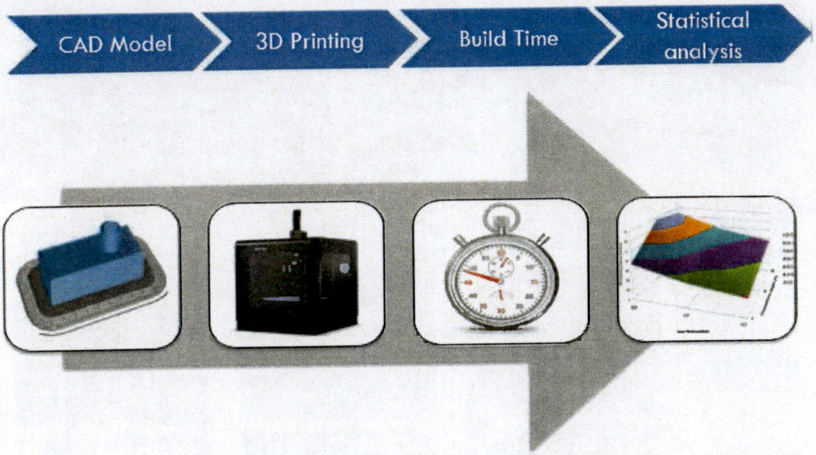

Fig. 2. Experimental procedure workflow.

Table 1. Variable parameters.

Process parameters	Level 1	Level 2	Level 3
Deposition angle (°)	0	15	30
Layer thickness (mm)	0.09	0.19	0.29
Infill ratio (%)	10	20	30
Infill pattern	Rectangular	Honeycomb	Octagonal

for printing the raft layers, the physical model and the support structure. The last two columns of Table 2 present experimental measurements of build time (min) and the calculated SN ratios of build time (db). The 'smaller-the-best' performance measure was selected as lower values of build time are desirable during 3D printing [20].

$$SN_{ratio} = -10Log\left(\frac{1}{n}\sum_{1}^{n} y_i^2\right) \quad (1)$$

Statistical analysis of the experimental results using Analysis of Means (ANOM) and Analysis of Variance (ANOVA) tools was conducted for optimizing build time of FDM 3D printing process.

ANOM was employed to identify the optimum levels for each FDM process parameter. The combination that minimizes 3D printing build time was found to be 0° deposition angle, 0.29 mm layer thickness, 30% infill ratio and honeycomb as infill pattern. It is observed from the ANOM diagram (Fig. 4) that deposition angle and layer thickness have a linear effect on the build time, while infill ratio and infill pattern showcase non-linear influence. ANOVA has been carried out in order to determine the influence of each control factor in the build time of FDM parts layer thickness (F = 50.87) and deposition angle (F = 45.52) have the highest effect on the build time. In contrast, the effect of infill

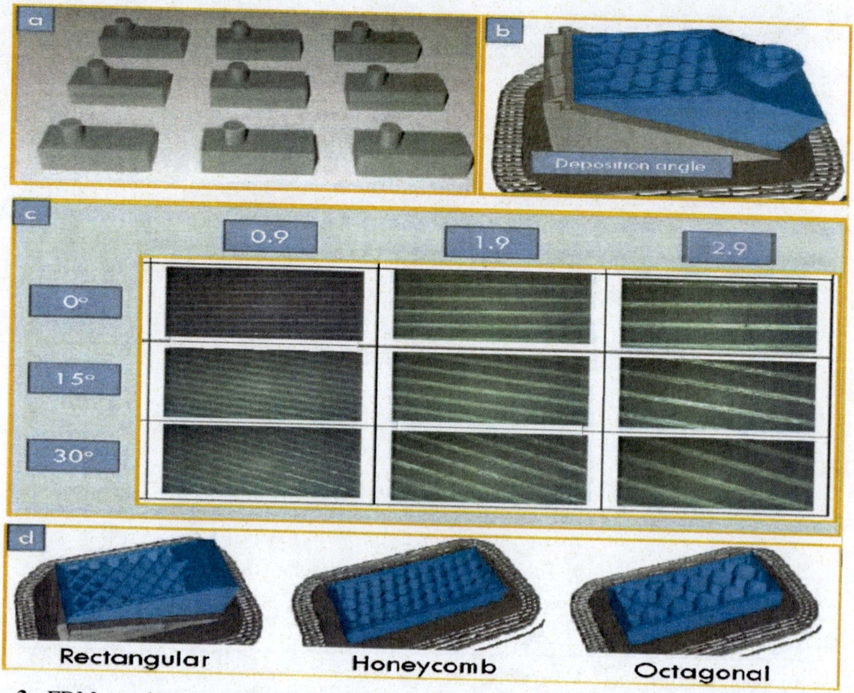

Fig. 3. FDM specimens **a** 3D printed parts, **b** deposition angle schematic representation (**c**), microstructure of specimens (XZ axis) under ×100 magnification, and **d** infill pattern.

Table 2. Experimental design and time measurements.

No. of exp	Deposition angle (°)	Layer thick (mm)	Infill ratio (%)	Infill pattern	Build time (min)	S/N ratio—Build time (db)
1	0	0.09	10	Rectangular	37	−15.682
2	0	0.19	20	Honeycomb	25	−13.979
3	0	0.29	30	Octagonal	19	−12.787
4	15	0.09	20	Octagonal	68	−18.325
5	15	0.19	30	Rectangular	40	−16.021
6	15	0.29	10	Honeycomb	29	−14.623
7	30	0.09	30	Honeycomb	63	−17.993
8	30	0.19	10	Octagonal	47	−16.721
9	30	0.29	20	Rectangular	37	−15.682

ratio (F = 1.004) and infill pattern (F = 0.996) can be attributed to error as they have a negligible impact on the build time inside the experimental region. Usually, increased values of the infill ratio parameter have a higher influence on the total printing time.

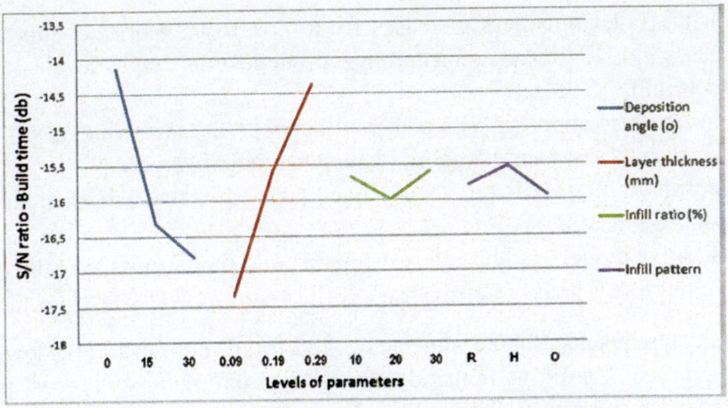

Fig. 4. Analysis of means according the SN ratios.

Layer thickness was identified as the most significant factor, having the highest contribution (51.70%) on the build time. An assessment of the interactions between the FDM process parameters revealed that "synergistic interactions" between the layer thickness and deposition angle, while the conflicting lines between layer thickness—infill ratio and layer thickness—infill pattern indicate an "anti-synergistic interaction". To validate the proposed optimization methodology, we developed an additive model in order to establish a relationship between the FDM process parameters and the build time.

$$n = m + (mDA1 - m) + (mLT3 - m) \pm e \quad (2)$$

A confirmation experiment was conducted by employing the optimum combination of factor levels, which is not included in the initial orthogonal array. The result of the verification experiment ($n = -13.01$ db) is within the two standard deviation confidence intervals [−12.33832 db, −13.17568 db], demonstrating the model's accuracy with an error of 5.70%.

4 Conclusions

In this research study, a systematic approach employing DOE—Taguchi robust design, Analysis of Means (ANOM), Analysis of Variance (ANOVA), interaction analysis and confirmation test was performed to optimize the FDM 3D printing process and determine the relationship between the FDM key process parameters and build time. Four critical FDM process parameters, each one consisting of three levels, were investigated using a Taguchi L9 orthogonal array. The main findings of this experimental research work can be summarized as follows:

(i) The levels 0° for deposition angle, 0.29 mm for layer thickness, 10 or 30% for infill ratio and honeycomb for infill pattern optimize and minimize the build time of the FDM process.
(ii) Layer thickness (51.70%) and deposition angle (46.26%) have the highest impact in the studied experimental area, while infill pattern and infill ratio have the lowest influence on the build time.

(iii) When the layer thickness increases from 0.09 to 0.19 or 0.29 mm, build time decreases for all the examined range of deposition angle levels ("synergistic interaction").
(iv) A non-linear relationship was observed between the layer thickness—infill ratio and layer thickness—infill pattern ("antisynergistic interaction").
(v) The result of the verification experiment is within the two standard deviation confidence intervals.
(vi) The additive model can accurately estimate build time using the optimal combination of the factor levels (5.70% error).

Overall, these results provide valuable insights for the optimization of low-cost FDM 3D printing process, resulting in the identification of optimum key process parameters, reduction of manufacturing time, energy and costs for a higher productivity. Investigating more process parameters and other performance quality indicators of the FDM process is proposed as a possible future work.

References

1. Daminabo, S.C., Goel, S., Grammatikos, S.A., et al.: Fused deposition modeling-based additive manufacturing (3D printing): techniques for polymer material systems. Mater. Today Chem. **16**, 100248 (2020). https://doi.org/10.1016/j.mtchem.2020.100248
2. Bikas, H., Stavropoulos, P., Chryssolouris, G.: Additive manufacturing methods and modelling approaches: a critical review. Int. J. Adv. Manuf. Technol. **83**, 389–405 (2016). https://doi.org/10.1007/s00170-015-7576-2
3. Ninikas, K., Kechagias, J., Fountas, N.A., Vaxevanidis, N.M.: A study of fused filament fabrication process efficiency: ABS vs PLA materials. IOP Conf. Ser. Mater. Sci. Eng. **1235**, 012007 (2022). https://doi.org/10.1088/1757-899X/1235/1/012007
4. Moza, Z., Kitsakis, K., Kechagias, J., Mastorakis, N.: Optimizing dimensional accuracy of fused filament fabrication using Taguchi design. Rec. Adv. Syst. Sign. Cont. Commun. Comp., pp. 110–114 (2015)
5. Kechagias, J.D., Ninikas, K., Salonitis, K.: An experimental study of laser cutting of PLA-wood flour 3D printed plates using a modified Taguchi design. Int. J. Exp. Des. Process Optim. **7**, 62–75 (2022). https://doi.org/10.1504/IJEDPO.2022.131230
6. Spahiu, T., Kitsakis, K., Kechagias, J.: Box-Behnken Design to optimise 3D printing parameters in applications for fashion products. Int. J. Exp. Des. Process Optim. **1**, 1 (2022). https://doi.org/10.1504/IJEDPO.2022.10053016
7. Pei, E., Shen, J., Watling, J.: Direct 3D printing of polymers onto textiles: experimental studies and applications. Rapid Prototyp J **21**, 556–571 (2015). https://doi.org/10.1108/RPJ-09-2014-0126
8. Choi, Y.-H., Kim, C.-M., Jeong, H.-S., Youn, J.-H.: Influence of bed temperature on heat shrinkage shape error in FDM additive manufacturing of the ABS-engineering plastic. World J. Eng. Tech. **04**, 186–192 (2016). https://doi.org/10.4236/wjet.2016.43D022
9. Spahiu, T., Canaj, E., Shehi, E.: 3D printing for clothing production. J. Eng. Fiber Fabr. **15**, 155892502094821 (2020). https://doi.org/10.1177/1558925020948216
10. Spahiu, T., Al-Arabiyat, M., Martens, Y., et al.: Adhesion of 3D printing polymers on textile fabrics for garment production. IOP Conf. Ser. Mater. Sci. Eng. **459**, 012065 (2018). https://doi.org/10.1088/1757-899X/459/1/012065

11. Isaac, C.W., Duddeck, F.: Current trends in additively manufactured (3D printed) energy absorbing structures for crashworthiness application—a review. Virtual Phys. Prototyp. **17**, 1058–1101 (2022). https://doi.org/10.1080/17452759.2022.2074698
12. Chaidas, D., Spahiu, T., Kechagias, J.D.: A Decorating 3D Printing on Cotton T-shirt Utilizing Fused Deposited Soft TPU Material (2022). https://doi.org/10.21203/rs.3.rs-2133402/v1
13. Gebler, M., SchootUiterkamp, A.J.M., Visser, C.: A global sustainability perspective on 3D printing technologies. Energy Policy **74**, 158–167 (2014). https://doi.org/10.1016/j.enpol.2014.08.033
14. Liu, Z., Jiang, Q., Zhang, Y., et al.: Sustainability of 3D printing: a critical review and recommendations. In: International Manufacturing Science and Engineering Conference, Vol. 49903, p. V002T05A004. ASME (2016)
15. Kechagias, J., Chaidas, D.: Fused filament fabrication parameter adjustments for sustainable 3D printing. Mater. Manuf. Process., pp. 1–8 (2023). https://doi.org/10.1080/10426914.2023.2176872
16. Kechagias, J.D., Zaoutsos, S.P.: Optimising fused filament fabrication surface roughness for a dental implant. Mater. Manuf. Process, pp. 1–6 (2023). https://doi.org/10.1080/10426914.2023.2176870
17. Fountas, N.A., Papantoniou, I., Kechagias, J.D., et al.: Modeling and optimization of flexural properties of FDM-processed PET-G specimens using RSM and GWO algorithm. Eng. Fail. Anal. **138**, 106340 (2022). https://doi.org/10.1016/j.engfailanal.2022.106340
18. Kechagias, J.D., Fountas, N.A., Ninikas, K., Vaxevanidis, N.M.: Kerf geometry and surface roughness optimization in CO_2 laser processing of FFF plates utilizing neural networks and genetic algorithms approaches. J. Manuf. Mater. Process. **7**, 77 (2023). https://doi.org/10.3390/jmmp7020077
19. Fountas, N.A., Zaoutsos, S., Chaidas, D., et al.: Statistical modelling and optimization of mechanical properties for PLA and PLA/Wood FDM materials. Mater. Today Proc. (2023). https://doi.org/10.1016/j.matpr.2023.08.276
20. Phadke, M.S.: Quality Engineering Using Robust Design. Prentice Hall PTR, Englewood Cliffs, New Jersey 07632 (1989)

The Effects of the Drill Microgeometry on the Induced Thrust Force and Cutting Torque

Anastasios Tzotzis[1](✉), Athanasios Manavis[1], Ioannis Sarris[2], and Panagiotis Kyratsis[1]

[1] Department of Product and Systems Design Engineering, University of Western Macedonia, 50100 Kila Kozani, Greece
a.tzotzis@uowm.gr

[2] Department of Mechanical Engineering, University of West Attica, 250 Thivon & P. Ralli Str, 12244 Egaleo Athens, Greece

Abstract. Microgeometry plays a key role in the performance of a drilling tool. In the present study, the performance of two twist drills is being evaluated in terms of the generated cutting force and torque. Both tools are ø8 TiAlN coated, carbide two-flute drills, with a point angle of 140°, as well as with equal flute length. A plate of Al7075 served as the workpiece and a CNC machine was utilized to perform two sets of holes, one for each of the tools. Despite the fact that both tools are coated, with identical helix and point angles, they performed in a dissimilar manner due to their different cutting profile. In specific, the comparison revealed that the tool with the larger cutting lip performed better in both output data, yielding lower force and torque values up to 23.1% and 17.3% respectively.

Keywords: Al7075 drilling · Drill bit · Tool microgeometry · Cutting edge · Rake face · Thrust force · Cutting torque

1 Introduction

One of the most applied manufacturing processes in the industry is drilling. Therefore, it has been widely studied in the past years, with studies focusing on the influence of several machining parameters on the generated cutting forces, torque, hole quality, surface roughness, burr formation, temperature distribution, tool wear and more [1–3]. Along the experimental testing, CAD-based techniques and the finite element method (FEM) [4, 5] are computational methodologies that are applied in such researches.

Chatha et al. [6] investigated the drilling of aluminium 6063 under dry, flooded, minimal quantity lubricated (MQL) and nano-fluid MQL conditions. The authors performed the experiments under low cutting speeds with 6 mm drills and reported thrust forces and torque in the magnitude of 20 N to 30 N and 0.1 Nm to 0.2 Nm respectively. Palanikumar and Muniaraj [7] studied the drilling of Al6061/15%SiC-4%Gr metal matrix composites with 4 mm, 8 mm and 12 mm carbide tools under dry conditions. The study revealed thrust forces within the range of 50–300 N approximately. Rajmohan and Palanikumar [8] experimentally investigated Al356/Sic composite with coated carbide drills. Thrust

G. Guxho et al. (Eds.): ITC-ICEE 2023, LNMIE, pp. 346–353, 2024.
https://doi.org/10.1007/978-3-031-48933-4_34

force was measured under different levels of cutting speed and feed, generating values between 500 N and 700 N approximately. Aamir et al. [9] worked on the Al2024-T3 drilling with uncoated twist drills. It was found that under the selected cutting conditions, the generated thrust forces varied between 100 N and 500 N. Similar studies [10–12] concentrate on the measurement and analysis of the thrust force and cutting torque, due to their strong influence on the tool wear, surface quality and power consumption.

Despite the advent of modern lightweight materials, aluminium alloys is still one of the top choices in the industry, especially when related to aerospace, vehicles and structures. This fact, coupled with the annually growing production of aircrafts, cars, trains and other vehicles, leads to the need for updated studies and models regarding machining of aluminium. Especially, the effects of the cutting microgeometry is significant, since most studies focus on the standard angles such as the helix and the point angle, as well as the tool diameter. In the light of the above, this work highlights the significance of the tool's micro-geometrical aspects during drilling. In specific, an experimental work is being presented, studying the effects of the microgeometry on the induced thrust force and cutting torque during Al7075-T651 drilling.

2 Experimental Work

Two sets of drilling experiments were carried out with the aid of a HAAS VF-1 CNC machine, each comprising of nine tests. The used workpiece is a plate with dimensions of 150 mm × 100 mm × 15 mm, made of Al7075 alloy. The selected temper of the alloy is the T651, which is widely used in machining. Only the cutting speed and the feed rate were taken into account in the study, in order to focus on the microgeometry of the tools. In addition, the performed holes were made "through holes". The cutting speed levels correspond to 50 m/min, 100 m/min and 150 m/min, whereas the equivalent levels for the feed rate are 0.15 mm/rev, 0.20 mm/rev and 0.25 mm/rev respectively. Table 1 presents the series of the drilling experiments with the corresponding cutting conditions.

Table 1. The experimental testing order and the cutting conditions.

Test no.	V_c (m/min)	f (mm/rev)
1	50	0.15
2	50	0.20
3	50	0.25
4	100	0.15
5	100	0.20
6	100	0.25
7	150	0.15
8	150	0.20
9	150	0.25

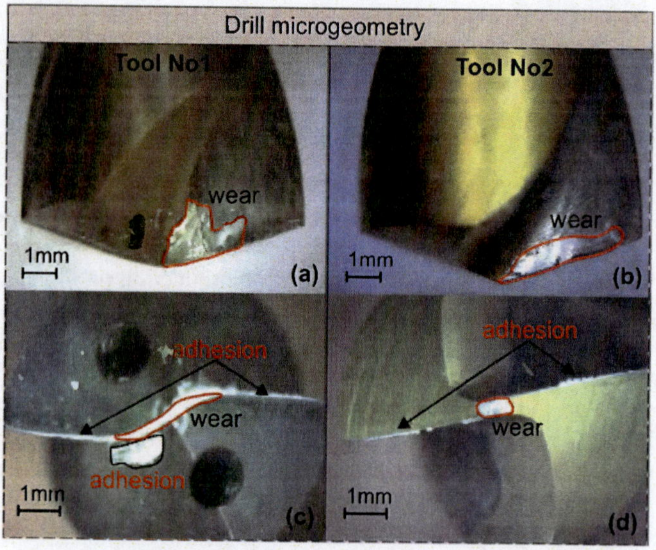

Fig. 1. The tools participated in the study.

The drilling tests were carried out by using the tools with reference number 12030800 (tool No. 1) and B041A08000 (tool No. 2), while feeding the cutting area with semi-synthetic coolant. Both tools are solid carbide, 3 × D two-flute twist drills and share equal characteristics such as diameter (8 mm), point angle (140°), cutting length (41 mm) and helix angle (30°). Moreover, both tools are intended to cut a variety of materials, including aluminium alloys. However, the key differences between the two tools are found in the microgeometry and specifically in the chisel edge and the cutting lip. The aforementioned geometrical aspects influence significantly the performance of the cutting edge as discussed in this paper. Figure 1 illustrates the physical models of the used tools, whereas Table 2 includes the most important characteristics of the drills, as given by the manufacturers and measured by utilizing enhanced images taken through a microscope. Moreover, Fig. 1 highlights the tool wear, as well as the material adhesion on the tool edges. Further discussion on the tool wear and material adhesion is made in Sect. 3.

To collect and measure the output data, a 4-component dynamometer was utilized, which is able to measure all three cutting forces and generated torque. An amplifier, coupled with a data acquisition card were implemented to amplify the signal to the required strength, as indicated by the manufacturer, and to store the measurement data to a computer. Figure 2a depicts the used equipment and part of the experimental work (Fig. 2b).

3 Results and Discussion

The present study focuses on the thrust force and cutting torque measurement, with respect to the cutting speed and feed. Furthermore, a short presentation of the tool wear areas, as well as the material adhesion locations is made. Figure 3 graphically presents

Table 2. Tool dimensions and characteristics.

Characteristic	Tool no. 1	Tool no. 2
Diameter (mm)	8	8
Point angle (°)	140	140
Flutes	2	2
Flute length (mm)	41	41
Helix angle (°)	30	30
Chisel edge (mm)	Pointed	0.35
Cutting lip (mm)	3	4.3

Fig. 2. The experimental equipment (**a**) and part of the drilling tests (**b**).

the comparison of the two drills in terms of the generated thrust force and torque. It is noted that both parameters are required to be as low as possible. According to Aamir [13] high cutting forces tend to rotate the axis of the spindle more, leading ultimately to low quality machined surface. In addition, higher values of thrust force affect the tool life negatively, whereas large torque indicates the existence of excessive friction between the tool and the workpiece, and thus leads to the generation of increased temperatures. By observing the charts of Fig. 3, it is derived that tool No. 2 produced lower forces compared to tool No. 1, by approximately 5–23.1%. The same pattern is observed for torque, where tool No. 2 produced higher torque in comparison with tool No. 1 only in three experiments (no. 1, no. 5 and no. 6), by approximately 0.7%, 2.4% and 2.7% respectively. The difference can be considered marginal, since tool No. 2 produced lower torque compared to tool No. 1 in six experiments, with percentage levels varying between 1.4% and 17.3%. The magnitude of the measured output are supported by the findings reported by similar researches for the 7075 alloy [10, 14].

The findings that yielded from Fig. 3, coupled with Fig. 1 can lead to conclusions related to the influence of the microgeometry on the output data. Specifically, Figs. 1a, b illustrates the wear on the rake face of the tool No. 1 and No. 2 accordingly. It can

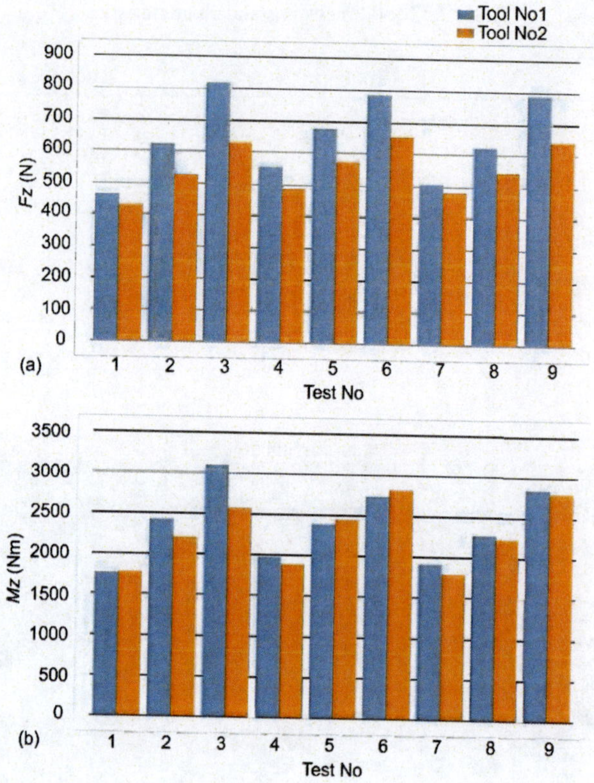

Fig. 3. The comparison charts for thrust force (**a**) and torque (**b**).

be observed that the wear on tool No. 2 is located across the edge, whereas on tool No. 1 closer to the core due to the adhesion of material that continuously rubs between the tool and the uncut material. Figure 1c proves the aforementioned claim, where a piece of adhered material is shown close to the core. Moreover, Fig. 1d illustrates the wear on the chisel edge of tool No. 2. Finally, it is shown that material is adhered on the cutting edges of both tools. However the quantity of material in tool No. 2 is fewer. The previously mentioned findings strengthen the fact that the narrower rake face formed by the microgeometry characteristics, probably is responsible for the accumulation of material that deteriorates the cutting process, and hence contributes towards the production of higher levels of forces and torque.

Concluding, Figs. 4 and 5 illustrate the interaction between the cutting speed and feed, and their combined effect on the output. In specific, Fig. 4 corresponds to the thrust force, whereas Fig. 5 to the torque. The plots indicate the same trend for both tools, where it is obvious that higher values of feed produce more thrust force and torque. The increase in cutting speed on the other hand, seems to have a different effect pattern. The lower level of cutting speed, coupled with the lower value of feed, are responsible for the production of the lower values of both F_z and M_z. The next level of V_c generated higher output values and finally, the higher level tends to produce lower output rates as to the medium one. A deviation from the pattern is visible with tool No. 1, at $f = 0.25$ mm/rev and $V_c = 50$ m/min, where it is seen that this combination is responsible for the production of the higher levels of F_z and M_z.

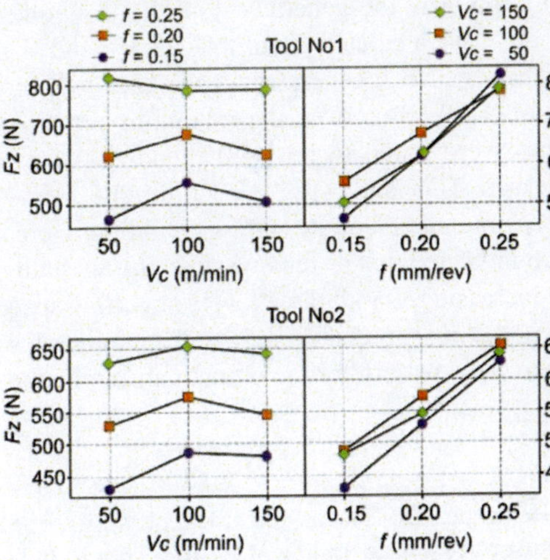

Fig. 4. The interaction plots of tool no. 1 and no. 2 for thrust force.

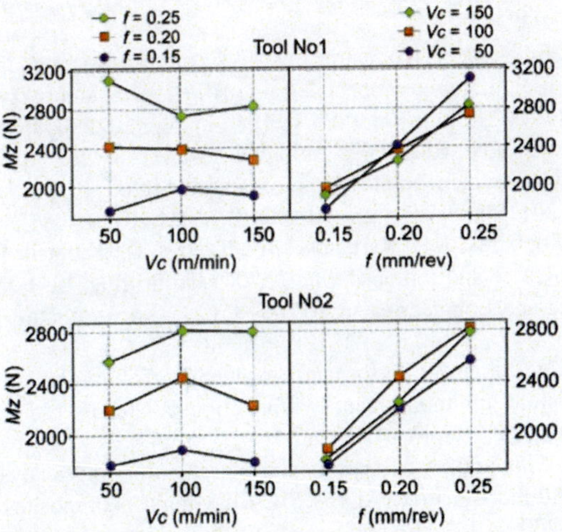

Fig. 5. The interaction plots of tool no. 1 and no. 2 for cutting torque.

4 Conclusions

The present study compared the performance of two twist drills with identical base characteristics but with different cutting microgeometry, during Al7075-T651 drilling. The results revealed that the microgeometry of tool No. 2 was responsible for the yield of lower values of thrust force in all nine experiments. In addition, the same tool performed better in terms of the cutting torque, generating lower values in six out of nine tests. The tool wear patterns, as well as the locations where material concentrated, suggest that the chisel edge of tool No. 2, as well as the broader rake face, prevented the material accumulation, reducing this way the generated forces and torque. The average value of the comparison levels was calculated equal to −14.1% and −3.8% for F_z and M_z respectively, meaning that tool No. 2 generated on average lower thrust force and torque in comparison with tool No. 1. Moreover, the minimum and maximum values of the comparison percentages were found to be −5.0% and −23.1% for the thrust force, whereas for the torque were 2.7% and −17.3% accordingly.

Concluding, the cross-interaction of the cutting conditions were highlighted with the interaction plots, which indicated that as feed was increasing, both the generated thrust force and torque were increasing, for all three levels of cutting speed. A different effect scheme was observed as cutting speed shifted, with the medium level producing higher output. Furthermore, the same pattern for the output data applied to both tools.

References

1. Tzotzis, A., García-Hernández, C., Huertas-Talón, J.-L., Kyratsis, P.: FEM based mathematical modelling of thrust force during drilling of Al7075-T6. Mech. Ind. **21**(4), 415 (2020). https://doi.org/10.1051/meca/2020046
2. Mondal, N., Mandal, S., Chandra, M.: FPA based optimization of drilling burr using regression analysis and ANN model. Measurement **152**, 107327 (2020). https://doi.org/10.1016/j.measurement.2019.107327
3. Babu, S.S., Dhanasekaran, C.: Comparative analysis of thrust force, roughness and roundness error in drilling of aluminium composites using RSM, ANN and fuzzy logic. Mater. Today Proc. **69**, 908–917 (2022). https://doi.org/10.1016/j.matpr.2022.07.368
4. Kyratsis, P., Tapoglou, N., Bilalis, N., Antoniadis, A.: Thrust force prediction of twist drill tools using a 3D CAD system application programming interface. Int. J. Mach. Mach. Mater. **10**(1–2), 18–33 (2011). https://doi.org/10.1504/IJMMM.2011.040852
5. Tzotzis, A., Markopoulos, A.P., Karkalos, N.E., Tzetzis, D., Kyratsis, P.: FEM based investigation on thrust force and torque during Al7075-T6 drilling. In: IOP Conference Series: Materials Science and Engineering, p. 012009 (2021). https://doi.org/10.1088/1757-899X/1037/1/012009
6. Chatha, S.S., Pal, A., Singh, T.: Performance evaluation of aluminium 6063 drilling under the influence of nanofluid minimum quantity lubrication. J. Clean. Prod. **137**, 537–545 (2016). https://doi.org/10.1016/j.jclepro.2016.07.139
7. Palanikumar, K., Muniaraj, A.: Experimental investigation and analysis of thrust force in drilling cast hybrid metal matrix (Al-15%SiC-4%graphite) composites. Meas. J. Int. Meas. Confed. **53**, 240–250 (2014). https://doi.org/10.1016/j.measurement.2014.03.027
8. Rajmohan, T., Palanikumar, K.: Experimental investigation and analysis of thrust force in drilling hybrid metal matrix composites by coated carbide drills. Mater. Manuf. Process. **26**(8), 961–968 (2011). https://doi.org/10.1080/10426914.2010.523915
9. Aamir, M., et al.: Effect of cutting parameters and tool geometry on the performance analysis of one-shot drilling process of aa2024-t3. Metals (Basel) **11**(6) (2021). https://doi.org/10.3390/met11060854
10. Ghasemi, A.H., Khorasani, A.M., Gibson, I.: Investigation on the effect of a pre-center drill hole and tool material on thrust force, surface roughness, and cylindricity in the drilling of Al7075. Materials (Basel) **11**(1) (2018). https://doi.org/10.3390/ma11010140

11. Singh, G., Sharma, V.S., Gupta, M.K.: Sustainable drilling of aluminium 6061–T6 alloy by using nano-fluids and Ranque-Hilsch vortex tube assisted by MQL: an optimisation approach. Int. J. Mach. Mach. Mater. **20**(3), 252–273 (2018). https://doi.org/10.1504/IJMMM.2018.093547
12. Kyratsis, P., Tzotzis, A., Davim, J.P.: Experimental and 3D numerical study of AA7075-T6 drilling process. In: 3D FEA Simulations in Machining, Cham: Springer International Publishing, pp. 63–75 (2023). https://doi.org/10.1007/978-3-031-24038-6_4
13. Aamir, M., Giasin, K., Tolouei-rad, M., Vafadar, A.: A review: drilling performance and hole quality of aluminium alloys for aerospace applications. J. Mater. Res. Technol. **9**(6), 12484–12500 (2020). https://doi.org/10.1016/j.jmrt.2020.09.003
14. Efkolidis, N., Hernández, C.G., Talón, J.L.H., Kyratsis, P.: Modelling and prediction of thrust force and torque in drilling operations of Al7075 using ANN and RSM methodologies. Strojniški Vestn. J. Mech. Eng. **64**(6), 351–361 (2018). https://doi.org/10.5545/sv-jme.2017.5188

CAD-Based Automated USV Hull Design

Anastasios Tzotzis[1](✉), Nikolaos Efkolidis[1], César García-Hernández[2], and Panagiotis Kyratsis[1]

[1] Department of Product and Systems Design Engineering, University of Western Macedonia, 50100 Kila Kozani, Greece
a.tzotzis@uowm.gr

[2] Department of Design and Manufacturing Engineering, University of Zaragoza, 50018 Zaragoza, Spain

Abstract. The unmanned surface vehicles (USVs) market is continuously growing due to the ever-increasing demand. Moreover, USVs are sophisticated products with a challenging design process. In this work, a CAD-based application is being presented that is able to automate the design process of a USV hull. Particularly, it is possible to either design a single piece or a batch of hull forms. The purpose of the batch design is to provide the user with a variety of test-pieces for comparing the most important geometric characteristics (buoyant volume and center of mass). By utilizing the methods that are available in the application programming interface (API) of SolidWorks™, the developed application can accept inputs in the form of a spreadsheet file, in order to automatically build a number of forms. The procedure is carried out according to variables that are used in the industry, such as the deadrise angles and the planning surface widths.

Keywords: API · CAD · CAD-based programming · Deadrise angle · Hull · Planning surface · SolidWorks · USV

1 Introduction

Unmanned vessels is a high-tech product that gains popularity day by day. In the past few years, a number of manufacturers worldwide have been building such products, intended for various purposes, such as leisure, photography, engineering, inspection and security, safety, as well as science. Despite the fact that the term "drone" usually refers to unmanned aerial vehicles (UAVs), ground drones, as well as water drones do exist. Especially unmanned surface vehicles (USVs), is a product that is oriented towards scientific tasks, such as water quality monitoring, wildlife observation and water structure inspection [1–3]. Most of the time, the design of the USV hull is a tedious task that requires a lot of time and effort, and the utilization of specialized software.

The automation level that the CAD-based programming offers is practically unlimited and can be applied to various engineering tasks. Kyratsis et al. [1] utilized the API environment of SolidWorks™ to automate the design process of a pneumatic actuator system. The authors developed an application that deals with both the design process

G. Guxho et al. (Eds.): ITC-ICEE 2023, LNMIE, pp. 354–362, 2024.
https://doi.org/10.1007/978-3-031-48933-4_35

and the engineering documentation automatically. Tzotzis et al. [2] implemented the methods that are available in a commercially available CAD system for the automatic assessment of the existing holes in a workpiece and the generation of the G-code for drilling operations. Similar studies applied in the manufacturing area of research are available in the literature, regarding the aspect of product design [3–5], the measurement and verification of elements [6, 7] as well as the integration of parametric cutting tool models in finite element analyses [8]. Off-line robot programming, simulation and path planning, extended to machine tool trajectories simulation, is another topic which benefits from the features that are available in a CAD programming environment. Neto and Mendes [9] worked on the robot motion representation in a CAD drawing, and the automatic extraction of the motion data, in order to realize data mapping from the CAD model to the physical environment. Vakondios and Kyratsis [10] presented a milling simulation, based on the ball-mill tools in microscale. The authors implemented the kinematics of the process in a CAD-based environment and captured the topomorphy results of the workpiece.

The present paper presents a CAD-based application that is able to automate the design process of typical USV hulls without the need for specialized software, by utilizing the application programming interface (API) methods of a commercially available CAD system. The application is oriented towards the quick build of hull form models, with respect to standard design aspects. It is possible to either build a single model or a batch of models for comparison purposes. This way, both the effort and required time during a trial-and-error model selection process is reduced significantly.

2 Materials and Methods

Typical characteristics of a USV hull, include the deadrise angle and the planning surface width [11], regardless of the hull type. Multi-hull, round, shallow or deep vee and flat-bottomed are hull types that are most commonly utilized in various USV designs. In the present study, ten design variables; three angles, three widths and four lengths, were taken into account during the parametric modelling of the hull, focused on a vee-bottomed design, which corresponds to the main hull of a trimaran design, intended for use in calm water such as small lakes, ponds and reservoirs, as well as to open water with light wind conditions. The aforementioned design aspects are considered critical to every known hull type, for both manned and unmanned vessels. Figure 1a illustrates a wireframe model of a generic hull, including the six out of ten variables, namely foreship transverse deadrise angle θ_f, midship transverse deadrise angle θ_m, aftship transverse deadrise angle θ_a, width of foreship planning surface L_f, width of midship planning surface L_m and width of aftship planning surface L_a. In addition, Fig. 1b illustrates the four lengths, the foreship, midship, aftship and stern lengths. Since the present paper focuses mainly on the CAD-based design process of the specific product, rather on the technical aspects of the product itself, a number of extra design parameters such as the longitudinal deadrise angles did not participate in this study. However, it is still possible to use the application to design a hull including all the previously mentioned variables.

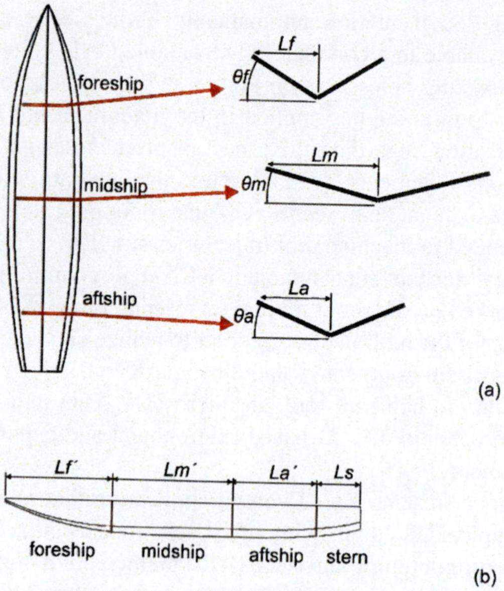

Fig. 1. The design parameters of the hull form model: bottom view (**a**) and side view (**b**).

2.1 Study Workflow

An initial form model of the hull was designed in SolidWorks™, according to the design principals found in the literature [11–13], in order to extract the level values of the design parameters. The design variables consist of three transverse deadrise angles, three planning surface widths and four longitudinal lengths, whereas the output parameters include two coordinates of the center of mass (COM) and the buoyant volume V_b. Next, the input parameter combinations for an arbitrary selected number of trials, were organized in a spreadsheet file, which was used to feed the developed design application. To link the input data with the design variables, first the Excel™ software was connected to the current SolidWorks™ session with the next lines of code [14].

- Dim xlApp As Excel.Application
- Set xlApp = New Excel.Application

To allow SolidWorks™ to access the input variables stored in the spreadsheet file, the next methods were used, which enable access to the active sheet (named "input_sheet") of the Excel™ file and then allow the extraction of a specific value (i.e. variable named "*Lf1*") in the sheet.

- Set input_sheet = GetObject(, "Excel.Application")
- Set Lf1 = inputsheet.ActiveSheet

With the design process that is described in Sect. 2.3, a total of 20 hull designs were produced, each with different COM and V_b. Using a similar method, the output results were recorded in the same spreadsheet file, leading to 20 results. The code that is responsible for the output data export is described in the next lines.

- Set swMass = swAssembly.Extension.CreateMassProperty
- COM = swMass.CenterOfMass
- excel_sheet.Range("A" & sheet_Row).Value = COM(0) * 1000
- excel_sheet.Range("B" & sheet_Row).Value = COM(1) * 1000
- excel_sheet.Range("C" & sheet_Row).Value = COM(2) * 1000
- excel_sheet.Range("D" & sheet_Row).Value = MassProp.Volume

Concluding, Fig. 2 summarizes the aforementioned design process of the hull forms.

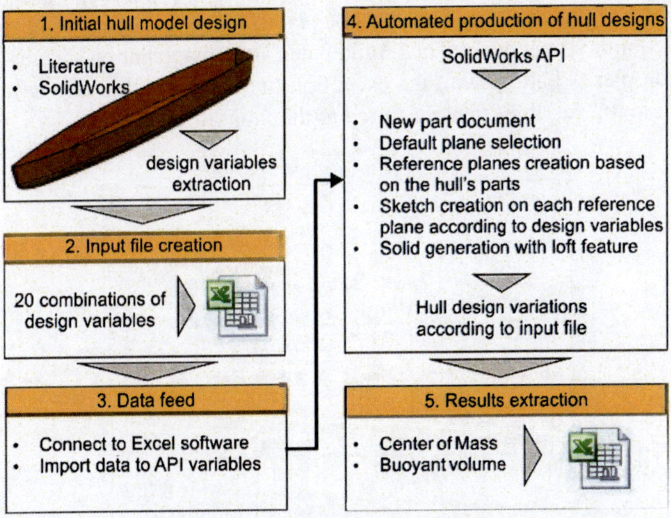

Fig. 2. The design process workflow.

To facilitate the hull design process, a user interface (UI) was designed within the SolidWorks™ API environment. Sections 2.2 and 2.3 describe the development of the UI, including the most important API methods used, as well as the design procedure.

2.2 User Interface Development

The UI of the application provides the user with the necessary tools to input the design variables in the code. Figure 3 illustrates the formation of the input fields and the command buttons that enable the application's functions. Specifically, it is possible to either use a spreadsheet file to input the design variables or use the equivalent input fields on the UI of the application. During the single piece design, the user must input a number of variables, which are divided into four categories: the deadrise angles in the transverse direction, the planning surface widths, and the longitudinal lengths of the hull, which essentially form the distances between the section planes of the hull. Both the transverse deadrise angles (θ_f, θ_m, θ_a) and the planning surface widths (L_f, L_m, L_a) refer to the foreship, the midship and the aftship parts of the hull. Lastly, the longitudinal lengths ($L_{f'}$, $L_{m'}$, $L_{a'}$, L_s) denote the lengths of the foreship, the midship, the aftship and the stern parts of the hull. By pressing the "Build single piece" command button the design process of the hull initiates according to the input parameters. The B-spline mesh of the loft command that is used to bond the profiles of the hull, can be controlled in the "Loft curvature" frame of the UI, as shown in Fig. 3. Concluding, the "Batch mode design" frame enables the mass production of hull form models, based on the instructions given in the input file, which must be pointed out to the application by providing the corresponding path. The input file must be in a spreadsheet format used in Excel™, such as ".xls". Moreover, it is important that the variables follow the same order every time, corresponding to the predefined columns and rows of the input file. For example, the transverse deadrise angles must be placed in the first column and each one of the foreship, midship and aftship values, must be placed in the first, the second and the third row of that column respectively. The "Build batch" command button is equivalent to the "Build single piece" button, with the exception that it will initiate the design process with respect to the variables that are stored to the input file.

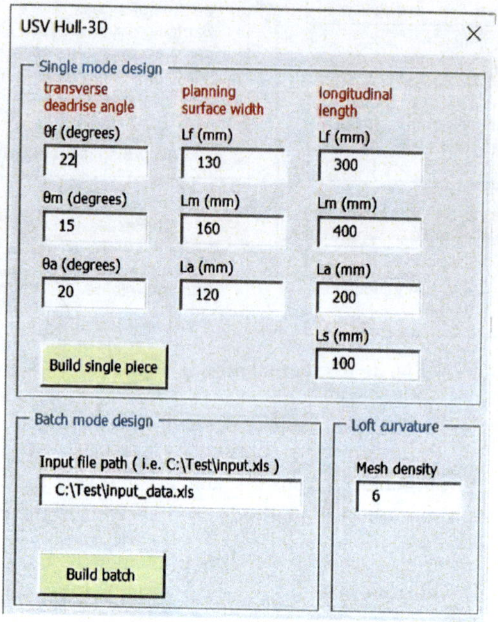

Fig. 3. The UI of the design application.

2.3 Design Process

The design process of the hull is simple and straightforward. A new part document is created with the "NewDocument" method and then five reference planes are created with respect to the default "Front Plane" of the part document in SolidWorks™, corresponding to the bow, foreship, midship, aftship and stern parts of the hull (Fig. 4a). The "SelectByID2" method is used to select the "Front Plane", whereas the "InsertRefPlane" method to create the reference planes. In addition, a sketch is created on each one of the six planes, forming the main sections of the hull (Fig. 4b). The sketch creation begins with the "Front Plane" and the code is repeated after each reference plane creation. The "ISketchManager" interface allows the sketch creation on the planes by utilizing the "CreateLine" method. In order to sketch the first line, its length, the inclination, as well as the starting coordinates are needed. Finally, the surfaces that derive from the sketches are bonded together with the "CreateLoftBody2" method, to form the final hull model. In detail, first the default "Front Plane" is selected and a new sketch is inserted. The sketch comprises of five lines. The line L', as shown in the detail A of Fig. 4b, is designed with respect to the given midship planning surface width L_m and the angle θ_m, with starting point of the line, the document's origin point. Thus, the length of line L' can be easily calculated with the cosine of θ_m. Equation 1 shown the calculation of the aforementioned line L', which essentially is the hypotenuse of the formed triangle.

$$L' = L_m \cos\theta_m \tag{1}$$

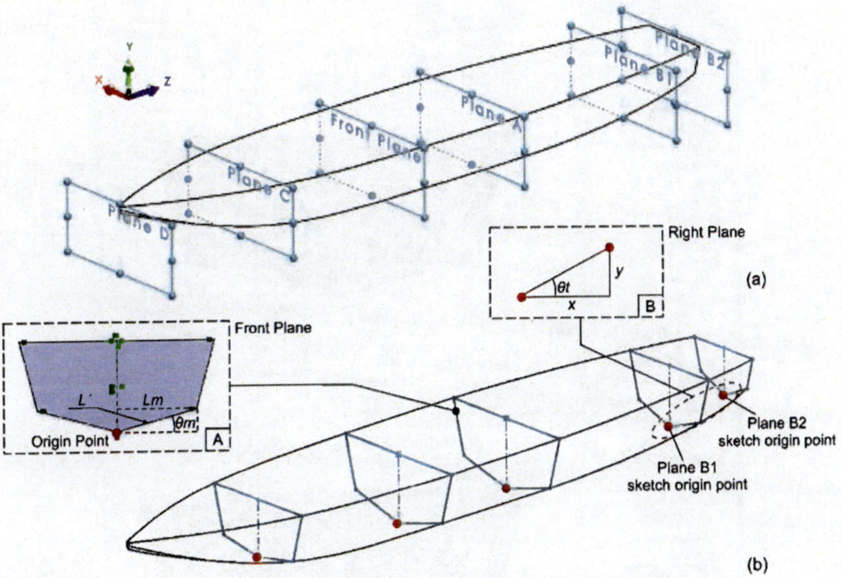

Fig. 4. Modelling process of the hull.

In a similar manner, the y coordinate of the origin points of the rest sketches, were determined with the Eq. 2. Where θ_t denotes the tangential deadrise angle for the equivalent part of the hull, and x is the horizontal distance (adjacent of the formed triangle) between two consecutive origin points, as shown in the detail B of Fig. 4, which practically is the distance between two neighboring planes that form the corresponding part of the hull.

$$y = x \tan\theta_t \tag{2}$$

3 Results and Discussion

Upon the completion of the design process, the generated models are stored in the specified folder where the input file resides, for the user's consideration. Moreover, the extracted output data (COM and V_b) are stored in a new spreadsheet file within the same folder. To realize the extraction of the COM coordinates, the "CenterOfMass" property is called and the coordinates are extracted in the form of an array. The boyant volume can be calculated by utilizing the "Volume" property. First, a reference plane is created with regard to the default "Top Plane" of the document, at a desired distance, which is usually equal to the draft of the vessel. Then, an internal region is created by intersecting the solid model on the reference plane, so that the byoant volume can be calculated. Figure 5 presents the result of the batch modelling, including two sample models out of a set of the 20 designed models along with their extracted geometric characteristics.

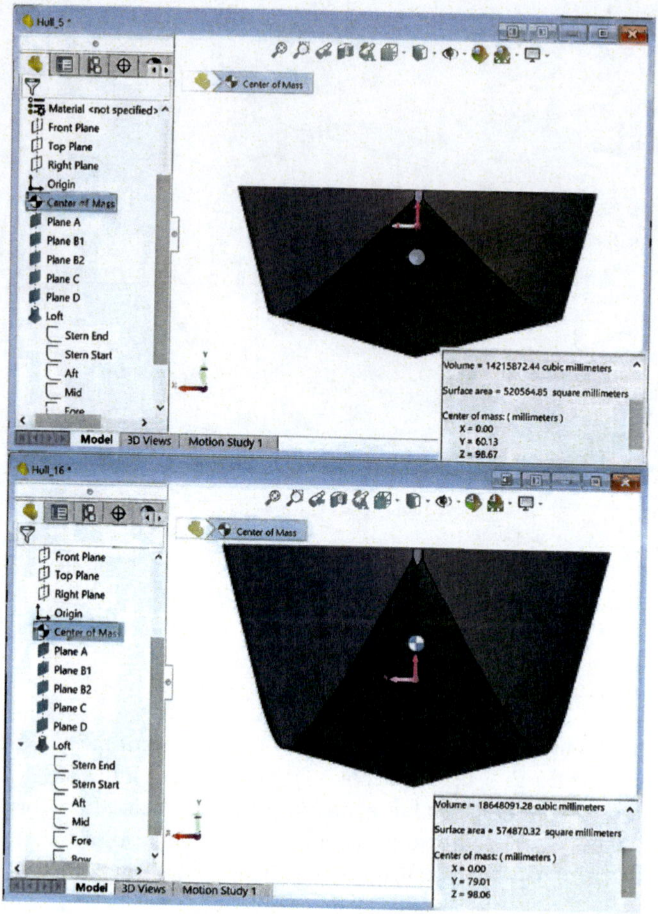

Fig. 5. Sample generated hull models.

Finalizing, it should be noted that it is possible to design a set of experiments according to a full or fractional factorial design in order to study the effects of the design variables on the output parameters and to optimize the design parameters according to the next goals: minimize the y coordinate of the COM, set the z coordinate of the COM equal to zero and finally, maximize the buoyant volume.

4 Conclusions

This study presented the development of a CAD-based programming application, which can enable the automated design of a USV hull. The presented application serves mainly two objectives: the automated parametric modelling of a USV hull, which can be extended to various other types, as well as the automatic comparison of multiple designs with respect to the COM and the buoyant volume. Finalizing, the following conclusions can be drawn:

- The design process yields an adequately accurate representation of the hull with minimum effort, proving the value of the application and the CAD-based programming in general.
- The structure of the application's code allows for the extension of the code's functionality in such a way, so that more hull types can be added to the application's reservoir.
- By utilizing a design of experiments, it is possible to study the effects of a variety of design parameters on the most important geometric characteristics of a USV hull. Therefore, the results can be used for the generation of the optimal solution regarding the hull design.

References

1. Kyratsis, P., Tzotzis, A., Tzetzis, D., Sapidis, N.: Pneumatic cylinder design using cad-based programming. Acad. J. Manuf. Eng. **16**(2), 107–113 (2018)
2. Tzotzis, A., Manavis, A., Efkolidis, N., Kyratsis, P.: CAD-based automated G-code generation for drilling operations. Int. J. Mod. Manuf. Technol. **13**(3), 177–184 (2021). https://doi.org/10.54684/ijmmt.2021.13.3.177
3. Kyratsis, P., Gabis, E., Tzotzis, A., Tzetzis, D., Kakoulis, K.: CAD based product design: a case study. Int. J. Mod. Manuf. Technol. **11**(3), 88–93 (2019)
4. Manavis, A., Tzotzis, A., Tsagaris, A., Kyratsis, P.: A novel computational-based visual brand identity (CbVBI). Machines (2022). https://doi.org/10.3390/machines10111065
5. Tapoglou, N., Efstathiou, C., Tzotzis, A., Kyratsis, P.: Study of the topography of face milled surfaces using CAD-based simulation. In: Kyratsis, P., Manavis, A., Davim, J.P. (eds.), Computational Design and Digital Manufacturing. Springer International Publishing, Cham, pp. 159–166 (2023). https://doi.org/10.1007/978-3-031-21167-6_8
6. Gella-Marín, R., Tzotzis, A., García-Hernández, C., Huertas-Talón, J.-L., Kyratsis, P.: CAD software integration with programming tools for modelling, measurement and verification of surfaces. In: Experiments and Simulations in Advanced Manufacturing, pp. 91–116 (2021). https://doi.org/10.1007/978-3-030-69472-2_5

7. García-Hernández, C., Gella-Marín, R., Huertas-Talón, J.L., Berges-Muro, L.: Algorithm for measuring gears implemented with general-purpose spreadsheet software. Measurement **85**, 1–12 (2016). https://doi.org/10.1016/j.measurement.2016.02.013
8. Tzotzis, A., García-Hernández, C., Huertas-Talón, J.L., Kyratsis, P.: CAD-based automated design of FEA-ready cutting tools. J. Manuf. Mater. Process. **4**(4), 1–14 (2020). https://doi.org/10.3390/jmmp4040104
9. Neto, P., Mendes, N.: Direct off-line robot programming via a common CAD package. Rob. Auton. Syst. **61**(8), 896–910 (2013). https://doi.org/10.1016/j.robot.2013.02.005
10. Vakondios, D.G., Kyratsis, P.: An innovative CAD-based simulation of ball-end milling in microscale. Adv. Comput. Des. **5**(1), 13–34 (2020). https://doi.org/10.12989/acd.2020.5.1.013
11. Guan, G., Wang, L., Geng, J., Zhuang, Z., Yang, Q.: Parametric automatic optimal design of USV hull form with respect to wave resistance and seakeeping. Ocean Eng. **235**(December 2020), 109462 (2021). https://doi.org/10.1016/j.oceaneng.2021.109462
12. Khan, S., Gunpinar, E., Sener, B.: GenYacht: an interactive generative design system for computer-aided yacht hull design. Ocean Eng. **191**(September), 106462 (2019). https://doi.org/10.1016/j.oceaneng.2019.106462
13. Zhang, P., Xiang Zhu, D., Hao Leng, W.: Parametric approach to design of hull forms. J. Hydrodyn. **20**(6), 804–810 (2008). https://doi.org/10.1016/S1001-6058(09)60019-6
14. Tzotzis, A., Garcia-Hernandez, C., Huertas-Talon, J.-L., Tzetzis, D., Kyratsis, P.: Engineering applications using CAD based application programming interface. In: MATEC Web of Conferences, pp. 1–7 (2017). https://doi.org/10.1051/matecconf/20179401011

Material Science

Quality Control of the Steel Bar Production for Engineering Applications

Klodian Dhoska[1,2,3](✉), Irida Markja[1,2], Elena Bebi[2,4], Brisilda Kaçorri[5], and Agus Pramono[6]

[1] Department of Production and Management, Faculty of Mechanical Engineering, Polytechnic University of Tirana, Mother Theresa Square No. 1, 1019 Tirana, Albania
kdhoska@fim.edu.al

[2] Albanian Mechanical Engineering Association (AMEA), Mother Theresa Square No. 1, 38954 Tirana, Albania

[3] Association of Talent under Liberty in Technology (TULTECH), Sopruse Pst 3, 10615 Tallinn, Estonia

[4] Energy Department, Faculty of Mechanical Engineering, Polytechnic University of Tirana, Mother Theresa Square No. 1, 1019 Tirana, Albania

[5] Quality Control Laboratory, Kurum International Ltd, Elbasan, Albania

[6] Department of Metallurgy, Universitas Sultan Ageng Tirtayasa, Kota Cilegon, Banten 42435, Indonesia

Abstract. Albania has one of the largest industries for several steel production facilities located in the Elbasan city. "Kurum" international ltd is the metallurgical industry that is focusing in production of the stainless steel. The production capacity is going around 700,000 of liquid steel every year. Most of the products have been imported to Bulgaria, Montenegro, Turkey, Kosovo, Serbia and Italy while 40% cover the needs of the domestic market. Due to the market needs, the scale and large production capacity has made the company to focused on production of the steel bar. For assuring the quality of the steel bar is needed quality control. Quality control is a crucial aspect of steel bar production to ensure that the produced bars meet the required standards and specifications. Based on it, our research work will be focused on the quality of control of the steel bar production.

Keywords: Metallurgy · Steel bar · Quality control · Tensile testing

1 Introduction

Nowadays the new economic growth target of Albanian government is focusing on the support of different metals manufacturing companies that are operating in Albania. Some of the important manufacturing companies are in the center part of Albania where they are focusing on the production of steel, aluminum, chromium, nickel and copper [1–3]. Furthermore, the largest sector of the metals production in Albania is focused in steel production which is located in Elbasan city. "Kurum international" is the metallurgical industry that is focusing in production of the stainless steel. The production capacity is

G. Guxho et al. (Eds.): ITC-ICEE 2023, LNMIE, pp. 365–371, 2024.
https://doi.org/10.1007/978-3-031-48933-4_36

going around 700,000 of liquid steel every year [4]. Due to the market needs, the scale and large production capacity has made the company to focused on production of the steel bar. Figure 1 shows the manufacturing scheme of the steel production that are using in "Kurum international" [4, 5].

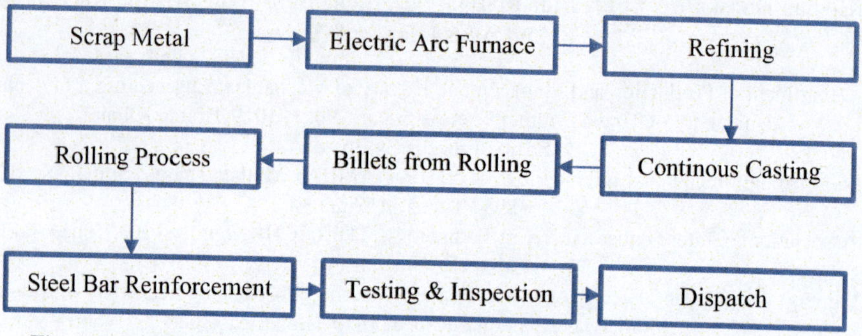

Fig. 1. Manufactured scheme of the steel bar production, "Kurum" International ltd.

In Fig. 1 it has been shown that there are different stages and methodologies to monitor and improve the quality of steel bar products. Some of the important factors are as follows:

- Raw material inspection
- Chemical composition
- Melting and refining
- Heat treatment
- Rolling and forming
- Testing and inspection.

Our research work will be focused on the inspection services for ensuring the quality of the manufactured steel bar. Quality control is a crucial aspect of steel bar production to ensure that the produced bars meet the required standards and specifications. In this paper, we will evaluate the mechanical resistance of the steel bars with diameters 12 and 14 mm by using calibrated uniaxial tensile strength. Furthermore, standard deviation has been used for investigating respectively the quality measurements of the uniaxial tensile strength and caliper through repeatability measurements of the applied force and diameters of the steel bars.

2 Materials and Methods

Two different chemical compositions of the steel bar production samples 1 and 2 that we have used in our research work are shown in Table 1 where respectively correspond to diameter 12 mm and diameter 14 mm. We have selected 5 random samples for each chemical composition to verify mechanical properties of the manufactured steel bar. The length for both samples has been prepared in cutting machine and the lengths varied from 551 mm until 556 mm.

Table 1. Chemical composition of the steel bar production for sample 1 and sample 2

Elements	Sample composition 1 (%)	Sample composition 2 (%)
Carbon	17	20
Silicon	13	15
Manganese	50	53
Phosphorus	0.3	0.1
Sulphur	0.4	0.2
Copper	20	21
Nitrogen	0.01	0.01
Equivalent Carbon	0.45	0.4

The measurement method has been focused on the tensile strength method by following standard ISO 6892-1 [6]. Tensile strength cannot be measured directly but should be calculated through the measurement of the applied load per unit area of the samples as can be seen in Eq. (1) [2, 7–9].

$$Rm = \frac{F}{A} \tag{1}$$

where Rm is tensile strength, F represent maximum applied force, and A is the cross-sectional area of the sample. Cross-sectional area of the samples is calculated by using following Eq. (2).

$$A = \frac{\pi d^2}{4} \tag{2}$$

where d is the diameter of the samples.

Figure 2 depict the calibrated tensile strength device which is located to "Kurum" International ltd laboratory for realization the measurements of the manufactured steel bar samples. Furthermore, in the Tables 2 and 3 are shown the measurements test results for the samples 1 and 2.

Due to the repeatability measurements for assuring the quality of the calibrated tensile strength device and caliper we have used Eq. (3) to calculate standard deviation of the force measurements and diameter measurements [10–15].

$$S = \sqrt{\frac{1}{\mathrm{n}-1} \cdot \sum_{i=1}^{n} (x_i - x_m)^2} \tag{3}$$

where S is standard deviation of the reading of applied force and diameters, x_i represent individual measurements, x_m is the mean value of individual measurements and n is number of individual measurements.

Afterward, we have used Eqs. (4) until (6) to analyzed the tensile test results of the samples 1 and 2.

$$Fracture = \frac{\mathrm{BL}}{\mathrm{A_0}} \tag{4}$$

$$PE = \left[\frac{(L_1 - L_0)}{L_0}\right] \cdot 100 \tag{5}$$

$$Ratio = \frac{\text{Fracture}}{R_m} \tag{6}$$

where *BL* is breaking load, *PE* is percentage elongation, L_1 is the measurement length after breaking loads, L_0 is the length before tensile testing and *Ratio* represent the ratio between the fracture per tensile strength.

(a) (b)

Fig. 2. Tensile strength device with manufactured steel bar, **a** before and **b** after measurements, Sample 1.

Table 2. Measurement test results of the sample 1.

Measurements no	Applied loads (N)	Breaking loads (N)	Diameters (mm)
1	76317.24	92223.11	12.02
2	76317.26	92223.18	12.03
3	76317.25	92223.14	12.02
4	76317.24	92223.12	12.01
5	76317.25	92223.13	12.02

Table 3. Measurement test results of the sample 2.

Measurements no	Applied loads (N)	Breaking loads (N)	Diameters (mm)
1	98470.12	111321.27	14.02
2	98470.11	111321.25	14.01
3	98470.11	111321.24	14.01
4	98470.12	111321.26	14.01
5	98470.12	111321.27	14.02

3 Measurement Results

Throughout the manufacturing process for ensuring the quality control of the manufactured steel bars in our research work we have been focused firstly on the results of the standard deviation of the force measurements and caliper. Tables 4 and 5 depict the results of the standard deviation of the applied force and diameters through repeatability measurements.

Table 4. Test results from repeatability measurements of the applied force.

Measurements	Mean value (N)	Standard deviation (N)
Applied force sample 1	76317.248	0.00836
Applied force sample 2	98470.116	0.00547

Table 5. Test results from repeatability measurements of the diameters.

Measurements	Mean value (mm)	Standard deviation (mm)
Diameters sample 1	12.020	0.00707
Diameters sample 2	14.014	0.00547

From the repeatability results it has been seen that our measurements devices ensure accuracy of the test results.

Afterward the tensile strength results of the manufactured steel bar samples are shown in the Table 6 and 7.

Results of the Sample 1 have shown that the ratio value varied from 1.14 until 1.21 which is in the accordance to the standard ISO 6892-1[6].

Results of the Sample 1 have shown that the ratio value varied from 1.13 until 1.15 which is in the accordance to the standard ISO 6892-1[6].

In general, it has been seen that all the experimental results fulfilling the requirements of the standard ISO 6892-1. The ratio values varied from 1.13 until 1.21 which respectively corresponds to the diameter 14.02 mm and 12.03 mm.

Table 6. Tensile testing results of the manufactured steel bar, Sample 1.

No	D (mm)	L_0 (mm)	L_1 (mm)	TS (N/mm^2)	Fracture (N/mm^2)	PE (%)	Ratio
1	12.01	551	571	674.01	768.37	3.63	1.14
2	12.03	552	576	671.77	811.78	4.35	1.21

Table 7. Tensile testing results of the manufactured steel bar, Sample 2.

No	D (mm)	L_0 (mm)	L_1 (mm)	TS (N/mm^2)	Fracture (N/mm^2)	PE (%)	Ratio
1	14.01	554	574	639.08	734.94	3.61	1.15
2	14.02	556	577	638.17	721.46	3.78	1.13

4 Conclusion

In this paper we have briefly describe the quality control of the manufacturing steel bar samples. Our research work has been focused on the inspection services for ensuring the quality of the manufactured steel bar. Firstly, repeatability results have shown that our measurements devices are accurate and ensure the accuracy of the test results. Secondly, all the experimental results fulfilling the requirements of the standard ISO 6892-1. The ratio values varied from 1.13 until 1.21 which respectively corresponds to the diameter 14.02 mm and 12.03 mm. The future research work would be focused on the investigation of the tensile strength with larger diameters and compositions of the reinforced steel bars.

Acknowledgement. The authors would like to thank manufacturing company "Kurum" International ltd for all the data and Polytechnic University of Tirana.

References

1. Dhoska, K., et al.: Manufacturing Process of the aluminum alloy AA6063 for engineering applications. J. Integr. Eng. Appl. Sci. **1**(1), 1–13 (2023)
2. Dhoska, K.: Tensile testing analysis of the HRB400 steel reinforcement bar. Int. J. Innov. Technol. Interdisc. Sci. **2**(3), 253–258 (2019)
3. Markja, I., Dhoska, K., Elezi, D., Moezzi, R., Petru, M.: Effect of the grain sizes on the ultrasonic propagation and attenuation on different types of steels microstructure during non-destructive testing. Ann. Chim.-Sci. Mat. **45**(4), 329–334 (2021)
4. Kurum International ltd Albania: https://www.kurum.al. Last accessed 21 July 2023
5. Dhoska, K., Lumi, D., Sulejmani, A., Koça, O.: Measurement uncertainty for mechanical resistance of manufactured steel bar. Pollack Periodica **17**(2), 104–108 (2022)
6. ISO 6892-1: 2016. "Metallic materials", https://www.iso.org/obp/ui/#iso:std:iso:6892:-1:ed-2:v1:en. Last accessed 23 July 2023

7. Feng, L., He, Y., Chang, J., Zheng, S., Frosolini, M.: The relation research of tensile strength and chemical components of HRB400 in China. J. Chem. Pharm. Res. **6**(6), 2358–2366 (2014)
8. Wijaya Hikmatullah, C., Setiawan, I., Pramono, A.: Environment effect on underwater wet welding process of API 5L X65 steel by coating resin-based E6013 electrode. J. Trans. Syst. Eng. **1**(1), 31–49 (2023)
9. Dhoska, K., Tola, S., Pramono, A., Vozga, I.: Evaluation of measurement uncertainty for the determination of the mechanical resistance of the brick samples by using uniaxial compressive strength test. Int. J. Metrol. Qual. Eng. **9**(12), 1–5 (2018)
10. Kuhinek, D., Zorić, I., Hrženjak, P.: Measurement uncertainty in testing of uniaxial compressive strength and deformability of rock samples, Meas. Sci. Rev. **11**(4), 112–117 (2011)
11. JCGM 100: 2018. Evaluation of Measurement Data—Guide to the Expression of Uncertainty in Measurement. https://www.iso.org/sites/JCGM/GUM-introduction.htm. Last accessed 12 Aug 2023
12. Koça, O., Sulejmani, A., Dhoska, K.: Pressure distribution on rolling-slide contact problem. Pollack Periodica **16**(1), 71–76 (2021)
13. Dhoska, K., Kübarsepp, T., Hermaste, A.: Uncertainty evaluation of angle measurements by using 3D coordinate measuring machine. In: Proceedings of the International Conference of DAAAM Baltic "Industrial Engineering", pp. 221–225 (2014)
14. Pramono, A., Dhoska, K., Markja, I., Kommel, L.: Impact pressure on mechanical properties of aluminum-based composite by ECAP-parallel channel. Pollack Periodica **14**(1), 67–74 (2019)
15. Dhoska, K., Hofer, H., Rodiek, B., López, M., Kübarsepp, T., Kück, S.: Improvement of the detection efficiency calibration and homogeneity measurement of Si-SPAD detectors. Springerplus **5**(1), 2065 (2016)

Asphalt Pavement Recycling Technology with Reclaimed Asphalt Pavement (RAP) According to the Standards EN 13108-8 and EN 13108-1

Irida Markja[1(✉)], Klodian Dhoska[1], Mentor Fraholli[2], Giorgio Mustafaraj[1], Blerta Ahmetaj[1], and Dervish Elezi[1]

[1] Department of Production and Management, Faculty of Mechanical Engineering, Polytechnic University of Tirana, Tirana, Albania
imarkja@fim.edu.al

[2] Laboratory Elba Ltd., Tirana, Albania

Abstract. Asphalt recycling technology is a method used to repair roads at a lower cost and with a lower environmental impact than conventional methods. The technology of recycling can reduce the use of new paving materials by recycling the recycled asphalt layer, called Reclaimed Asphalt Paving ("RAP") which is generated when roads are repaired, to produce recycled asphalt mixtures and use these mixtures again for road paving. RAP material is the main component of new recycled asphalt mixes, the characteristics of this material must be in accordance with the National and European specifications (EN 13108-8). One of the characteristics of RAP material is its heterogeneity. Thus, the RAP material used in the laboratory is reduced by quaternization, according to EN 932-1 and EN 932-2 standards, in order to obtain samples that accurately represent the studied RAP. In this work, it is described the recycling of asphalt, where its improvement has the purpose of re-use in road paving projects. The set of laboratory analyses control performed gives us the necessary data which determine the characteristics of the reclaimed asphalt pavement. With the obtained mass of the recycled asphalt, we performed the tests for the evaluation of density, stability and Marshall Flow test. Form the results that we compared to the existing asphalt, we can conclude that the Reclaimed Asphalt Paving ("RAP") processed asphalt is a very good and profitable alternative for road paving projects.

Keywords: Asphalt recycling technology · Reclaimed asphalt paving · Stability · Marshall and flow · Graulometry analysis

1 Introduction

In the Republic of Albania, there are practices of recycling existing asphalt. We are focusing on the study in the experimental phase on the harmonization of the old Albanian Standard internship (STAZH Standard) with the European EN standards, which are now harmonized as SSH EN.

G. Guxho et al. (Eds.): ITC-ICEE 2023, LNMIE, pp. 372–387, 2024.
https://doi.org/10.1007/978-3-031-48933-4_37

The Albanian standard has worked with bitumen quantities much higher than the European standards, therefore we consider this experimental study reasonable, which is important due to the fact that, in addition to the recycling of asphalt, which reduces environmental pollution, since we do not throw it away, we reuse it even we obtain a larger amount of bitumen, which gives us the opportunity to also obtain a larger amount of asphalt, which reduces the cost of construction, since we only add aggregates without adding the bitumen binder [1, 2, 11].

We will express it above with the asphalt concrete tests carried out near the Laboratory ELBA Ltd., which is an accredited laboratory in our country, near the Albanian Accreditation Directorate.

2 Materials, Methods and Standards Used and Related Calculations

The materials used in this study have been focused on existing asphalt and new asphalt, comparing the binder with the required properties according to the European standards.

The methods that we have used to analyze the asphalts are the percentage of bitumen, the granulometry analyses, Marshall flow and determination of percent voids residual according to the EN standard and technical specification of European roads. The existing asphalt that we have analyses in this study was taken during the works on the street with the address: Road Dora D'istria, Tirana, Albania (Figs. 1 and 2).

Fig. 1. The view from the shredding

2.1 Existing Asphalt (Wearing Course)

The sample taken from the existing asphalt was treated in the oven in order to melt it. In this way, one part of the sample was prepared for laboratory extraction and the other for Marshall Stability.

Fig. 2. The sample taken for study of the existing asphalt.

The extraction test serves us to calculate the content of the binding material (bitumen) as well as the granulometry of the asphalt component aggregates taken in the road (in our case *Surface layer*) [2]. This procedure was performed according to the standard SSH EN 12697-1 to evaluate the percentage of bitumen (*Soluble binder content*) and the standard SSH EN 12697-2:2015 + A1:2019 for granulometric analysis (*Determination of particle size distribution*) [4, 8] (Figs. 3, 4, 5, 6 and 7).

Fig. 3. Asphalt-filled pit and cylinder

2.1.1 Centrifuge Extractor Method for Bitumen

The following table (Table 1) shows the data obtained during the bitumen test, where the percentage of bitumen in the mixture is **6.75%** and the percentage of bitumen in the aggregate is **7.23%** [4].

Fig. 4. Cylinder assembly assembly tetrachlorethylene before the test.

Fig. 5. The cylinder with solvent content

Fig.6. Extraction unit bowl with agregates

Fig. 7. Mixing of aggregates at the end of extraction, after drying

Table 1. The data obtained during the bitumen extraction test

Referenced standard	SSH EN 12697-1:2020	
Description	Unit	M
Weight of mix + weight of tare before extraction	g	3646.70
Weight of tare	g	2367.50
Weight of mixtures before extraction	g	1279.20
Weight of bowl	g	195.30
Weight of bowl + filler	g	202.70
Weight of Filler in bowl	g	7.40
Weight of mix + weight of tare after extraction	g	3553.00
Net weight of aggregates	g	1185.50
Weight of bitumen	g	86.30
Percentage of bitumen to mixtures	%	**6.75**
Percentage of bitumen to aggregates	%	*7.23*

Granulometric Analysis was performed according to the standard SSH EN 12697-2:2015 + A1:2019 and the data obtained from granulometric analysis are presented at the table below (Table 2) [5]. We can see that the curve is within the standard limits, because the selected granulometry curve passes between the curves according to the standard [5, 11, 13]. As can be seen in the graph below Fig. 8, the factice curve which shown the sieves dimension in *mm*, is within the standard [5].

Graph of aggregate granulometry curve obtained at the end of the bitumen extraction of existing asphalt whereas with the rest of the hot sample, we perform laboratory tests of Marshall Stability [5, 10–13]. This is based on the standard SSH EN 12697-34:2020, SSH EN 12697-30:2020 [9, 10]. Four sample are filled with hot asphalt, before they are put into the compactor, where they receive 75 blows. Then the sample return and are

Table 2. The data obtained from the grain size analyses.

Standard sieve size [mm]	Lower passing limit [%]	Upper passing limit [%]	Progressive passing [%]	Partial retained weight [g]	Progressive retained weight [g]	Progressive retained weight [%]	Progressive passing [%]
20.000	*100*	*100*	**100.00**	0.00	0.00	0.00	100.00
16.000	*100*	*100*	**100.00**	0.00	0.00	0.00	100.00
12.500	*100*	*100*	**99.72**	3.30	3.30	0.28	99.72
10.000	*90*	*100*	**98.39**	15.90	19.20	1.61	98.39
6.300	*50*	*85*	**79.38**	226.80	246.00	20.62	79.38
4.000	*35*	*70*	**55.44**	285.60	531.60	44.56	55.44
2.000	**15**	**60**	**37.36**	215.60	747.20	62.64	37.36
1.000	*8*	*50*	**26.26**	132.50	879.70	73.74	26.26
0.500	*6*	*40*	**18.92**	87.50	967.20	81.08	18.92
0.250	*5*	*30*	**11.89**	83.90	1051.10	88.11	11.89
0.125	*4*	*20*	**6.37**	65.80	1116.90	93.63	6.37
0.063	*2*	*12*	**2.80**	42.60	1159.50	97.20	2.80
0.000				33.40	33.40		
Total weight				**1192.90**	**1192.90**		

given another 75 blows from the other side. Then the mold is left to cool for more than 12 h and the cooled samples are taken out of the mold [3] (Figs. 9, 10, 11, 12, 13 and 14).

Below at Table 3 are shown the summary calculations of the Marshall density for the existing asphalt with the hydrostatic method [6, 10, 11].

After calculating the density, the samples are placed in a warm bath 60 °C for 45 min, in order for the samples to be ready for the Marshall test for the calculation of stability and flow according to SSH EN 12697-34:2020 [10, 11, 13] (Figs. 15 and 16).

Below at Tables 4 and 5 are shown the data and the summary calculations of the Marshall flow test for the existing asphalt [7, 8].

Related to the Marshall porosity, calculations were made for the porosity of the sample for the existing asphalt.

2.2 New Asphalt (Wearing Course)

Studying the data obtained from the existing asphalt tests, we find it necessary to add filler and aggregate to improve the properties according to EN standards. Under these conditions, we repeat the above tests for the new asphalt with the addition of aggregates. We melted the existing asphalt in the oven, and mixing it, we added according to the

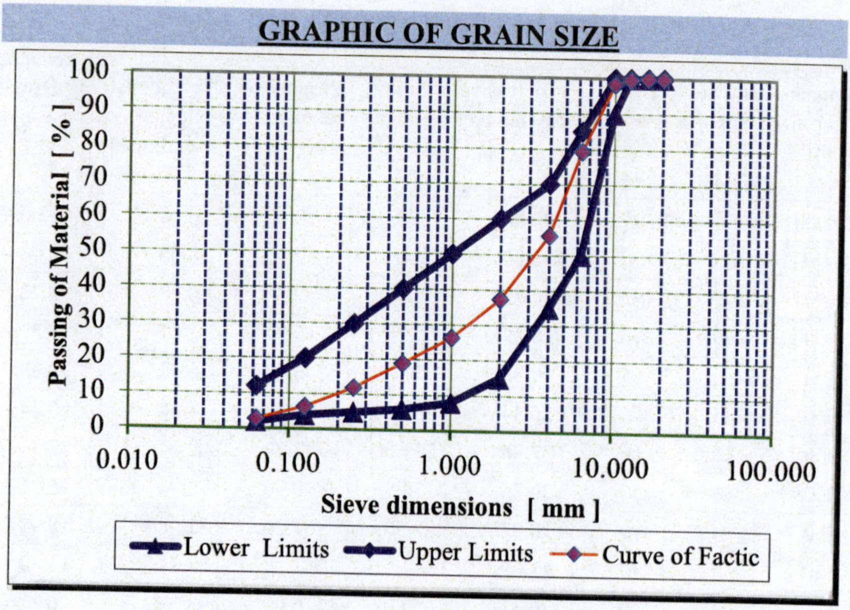

Fig. 8. The graph above shows the grain size analyses of the exciting asphalt. This curves are obtained from the values of sieve dimensions and values of percentage of passing.

Fig. 9. Marshall compactor

calculations of dust (filler) and mountain sand. By constantly mixing it, we manage to obtain an increased amount of asphalt, but with 5.69% bitumen by weight (Fig. 17).

The mixture is constantly stirred until a homogeneous mass of asphalt is obtained. The new granulometry curves with the addition of sand and filler is as follows (Fig. 18 and Table 6).

Below at Table 7 are shown the summary calculations of the Marshall density for the new asphalt with the hydrostatic method.

From the values obtained, we say the density of new asphalt has a decrease, from the value of existing asphalt [14, 15]. Specifically, before processing it was 2364 kg/m^3,

Fig. 10. The mold preheat.

Fig. 11. Compacted samples

Fig. 12. Remove of the samples.

while after processing it is 2331 kg/m^3. This phenomenon comes as a result of the increase in the amount of voids in the recipe as a result of adding aggregates to it [14, 16] (Table 8).

Related to the Marshall porosity, calculations were made for the porosity of the sample for the new asphalt (Table 9).

Fig. 13. Sample measurement

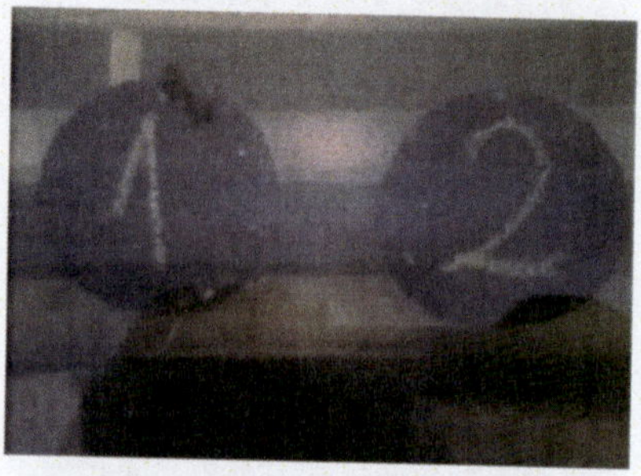

Fig. 14. Hydrostatic pressure of samples.

Table 3. Calculation of bulk density of bituminous samples for existing asphalt.

Referenced standards	SSH EN 12697-6:2020					
Number of sample	Identification	Unit	1	2	3	4
Weight of sample before saturation	m_1	g	1181.0	1282.5	1202.5	1189.3
Weight of sample in water	m_2	g	684.5	743.3	696.6	695.5
Weight of sample after saturation	m_3	g	1183.7	1284.2	1204.3	1197.6
Temperature of water	t	°C	23.2			
Density of water	ρ_w	kg/m^3	998.0			
Density Marshall (apparent)	ρ_{th}	**kg/m^3**	**2361**	**2366**	**2364**	**2364**
Average density (apparent)	ρ_{th}^{mes}	**kg/m^3**	**2364**			

It can be noted that compared to the existing asphalt, the fluidity value has moved from **3.9** to **3.8** mm, while the stability value has increased from **1741** to **2357** kg.

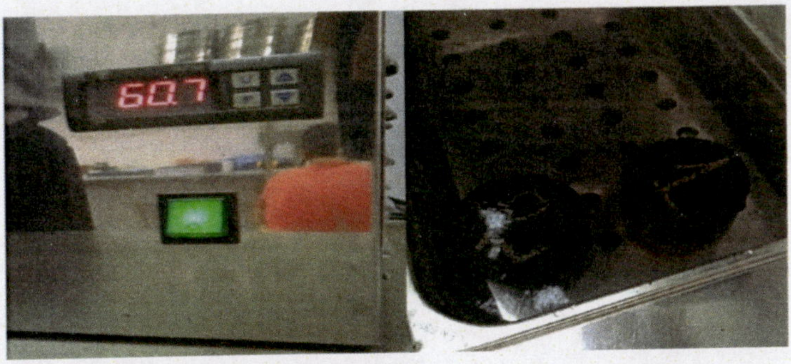

Fig. 15. Samples during the treatment in the water bath.

Fig. 16. Development of the Marshall test.

3 Conclusions

In the table below are show the average data of the results of the existing and new asphalt according to the standard and also compare to the technical specifications of European roads [2, 14, 15].

Referring to the Table 10, we arrive at some results obtained for both layers such as:

The percentage of bitumen in the existing asphalt is 7.23%, while in the new asphalt it is 5.69%, which shows a decrease in the amount of bitumen, which has a high cost in asphalt compared to inert, about 15.4 L/ton or about 1400 lek/ton, which means for 1 km of road with a width of 6m and a layer thickness of 6 cm, the profit is about 900,000 lek/km.

Stability, density and Marshall flow, in both layers (existing asphalt and the new one) are within the technical conditions according to technical specifications of European roads.

Table 4. Determination of stability and flow, data processing at the end of the Marshall test for the existing asphalt sample.

Determination of stability and flow					
Referenced standards		SSH EN 12697-34:2020			
Sample	Nr	1	2	3	4
Height sample	mm	62.50	67.50	63.50	64.10
Dial reading	[μm]	754	905	780	806
Carico/load	–	16.2883	19.5503	16.8500	17.4117
Load correction	[kg]	1660	1993	1718	1775
Correlation ratio	–	1.02	0.91	1.00	0.98
Marshall stability	[kg]	1694	1814	1718	1739
Average stability	**[kg]**	**1741**			
Marshall flow	[mm]	3.80	3.90	3.80	3.90
Average flow	**[mm]**	**3.85**			
Stiffness	**[kg/mm]**	**440**	**465**	**452**	**446**

Table 5. Determination of percent voids residued, data processing at the end of the Marshall test for the existing asphalt sample.

Determination of percent voids residued		
Referenced standards	SSH EN 12697-8:2018	
Air voids content (Va)	%	1.89
Voids in mineral aggregate (percent of bulk volume) (VMA)		18.6
Percent of the voids in the aggregates filled with binder (VFA)	%	89.8

While the porosity in the existing asphalt is below the minimum allowed value of 1.89% from 3–5%, while the new asphalt has reached the required parameters of 3.24%.

Comparing the values obtained from the test results, we can see that the percentage of bitumen in the existing asphalt is quite high, compared to the European technical specifications.

As for density, Marshall stability and flow, the parameters are in accordance with the European standards EN 13108-1:2016 & EN 13108-8:2016.

The granulometric analysis, regardless of the fact that it is within the limits of the European standard, there is a deviation which has to do with the dust under the sieve 0.063 mm, as it does not respect the bitumen powder ratio which should vary from 0.6–1.4, i.e. the old asphalt $2.8/7.23 = 0.4 < 0.6$, while new asphalt $6.21/5.69 = 1.1 < 1.4$.

The porosity of the sample of the existing asphalt is in values below the normal value, which shows that the existing asphalt, in addition to having low rates of porosity,

Fig. 17. Adding sand and filler to the hot mix of existing asphalt.

Table 6. Evaluation of the sieve passing by comparing it with the standard curves according to the reference standard SSH EN 13108-1:2016.

Referenced standard	SSH EN 13108-1:2016		
Standard sieve size [mm]	Lower passing limit [%]	Upper passing limit [%]	Progressive passing [%]
20.000	*100*	*100*	**100.00**
16.000	*100*	*100*	**100.00**
12.500	*100*	*100*	**99.76**
10.000	*90*	*100*	**98.63**
6.300	*50*	*85*	**82.47**
4.000	*35*	*70*	**62.03**
2.000	*15*	*60*	**44.76**
1.000	*8*	*50*	**33.14**
0.500	*6*	*40*	**24.24**
0.250	*5*	*30*	**15.98**
0.125	*4*	*20*	**9.90**
0.063	*2*	*12*	**6.21**

which is due to its obsolescence from the deformation processes, shows us that the time of removal has come of this layer after it has lost its deforming properties and has gone towards a rigid and non-elastic layer.

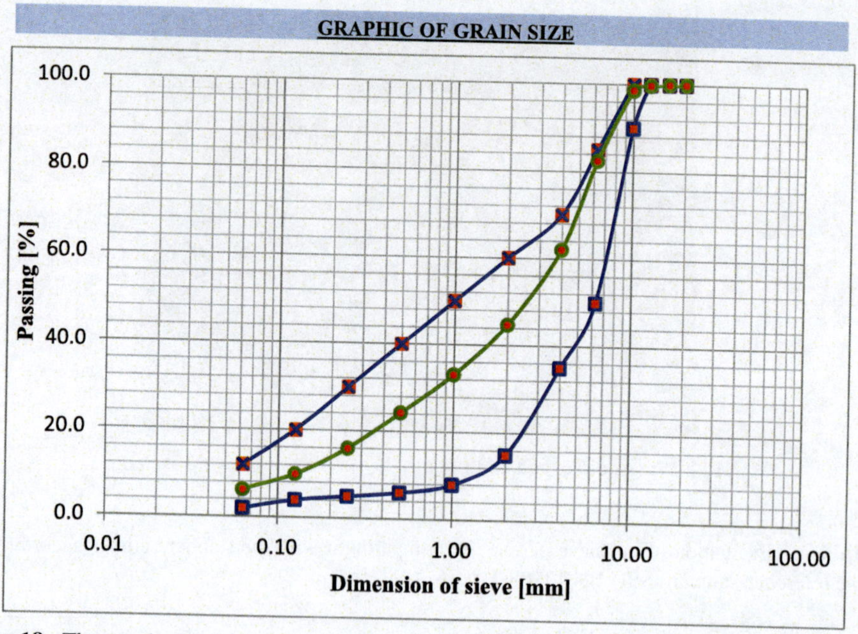

Fig. 18. The graph above shows the grain size analyses of the new asphalt. This curves are obtained from the values of sieve dimensions and values of percentage of passing.

Table 7. Determination of bulk density of bituminous specimens according to the standard SSH EN 12697–6:2020.

Referenced standards	SSH EN 12697-6:2020					
Number of samples	Identification	Unit	1	2	3	4
Weight of sample before saturation	m_1	g	1143.5	1138.5	1182.5	1154.2
Weight of sample in water	m_2	g	661.9	659.8	685.0	669.8
Weight of sample after saturation	m_3	g	1151.6	1147.4	1190.6	1163.9
Temperature of water	t	°C	21.8			
Density of water	ρ_w	kg/m^3	997.8			
Density Marshall (apparent)	ρ_{th}	**kg/m^3**	**2330**	**2330**	**2334**	**2331**
Average density (apparent)	ρ_{th}^{mes}	**kg/m^3**	**2331**			

Table 8. Determination of stability and flow, data processing at the end of the Marshall test for the new asphalt sample.

Determination of stability and flow					
Referenced standards	SSH EN 12697-34:2020				
Sample	No	1	2	3	4
High sample	mm	61.00	61.10	63.50	64.10
Dial reading	[μm]	1125	1098	1101	1090
Carico/load	–	24.3029	23.7196	23.7844	23.5468
Load correction	[kg]	2477	2418	2425	2400
Correlation ratio	–	0.95	0.95	1.00	0.98
Marshall stability	[kg]	2353	2297	2425	2352
Average stability	**[kg]**	**2357**			
Marshall flow	[mm]	3.70	3.80	3.70	3.80
Average flow	**[mm]**	**3.8**			
Stiffness	**[kg/mm]**	**635**	**605**	**655**	**619**

Table 9. Determination of percent voids residued, data processing at the end of the Marshall test for the new asphalt sample.

Determination of percent voids residued		
Referenced standards	SSH EN 12697-8:2018	
Air voids content (Va)	%	3.24
Voids in mineral aggregate (percent of bulk volume) (VMA)		15.34
Percent of the voids in the aggregates filled with binder (VFA)	%	81.42

Table 10. Data comparison of existing and new asphalt according to the technical specifications of European roads.

Description	Existing asphalt	New asphalt	Technical specifications of European roads
Sieve analysis	According to the standard	According to the standard	
Percent of bitumen	7.23%	5.69%	(4.5–6.0)%
Marshall stability	1741 kg	2357 kg	>1000 kg
Marshall density	2364 kg/m^3	2331kg/m^3	>2300 kg
Flow	3.9 mm	3.8 mm	(2–4) mm
Air voids	1.89%	3.24%	(3–5)%

Acknowledgment. The authors would like to thank Laboratory Elba Ltd. for all the measurement data and Polytechnic University of Tirana for supporting this research.

References

1. EN 12697-27, Bituminous mixtures—Test methods for hot mix asphalt — Part 27: Sampling
2. EN 13108-8, Bituminous mixtures — Material specifications — Part 8: Reclaimed Asphalt
3. EN 13108-1, Bituminous mixtures—Material specifications—Part 1: Asphalt Concrete
4. EN 12697-1, Bituminous mixtures—Test methods for hot mix asphalt—Part 1: Soluble binder content
5. EN 12697-2, Bituminous mixtures—Test method for hot mix asphalt—Part 2: Determination of particle size distribution
6. EN 12697-5, Bituminous mixtures—Test methods for hot mix asphalt—Part 5: Determination of the maximum density
7. EN 12697-6, Bituminous mixtures—Test methods for hot mix asphalt—Part 6: Determination of bulk density of bituminous specimens
8. EN 12697–8, Bituminous mixtures—Test methods for hot mix asphalt—Part 8: Determination of void characteristics of bituminous specimens
9. EN 12697–30, Bituminous mixtures—Test methods for hot mix asphalt—Part 30: Specimen preparation by impact compactor
10. EN 12697-34, Bituminous mixtures—Test methods for hot mix asphalt—Part 34: Marshall test
11. Markja, I., Dhoska K., Ahmeta B., Fraholli M., Sita E., Sakaj, E., Elezi, D.: Evaluation of granulometry, stability and flow of asphalt recipes. In: Proceedings of the 9th International Textile Conference & 3rd International Conference on Engineering and Entrepreneurship (9th ITC & 3rd ICEE 2021), pp. 167–174 (2021)
12. Markja, I., Dhoska, K., Elezi, D., Moezzi, R., Petru, M.: Effect of the grain sizes on the ultrasonic propagation and attenuation on different types of steels microstructure during non-destructive testing. Journal Annales de Chimie-Science des Matériaux **45**(4), 329–334 (2021)
13. Markja, I., Fraholli M., Dhoska K., Sita E., Sakaj, E., Elezi, D.: Performance evaluation analyses of recycled concrete in albania by using ultrasonic pulse velocity. Int. J. Innov. Technol. Interdiscip. Sci. **4**(4); 784–791 (2021)

14. Dhoska, K., Markja, I.: Recycled tire rubber modified bitumens for improving the quality of the road constructions in Albania. J. Trans. Motauto World **4**(1), 34–35 (2019)
15. Dhoska, K., Markja, I., Pramono, A.: Analysis of recycled tire rubber modified bitumen in Albania for quality of the road construction. IOP Conf. Ser. Mater. Sci. Eng. **673**(1), 012126 (2019)
16. Pramono, A., Markja, I., Dhoska, K., Milo, P.: Fabrication of polymer matrix composites by bagasse based on Yukalac polyester resin. J. Pollack Periodica **18**(1), 138–143 (2023)

Mechatronics

Path Planning Algorithms of the Mobile Robots Control

Arbnor Pajaziti[1](✉), Lavdrim Panxha[1], Xhevahir Bajrami[1], Shaban Buza[2], and Mirlind Bruqi[3]

[1] Faculty of Mechanical Engineering, Department of Mechatronics, University of Prishtina, Prishtina, Republic of Kosovo
{arbnor.pajaziti,xhevahir.bajrami}@uni-pr.edu, lavdrim.panxha1@student.uni-pr.edu

[2] Department of Engineering Design and Vehicles, University of Prishtina, Prishtina, Republic of Kosovo
shaban.buza@uni-pr.edu

[3] Department of Industrial Engineering and Management, University of Prishtina, Prishtina, Republic of Kosovo
mirlind.bruqi@uni-pr.edu

Abstract. The purpose of this paper is to provide a complete analysis and comparison of the most used algorithms to plan the trajectories of mobile robots. By examining the principles, functions, and applications of these algorithms, it is intended to see their particular advantages and limitations. Each algorithm is analyzed in detail, including basic principles, mathematical fundamentals, and applications in the real world. Further, a comparative analysis has been developed, assessing algorithms based on various criteria, such as performance, applicability, limitations, computational requirements, and suitability. By examining these factors, we aim to provide a thorough understanding of the advantages and disadvantages of each algorithm. There are a wide range of algorithms and models available for controlling mobile robots. Depending on the specific application needs, various algorithms such as A*, RERT, Pure Pursuit, MPC, etc., as well as mathematical and Artificial Intelligence-based models for robot control, object tracking, obstacle avoidance, and many more can be used.

Keywords: Path planning · Mobile robot · Control · Simulation · Algorithm

1 Introduction

In recent years, the field of robotics has experienced a rapid evolution marked by significant strides in automation and autonomous systems. An essential driver of this progress is the development and integration of algorithms that enable robots to navigate and function in intricate and dynamic environments. These algorithms are pivotal in tasks encompassing robotic control, path planning, obstacle avoidance, and object recognition, among others. Evaluating the merits and drawbacks of these algorithms is pivotal in crafting efficient and dependable robotic systems.

G. Guxho et al. (Eds.): ITC-ICEE 2023, LNMIE, pp. 391–400, 2024.
https://doi.org/10.1007/978-3-031-48933-4_38

This paper's primary objective is to conduct a comprehensive analysis and comparison of the most prevalent algorithms in the realm of robotics. By scrutinizing the principles, functions, and applications of these algorithms, we aim to elucidate their specific strengths and limitations. Our methodology involves a meticulous algorithm selection process based on their ubiquity, relevance, and impact in robotics. Furthermore, we concentrate on algorithms that address fundamental challenges in robotics, including control, path planning, localization, perception, and learning.

Each chosen algorithm undergoes an in-depth examination encompassing its underlying principles, mathematical foundations, and real-world utility. Additionally, we perform a comparative analysis, assessing the algorithms across various criteria, such as performance, applicability, limitations, computational demands, and adaptability. Through this comprehensive evaluation, we endeavor to provide a profound comprehension of the advantages and disadvantages associated with each algorithm.

1.1 Types of Algorithms Used in Advanced Robotics

Recent years have seen remarkable advancements in robotics, enabling their application in diverse and complex scenarios, largely attributed to the utilization of advanced algorithms. These algorithms unify mathematics, engineering, AI, and computer science [1].

This paper offers a comprehensive overview of advanced robotics algorithms, elucidating their impact on enhancing robot capabilities and performance. Subsequently, we will delve into an assessment of widely utilized algorithms in the realm of advanced robotics, considering their applications [2, 3].

2 Robot Control Algorithms

Control algorithms are pivotal in directing and managing a robot's behavior by utilizing information from its sensors to determine actions required to meet predefined objectives. These algorithms find application in tasks such as executing precise movements, object manipulation, and trajectory planning. The following sections present evaluations of different algorithm types, highlighting their respective pros and cons.

2.1 PID Control

Various robot control algorithms are employed in advanced robotics, with the PID (Proportional-Integral-Derivative) algorithm being among the most prevalent. PID, a classic control system, integrates three components: the Proportional (P), Integral (I), and Derivative (D) terms [4].

Utilizing feedback from the controlled system, the PID algorithm computes and modifies control inputs, such as motor position or speed (Fig. 1). This algorithm comprises three components: the proportional component, dependent on the current error; the integral component, reliant on the accumulated error over time; and the derivative component, contingent on the error's rate of change. By fine-tuning these component parameters, the PID controller can achieve precise control behavior.

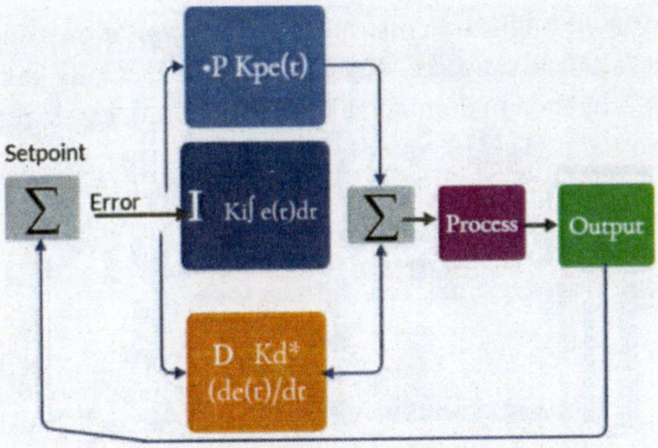

Fig. 1. Block diagram of PID [4].

PID represents a feedback-driven control algorithm that adapts control inputs in response to the current error, accumulated error over time, and the error's rate of change. While it offers simplicity in implementation and adjustment, it may encounter challenges when handling intricate systems and constraints.

2.2 MPC Control

Model Predictive Control (MPC) is a control system that employs a mathematical model of the controlled system to forecast future behavior. In the realm of robotics, MPC finds application in trajectory control, where the objective is to precisely track a predefined path (Fig. 2). This algorithm relies on a dynamic mathematical model of the system to anticipate its future evolution [5].

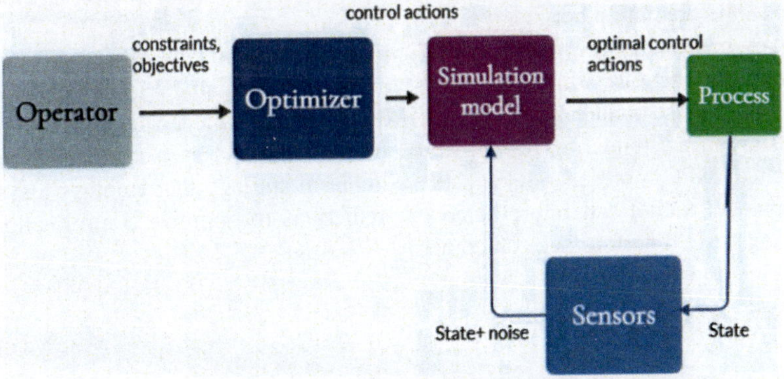

Fig. 2. Basic scheme of MPC [5].

MPC involves addressing an optimization challenge with the aim of minimizing an objective function that encapsulates the execution of control actions to meet predefined conditions and objectives. This optimization problem's resolution is influenced by control conditions, including action restrictions, load limitations, and trajectory requirements. MPC stands as an advanced control approach, especially well-suited for systems entailing constraints, as seen in domains like chemical processes, power generation, and industrial operations (Fig. 3).

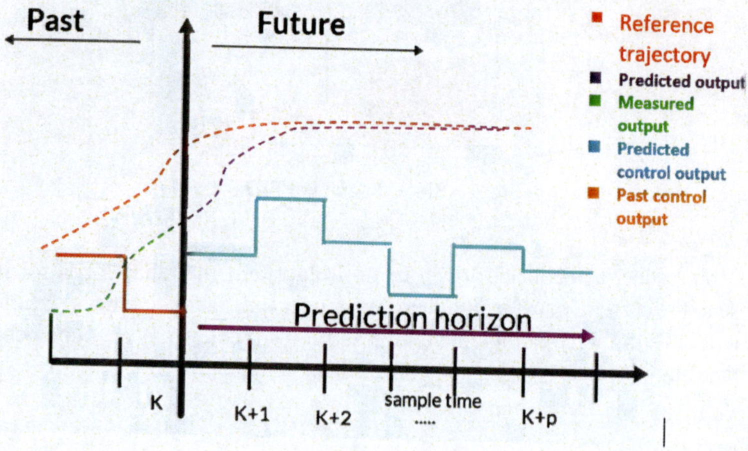

Fig. 3. MPC action method [5].

In MPC control, a process model is harnessed to forecast the future performance of the system, and these predictions guide the optimization of control actions to attain a desired level of performance. The model relies on the physical attributes of the system and aligns with the control objectives, and it can assume either linear or non-linear form. The MPC algorithm leverages this model to optimize control actions across a time span known as the prediction horizon [6].

A key benefit of MPC lies in its ability to handle constraints, encompassing input and output restrictions, while optimizing control actions to attain desired performance levels within these boundaries. Moreover, MPC exhibits resilience to model uncertainty and disturbances, rendering it well-suited for intricate and ever-changing processes. In essence, MPC represents a sophisticated control strategy that employs a system's mathematical model and future behavior predictions to optimize control actions for achieving desired performance outcomes.

2.3 Fuzzy Logic Control

Fuzzy Logic serves as a method for modeling information that accommodates a spectrum of values, not limited to binary "yes" or "no" distinctions. In the field of robotics, Fuzzy Logic has found application in addressing intricate challenges linked to robot perception, operation, and control within environments characterized by ambiguity or complexity (Fig. 4) [7].

In contrast to traditional binary logic, which relies on "yes" or "no" cutoff values to denote statement accuracy, Fuzzy Logic employs variables representing the degree of

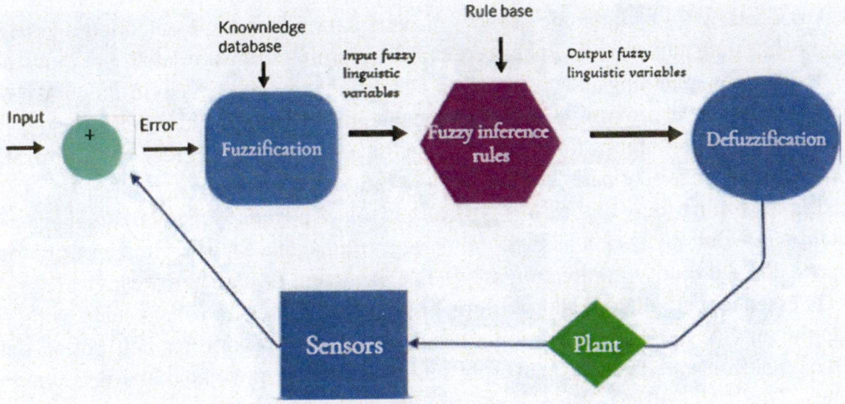

Fig. 4. A platform with Fuzzy Logic Controller [7].

truth or certainty in a statement. These variables manifest as functions that describe the connection between input and output values.

In the realm of robotics, Fuzzy Logic proves valuable for modeling robot perception, decision-making, and behavior control. A Fuzzy Logic Controller (FLC) can intake data from diverse sensors, such as ultrasonic sensors for obstacle distance detection and global positioning systems for determining the robot's position. Through fuzzy inference, the FLC generates an output that enables the robot to distinguish environmental changes and execute designer-desired actions. For instance, a wheeled robot may adjust its rotation speed to navigate corners and avoid obstacles, while multi-legged robots could utilize inputs like roll angle or forward speed.

Leveraging Fuzzy Logic in robotics offers numerous benefits. It facilitates the management of ambiguous and unstructured information, enhancing a robot's adaptability in challenging environments. Furthermore, FLC can handle intricate, non-linear relationships between inputs and outputs, enabling the execution of sophisticated behaviors.

3 Path Planning and Obstacle Avoidance

A path is a sequence of poses connecting a start configuration to an end configuration while adhering to specific requirements, such as obstacle avoidance or force profiles. It differs from a trajectory, which represents the execution of a path over time.

Path planning and obstacle avoidance are critical facets of robot control within a given environment, ensuring safe and efficient robot movement by finding optimal paths while avoiding obstacles.

Path planning entails identifying the best route to a destination. The robot utilizes environmental data and its goal to determine the most appropriate path, considering factors like open spaces, obstacles, and employing search algorithms to optimize based on criteria such as speed, length, or convenience.

Obstacle avoidance focuses on detecting and evading obstacles in the robot's path, safeguarding the robot's integrity and enabling obstacle-free navigation. Sensors like cameras, LIDAR, or ultrasonic sensors provide data for obstacle detection and trigger actions like path adjustments, speed reduction, or turning to ensure safe robot movement.

During path planning and obstacle avoidance, the robot utilizes environmental inputs, including obstacle proximity, type, movement limitations, and the defined goal. The choice of planning algorithm can lead to paths with varying degrees of optimization, considering factors like path length, turns, or braking.

Implementing these algorithms typically involves advanced mathematical and computational techniques, such as pathfinding algorithms and AI. The choice of algorithm depends on the robot's application and the environment's characteristics.

In essence, path planning algorithms find optimal routes for robots, addressing path mapping and obstacle identification. Obstacle avoidance algorithms help robots detect and navigate around obstacles, ensuring safe and efficient movement in diverse environments. These algorithms leverage environmental data to determine suitable paths and effectively avoid obstacles.

3.1 Shortest Path Search Algorithm

The shortest path search algorithm is a technique utilized to discover the most efficient route between two defined points within a graph. In the context of robotics, this algorithm is instrumental in determining the quickest path for a robot to reach a specified destination while avoiding obstacles in its course. This algorithm plays a pivotal role in robotics by enabling robots to pinpoint the shortest and most efficient way to reach their goals. Its application proves especially valuable in planning a robot's route through diverse environments while adeptly navigating around obstacles.

Multiple variants of the shortest path search algorithm exist, including:

Dijkstra's Algorithm: Dijkstra's algorithm is a renowned and straightforward method used to identify the shortest path within a graph that features diverse weights on its nodes and edges (Fig. 5) [8]. It commences at the initial points, computes the shortest distance to all other points in the graph, and through successive iterations, selects the point with the smallest distance. This process continues until the destination point is reached.

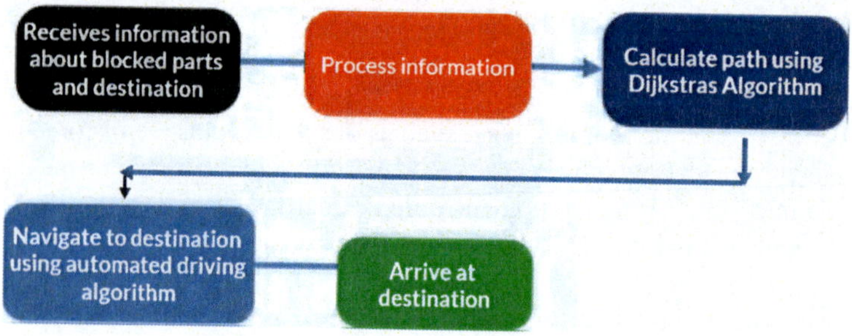

Fig. 5. Dijkstra's algorithm [8].

Bellman-Ford Algorithm: It finds the shortest path in graphs with negative weights by iteratively updating distances between points. It initializes distances, then updates them by comparing new distances to the existing ones. It can detect negative cycles and is used in robotics for path planning and obstacle avoidance.

A* Algorithm: This algorithm combines graph search with heuristics to find the shortest path. It maintains open and closed lists of visited points and selects points based on a function combining distance and a heuristic estimate. A* is valuable in robotics for efficient path planning in complex environments [9].

Rapidly-Exploring Random Tree (RERT) Algorithm: RERT explores a robot's configuration space by building a spanning tree through random point selection and expansion. It efficiently plans paths in complex spaces, making it suitable for various robotic applications.

Pure Pursuit Algorithm: Pure Pursuit guides autonomous robots along planned paths by determining waypoints, calculating angles, linear velocity, and angular velocity, and implementing control commands. It adapts to changes in speed and orientation [10].

Algorithm with LIDAR: LIDAR-based algorithms use LIDAR sensor data for tasks like object segmentation, path identification, determining target points, and integrating with the Pure Pursuit algorithm for path following. They enable robots to adapt to changing environments, including obstacle detection [11].

Probabilistic Roadmaps (PRM): PRM is used in robotics for motion planning in obstacle-filled spaces. It samples robot configurations, checks for collisions, connects collision-free configurations, and constructs a roadmap of potential paths through the environment while avoiding obstacles.

4 Simulations of Mobile Robot and Discussions

In this section we will present simulations of the movement of mobile robots by comparing the algorithms described above [12, 13].

In the first case, a comparison was made of the efficiency of the algorithms: RERT and LIDAR for following the route and target points while avoiding various obstacles.

The RERT algorithm is a tree-based algorithm that explores space.

This algorithm helps in finding the optimal path to reach the target point efficiently. By using the RERT algorithm, the robot can find the shortest and most optimized path to reach the target point by avoiding obstacles and using only the available information.

The LIDAR algorithm is a technology to detect obstacles and help track the road. With the help of LIDAR, the robot can detect objects and obstacles in its surroundings and react efficiently to avoid them. The use of LIDAR can provide a more detailed view of the surrounding environment and help track the route and target points with greater precision.

After testing and analyzing both algorithms under different conditions, the results showed that the RERT algorithm was more efficient and had better results in tracking the path and target points. This algorithm provided a more stable, faster, and more accurate tracking, minimizing the distance traveled and the time required to reach the target point. However, it has been estimated that the use of LIDAR as an additional tool can improve the results and increase the level of safety in following the road, detecting objects and obstacles efficiently.

According to the performance analysis and evaluation, the RERT algorithm was more efficient and provided better results in tracking the path and target points.

The results achieved and comparisons in the following Figs. 6 and 7.

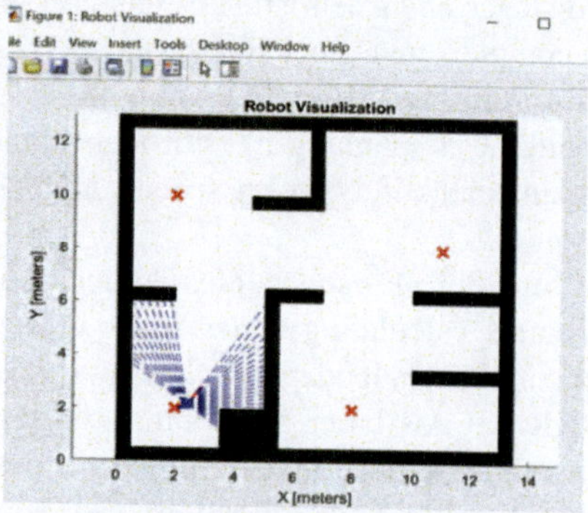

Fig. 6. Robot movement with LIDAR algorithm.

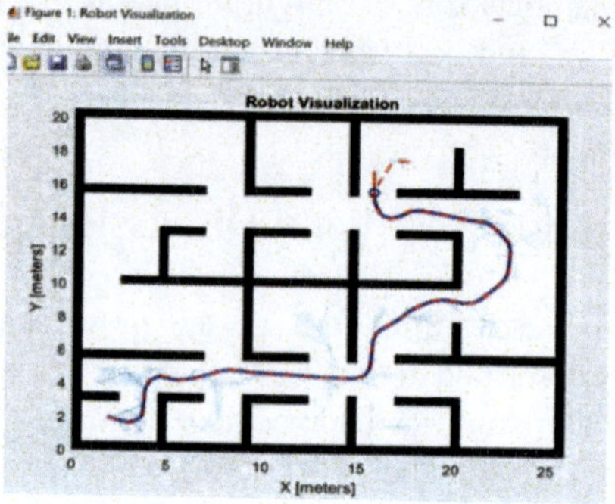

Fig. 7. Robot movement with RERT algorithm.

In the second case, a comparison was made of the efficiency of the algorithms: PP and MPC for following the path and target points. The PP algorithm is a method that follows the path using a target point as a target. This algorithm uses a defined trajectory and finds the closest position of the robot relative to this trajectory. After that, use a strategy to move towards the target point using direct control. The PP algorithm is relatively simple to implement and requires a limited responsibility in data processing.

MPC is an algorithm that aims to find an optimized trajectory given a system model and a set of defined rules. This algorithm makes continuous predictions of the system's behavior in a future period and makes control decisions in an optimized way to follow the intended path. MPC requires an accurate system model and continuous computation to generate control decisions.

After testing and analyzing both algorithms under different conditions, the results showed some fundamental differences. The PP algorithm provided good performance on the directed path to the target point and was simple to implement. However, it can have problems when disruptions occur in the road, such as obstacles or unforeseen interruptions, requiring a quick adaptation to follow the new path.

On the other hand, MPC showed a good performance in complicated trajectories and unpredictable situations. Through the system model and continuous predictions, the MPC can adapt and determine control decisions in an optimized way to reach the target point. Although MPC requires an accurate system model and continuous computation, it offers a high level of performance and precision in path tracking.

According to the performance analysis and evaluation, the PP offered a simple and efficient solution for tracking the route and target points, while the MPC offered an advanced performance suitable for complex trajectories and unforeseen situations. The choice of the most appropriate algorithm will depend on the specific system requirements and application conditions.

The results achieved and comparisons in the following Figs. 8 and 9:

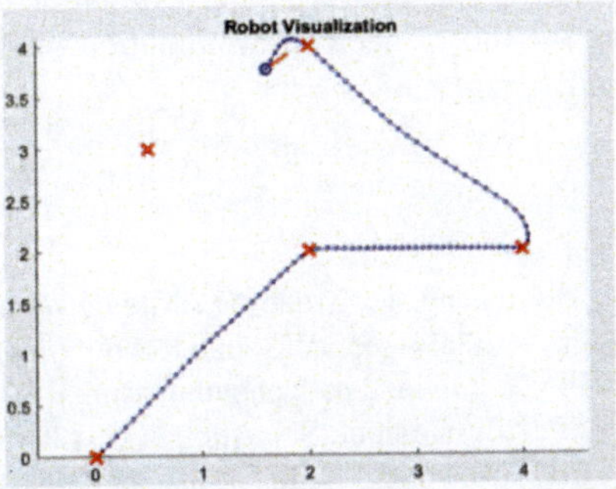

Fig. 8. Robot movement with MPC algorithm.

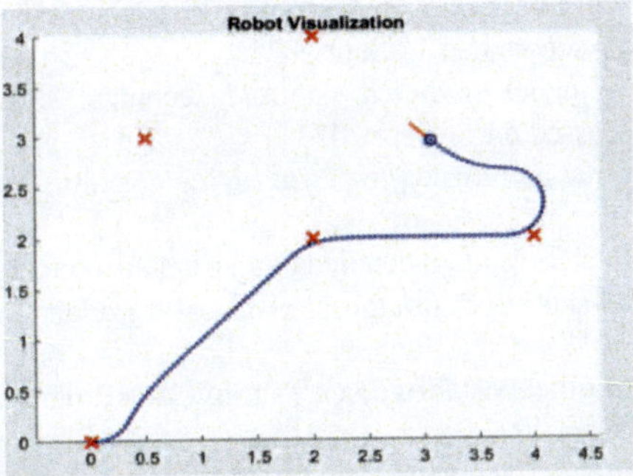

Fig. 9. Robot movement with PP algorithm.

5 Conclusions and Future Work

In the realm of mobile robot control, a multitude of algorithms and models stands ready to serve diverse purposes, from tracking objects to evading obstacles. The effectiveness of these solutions hinges on environmental conditions, path intricacies, tracking precision, robot constraints, and computational speed. Furthermore, the selection of sensors, be it LIDAR, cameras, or GPS, and the data they provide, including size, direction, and distance, play pivotal roles in determining the suitability and performance of these algorithms and models. Choosing the right algorithm depends on the robot's intended objective. For instance, the Pure Pursuit (PP) algorithm suits trajectory tracking, while the A* algorithm excels at navigating obstacle-laden terrain. Evaluating these options through rigorous testing enables the identification of strengths and weaknesses, driving enhancements and superior solutions for mobile robotics. Future decisions regarding mobile robot control should factor in application specifics, environmental nuances, available resources, and defined objectives. Meticulous evaluation and results-driven testing pave the way to selecting the most effective combination of algorithms and models to optimize mobile robot performance.

References

1. Borenstein, J., Everett, H.R., Feng, L., Wehe, D.: Mobile robot positioning: sensors and techniques. J. Robot. Syst., Special Issue on Mobile Robots **14**(4), 231–249 (1997)
2. Noreen, I., Khan, A., Habib, Z.: Optimal path planning using RRT based approaches a survey and future directions. Int. J. Adv. Comput. Sci. Appl. **7**(11) (2016)
3. Maulana, M.J.S., Riky, D.P., Wahyu, C.: Artificial potential field path planning algorithm in differential drive mobile robot platform for dynamic environment. Int. J. Robot. Control Syst. **3**(2), 161–170 (2023)
4. https://github.com/ivmech/ivPID. Accessed on 11 Sept 2023
5. https://towardsdatascience.com/ai-for-industrial-process-control-intro-to-control-strategies-part-1-b855fb6df61d. Accessed on 11 Sept 2023
6. Agrawal, M.: What is model predictive control? Technical Article, Control Automation, August 10 (2020). Accessed on 11 Sept 2023
7. https://www.tutorialspoint.com/fuzzy_logic/fuzzy_logic_control_system.htm. Accessed on 11 Sept 2023
8. Dijkstra's shortest path algorithm - a detailed and visual introduction. https://www.freecodecamp.org/news/dijkstras-shortest-path-algorithm-visual-introduction/. Accessed on 11 Sept 2023
9. Introduction to A*. http://theory.stanford.edu/~amitp/GameProgramming/AStarComparison.html. Accessed on 11 Sept 2023
10. Introduction to pure pursuit tracking algorithm. https://vinesmsuic.github.io/robotics-purepersuit/. Accessed on 11 Sept 2023
11. Raber, R.B.: Lidar guidebook: concepts, project design, and practical applications. https://www.urisa.org/clientuploads/directory/Documents/Books%20and%20Quick%20Study/lidar_guidebook.pdf. Accessed on 11 Sept 2023
12. Conduraru, I., Doroftei, I., Conduraru, A.: Localization methods for mobile robots - a review. Adv. Mater. Res. (2014)
13. Matlab & Simulink: https://www.mathworks.com/products/simulink.html. Accessed on 11 Sept 2023

Design and Fabrication of the Remote-Control Field Mobile Robot

Arbnor Pajaziti[1(✉)], Faruk Selmani[1], Shaban Buza[2], Xhevahir Bajrami[1], Ramë Likaj[1], and Afrim Gjelaj[3]

[1] Department of Mechatronics, Faculty of Mechanical Engineering, University of Prishtina, Prishtina, Republic of Kosovo
{arbnor.pajaziti,xhevahir.bajrami,rame.likaj}@uni-pr.edu, faruk.selmani@student.uni-pr.edu

[2] Department of Engineering Design and Vehicles, Faculty of Mechanical Engineering, University of Prishtina, Prishtina, Republic of Kosovo
shaban.buza@uni-pr.edu

[3] Department of Manufacturing and Industrial Engineering with Management, Faculty of Mechanical Engineering, University of Prishtina, Prishtina, Republic of Kosovo
afrim.gjelaj@uni-pr.edu

Abstract. The purpose of this paper is to build a mobile robot for spraying trees with pesticides and mowing grass remotely controlled, based on the analysis of international best practices argued. To carry out this work, the following three phases have been applied. The first phase deals with the selection of the literature was to determine the field of study. The selection of literature is spread over a considerable number of different authors. The second phase includes the data collection from references by different authors. The third phase includes data analysis. To achieve the objectives set in this work, we have relied on the method of analysis and synthesis, where the collected data is structured according to the work's requirements and then processed and we have reached the objectives of the paper.

Keywords: Design · Fabrication · Mobile robot · Remote control

1 Introduction

Nowadays, technology is developing rapidly, especially when it comes to Artificial Intelligence (AI) and Robotics.

AI is not something extraordinary that is achieved with the help of a computer to act like a human being which includes a large number of applications such as software simulators and robotics.

A robot is an electro-mechanical device that replaces a human in performing various tasks. These tasks are performed by the robot by itself or directly guided by the human. Robots can be mobile or positioned in one place. Today they are mainly used for performing tasks that are very dangerous (e.g. in the cleaning of atomic waste), and heavy or difficult for humans to perform. Robots today perform almost more than half of

G. Guxho et al. (Eds.): ITC-ICEE 2023, LNMIE, pp. 401–411, 2024.
https://doi.org/10.1007/978-3-031-48933-4_39

the current tasks in workplaces ranging from heavy industries, agriculture to household chores (cooking, cleaning, yard maintenance, etc.) [1–5].

The purpose of this work is, through theoretical and practical knowledge previously studied, research, analysis and international best practices argued, to build a prototype of a mobile robot with remote monitoring and control, where its task will to spray trees with pesticides and at the same time mow the grass in plantations, greenhouses, gardens, etc.

After determining the goal for the work of this project, it is also important how we will approach the forms to realize the goal. The objectives set before us are as follows:

- First objective: Literature review related to the design, modeling and manufacture of a mobile robot for the given field.
- Second objective: Analyzing and reviewing similar projects.
- Third objective: The prototype is experimentally functional for defined purposes.

1.1 General Concepts

The mobile spraying and mowing robot is a type of 2-in-1 automated machine that is designed for spraying trees and mowing grass. These mobile robots typically use sensors and navigation systems to move around a lawn, spraying trees and cutting grass as they move, simultaneously. They can be programmed to spray and cut the grass at specific intervals and can also be controlled remotely. Some models also come with features such as obstacle detection and the ability to return to their charging stations when the battery is low.

In this paper we have built an automatic robot for spraying and mowing using Arduino as microcontroller components. The robot can do tree spraying and grass cutting on the pre-defined path.

2 Design of the Electronic Components of the Robotic Prototype

For an experimental system, built and controlled for a 2-in-1 robotic prototype for spraying and mowing, in order to understand its functionality according to the defined task, the followingcomponents are needed:

1. 1 × Arduino UNO Board
2. 1 × L293D Motor Drive Module
3. 1 × Ultrasonic Sensor HC-SR04
4. 1 × Ultrasonic Sensor Case/Holder
5. 4 × DC Gear Motors
6. 1 × BLDC Motor 100 kV
7. 1 × Servo Motor SG-90
8. 1 × ESC Module
9. 1 × Servo Motor Tester
10. 1 × 3-Pin Slide Switch
11. 1 × X Type Cross Holder
12. 1 × Robot Chassis
13. 1 × 11.1 V Lipo Battery

14. 1 × Water pump
15. 2 × Sprayers
16. Drilled PCB
17. Breadboard
18. Wires.

2.1 Arduino UNO Microcontroller

Arduino UNO is an open source microcontroller board based on ATmega328P, Fig. 1. There are 14 digital I/O pins of which 6 can be used as PWM (Pulse Width Modulation) outputs, 6 analog inputs, a 16 MHz resonator, a USB connection, a power source, an ICSP (In Circuit Serial Programming) header and a reset button.

The power source is powered by the USB cable or an external 9 V battery, although it accepts voltages between 7 and 20 V.

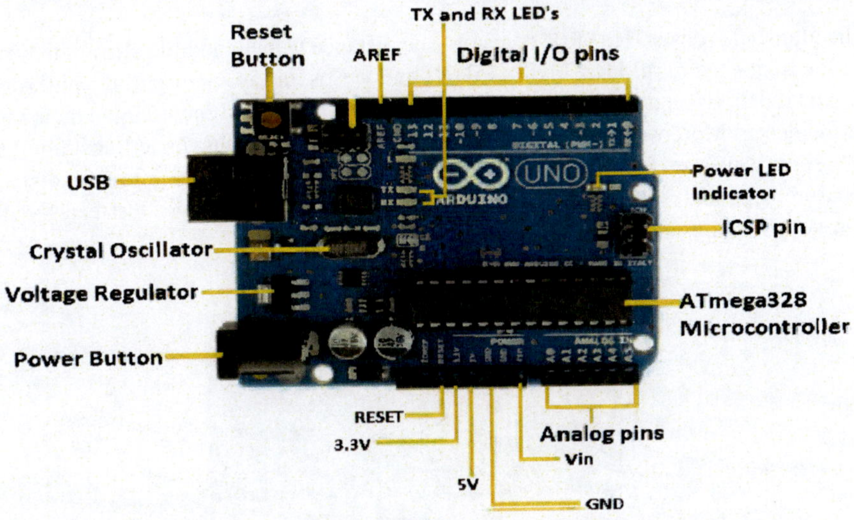

Fig. 1. Arduino UNO - Microcontroller board with the composition of its parts [6].

2.2 DEVMO L293D Motor Drive Module

The L293D is an integrated, high-voltage, high-current, 4-channel driver, Fig. 2.

Basically this means that by installing this chip we can control DC motors and with power supplies up to 36 V. These are quite large motors and the chip can deliver a maximum current of 600 mA per channel, the L293D chip being known as a type of H-Bridge. The H-bridge is typically an electrical circuit that enables a voltage to be applied to a load in either direction of an output, e.g. motor.

This module has the ability to control four DC motors, two servo motors and two stepper motors.It consists of:

- Interface for 5 V Servo connected to dedicated timer with high resolution,
- External terminal power interface, for separate logic/motor supplies.

Fig. 2. DEVMO L293D - Motor drive module [7].

2.3 Ultrasonic Sensor HC-SR04

The ultrasonic sensor HC-SR04 measures the distance using inaudible ultrasonic waves with a frequency of 40 kHz, Fig. 3. Like sound waves, ultrasonic waves travel through air and if there is an obstacle in front of them, they are reflected according to their angle of incidence. Moreover, if an object is placed parallel to an ultrasonic transmitter, the ultrasonic waves are reflected at an angle of exactly 180 degrees. Therefore, for distance measurement with the HC-SR05 sensor, we place the object under test exactly in a parallel position with an ultrasonic sensor.

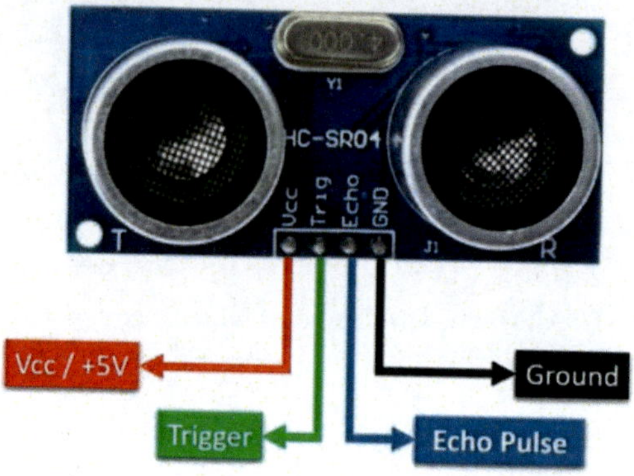

Fig. 3. Connection diagram for the HC-SR04 ultrasonic sensor [8].

2.4 DC Gear Motors HC

A DC motor is a class of rotary electric motor that converts direct current (DC) electrical energy into mechanical energy. In the following figure, Fig. 4, the connection of the DC Gear Motors to the motor drive module and plate with Arduino sensor shield is shown.

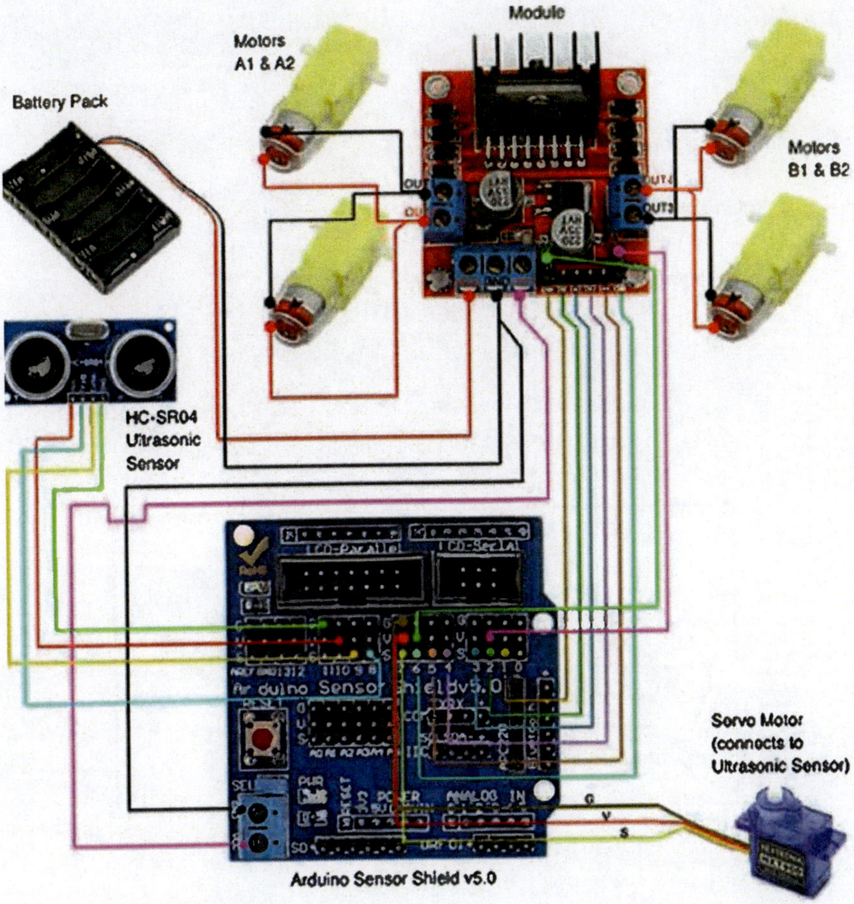

Fig. 4. Connecting DC gear motors to the motor drive module and other interface components [9].

2.5 Servo Motor SG-90

The SG90 servo motor is an electromechanical device which can be used during the design of a controlled robot and in other industrial projects, Fig. 5.

This device has a great ability to control its location from any position in an operating range and works with a control card that has three wires.

2.6 Potentiometer

The potentiometer can be used as a sensor, since the angular movement exerted on it will cause the resistance and therefore the voltage will change, Fig. 6. By calibrating the system and measuring the values it can be determined how much it has moved, but it can also be used as a control element.

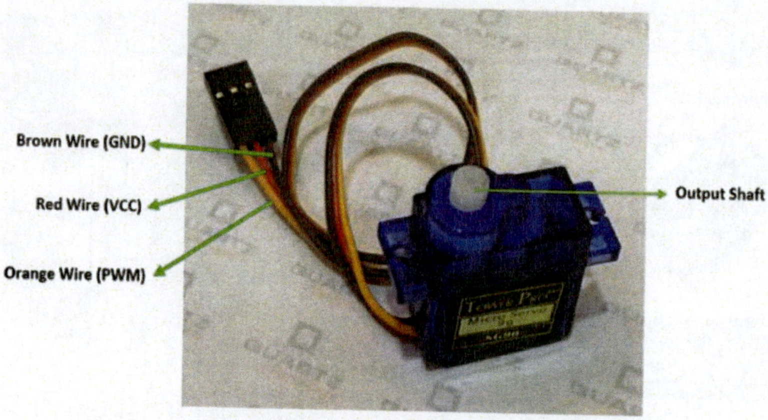

Fig. 5. Servo motor SG-90 with terminals for connection [10].

Fig.6. Potentiometer [11].

2.7 3-Pin Slide Switch

This switch's role is that the current flow does not go directly from the power source to the corresponding device, but it is controlled and used only when that device must be in operation in the task, Fig. 7.

2.8 Robot Chassis

The chassis is the structural component of the robot, which contains the drive motor and allows the robot to be mobile using wheels or some other method, Fig. 8. The chassis of a robot is often referred to as the robot frame.

2.9 Brushless DC Electric Motor - BLDC Motor

A brushless DC electric motor (BLDC), also known as an electronically commutated motor, is a synchronous motor that uses a direct current (DC) supply, Fig. 9. It uses an electronic controller to pass DC currents to the motor windings to produce magnetic fields and that effectively rotates in space followed by the permanent magnet rotor.

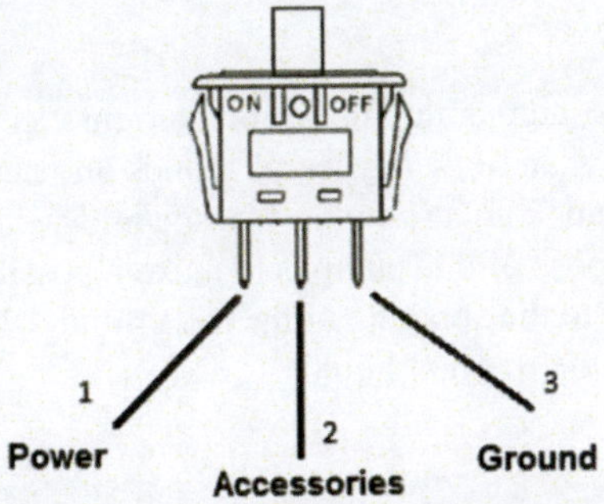

Fig. 7. 3-pin slide switch terminals [12].

Fig. 8. The chassis of the robot [13].

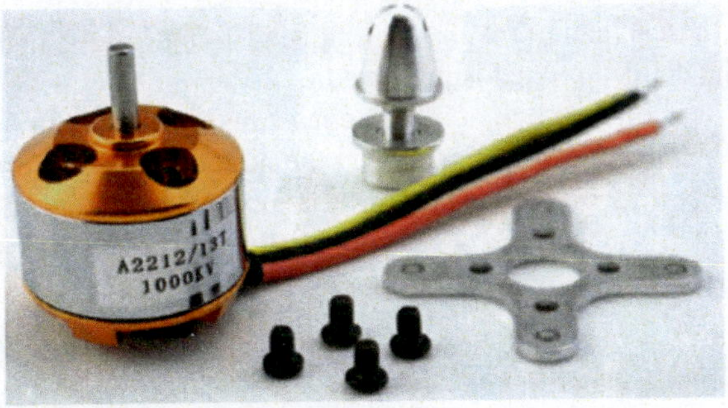

Fig. 9. Brushless DC electric motor [14].

2.10 DC Water Pump

DC water pump refers to pumps that use direct current electricity as a power source, Fig. 10. DC water pumps and other types of pumps are shaped like a tube and their purpose is to convey liquid from a reservoir to a specific destination.

In this paper, the purpose of this pump is to make it possible to convey the liquid of the Pesticide preparation to the spraying of the trees through it and the conveying pipes from the tank filled with the relevant liquid.

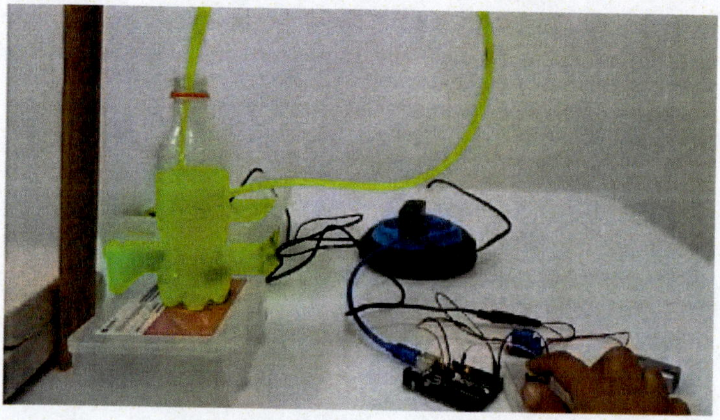

Fig. 10. The process of liquid flow through the pump controlled with the Arduino microcontroller.

3 Design of Electronic Circuit

The electric circuit of the field robot consists of the electric source, the wires for connecting the elements, the switchand the consumers.

In addition, theschematicform in the following figure, Fig. 11, the entire circuit connection for the defined project is presented.

Figure 12 a) and b) shows the circuit connection of these components as well as the assemblyof the electronic components on the mobile robot platformrealized in the Mechatronics Laboratory of the Facultyof Mechanical Engineering.

4 Software

In the realization of this experimental project before the start of the hardware implementation, we have previously used the Software, specifically the Tinkercadsoftware, which in a virtual and demonstrative way has given the opportunity to understand and practically realize the connections of an electric circuit, reading signals, their measurements, etc. Fig. 13.

The use of this software has made the hardware realization of the project much easier and understandable.

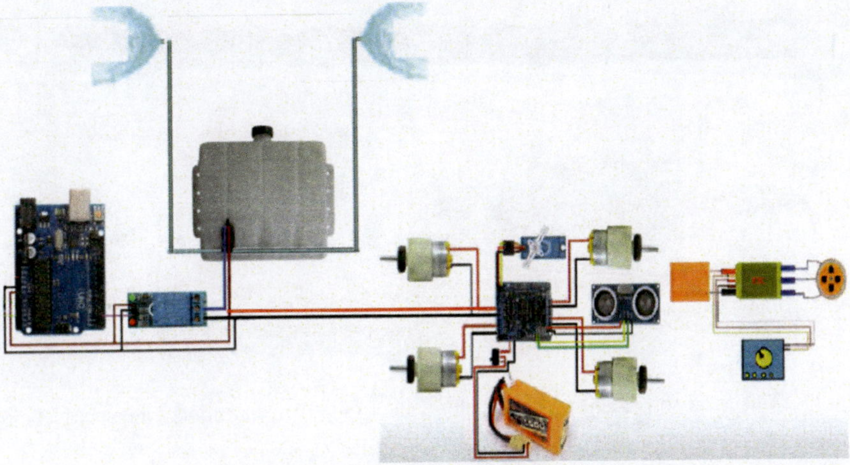

Fig. 11. Circuit diagram for the entire project defined.

(a) **(b)**

Fig. 12. **a**, **b.** Assembly view of the electronic components of the mobile robot.

5 Experiments with Mobile Robot and Discussions

The mobile robot has been tested to perform simultaneous spraying and mowing operations.

Activation of the robotic platform is done manually. First the mower is activated, then the sprayer and finally the robotic platform to move according to the defined trajectory. During the movement of the platform, the ultrasonic sensor is installed in the front part, which rotates through the servo motor, so that if any static or dynamic obstacle appears, the robot will stop and move, bypassing the obstacle, Fig. 14.

The diameter of the mower is 12 cm and it rotates with the help of a brushless motor. During mowing, if the mower encounters strong obstacles, then it stops.

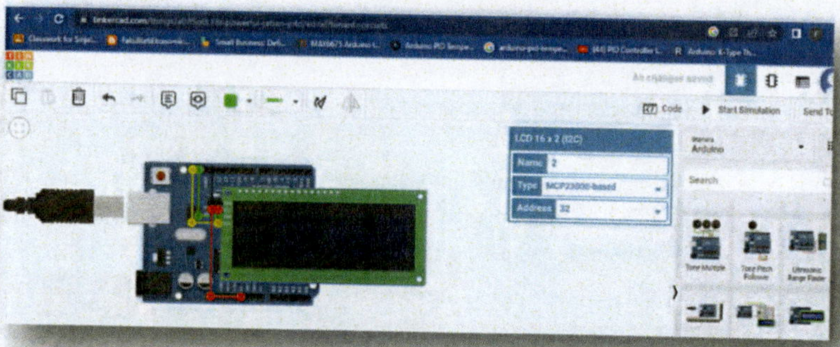

Fig. 13. Tinkercad software used to virtually demonstrate the connectionof an electrical circuit [15].

The spraying device is installed on the back side of the robotic platform. The pump, which is connected to the distributor, supplies both sides of the sprinklers with the liquid coming from the reservoir.

The LIPO battery with a power of 3 A supplies the brushless motor of the mower, while three 3.7 V batteries connected in series supply the pump motor.

Two 3.7 V batteries connected in parallel supply the Arduino Uno microcontroller while four DC motors drive the platform.

The Motor shield is housed in the Arduino Uno microcontroller and controls the movement of the wheels in pairs.

The pump motor is not controlled by the microcontroller, while the brushless motor is controlled by the BLDC Motor 100 kV controller.

(a) **(b)**

Fig. 14. **a**, **b**. Views of the field mobile robot prototype.

During the testing of the prototype, it was noticed that we have three sources of energy for the respective devices, so the system in the future should be replaced with a common source by using the necessary DC/DC converters.

In order for the device to work autonomously, in the future it is necessary that the direction and control of the robotic platform be done automatically in the defined space.

6 Conclusions and Future Work

Based on the analysis of algorithms and models used in mobile robots for their control, we reach some important conclusions and future work recommendations.

The goal of this project has been successfully realized from the software and hardware aspect, from which a complete system of a prototype of a robot for spraying trees with pesticides and mowing the grass, controlled from a distance, has been designed, constructed and implemented. ARDUINO, and after the tests and experiments done, it turned out to be functional in the intended purpose.

During the testing of the prototype, it was noticed that we have three sources of energy for the respective devices, so the system in the future should be replaced with a common source by using the necessary DC/DC converters.

In order for the device to work autonomously, in the future it is necessary that the direction and control of the robotic platform be done automatically in the defined space.

References

1. Rehg, J.A., Sartori, G.J.: Programmable Logic Controllers, 2nd edn. Pearson Education, Upper Saddle River, New Jersey (2007)
2. Geisllari, K.: Programmed Automata – PLC. SHBLU, Tirana (2008)
3. Buza, S.: CAE and Digital Factory, WUS Austria and University of Prishtina, Prishtina (2010)
4. Gashi, A.V.: Electrical Measurements. University of Prishtina, Prishtina (2014)
5. Bishop, R.H.: Mechatronics an Introduction. University of Texas at Austin, Taylor & Francis Group (2006)
6. https://www.javatpoint.com/arduino-uno. Accessed on 17 Sept 2023
7. https://microcontrollerslab.com/arduino-l293d-motor-driver-shield-tutorial/. Accessed on 17 Sept 2023
8. https://circuitdigest.com/microcontroller-projects/arduino-ultrasonic-sensor-based-distance-measurement. Accessed on 13 Sept 2023
9. https://forum.arduino.cc/t/arduino-code-for-controlling-4wd-motors/462846. Accessed on 17 Sept 2023
10. https://quartzcomponents.com/products/tower-pro-sg90-servo-9-gms-mini-micro-servo-motor. Accessed on 17 Sept 2023
11. https://techzeero.com/arduino-tutorials/how-to-use-potentiometer-with-arduino/. Accessed on 17 Sept 2023
12. https://www.pinterest.com/pin/rocker-switch-diagram--731835626782227049/. Accessed on 17 Sept 2023
13. https://eu.robotshop.com/collections/4-wheeled-development-platforms. Accessed on 17 Sept 2023
14. https://nyerekatech.com/shop/xxd-a2212-kv1000-brushless-motor/. Accessed on 17 Sept 2023
15. https://forum.arduino.cc/t/. Accessed on 17 Sept 2023

Textiles (Garments, Footwear, and Leather)

Resource Efficiency and Clean Production Assessment for the Albanian Textile Manufacturing Sector

Ermira Shehi[1](✉), Elmira Dumishllari[1], and Tatiana Chernyavskaya[2]

[1] Textile and Fashion Department, Tirana, Albania
{eshehi,edumishllari}@fim.edu.al
[2] UNIDO, Vienna, Austria
tchernyavskaya@unido.org

Abstract. There are different methods and tools for the Resource efficiency and Cleaner Production measurements introduced. The resource efficiency is defined as the ratio of a benefit to its use of natural resources. The benefit can be a specific product or service or described by a function or functional unit. Natural resources include renewable and non-renewable primary raw materials, energy, air, water, surface area and soil, biodiversity and ecosystem services. The natural resources of primary raw materials, energy, water and surface area are predominantly relevant for businesses. The proportion of materials cost for the textile sector in Albania, is 44.8% compared to 41%, the proportion of the material cost in the manufacturing sector in general, according to the Federal Statistical Office of Germany. This illustrates that enterprises are significantly impacted by their material costs and economic success hinges on price increases in the raw material market. The resource efficiency assessment under this paper has been completed under the one and half year project of the Textile and Fashion Department with UNIDO and Swedish Embassy. The procedure for implementing resource efficiency measures is divided into five steps analysis, solution development, evaluation, implementation and control. Six manufacturing enterprises in the garment and footwear production have been identified and assessed with some important KPIs. 29 measures have been suggested for the six companies giving the savings amount and the investment payback period.

Keywords: Resource efficiency · Cleaner production · Raw material · Energy · Water · Key performance indicators (KPIs)

1 Overview of the Current Situation of Textile Sector in Albania

1.1 General Overview

Albania at present is implementing important structural reforms that will support equitable growth, raise productivity and competitiveness in the economy, create more jobs, and improve governance and public service delivery. Enhanced regional connectivity and access to regional and global markets, coupled with export and market diversification, can also help promote faster growth [1].

G. Guxho et al. (Eds.): ITC-ICEE 2023, LNMIE, pp. 415–424, 2024.
https://doi.org/10.1007/978-3-031-48933-4_40

The Government's focus has been on recovery from the pandemic and the devastating earthquake of November 2019. Sustaining the pace of reforms under these circumstances has been challenging.

A robust recovery took place in 2021 in Albania thanks to a policy stimulus and the resurgence of travel, construction, and extractive activity. Private investment, consumption, and public spending drove growth, while public debt remained high. Rising inflation and the war in Ukraine threaten economic and poverty prospects in 2022.[1] The World Bank and other partners are cooperating to support the Government in overcoming these challenges and implementing the country's longer-term vision.

1.2 Economic Outlook

Economic activity is projected to expand at an average of 2.7% through 2024, following global conditions and persistent supply side shocks. Enduring geopolitical tensions could further increase inflation, disrupt supply chains, and disturb financial markets, all of which could further dim Albania's growth prospects. In turn, a sluggish job market combined with diminished purchasing power could dampen progress on poverty reduction.

Public debt is expected to decline to 68.9% of GDP in 2022, and more significantly over the medium term. This is based on the assumption that the primary balance turns positive in 2024 in line with the fiscal rule. However, the fiscal balance could further deteriorate in a worsening international context, which may force the government to cut public spending to prevent a hike in the debt-to-GDP ratio. Given Albania's growing reliance on external financing, interest rate and refinancing-related risks remain elevated.

In the medium term, private consumption is projected to return as a key driver of GDP growth. Private investment could provide further support to growth if business climate reforms are implemented. After a significant reduction in 2021, poverty is expected to continue declining in 2022, but persistent inflationary pressures could hamper further poverty reduction [1].

On July 2022 the value of exports was ALL 39 billion, increasing by 23.0% compared with the same period of previous year and decreasing by 14.3% compared with June 2022. The value of imports was ALL 82 billion, increasing by 14.1% compared with the same period of previous year and by 0.6% compared with June 2022. This month the trade deficit is ALL 42 billion, increasing by 6.9% compared with July 2021 and by 20.3% compared with June 2022 [2]. In the annual increase of exports by 23.0% this month, a positive contribution was given by the groups: Textile and footwear" with + 6.6%, "Construction materials and metals" with +5.2% and "Mineral, fuels, electricity" with +5.0%.

1.3 Overview of Textile, Clothing, Leather and Footwear in Albania

The annual growth of goods exports in 2021 resulted in the level of 35.6%. The main contributions to this increase were given by the category "Construction materials and metals" and the category "Minerals, fuels, electricity". In both cases, an important growth factor is estimated to have been the increase in the international prices of these products, especially during the second half of the year. The category "Textiles and shoes", which occupies the largest weight in the export of goods, has continued to encounter difficulties in recovery throughout 2021, remaining below the levels before the pandemic [3].

The export of textiles and shoes, which constitutes an important item of Albania's foreign trade, continued to recover during the first five months of 2022, accounting for 27% of the volume of exports [4]. The exports of this group were higher even compared to the first five months of 2019, which coincides with the period before the pandemic.

The leather and footwear production in Albania is an important sector with high potential for the country's economy. At the end of 2021, there are approximately 236 footwear manufacturing factories across the country, with a total of more than 25,000 employees, or about 6% of the total employment in the country [5].

The leather and footwear manufacturing industry in Albania accounts for 20% of total exports, with about 500 million euro per year. Whereas, for 2021, the weight occupied by the import of raw and processed leather alone was 6,906 tons. The main markets are Italy, Germany, Greece, France and the Netherlands [6].

Periods of crisis such as Covid-19 and the War in Ukraine can be believed to bring about the displacement of the production activities of Western investors from their old markets and Albania can be one of the countries that would potentially benefit from this displacement, due to the textile and footwear sector, which is constantly growing and an important contributor to the Albanian economy. Therefore, it is urgently required that this sector becomes stable and competitive for global markets, showing high productivity, flexibility, product diversification and cooperation partners.

1.4 Sustainable Development Goals in Textile Sector in Albania

The SDGs are a global call to action to end poverty, protect the earth's environment and climate, and ensure that people everywhere can enjoy peace and prosperity. Through a Sustainable Development Cooperation Framework with the Government of Albania, the UN fully supports and works towards the complementary agendas of Albania's Goal of accession to the European Union (EU), national priorities, as well as harmonization and aid effectiveness. This includes Albania's commitment to achieving the SDGs [7].

Albania's baseline report underscores that the national strategic policy framework is most harmonized with the SDG targets in SDG 3 (good health and well-being); SDG 7 (affordable and Clean Energy); SDG 8 (good jobs and economic growth); and SDG 9 (industry, innovation and infrastructure), while the least harmonization is with SDG 14 (life below water). Albania's actions towards SDGs with an environmental dimension focus primarily on SDGs 3, 7, 8, 9, 11, 14 and 15.

The main objective of the UNIDO Project under which the actual situation and perspective of the textile sector in country has been deeply analyzed, was to see the actual of the SDGs implementation level and suggest under a pilot project the RECP measures implementation.

1.5 Methodology

Adesk research on the actual situation of the textile sector was conducted, followed by a survey with the manufacturing companies operating in clothing and footwear sectors. A questionnaire has been developed with the target defined as follows:

Contribute to the development of methodology for a survey of Albanian enterprises from clothing and footwear sector to identify perception, potential and best practices from application of SDGs/EU Green agenda for ESG improvement of the businesses. Include identification of the status of the manufacturing processes at enterprises from clothing and footwear producing companies along their supply chains into the survey.

The questionnaire related to targets and indicators of SDGs 8 (Decent work and economic growth), 9 (Industry, innovation, and infrastructure), and 12 (Responsible production and consumption) and along their supply chains.

The survey aimed is to get a clear view and perception of the current situation of the garment and footwear manufacturing industry in Albania; to identify the gaps this sector has toward productivity, human resources, equipment technology, digitalization, OSH standards, lean manufacturing with the final objective to work together towards reduction of manufacturing waste, increasing awareness for lean manufacturing, recycling, circular economy, green production, energy consumption.

According to INSTAT data there was a total of 804 textile. Other sources of information give other approximate and similar numbers. Important is to know that about 700 are active companies. Clothing, leather and footwear companies in 2020 in Albania, 581 textile and clothing companies and 226 footwear and leather companies (INSTAT).

1.6 Selection Criteria for the Companies Participating in Survey

1. The size of companies (suggestion: big companies, applying at least Cut/make/trim as a business model.
2. Preferably companies producing jeans or companies applying finnishing processes such as washing, bleaching, dyeing.
3. Companies that are doing sourcing of raw materials.
4. Membership in the Textile association of Albania, ProExport.

1.7 Sample Size for the Survey

The number of ProExport Association membership actually is 185 companies. 50 companies have been selected from the main database of the companies members of the textile association fulfilling the selection criteria, 27% of the members. The respondent rate was 56% with 28 companies. As a limitation of the survey was the extremely short time at our disposal for conducting the survey (2 weeks).

1.8 Survey Results

The results are presented below according to the categories of questions in the questionnaire. There are 7 main categories of questions:

1. Information on company

17.9% of the companies work in a joint ownership business, 82.1% in Albanian ownership; 10 companies have 50–100 employees, 6 companies between 100–200 employees; 90% of the employees are females; 22% of the companies work for more than 10 clients/suppliers, 64% of them for up to five clients and 14% for 6 to 10 clients.

The distribution of the participants in the survey correlates with the real distribution of the companies of this sector in country. The main cities where this activity is located mostly are Tirana, Shkoder, Durres, Kruje, Berat, Fier. 32% of the companies are old in the sector, created before 1995.

2. Sustainable Development Goals (SDGs)

3.6% of the respondents say they participated in this survey because the association asked them, the rest answer because they want to improve their name and reputation, they want to benefit from the incentives and the tax, they want to attract new workers and for cost efficiency; 11.1% of them make a quality control/inspection only at the end of the production line; 18.5% of them do not have written procedures for new staff training programs; 64.3% do not have R&D staff or sector; about their needs for qualified staff the results are: 54% need sewing operator, 10% textile engineers, 10% production technicians and managers, others need graphic designer, machine technicians.

3. Management System

60.7% of respondents do to have Written procedures for processes, maintenance, handling and storage of chemicals; 53.6% answer they have protection measures for operators against hazardous chemicals; 35.7% do not have any internal procedure about maintenance of spare parts or unused machines.

4. Products

About products and production: the distribution of the production markets: 40% export to Italy, 16% of them export to Germany, 28% other European countries, 4% to China, 4% to Greece and 4% for home market. About the monthly volume of production (in pieces) 27% of companies produce 10000–30000 pieces/month, 15% of them 50000–100000 pieces/month, 11% 30000–50000 pieces/month.

21.4% face difficulties in preparing new styles for production; only 25.9% make product design by their own; 40.7% do not prepare technical specification for production lines; 70% do not make any material testing.

5. Processes

48% of companies work with Full Package or Full cycle of production, adding value to the production chain; 22% apply Cut Make and Trim model of their business including cutting sector and pattern making themselves; 11% of the companies participating in the survey have their Own Brand at least in one production line; 7% Cut & Make; 8% only Make and 4% is atelier category of company with 1–5 employees; Two out of 28 companies say they do not have automatic or semi-automatic in production; Despite the fact that almost 80% of companies apply CMT, FP and OB business model, not all of them have design sector but some of them receive patterns already designed by the clients.

6. Materials

Only 32.1% of the companies make raw material sourcing; 10.7% do not have any raw materials/accessories storage policies; 57.1% do not have any acclimatization standards for storage; Almost half of companies are interested in finding new suppliers for finding raw materials; 64.3% do not apply any recycling or separation of waste; 50% of companies do not make any material consumption analysis or only if clients require.

7. Environment

26.9% do not have any environmental management system implemented; about 61% do not select clients according to environment protection criteria; 61% of companies do not have requirements from suppliers for organic material use; 58% do not have implemented any water recovery projects; 14.3% do not have any energy saving plans.

1.9 Intervention

As a first phase of the project, a piloting step was foreseen selecting 6 companies for the Resource Efficiency and Cleaner Production Implementation. 12 local experts have been trained and certified by an international expert on RECP methodology for businesses assessment following by companies visits and data collection, gaps identification, options and suggestions identification for the companies.

40 suggestions for 6 companies have been given in the form of a report including mainly: optimization of compressed air system, new heating tunnels, infrared insulation, renewable energy, photovoltaic plants, use of water softener especially for ironing sector, CAD/CAM system for cutting sector, use of LED lighting, environment management system in place etc.

The RECP analysis for each company consisted of:

1. Environmental and health policy statement
2. Technology used
3. Production process flow chart
4. Selection of priority flows or areas
5. Results of detailed analysis
6. Material balance conclusions
7. Water balance conclusions
8. Energy analysis conclusions

A thorough data analysis was done for each company and each indicator. An example is given below for the energy consumption (Figs. 1, 2, 3 and 4; Table 1).

After the data analysis the selection of priority areas has been completed for each company and also a root – cause analysis (Tables 2 and 3).

A saving catalogue was produced for all the suggested RECP measurements for each company, including the feasibility of the selected RECP Measures (Tables 4 and 5).

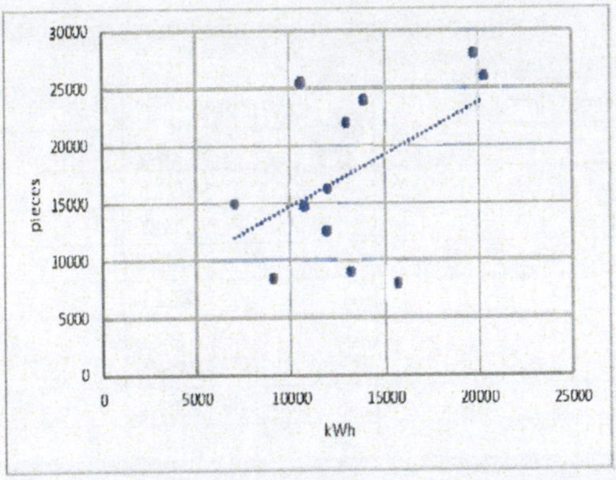

Fig. 1. Production versus energy consumption for 12 months.

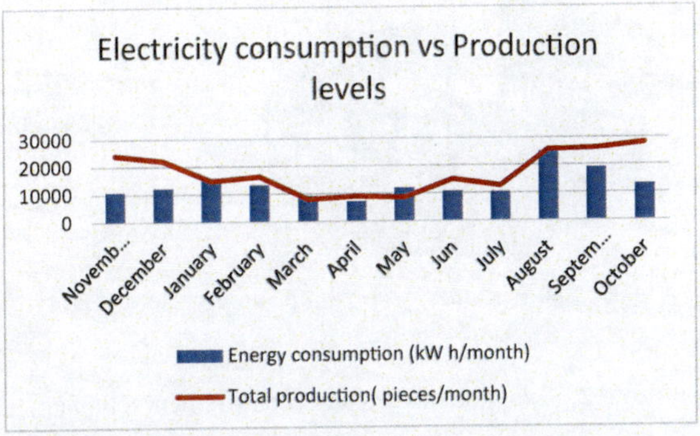

Fig. 2. Energy consumption versus production levels.

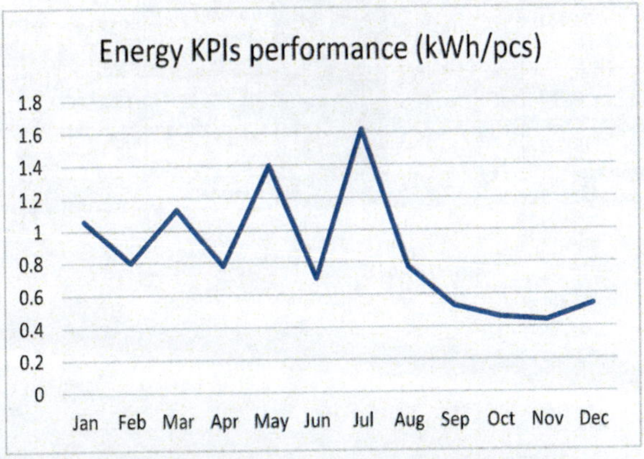

Fig. 3. Monthly energy consumption in 2023

Table 1. Energy consumption according to machine category.

Category	kWh/year	%
Lighting	36288	20.8
Air Compressor	16621	9.5
Sewing machine	56700	32.5
Boiler	28080	16.1
Vacuum	26244	15
Iron	8950	5.1
Cutting table	1350	0.8
TOTAL	**174233**	**100**

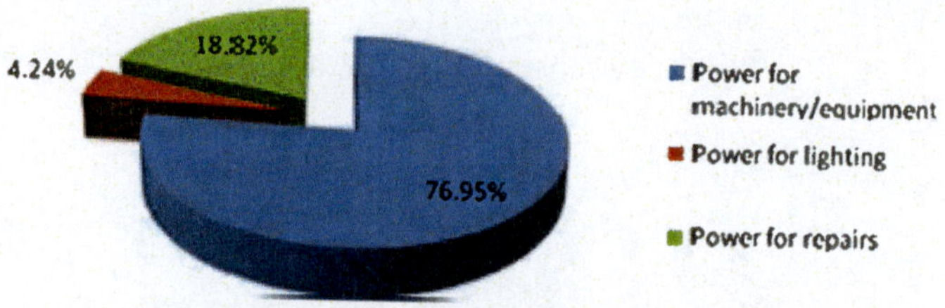

Fig. 4. Share of energy consumption per user category

Table 2. Example of indicator measurement template

1.5. Energy (use kWh)				3.4. Air Emissions (tons)		
Electrical energy	174,151			Emission from electrical energy	4.179,6	
LPG Gaz	8,283			Emission from LPG	1.822,26	
Diesel	75,627			Emission from Diesel	20.419	
Subtotal	**258,061**	**0**		**Subtotal**	**26.42**	**0**

Table 3. Priority flow and measure to be taken.

Priority flow	Measure
Waste	Consider training software users
Electricity	Install PV plant
Lights	Use presence and daylight sensors
Ironing	Use demineralized water instead of hard water
Cutting	Train operators, monitor and control cutting waste
Cutting	Consider recycling polyester fibres
Ironing	Support extraction with self-propelling extractors

Table 4. Catalogue for RECP measurements.

ID	RECP Measure	Cost saving [EUR/yr]	Investment [EUR]	Payback [years]	Energy saving	
					kWh/yr	%
1	Installation of LED luminaires with daylight and presence sensors	1 500	6 000	4	7 500	
2	Check efficiency of sewing machines and upgrade to VSD	Up to 50% or 1 000 kWh per machine, or 200 EUR	1 500 per machine	7 years	1 000 per machine	
3	Compressed air system optimisation	1 000	1 000	1	5000	
4	Installation of a PV plant	17 500	50 000	4	75 000	

Table 5. Example of RECP recommendations template

ID number	1
Title of measure	Luminaires replacement with LED, Install daylight and presence sensor
Cause	Actual luminaires have high electricity consumption. They are left switched off even when not necessary.
Description	The installation of daylight and presence sensors will help to make better use of sunlight, reducing electricity costs. The replacement of the existing lamps with LEDs, which have minimal energy consumption (4W/m), will also contribute in this direction. There is expected a reduction of 96.6% of the energy consumption for lightening. The sector that needs installation of sensors is mainly the storage sector. Since it is positioned to the south-east direction there is sufficient day light. Furthermore, it is not constantly frequented.
Technical feasibility evaluation	Easy to install and do not affect the production process.
Economic savings	The economic savings from the installation of the sensors at the storage sector are 600 EUR/year. Installation of LED luminaires will yield to economic savings of 1818 EUR/year, 87.7%.
Environmental savings	The environmental saving is the reduction of the CO_2 emissions due to total energy consumption.
Other benefits	-
Potential Supplier	Mr.Light, https://mrlight.al/
Investment (estimate)	6355.8 EUR
Payback	3.49
NPV	
Annexes	

References

1. https://www.worldbank.org/en/country/albania/overview
2. http://www.instat.gov.al/
3. Bank of Albania – Annual report, 2021
4. INSTAT, Albania
5. https://ata.gov.al/
6. https://www.aida.gov.al/
7. https://ec.europa.eu/info/strategy/international-strategies/sustainable-development-goals/eu-and-united-nations-common-goals-sustainable-future_en

The Perception of Athleisure Trend for a Denim Brand: The Case of Mavi

Özgür Ceylan[1(✉)], Nesligül Kiliç[1], Özgen Özturan[2], Cihad Sünbül[2], and Özge Polat[2]

[1] Eskisehir Technical University, Eskisehir, Turkey
ozgurceylan@eskisehir.edu.tr
[2] Department of Design, Mavi, Istanbul, Turkey
{ozgen.ozturan,cihad.sunbul,ozge.polat}@mavi.com

Abstract. The pandemic period and its aftermath have led to a shift in the interaction between work, home, and social activities, resulting in the adoption of a more flexible lifestyle. Consequently, this change has also influenced daily clothing preferences. Furthermore, an increased tendency towards engaging in physical activities in outdoor environments has been observed following prolonged periods spent indoors for work and home life. This has facilitated the emergence of comfort, functionality, and multitasking features as fundamental needs in clothing preferences. Additionally, the growing inclination towards sports and physical activities has necessitated the incorporation of protective qualities in garments against environmental conditions. Thus, the concept of Athleisure clothing has emerged as a combination of daily, sportswear, and active attire, meeting the requirements of comfort and style for both sports and casual wear. Athleisure clothing represents a mega trend, which offers versatile garments that can be worn in various life domains. The primary objective is to provide users with clothing items that allow multiple applications. Thus, functionality takes precedence in clothing products. In this context, the purpose of this study is to conduct an in-depth examination of the design, fabric, and accessory characteristics of the collections created for Mavi, a denim brand, drawing inspiration from the Athleisure mega trend. Within the scope of the research, the innovative aspects, design qualities, as well as research and development studies of outerwear, knitwear, woven, and activewear product groups have been examined in line with the determined objectives. The necessary data for the research were obtained through source (visuals, text, products) review, and an observation-based content analysis method was employed. The products produced by Mavi within the context of the Athleisure trend were visually examined, and the details of the product development process were conveyed. The study will provide valuable insights into how leading fashion brands incorporate current trends into their collections.

Keywords: Athleisure · Functional clothing · Active wear

G. Guxho et al. (Eds.): ITC-ICEE 2023, LNMIE, pp. 425–433, 2024.
https://doi.org/10.1007/978-3-031-48933-4_41

1 Introduction

Athleisure, emerging as a combination of two concepts namely "athletics" and "leisure," refers to casual clothing designed to be used for both exercise and daily use. The introduction of form-fitting yoga pants into the fashion world sparked interest in Athleisure, and since then, it has evolved into a dominant fashion trend of the 21st century [1]. In modern times, functional sportswear intended for non-work times and physical activities gradually transitioned into more extended daily wear. The rise of "fast fashion," characterized by frequent production of new models, has further fuelled the growing inclination towards sportswear in fashion brands [2]. According to the Global Industry Analysis Reports of 2017, the increasing health consciousness and participation in sports activities are expected to drive the fitness and sportswear market to reach 231.7 billion dollars by 2024. Acknowledging these projections, global luxury fashion brands are also focusing on Athleisure clothing collections. Particularly, after the 1990s, the concept known as "new luxury" has combined the features of sport-specific clothing with designs intended for more formal work and non-work settings. This fusion has given rise to a new fashion clothing trend termed Athleisure, which can be defined as "daily sporty-chic attire/recreational clothing". The trend incorporates mass-produced characteristics in terms of design, manufacturing, distribution, and pricing strategies while providing consumers with a more comfortable wearing experience through positioning and promotional strategies. Consequently, this has led to the formation of new consumption and production behaviours compared to the past [3].

Athleisure garments refers to clothing designed not only for athletic and workout purposes but also for leisure and everyday wear, extending beyond garments intended solely for gym or exercise activities [4]. These versatile garments can be worn in various social settings, including work environments. Athleisure has become a significant trend in the fashion industry as it combines casual style with a healthy and vibrant appeal. In 2018, the global athleisure market was estimated to be worth over 300 billion USD [5]. The increasing interest in yoga and thus yoga-specific clothing, aimed at supporting mental and physical well-being while reducing stress and anxiety, has also sparked awareness in the global fashion market. This has led to a remarkable rise in sales of high-quality elastic leggings and sports bras, resulting in a considerable 55% annual increase in demand for sportswear across various fashion brands. With the continuous growth of these trends, the Athleisure category is expected to generate a worldwide revenue of 547 billion USD by 2024 [6].

Brands aiming to adapt to the Athleisure trend need to develop insights into how consumers, especially female consumers, utilize and integrate sportswear in both athletic and non-athletic contexts, and how their experiences align with their personal values [7]. Values represent an individual's enduring beliefs, what they consider important in life, and what they strive to pursue as an ideal state of being. Understanding consumer values is crucial in marketing research as it reflects their fundamental needs, and the primary objective of marketing is to assist consumers in fulfilling these needs through product consumption. In this context, this research entails a comprehensive examination of the Athleisure collection prepared by the brand Mavi for the FW22 season. Mavi, incorporated in 1991 in Istanbul, is a global lifestyle brand with strong denim roots. Company has investments in 33 different countries including Turkey, USA, Canada, Germany,

Russia and Australia. Mavi has 439 mono-brand stores and 4500 points of sale and ranks among the world's leading denim brands. Mavi has a unique brand position with fashion-savvy, young adult customers and collaborates with top local and international celebrities in its marketing communications. The study will investigate the design and product development process of the Athleisure collection extensively and will provide valuable insights into the creative process of fashion design inspired by a global trend. This will contribute to the better understanding of inspiration sources used in creative fashion design process.

2 Collection Development Process

The primary objective in preparing a clothing collection is to present visual and aesthetic elements in a cohesive manner. Given its universal market structure, the clothing industry develops designs within the framework of annually determined fashion trends to establish a common language. Before designing garments, they are classified based on the target users and usage purposes [8]. Socioeconomic events significantly influence the flow of daily life and results in fundamental changes and developments in the clothing industry. Fashion trends, influenced by emerging diverse needs resulting from these developments, have a considerable impact on the process of collection development. Decisions on the fits, fabrics, colors, and patterns to be used in garments are made in line with trends, giving rise to tangible design details [9]. The preparation of a clothing collection involves completing preliminary preparations for design, making decisions on necessary accessories, finalizing product pattern, procuring external materials, conducting research and development efforts, and constructing the sample garment (Fig. 1) [10].

3 Method

The primary objective of this study is to investigate the design and product development process of a fashion collection inspired by the prominent fashion trend namely, Athleisure. Employing a qualitative approach, the study adopted a case study design to gain insight into the collection development process of the fashion brand under examination. The data for this investigation were gathered through participant observation and document analysis. Document analysis involves the scrutiny of written and visual materials pertinent to the phenomenon under investigation. In qualitative research, document analysis can serve as a standalone data collection method or be employed to corroborate findings obtained through other data collection techniques [11]. In the context of qualitative case studies, document analysis is particularly relevant as an additional means of gathering data, and thus, it was used to complement the data collected through participant observation [12].

The utilization of participant observation as a data collection method was justified due to three of the authors' involvement in the design team at Mavi Design Center. This proximity enabled them to observe and participate in the collection design and development process in a natural setting, fostering a deeper understanding of the activities

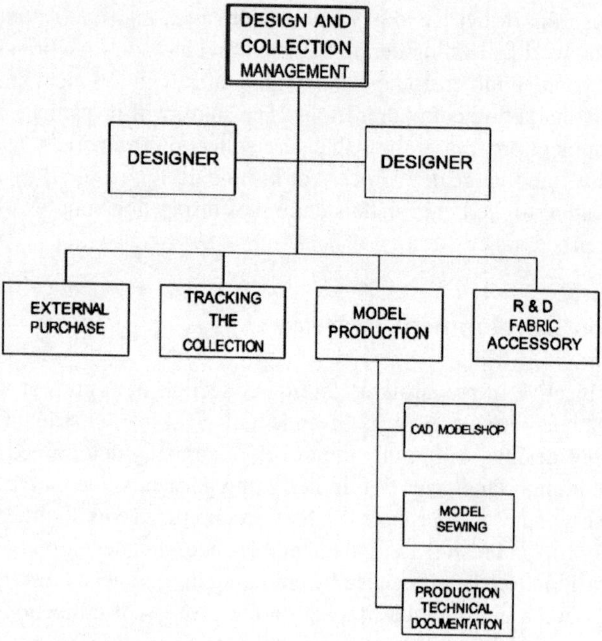

Fig. 1. Structure of design and collection

involved. The data collection process commenced with an analysis and review of all relevant documents pertaining to the target collection, encompassing photographs, textual materials, news articles, announcements, and any other pertinent information utilized by the designers. As a result of examining written texts, themes or categories are identified, whereas in the examination of visual data, detailed interpretations are conducted to obtain analytical outcomes [13]. Notably, the observation period focused solely on aspects relevant to the Athleisure collection. The findings derived from the combination of the two data collection methods were subsequently summarized in the Results section.

4 Results and Discussions

Mavi conducted an in-depth collection study for the FW22 season by examining the diverse needs of the increasingly influential Athleisure lifestyle and performed an extensive research on models, fits, fabrics, and accessories to address these needs while achieving a balance between performance and quality expectations. Through its design and product development efforts, the brand identified outerwear, knitwear, woven, and activewear product groups as suitable groups for Athleisure products. The collection includes various designs where functionality takes precedence, with variations in fabrics, fits, and technical specifications based on the product group (Table 1).

As a result of trend research, the brand has created a distinct design narrative by incorporating athleisure details and designs into its current fashion collections, thereby offering a fresh perspective on the combined use of casual, sport, and active wear. In

Table 1. Product group and specifications

Product group	Product specifications		
Outerwear	Water-repellent	Seam seal tape	TPU film membrane
Knitting	Modal/cupro fabric	Scuba fabric	New fit (sport corset)
Weaving	Micro-parachute	Micro-poplin	High performance
Activewear	High spandex	Polyamide fabric	New fit

this context, the fundamental approach of the narrative is to respond to customer needs with robust and functional product designs. The product development details of the FW22 season Athleisure collection, characterized in Table 1, are provided below for each product group:

1. Outerwear

Polyester and microfiber fabrics were utilized in the product range to impart water-repellent properties to the fabric surfaces, to prevent the transfer of water to the body. While introducing this feature in the products, comfort characteristics were also emphasized. Novel techniques were explored for the accessories and other details used in the raincoat. In designing the adjustable hood, adjustable elastic drawstrings and stoppers were incorporated at the hem. Seam Seal Tape was applied to specific areas of the garment (hood, nape, shoulders) to prevent rainwater from permeating the jacket (Fig. 2).

Fig. 2. FW22 athleisure windbreaker/relaxed fit/windproof

2. Knitting

In the knitted apparel group prepared for the Athleisure theme, the primary objective was to offer a new and dynamic collection. Accordingly, the products have been reinterpreted with theme-fitting colours and styles. Emphasis was on utilizing cotton, polyester, 3-ply, and lycra-blended supreme fabrics, as well as rib qualities, to ensure the necessary performance in the products (Fig. 3).

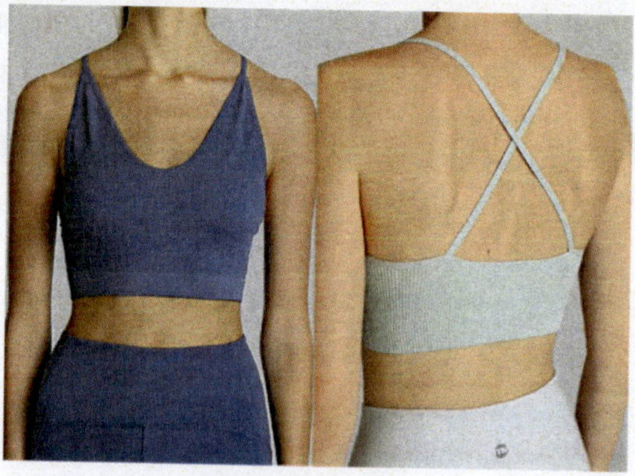

Fig. 3. FW22 athleisure sports bras

During the fit development of knitted apparel models, sporty corset trials were included. To achieve a more athletic appearance and to accommodate an important trend of the FW22 season, which involves the combined use of knitwear and knitted details, development efforts were undertaken in collaboration with two different department teams. Within the knitted apparel group, approximately 20 designs were created to ensure freedom of movement. These designs include long sleeve, zippered fitted hoodies, polo t-shirts, polo sweatshirts, and colour-block sweatshirts, employing fabrics such as cotton, supreme, 3-ply, and high elastane content. The integration of sporty micro-weave fabrics in the designs contributed to integrity of the collection (Fig. 4).

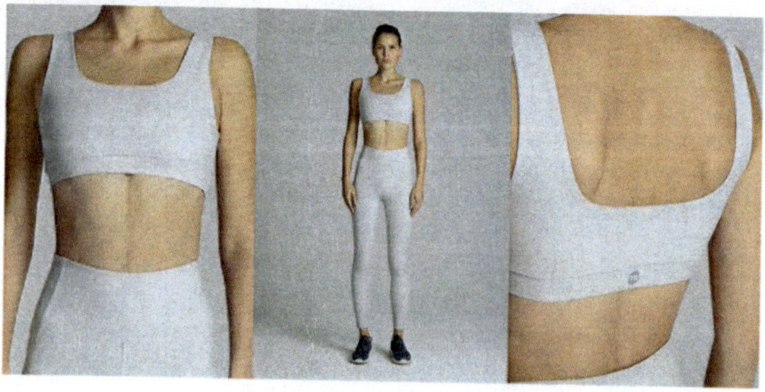

Fig. 4. FW22 athleisure sports bra/body-fitting

3. Weaving

For the woven product group, particular emphasis was given on evaluating performance fabrics, specifically micro-parachute, belonging to the Activewear category. When using such fabrics, attention was given to ensure similar properties to commonly used fabrics in the collection, such as Tencel-poplin. For models incorporating combinations of micro and poplin fabrics, research and development efforts were undertaken to prevent any post-wash alterations due to content differences (synthetic-natural). In this product group, 5 sweatshirt and shirt designs were created, utilizing performance fabrics (micro-parachute) (Fig. 5).

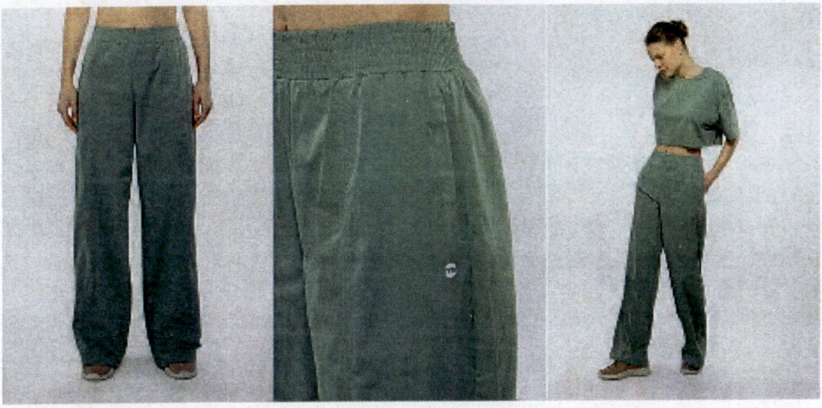

Fig. 5. Parachute fabric pants/wide leg

As a distinct detail, sequenced, wide-step decorative stitches were incorporated in these designs, deviating from the previously used stitching details. Fabrics, accessories, and details typically used in knitted and activewear groups were also integrated into the woven collection, resulting in new looks for novel designs. The inclusion of Athleisure themed woven models facilitated the ease of wearing these products in daily life.

4. Activewear

In the Activewear product group, medium-support bras (sports bras) were designed to provide increased comfort during exercise. The collection was diversified, taking into account disciplines such as yoga and pilates. Approximately 25 designs were developed for the Activewear product group, focusing on durable and high-support fabrics (elastane, polyamide, etc.). The aim was to create a new Mavi sporty female silhouette through the use of these fabrics, stitching techniques, and newly developed fits. Various shoulder and back details were incorporated into the upper body products to ensure designs that cater to individual comfort preferences. For the lower body product group, different cutting details were explored to ensure optimal fit around the waist during movement, with a preference for fabrics with high elastic content (Fig. 6).

The Athleisure trend is expected to have a long-term impact on Mavi's growth projections. It is anticipated that this trend will not be merely a short-lived fashion trend but will persist over time. The main reason for this expectation lies in the unique sales

Fig. 6. FW22 athleisure activewear group/high elastic/comfortable

potential of sportswear, not only due to its fashionable appearance but also because of the comfort it provides to users, stemming from innovations in materials and new production processes, and the increasing acceptability of wearing such garments in various contexts [14].

5 Conclusions

The main objective of this study was to examine extensively the impact of the ongoing Athleisure mega trend on the Mavi collections, specifically in the context of the FW22 season, through detailed analysis of the research and development efforts. The period from 2020 to 2022, which was marked by the pandemic, witnessed increased interest in social life and sports activities, leading to a rise in sales of sportswear categories in 2021. Consequently, global clothing brands have increased their production of activewear to capitalize on the value growth in the sportswear market.

The interaction between work and social life has facilitated the emergence of new fashion trends. In line with the emerging needs in clothing, priority has been given to the production of products that combine sportiness, classic style, and elegance, while providing comfort within a flexible lifestyle. The Athleisure trend, which forms the basis of clothing products preferred for active sports groups and yoga, supporting healthy and social living, is predicted to be among the rising trends from 2018 to 2026.

In line with this expectation, it has been observed that research and development efforts of Mavi for the FW22 Athleisure collection significantly meet these expectations. It is anticipated that the designed Athleisure collection will create value for future market endeavours.

References

1. Lipson, S.M., Stewart, S., Griffiths, S.: Athleisure: a qualitative investigation of a multi-billion-dollar. Body Image **13**, 5–13 (2019)

2. Çileroğlu, B., Kıvılcımlar, İ.: Fast fashion koleksiyonlarında yer alan spor-casual giysilerin tasarım özelliklerinin belirlenmesi. Int. J. Sci. Culture Sport **1**, 623–633 (2015)
3. Çakmak, Ş., Çeğindir, N.Y.: Farklı pazar özelliklerine göre rekreatif giyimden athleisure giyime gelişimin incelenmesi. Int. J. Human Sci. **3**, 799–815 (2019)
4. Webster, M.: Electronic resource (2021). Available: https://www.merriam-webster.com
5. Dams, L.: The strategic challenges of under armour ın 2021 with an analysis of the athleisure market. Nova School Bus. Econ. (2021). Available: https://run.unl.pt/handle/10362/144881
6. Marci, K.: Activewear market analysis (2021). Available: https://blog.edited.com/blog/resources/activewear-market-analysis
7. Kennedy, P., Kahle, L.R.: Using the list of values (LOV) to understand consumers. J. Serv. Mark. **2**, 49–56 (1988)
8. Alpan, E.: Hazır giyim sanayiinde koleksiyon hazırlama ve geliştirme yöntemleri. Marmara Üniversitesi Fen Bimleri Enstitüsü, İstanbul (2005)
9. Mehrali, N.: Study on the effects of fashion trends on occurrence process of clothing collection, p. 117. Istanbul Arel University, Fashion and Textile Design Art Department (2015)
10. Gürpınar, M.: Hızlı moda'da koleksiyon tasarımı ve yönetimi, Türkiye uygulaması, p. 130. Marmara Üniversitesi, Fen Bilimleri Enstitüsü. İstanbul (2010)
11. Yıldırım, A., Şimşek, H.: Nitel araştırma yöntemleri [Qualitative research methods], p. 189. Seçkin Yayıncılık, Ankara (2008)
12. Bowen, G.A.: Document analysis as a qualitative research method. Qual. Res. J. **9**(2), 27–40 (2008)
13. Gül, S.S., Nizam, Ö.K.: Content and discourse analysis in social sciences. Pamukkale Univ. Journal of Soc. Sci. Inst. **42**, 181–198 (2020)
14. Vajen, L.: Capitalizing on the athleisure trend. Nova School Bus. Econ., 1–31 (2015)

The Use of Innovative 3D Simulation Tools in Fashion Boutiques

Nesligül Kiliç and Özgür Ceylan(✉)

Eskisehir Technical University, Eskisehir, Turkey
ozgurceylan@eskisehir.edu.tr

Abstract. Technological advancements occurring worldwide have also influenced the fashion and textile industry. This rapidly evolving landscape, which is readily reflected in design and production processes, indicates the industry's openness to innovative developments. Particularly in garment design, traditional hand drawings have been replaced by tablets and computers. Computer-Aided Design (CAD) and Computer-Aided Manufacturing (CAM) software are extensively utilized in the sector. CAD/CAM software can be categorized into two-dimensional systems (such as Adobe Illustrator, Adobe Photoshop, CorelDraw) and three-dimensional systems (such as Optitex, Vidya (Assyst), V-Stitcher/Gerber, Modaris/Lectra). Additionally, there has been a recent emergence of three-dimensional garment simulation software products (such as CLO 3D, Marvelous Designer, Style 3D). The utilization of these software applications minimizes the time spent in the design process and reduces error margins by enabling pre-production checks of designs. The primary objective of this study is to investigate the utilization of 3D simulation software in the fashion and textile sector within the context of personalized production carried out by small-scale enterprises. Considering that the design and production processes of these boutique enterprises significantly differ from those of ready-to-wear businesses, this research will provide valuable insights into the integration of 3D simulation software into their relevant operations.

Keywords: CAD · CAM · 3D simulation · Fashion boutique

1 Introduction

The relationship between humans and the environment has generated public concern, leading to the emergence of sustainability concept. The exponential increase in consumption, along with associated production intensity, contributes to environmental issues [1]. The textile and fashion industry is one of the most harmful industries to the environment. Within the scope of fast fashion, the perception that resources are unlimited contributes to exacerbating this effect [2].

For fashion industry, sustainability can be defined as the reduction of environmental and socio-economic impacts of produced goods. This concept involves a continuous effort to enhance all stages of a product's lifecycle, ranging from its design phase to

G. Guxho et al. (Eds.): ITC-ICEE 2023, LNMIE, pp. 434–441, 2024.
https://doi.org/10.1007/978-3-031-48933-4_42

eventual recycling/upcycling [3]. The environmental and social issues stemming from excessive consumption necessitate greater preventive measures for the industry. Another significant concern in this regard is the initiation of the phases of reduction and deceleration [4]. The notion of a "slow culture" has become an integral component of the paradigm shift towards sustainability in the fashion industry [5]. Slow design processes are characterized by their open-source nature and a collaborative approach rooted in transparency [6]. Personal production in fashion can also be categorized within the scope of these principles.

With the advent of technology, personal production in fashion have gained prominence in recent years. Personal productions, distinct from mass production, are built upon a model wherein garments are uniquely tailored to fit an individual customer's body measurements and involve the customer in the design process. In personal production, garments are designed with specific fabrics and accessories as per the customer's needs, thereby yielding variable costs and obviating the necessity for stockpiling during production [7]. Furthermore, quality tends to increase, and product life cycles extend while quantities decrease in personalized products catering to specific target groups. The current century and ongoing advancements exhibit a supportive disposition towards personal production. Trends favoring individuality foster an escalation in demand for personalized products [8]. Within this context, it is conceivable to approach all stages of product development through the user experience concept. Often, users sharing the experience of a product overshadow the product itself [9]. The benefit of this experience lies in its encompassing of an entire spectrum of connection between the product and the user [10].

Computer-aided design programs can be seen as technologies that facilitate these activities [11]. The trend of digitization within the industry possesses a facilitating quality for personal production. Within the scope of this study, based on the studies related to CAD systems, 3D design programs, and personal production in fashion, semi-structured interviews were conducted to gather information about the operations of boutique ateliers engaged in personal production and their perspectives on the software employed in the fashion industry. In line with the conducted research, a garment model created using CLO 3D, one of the three-dimensional clothing simulation programs, was produced.

2 Computer Aided Design Programs

Computers, due to their extensive capabilities, are very suitable to be employed in many stages of the design process. In the field of clothing, computers began to be used in production-oriented applications in the 1960s and have progressed to encompass design-related processes such as pattern design and making, thus advancing considerably. With the advancement of computer technologies, Computer-Aided Design (CAD) and Computer-Aided Manufacturing (CAM) software have actively started to be used in the industry [12]. Additionally, three-dimensional simulation programs, which play a significant role in the production process, have also started to develop, making sample preparation much easier [13]. As two-dimensional graphic programs, (Adobe) Illustrator, (Adobe) Photoshop, and CorelDraw can be given while examples of three-dimensional clothing design programs include Optitex, Vidya (Assyst), V-Stitcher (Gerber), and Modaris (Lectra) [14].

Three-dimensional simulation systems, in addition to visualizing real-time interaction between two-dimensional models, can also use the same materials as final garments. Thus, corresponding systems have the potential to minimize the time and cost involved in the development process of a model [15].

Personalization in clothing occurs in two consequent stages. The first is the phase of ensuring the fit of the garment to the body, and the second involves determining details such as the model, fabric, color, and accessories [16]. In a study conducted on the options that should be available on the website for a T-shirt design application in the field of online mass customization, the stages of customization have been prioritized in order of importance as in Table 1 [17].

Table 1. Website options required for online t-shirt design

Precedency	Requested option on website
1	Design by body measurements
2	Height
3	Original design
4	Collar type
5	Fabric color
6	Cut (narrow, wide cut, etc.)
7	Sleeve length
8	Fabric type
9	Button, zipper etc. auxiliary material
10	Text printing
11	Multiple colors
12	Pocket size and place
13	Photo print
14	Sewing thread color

3 Materials and Methods

The study was conducted within the framework of a qualitative research method. Qualitative research encompasses three types of data collection methods based on environment, process, and perceptions. These methods are supported by data collection techniques such as interviews, observations, and the examination of written materials [18]. In this research, the design and production processes of the personal clothing production were investigated. Semi-structured interviews were conducted with two boutique ateliers located in Eskişehir. The semi-structured interview audio recordings were transcribed into text using 'Sonix.' The obtained texts were analyzed using descriptive statistical methods, and the results were presented in the corresponding section. Descriptive statistics is an approach classified as a statistical method that allows for the explanation and summarization of collected data [19].

The study utilized data obtained from literature review and interviews to create avatars in the CLO 3D garment simulation program based on individual measurements. A sewing simulation of the garment, created according to the obtained measurements, was performed on the avatar. The digital design and production processes were carried out in a digital environment, and the actual production of the garment model was conducted, allowing for quality control on a dummy.

4 Results and Discussions

Within the scope of this study, semi-structured interviews were conducted with boutique ateliers specializing in personalized production. The content of the questions posed to workshop managers during the interviews included topics such as the process they followed before commencing production, how many fittings they conducted during production, the challenges of preparing orders without fittings, and their thoughts on digitalization in the textile industry. In this context, below is a compilation of some of the responses provided by the participants and the data obtained from the interviews:

Participant 1

- "I engage in both with-fitting and without-fitting production."
- "For online orders, I take measurements through video calls."
- "I don't work with ready-made patterns; I perform custom measurements."
- "If we can input individuals' body measurements into the computer program and see how a designed garment fits on a model, and how the contours will appear, I believe we can easily produce without fittings in a comfortable and problem-free manner."

Participant 2

- First, we prepare a collection. Based on that collection, we tailor garments according to the measurements of customers requesting models.
- For custom clothing orders, we engage in 80% without-fitting production and 20% with-fitting production.
- It depends on how well it's digitized. In other words, if the sewing details and body posture are very accurate and the presentation to the customer is very close to reality, it certainly can facilitate the process.

In both interviews, it was found that the companies predominantly employ traditional methods for customized production with fittings and without fittings. After taking customer measurements, patterns are hand-drawn. Participant 1 mentioned conducting at least one fitting for products that require it. However, they also expressed that for designs created using 3D simulation software and patterns prepared in the digital environment, production without fittings could be achieved without any issue. Participant 2 stated that if the image quality of the digitally transferred product is very good, it would facilitate customized production.

Based on the data obtained from the literature and semi-structured interviews, a model was developed within the scope of the study using the CLO 3D simulation software. After making fabric selection and forming the design concept, an avatar was created in the program according to the specified body size (38). Subsequently, the 2D pattern of the designed model was prepared (Fig. 1).

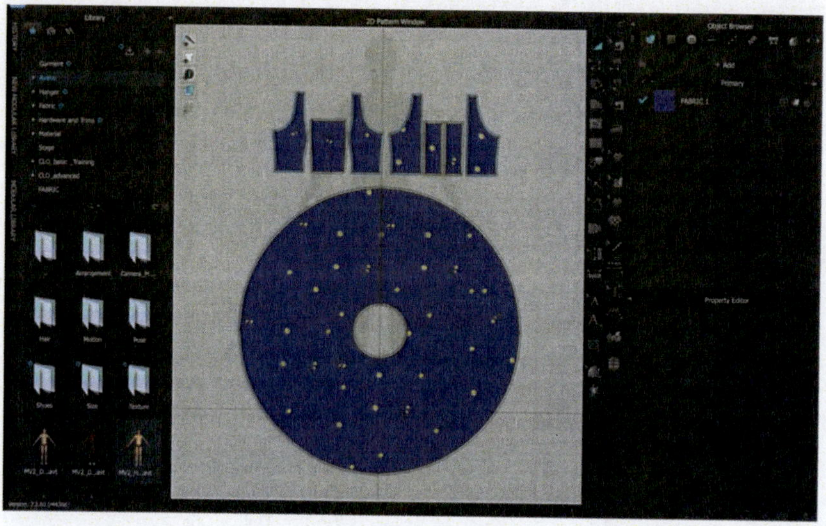

Fig. 1. Clo 2D field mold drawing

To integrate the selected fabrics into the 3D program, photographs of the fabrics were taken, and for the patterned fabric, a pattern report was created using Adobe Photoshop program (Fig. 2).

Fig. 2. Fabric photo and pattern report

At this stage, the model, for which pattern drawing, fabric color, and print design were conducted, was dressed to the avatar in the 3D environment within the program (Fig. 3).

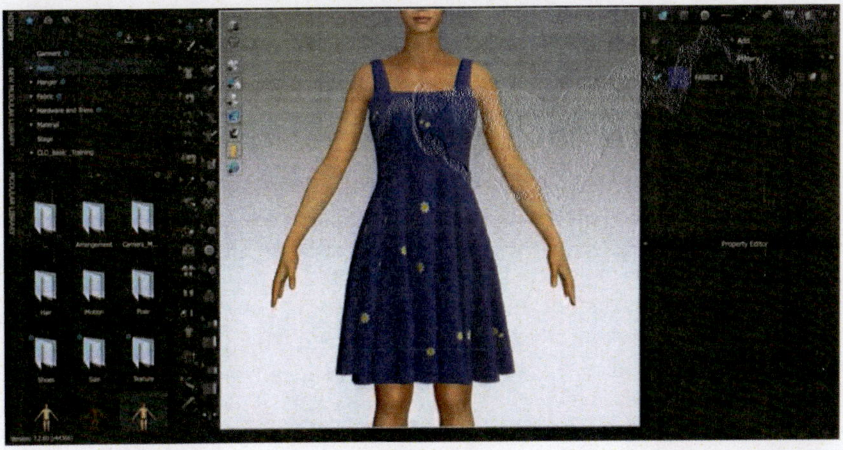

Fig. 3. CLO 3D garment sewing

The digital design and production process were finalised for the model, and the pattern drawings were transferred to the fabric for cutting and sewing operations. The sewn model was subjected to fit and measurement controls were performed on a dummy (Fig. 4). No fitting was performed on a human model during the production process. Upon completion of garment construction and the subsequent quality checks, it was observed that with correct measurements and accurate avatar construction within the program, the error rate in personal production with no fittings could be close to zero.

Fig. 4. Fit check on product sewing and model

5 Conclusions

The primary aim of this research was to investigate whether digital programs particularly 3D simulation ones used in the textile and fashion industry are effective in the context of personal production and to conduct a practical application in this regard. In the last decades, the concept of fast fashion has gained momentum due to changes in supply chain, production and consumption habits. While this drastic change has been considered an advantage by certain customer segments due to the variety of products available, it has also led some consumers to lean towards personalized clothing, driven by the desire for individuality. Just as in any other industry, it is observed that with the influence of technological advancements in textile and fashion industry, there is an increasing number of software applications to support the design and production processes. Research has shown that CAD/CAM and 3D garment simulation softwares facilitate personal production processes.

From the conducted interviews, opinions support the idea that novel softwares enabling digital design and production would be conducive to the practice of personal production without the need for fittings. The personal production carried out with the CLO 3D garment simulation program in the scope of this study corroborates also this perspective. Thus, it is believed that the research conducted will add value to the personal clothing sector and pave the path for future studies in this field.

Acknowledgment. The 3D clothing model application included in this study was supported by Masterkey, the Türkiye distributor of CLO 3D.

References

1. Cengiz, G.: Sürdürülebilir tüketim ekseninde sürdürülebilir moda pazarlaması uygulamaları. İktisadi ve İdari Yaklaşımlar Dergisi **3**(2), 131–144 (2021)
2. Mangır, A.F.: Sürdürülebilir kalkınma için yavaş ve hızlı moda. Selçuk Üniversitesi Sosyal Bilimler Meslek Yüksekokulu Dergisi **41**, 143–154 (2016)
3. Brydges, T., Retamal, M., Hanlon, M.: Will covid-19 support transition to a more sustainable fashion industry? Sustain. Sci. Pract. Policy **1**, 298–308 (2020)
4. Kaya, E.: From crisis of the speed to slowing: fast-fashion, pp. 1–16. İstanbul University (2021)
5. Fletcher, K.: Slow fashion: an invitation for systems change. Fash. Pract. **2**, 259–265 (2010)
6. Strauss, C.F., Fuad-Luke, A.: The slow design principles. Changing the Change, 1–14 (2008)
7. Bushra, A.: Mass Customization: A Study on the Textile Industry from the Production Management Perspective, pp. 1–74. Pamukkale University, Business Administration (2021)
8. Pazarcık, G.: Method of Mass Customization and an Implementation, pp. 1–227. Yıldız Teknik University, Institute of Social Science (2009)
9. Başer, E., Akıncı, S.: Kullanıcı deneyimi ve kişiselleştirme bağlamında bir dijital platform incelemesi. Selçuk İletişim Dergisi **2**, 866–897 (2020)
10. Yenilmez, F.: Ambalaj tasarımında kullanıcı deneyimi: bir zeytinyağı ambalajı üzerinden değerlendirme, Yayınlanmamış Yüksek Lisans Tezi. İstanbul: İstanbul Teknik Üniversitesi, Fen Bilimleri Enstitüsü (2012)
11. Tabaklı, G.: Kişiye özel giysi üretimi. Yayınlanmamış Yüksek Lisans Tezi. İzmir: Ege Üniversitesi, Fen Bilimleri Enstitüsü (2018)

12. Demirel, M.Y., Karaağaç, İ.: An overview of computer aided manufacturing process. Mühendis ve Makina **652**, 51–61 (2014)
13. Servi, Ş.: Türkiye'de bilgisayar destekli kalıp tasarım sistemlerinin hazır giyim endüstrisi ve moda tasarımı eğitimlerine uyumlama süreçleri. Yayınlanmamış Yüksek Lisans Tezi. İstanbul: Haliç üniversitesi, Lisansüstü Eğitim Enstitüsü (2019)
14. Öğülmüş, E.: Giysi tasarımında tekstil yüzeylerinin 3D program uygulamaları ile örneklendirilmesi. Yayınlanmamış Yüksek lisans Tezi. Eskişehir: Anadolu Üniversitesi, Fen Bilimleri Enstitüsü (2016)
15. Choi, K.H.: 3D dynamic fashion design development using digital technology and its potential in online platforms. Fashion and Textile **9**, 1–28 (2022)
16. Vuruşkan, A.: Developing a New Approach on Customized Clothing Considering Fit and Design Parameters. Dokuz Eylül University, Institute of Science, pp. 1–201 (2010)
17. Özbek, A., Topaklı, A.: Online mass customized t-shirt design. Vocational Education (NWSAVE) **3**, 1–10 (2019)
18. Karataş, Z.: Sosyal bilimlerde nitel araştırma yöntemleri. Manevi Temelli Sosyal Hizmet Araştırmaları Dergisi **1**, 62–80 (2015)
19. Süleymanova, G.: Betimsel istatistik ve histogram karma metodu. Kırgızistan Türkiye Manas University, Institute of Science, pp. 1–51 (2015)

Encouraging 3D Virtual Design in Fashion Education: Best Practices

Evridiki Papachristou[1,2](✉) and Nefeli Zolota-Tatsi[1]

[1] Creative Design & Clothing, International Hellenic University, Thermi, Greece
evridikipapa@ihu.gr

[2] School of Production Engineering & Management, Technical University of Crete, Chania, Greece

Abstract. This paper presents the outcomes obtained from a first interaction experience, in which third year fashion students were invited to explore and use 3D virtual prototyping technology tools to develop their ideas. Tools like 2D CAD pattern design and 3D clothing design software were introduced and specific fashion-design tasks were performed during their learning process throughout a spring semester of twelve weeks. Participants were then asked to resolve these tasks resorting to the tools presented and integrate them into the fashion design project by exploring and experimenting their potential uses. The result demonstrated that fashion students acknowledged the potentiality of these technology tools for their subject's project, showed dedication, perseverance and teamwork, overcame the barriers and challenges of the lack of efficient use in 3D virtual prototyping and did not resort to traditional methods of developing and communicating their projects with pen/marker and paper in two dimensions. Lectures' structure was tailored to students' strengths and interests with an aim to ignite curiosity. They were encouraged to leverage the use of any available 3D digital design tool by questioning the presented-during lectures-advantages and by building their existing skills with new ones in digital environments. A selected best practice of a fashion collection development through 3D-design and visualisation is presented in this paper.

Keywords: 3D virtual design · Fashion education · Education 4.0 · Digital transformation · Digitalisation

1 Introduction

The rapid progress of technology and the growing need for digitalization are leading the fashion industry to change. As stated by Avădanei et al. [1], digital transformation can be perceived as a process of redefinition of a business, digitization of its operations, restructuring, and expansion of its value chain to allow companies to explore new possibilities and benefit from the full potential of technology.

According to Nissen and Nerup [2], one of the main characteristics of Industry 4.0 is the integration of the digital world with physical production and they state that it is going to vastly change (up to 50%) the way of working and producing. That explains why nowadays the enterprises of the fashion industry, such as industry garment factories,

G. Guxho et al. (Eds.): ITC-ICEE 2023, LNMIE, pp. 442–452, 2024.
https://doi.org/10.1007/978-3-031-48933-4_43

ateliers, design studios, etc., are looking for multi-field specialists who are capable in the areas of design, construction, and clothing technology and have skills in various natures [3, 4].

A big part of that change is the use of 3D technologies in all the steps of the circle of life of clothing, from design and prototyping to marketing and merchandising. In particular, 3D prototyping is becoming more and more important in the apparel industry for a plethora of rea-sons. In terms of design visualisation and communication, digital prototypes provide a more realistic representation of the final product compared to sketches or flat drawings. They allow designers to see their designs in three dimensions, simulate fabric draping, and visualise how the garments will look on a human body [5]. This helps designers better communicate their ideas to clients, manufacturers, and other stakeholders involved in the production process and cooperate with pattern makers more efficiently [6]. Moreover, it can reduce costs and be more time efficient. By virtually testing the fit, it can reduce the average sample iterations, raw material, and labour costs. This is also important for the sustainability goals of the industry [7]. Yet, although 3D technology is firmly established among other industries, apparel is at the beginning of discovering its possibilities and integrating it into the production workflow [8].

In this shifting environment, the way fashion is studied changes. Crittenden et al. [9] suggest that higher education should be ready not just to impart knowledge but also to assist students in acquiring different ways of working. It is essential to guarantee that students receive adequate exposure to crucial technologies and cultivate their critical thinking skills, enabling them to discern and differentiate among the plethora of available information sources. Overall, universities ought to prepare students for their future working environment and help them be-come multi-skilled professionals.

This paper presents the outcomes obtained from a first interaction experience, in which third year fashion students were invited to explore and use 3D virtual prototyping technology tools to develop their ideas. Tools like 2D CAD pattern design and 3D clothing design software were introduced and specific fashion-design tasks were performed during their learning process throughout a spring semester of twelve weeks. Participants were then asked to resolve these tasks resorting to the tools presented and integrate them into the fashion design project by exploring and experimenting their potential uses. The result demonstrated that fashion students acknowledged the potentiality of these technology tools for their subject's project, showed dedication, perseverance and teamwork, overcame the barriers and challenges of the lack of efficient use in 3D virtual prototyping and did not resort to traditional methods of developing and communicating their projects with pen/marker and paper in two dimensions. Lectures' structure was tailored to students' strengths and interests with an aim to ignite curiosity. They were encouraged to leverage the use of any available 3D digital design tool by questioning the presented-during lectures-advantages and by building their existing skills with new ones in digital environments. A selected best practice of a fashion collection development through 3D-design and visualisation is presented in this paper. Subsequent paragraphs, however, are indented.

2 Methods

A previous research [10] tried to teach students 2D pattern-making software for creating garments and 3D garment simulation software for analyzing fit on a virtual model. During the course, the instructor observed that students found the 3D program more challenging than the 2D program. While students worked diligently and remained on task with both programs, they exhibited more independent work with the 2D software compared to the 3D software. In contrast, the 3D program prompted more peer interactions and small-group settings. Students also had more instructional interaction with the professor when using the 3D program, yet still interacted positively with both CAD programs compared to traditional manual methods. Overall, the project was found successful in preparing students to use different software in the fashion industry.

In a separate study [11], a similar approach was undertaken, with a specific emphasis on 3D design. The project was divided into three distinct phases, and industry experts were involved throughout the process. Students were allocated a period of two weeks to familiarize themselves with the design software.

The outcomes of the study were remarkably positive, with participants reporting significant benefits from the experience. The only major concern expressed by the students was that the allocated time proved insufficient to fully explore the potential of the design software adequately. Despite this limitation, the overall feedback indicated that the incorporation of 3D design in the project was highly beneficial and well-received by the participants.

The described experimentation of this study was performed during the subject Virtual Prototyping which is taught in the third year (sixth semester) of a four year bachelor course on Creative Design & Clothing. In this course, students explore the innovative use of computer simulations and digital models to design and test products and systems, gaining hands-on experience in creating virtual prototypes to revolutionise the fashion industry. During the theoretical part of the class, students are introduced to the concept of 3D design and its potential use across the industry. New technological terms, such as AI (Artificial Intelligence), VR (Virtual Reality) and AR (Augmented Reality), blockchain, and digital twins, are explained and studied and students are encouraged to share their opinions and thoughts about the integration and implementation of the mentioned technologies in the apparel industry and its future evolution.

During practical experimentation with 3D virtual prototyping for fashion, students employ cutting-edge software to create and manipulate digital garment designs, enabling them to visualise and refine clothing concepts in a virtual environment before actual production, thereby streamlining the fashion design process and fostering creative exploration. During the spring semester of 2023, students were introduced to VStitcher by Browzwear. At the end of the semester, students were able to:

- Create and edit 2D clothing patterns in V-Stitcher
- Construct and sew virtual garments
- Assign fabrics according to their physical properties and apply all-over pattern prints
- Use hardware and trims (zippers, buttons, types of seams, etc.) and apply artworks
- Edit avatars
- Render for the maximum quality of the final result

At the end of their learning process, students were assigned an open project for their final evaluation. The subject of the project was not specific in order to not restrict their imagination and creative thinking. They were granted the opportunity to apply the provided knowledge in any manner they deemed appropriate and explore the potential of 3D through experimentation. Additionally, there were no constraints imposed on the choice of 3D garment software they could employ, although the official academic software for the University was specific [12].

The students were allotted a one-month timeframe to diligently prepare the project, and upon its completion, they were required to deliver a formal presentation during the class session [13, 14]. Throughout this designated period, the course instructor provided guidance and assistance in resolving any technical challenges that arose. However, the students were also actively encouraged to independently seek out solutions to the obstacles encountered. Despite the project's intended individual nature, the students collaborated, addressing queries collectively, and providing mutual assistance during the preparation phase, strengthening the spirit of cooperation among them [15, 16].

The project yielded several noteworthy outcomes that significantly impacted the students' development and academic journey. Notably, it fostered the cultivation of critical thinking abilities, honing their aptitude for effective problem-solving. Through engaging in experiential learning, the students acquired a deeper understanding of pattern-making and 3D technologies, equipping them with practical skills relevant to their fields of study.

Moreover, the project served as a platform for students to unleash their creative vision, granting them the freedom to explore and express their imaginative ideas. This opportunity to delve into the realm of creativity resulted in an evident boost in their confidence levels, as they gained assurance in their own capabilities and potential.

Beyond individual growth, the project also played a pivotal role in enhancing the bonds within the academic community. Collaborative efforts and shared experiences fostered a sense of unity and cooperation among the students, strengthening the ties that bind them together.

In conclusion, the project's outcomes encompassed an array of critical skills and personal developments, contributing to the students' holistic growth and enriching their educational journey.

3 Results

A selected best practice of a fashion collection development through 3D-design and visualisation is presented in this paper. All students are well familiar with the steps of developing a fashion collection, pattern making on paper, physical properties and behaviour of fabrics and constructing a physical garment. Equipped with this existing knowledge, students are able to enrich their skills and abilities with assistance of technology. However, the specific and presented practice on virtual prototyping was chosen among others due to its exceptional ability to seamlessly integrate traditional design principles with advanced 3D modelling techniques, resulting in innovative and highly realistic virtual garments that demonstrated both aesthetic appeal and functional viability. The programs the student adeptly utilised for this purpose were:

- Polypattern by Polytropon[1], for 2D pattern making
- Photoshop and Illustrator by Adobe[2], for creation of needed boards (moodboard, colour board, trend board etc.) and for designing all-over prints and stamps
- VStitcher by Browzwear[3], for creating virtual garments

The selected presented work was selected from a ready-to-made fashion collection with the title "*Fragments of humanity*". The conceptual framework of this collection finds its origins in the art movement of Surrealism, with an emphasis on the portrayal and exploration of the human body and facial features.

At the beginning, some rough and improvised garments were created in the software to obtain an initial impression and visualization of the designs. Later, during the phase of carefully crafting the final 3D garments, the student meticulously introduced alterations to the forms and colours originally depicted in the initial sketches. This process allowed the designer to witness real-time aesthetic changes, ultimately leading to the enhancement of the final design compositions. A colour palette was derived from the concept, taking also into consideration the trends for the Spring/Summer 2024 season. The software used to develop prints was Adobe's Photoshop and Illustrator. Those files were later applied on the virtual garments. Following that, the designer proceeded in generating the basic clothing patterns utilising Polypattern software, and subsequently exported them as DXF files to facilitate their utilisation and modifications in V-Stitcher according to her preferences. The fashion collection comprises six distinct outfits, with each outfit offering two different colour variations.

1st Outfit:

The outfit (Fig. 1.a., b.) is composed of a sleeveless top, where a face profile is intricately integrated into the pattern, along with patch-worked pants. The top features a cut-out design for the eye and puffed lips, employing poplin fabric for the simulation. The student experimented with various properties, layering, and thickness of fabric to achieve the desired effect on the "lips," incorporating seam puckering as part of the process. For the trousers, the designer deliberately simulated the characteristics of denim fabric.

2nd Outfit:

The second ensemble (Fig. 2.a., b.) comprises a mini dress with distinct components. The bodice is virtually crafted from viscose fabric, while muslin is utilised for the hem and sleeves for the simulation. The bodice features four cut-outs designed to emulate eyelashes. To create the cut-outs, the designer used the pen tool for creating interior lines and changed their form to holes. Moreover, an all-over pattern was imported into the software and applied to the hem and sleeves of the dress, with the assign tool.

3rd Outfit:

The third ensemble (Fig. 3.a, b.) is a sleeveless, high-low dress. Viscose fabric was selected for the simulation. The designer experimented with the physical properties of the fabric, specifically adjusting settings to achieve a translucent effect on the shoulders and

1 https://www.polytropon.com/en/polypattern/cad-pattern-design-software.

2 https://www.adobe.com/gr_el/.

3 https://browzwear.com/products/v-stitcher/, 2022.2 Edition.

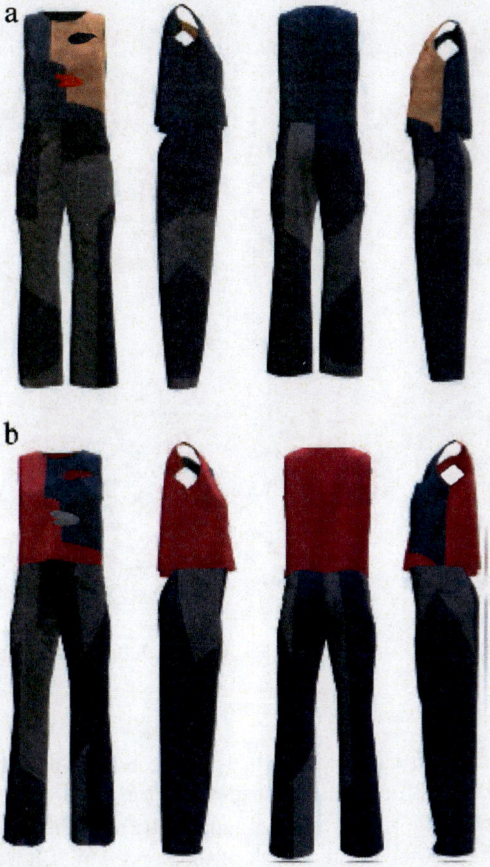

Fig. 1. **a**. Outfit 1/colourway 1 (front, right, back, left). **b**. Outfit 1/colourway 2 (front, right, back, left)

at the bottom of the dress. Furthermore, an intricately designed all-over print, produced in Illustrator, was applied to the main bodice of the dress.

4th Outfit:

The fourth outfit (Fig. 4.a., b.) showcases a strapless maxi dress simulated with a combination of denim for the bodice and main skirt, and muslin for the bottom portion. An artwork, which was created using Illustrator and Photoshop, was applied on the left side of the dress, while deliberate top stitches have been applied to replicate the stitching typically associated with denim fabric.

5th Outfit:

The ensemble comprises a sleeveless mini dress adorned with a "Peter Pan" collar and a two-layered circle skirt. The initial version of the dress showcases an intricately integrated lip design within the pattern (Fig. 5.a.), while the subsequent edition features

Fig. 2. a. Outfit 2/colourway 1 (front, right, back, left). **b.** Outfit 2/colourway 2 (front, right, back, left)

applied artworks on the front of the bodice (Fig. 5.b.). Poplin fabric was chosen to simulate the desired appearance. In the case of the top layer of the skirt, the designer deliberately modified the fabric's properties and opacity to achieve the intended outcome.

It is worth noting that the initial sketch of the dress depicted a different skirt style, characterised by an A-line silhouette with only one layer. However, during the process of creating the virtual garment, the designer expressed dissatisfaction with the aesthetic outcome. As a result, the user engaged in iterative experimentation with the dress's style within VStitcher software until she successfully attained the desired result.

6th Outfit:

The final outfit comprises a maxi dress with shoulder pads and wide-leg, high-waist trousers. The bodice of the dress is adorned with two assigned artworks, while the skirt displays an all-over print. To achieve the desired simulation, double jersey fabric was chosen for the bodice, and elastic tulle and denim fabric were utilised for the skirt and trousers, respectively.

In order to achieve the transparent effect of the skirt, the designer adjusted the opacity of the fabric. For the creation of the shoulder pads, the student designed new patterns and skillfully sewed them internally, while also altering the fabric's thickness to achieve the desired appearance (Fig. 6).

At the conclusion of the project, the student expressed how working on own designs not only facilitated the familiarity with design software and the fundamentals of 3D design but also sparked creativity and heightened the engagement with the project. The

a

b

Fig. 3. A. Outfit 3/colourway 1 (front, right, back, left). **b**. Outfit 3/colourway 2 (front, right, back, left)

virtualization of sketches and the ability to instantly change and modify patterns, colours, and other elements helped visualise ideas effectively and enhance the aesthetic outcomes of student's work.

Furthermore, it was highlighted that the presentation of the collection became more impactful through the incorporation of 360-degree rotated gifs, enabling the demonstration of all facets of the garments. Overall, the knowledge and application of 3D design, along with the practical experience gained during the project, significantly contributed to the user's growth as a fashion designer. The newfound abilities allowed the student to refine the designs, create more immersive presentations, and ultimately elevate skills and craftsmanship in the field of fashion design.

Fig. 4. **a**. Outfit 4/colourway 1 (front, right, back, left). **b.** Outfit 4/colourway 2 (front, right, back, left)

Fig. 5. **a**. Outfit 5/colourway 1 (front, right, back, left). **b.** Outfit 5/colourway 2 (front, right, back, left)

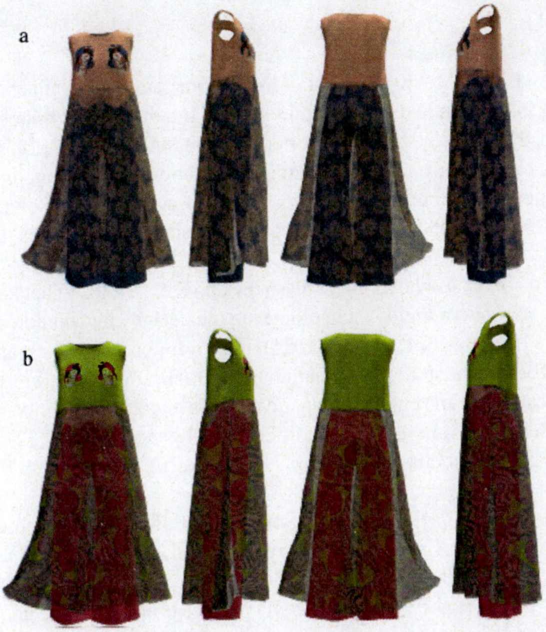

Fig. 6. **a**. Outfit 6/colourway 1 (front, right, back, left). **b.** Outfit 6/colourway 2 (front, right, back, left)

4 Conclusion

The integration of 3D virtual prototyping in fashion education has proven to be a valuable approach. The case study showcased how students effectively utilised 3D technology tools to develop their fashion collections, bridging the gap between traditional design principles and cutting-edge 3D modelling techniques. By embracing this transformative technology, fashion students demonstrated the potential for enhanced design visualisation, efficient communication, and sustainable practices in the industry. Moreover, the project's impact extended beyond technical proficiency, nurturing critical thinking and encouraging creative expression among students. As the fashion industry continues to evolve, the adoption of 3D virtual prototyping in fashion education becomes crucial to prepare future professionals for a digitalized and innovative landscape.

Funding. The author(s) disclosed receipt of the following financial support for the research, authorship, and/or publication of this article: This research has been co-financed by the European Regional Development Fund of the European Union and Greek national funds through the Operational Program Competitiveness, Entrepreneurship and Innovation, under the call RESEARCH—CREATE—INNOVATE (project code: T2EDK-04770).

References

1. Avădanei, M., Blaga, M., Kazlacheva, Z.: Best practices of sustainable product development through 3D-design and visualisation. In: 7th International Symposium on TTPF (Iasi, Romania), pp. 227–235 (2021). https://doi.org/10.2478/9788366675735-037
2. Nissen, B.K., Nerup, L.: Using 3D apparel visualisation and teaching to develop new skills and possibilities. In: Simonsen, R.R. (ed.) Fashion at a Turning Point: Implementing the Sustainable Development Goals in Education and Business Copenhagen, Denmark, pp. 55–61 (2020)
3. Kolosnichenko, M.V., Yezhova, O.V., Pashkevich, K.L., Kolosnichenko, O.V., Ostapenko, N.V.: The use of modern digital technologies in the design and technology VET in Ukraine. J. Tech. Educ. Training **13**(4), 56–64 (2021). https://doi.org/10.30880/jtet.2021.13.04.005
4. Kampakaki, E., Papahristou, E., Tarnanidis, T., Bilalis, N.: What are the most important digital qualifications of the employees in the clothing and fashion companies? In Proceedings of the 12th Hellenic Conference on Artificial Intelligence (SETN '22), Article 63, pp. 1–7. Association for Computing Machinery, New York, NY, USA. https://doi.org/10.1145/3549737.3549804
5. Spahiu, T., Manavis, A., Kazlacheva, Z., Almeida, H., Kyratsis, P.: Industry 4.0 for fashion products—case studies using 3D technology. IOP Conf. Ser.: Mater. Sci. Eng. (2020). https://doi.org/10.1088/1757-899X/1031/1/012039
6. Jevšnik, S., Stjepanovic, Z., Rudolf, A.: 3D Virtual prototyping of garments: approaches, developments and challenges. J. Fiber Bioeng. Inf. **10**(1), 51–63 (2017). https://doi.org/10.3993/jfbim00253
7. McQuillan, H.: Digital 3D design as a tool for augmenting zero-waste fashion design practice. Int. J. Fashion Des. Technol. Educ. **13**(1), 89–100 (2020). https://doi.org/10.1080/17543266.2020.1737248
8. Papachristou, E., Bilalis, N.: Can 3D virtual prototype conquer the apparel industry? J. Fashion Technol. Textile Eng. **1**, 1–6 (2016). https://doi.org/10.4172/2329-9568.1000141
9. Crittenden, W.F., Biel, I.K., Lovely, W.A.: Embracing digitalization: student learning and new technologies. J. Mark. Educ. **4**(1), 5–14 (2019). https://doi.org/10.1177/0273475318820895
10. Hwang, J.Y.: A case study of 2D/3D CAD virtual prototype simulation programs to enhance student performance in student-centered fashion design education. J. Textile Eng. Fashion Technol. **3**(1) (2017). https://doi.org/10.15406/jteft.2017.03.00088
11. Hodges, N., Watchravesringkan, K., Min, S., Lee, Y., Seo, S.: Teaching virtual apparel technology through industry collaboration: an assessment of pedagogical process and outcomes. Int. J. Fashion Des. Technol. Educ. **13**(2), 120–130 (2020). https://doi.org/10.1080/17543266.2020.1742388
12. Kokotsaki, D., Menzies, V., Wiggins, A.: Project-based learning: a review of the literature. Improv. Sch.. Sch. **19**(3), 267–277 (2016). https://doi.org/10.1177/1365480216659733
13. Girard, T., Altoona, P., Trapp, P.: An exploraty study of class presentation and peer evaluations: do students perceive the benefits? Acad. Educ. Leadership J. **15**(1), 77–94 (2011)
14. Živković, S.: The importance of oral presentations for university students. Mediterranean J. Soc. Sci. **5**(19), 468–475 (2014). https://doi.org/10.5901/mjss.2014.v5n19p468
15. Fry, H., Ketteridge, S., Marshall, S.: Supervising projects and dissertations. In: A Handbook for Teaching and Learning in Higher Education: Enhancing Academic Practice, 3rd edn., chap.11, pp. 150–164. Routledge, New York, U.S.A. (2003)
16. Guo, P., Saab, N., Post, L.S., Admiraal, W.: A review of project-based learning in higher education: student outcomes and measures. Int. J. Educ. Res. **102**, 1–13 (2020). https://doi.org/10.1016/j.ijer.2020.101586

Enhancing Realism in Cloth Simulation: A Review of Methods and Issues

Lilia Sabantina(✉)

Berlin School of Culture + Design, Department of Apparel Engineering and Textile Processing, Berlin University of Applied Sciences - HTW Berlin, 12459 Berlin, Germany
lilia.sabantina@htw-berlin.de

Abstract. In the pursuit of high-quality clothed model animations, the accurate assessment of simulated garment and fabric properties is important. This task is particularly challenging due to the complexity of cloth as an anisotropic material, resulting in a range of evaluation approaches. Moreover, the physical material parameters must align with simulation program requirements. Recent advancements in simulation systems have enabled the representation of dynamically moving garments, meeting industry expectations. However, these developments have posed new challenges for fabric characterization, necessitating enhancements in fabric measurement methods and parameter derivation. This review explores the evolving field of 3D simulation tools and their pivotal role in visualizing and manipulating virtual garment representations. These simulations strive to replicate real-world behavior by considering various physical parameters, such as tensile strength and flexural stiffness, bridging the gap between traditional material testing methods and advanced simulation software.

Keywords: 3D simulation tools · Physical material parameters · Apparel · Fabric testing · Tensile test · Flexural stiffness · Drape test

1 Introduction

Achieving high-quality animations of clothed models necessitates accurate and tailored evaluation of the simulated garment's and fabric properties. Given cloth's complexity as an anisotropic material, evaluating its properties is challenging and various approaches exist. The determined physical material parameters must also match the requirements of the simulation program. Simulation systems have advanced to the point where they can simulate not only simplified static garments but also complex dynamically moving ones, meeting the clothing industry's expectations. This evolution has introduced new challenges for fabric characterization methods, necessitating advancements in fabric measurement methods and parameter derivation. Developments in the field and use of 3D simulation tools in various sectors have been steadily increasing. The use of 3D simulation tools in apparel industry enables designers, manufacturers and stakeholders to visualize and manipulate virtual representations of garments in a way that was previously unimaginable [1]. 3D simulation attempts to replicate real-world behavior and results in

G. Guxho et al. (Eds.): ITC-ICEE 2023, LNMIE, pp. 453–460, 2024.
https://doi.org/10.1007/978-3-031-48933-4_44

a digital environment. This requires a thorough understanding of the intricate interplay of various physical parameters that affect the properties of a material, such as tensile strength, flexural stiffness, or drapability [2]. The convergence of traditional material testing methods and sophisticated simulation software forms a synthesis between the real and virtual worlds.

2 Material Testing and Simulations

The geometric representations can serve as a basis for the study of forces and deformations to which the textile material is exposed. The use of finite element (FE) simulation programs in various specialized software complexes such as ANSYS, LS-DYNA, ABAQUS, NASTRAN etc. allows the analysis of complex structural parameters that are difficult to determine using conventional methods. Software such as TexGen, ESI etc. can provide support and are used for simulations of textiles [3, 4]. For example, the use of TexGen and LS-DYNA was presented in the study by Lang et al. Mesoscale FEM models were created to support the development of gradient fabrics. TexGen software generates realistic material geometry, which is then converted to beam element models using a Python script. The results of real and virtual tensile tests show a clear correlation between the crimping of the threads in the fabric structures and the resulting elongation [5]. Figure 1 illustrates the generation of models for angle-through-thickness weave using TexGen and LS-DYNA.

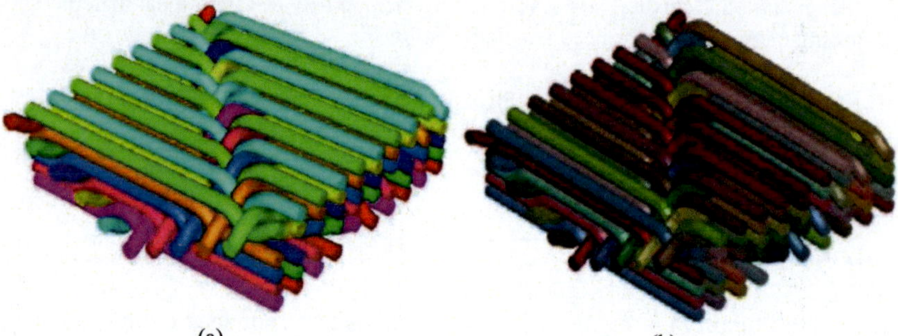

(a) (b)

Fig. 1. Model generation for angle-through-thickness weave, as illustrated in the following figures: (a) TexGen model, and (b) beam element model created in LS-DYNA. Adapted from [5], originally published under a CC-BY license.

Stephan et al. conducted a finite element investigation into how the orientation of fibers affects the response of glass and carbon fiber-epoxy composites under high-velocity impact conditions using the ANSYS LS-Dyna simulation tool [6]. Orlik et al. utilized the TexMath module MeshUp to digitally capture spacer fabric in its initial state as it emerges from the machine. Subsequently, they virtually expanded and compressed the fabric within the same TexMath module FiberFEM, designed for mechanical textile simulations [7]. Other examples of virtual simulation programs include CLO 3D, OptiTex™ 3D Runway Designer, Lectra-Modaris, and Assyst-Vidya systems. For example,

Shim et al. analyzed various nylon fabrics, measuring their physical properties using Kawabata systems and CLO fabric kits. It focused on the correlation between physical properties and drape characteristics, using the CLO fabric kit results for 3D virtual clothing simulations. The study identified key factors influencing drape properties and aimed to reduce disparities between real and virtual fabrics. Bending and shear properties showed the highest correlation with drape ratios. Integrating Kawabata system and CLO fabric kit data into a database reduced differences between real and virtual fabrics, including drape variations in different fabric thicknesses in CLO 3D simulations [8]. Figure 2 illustrates the stages, from diverse material characteristics integrated into 3D simulation, leading to 3D clothing simulation.

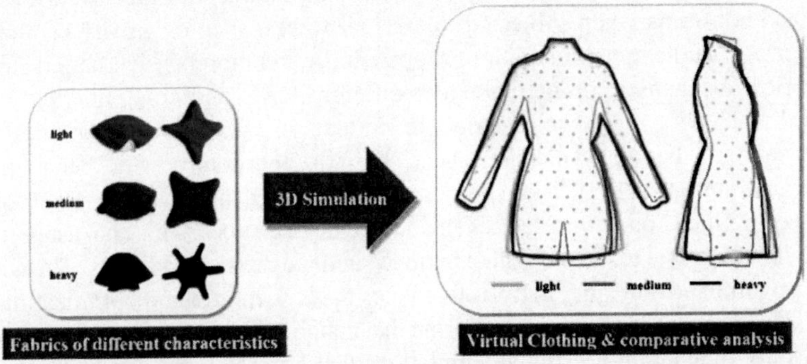

Fig. 2. A schematic representation of the stages involving materials with diverse characteristics, implemented in 3D simulation, leading to 3D clothing simulation with a focus on comparative analysis. From [8], originally published under a CC-BY-NC-ND 4.0. License.

3D Runway Designer is advanced cloth simulation software utilizing CAD patterns and real fabric properties. The OptiTex™ 3D Garment Draping and 3D Visualization software enables individuals involved in fashion and clothing design to instantly visualize pattern adjustments in full 3D. It enables instant 3D visualization of pattern modifications, while OptiTex Runway™ 3D allows for fabric and stitch customization in simulations, covering gathering, shrinkage, strength, width, and texture [9]. VStitcher's (Browzwear) virtual stitching technology translates 2D patterns into realistic 3D garment simulations with multisize grading on a virtual body, while texture mapping enables high-quality representation of fabric, seams, prints, and colors. Measurement of fabric properties such as bending, stretching and slippage are possible. Additionally, it provides essential data such as weight and thickness to support fabric modeling within the program. For instance, fabric extensibility, which quantifies elongation in both the warp and weft directions, is represented as 'stretch rigidity' within the software. 3D modeling software allows for the creation of virtual garment models and simulation of draping based on physical fabric conditions. The emergence of tools like Browzwear's Fabric Analyzer (FAB) serves as an instrument that meticulously captures the physical attributes of fabrics and translates these data into a format compatible with 3D design software, such as the VStitcher software. By accurately quantifying material properties like mass, thickness, bending, stretching, and shearing, FAB empowers designers and researchers to integrate these parameters seamlessly into their simulations, thereby enhancing the fidelity of their virtual models [10]. However, the primary challenge in modeling textiles, especially in non-crimp situations, is accounting for yarn contact, which significantly alters the fabric's macroscopic behavior. Accurately computed fabric characteristics are essential for representing the fine qualities of garments, making their correct derivation from real fabric characteristics crucial.

3 Textile Behavior and Drape Simulations

One of the pivotal aspects of textile behavior that 3D simulations aim to capture is the phenomenon of drape. Drape, in essence, refers to the manner in which a fabric cascades, folds, and conforms when subjected to external forces, such as gravity or movement. Achieving an authentic simulation of drape entails a comprehensive understanding of the fabric's mechanical properties under different loads and conditions [11, 12]. To model the behavior of textile materials in simulations, mechanical properties such as tensile strength, bending behavior, shear strength, and compression behavior are of significant importance [13]. These properties are observed under varying levels of stress, from the minimal loads that fabrics encounter during processing and usage to more substantial loads that textiles might experience under extreme conditions. This approach offers a twofold advantage. Firstly, it facilitates process control and optimization during manufacturing and design phases, ensuring that materials perform as intended. Secondly, it forms the foundation for digital simulations that accurately predict how fabrics will behave in different contexts, ultimately contributing to enhanced design precision and better-informed decision-making.

The research conducted by Rudolf et al. delves into the effects of simulating parameters in 3D garment drape simulations, shedding light on their significance in this context. A comparison was made between the orthogonal projections of real fabric drapes based on the Cusic method and the appearances and dimensions of real garments and their 3D virtual counterparts. The results indicated that simulation parameters had an impact on 3D virtual fabric and garment drape simulations, which varied depending on fabric properties and pattern piece sizes. However, the study identified a fundamental challenge related to incorporating real fabric physical and seam properties into garment drape simulations. Therefore, further comprehensive research is necessary to address these limitations and explore the synergistic effects of simulation parameters on 3D garment drape simulations [14]. Meanwhile, Han and Chang contribute to the field by conducting draping simulations on carbon/epoxy plain weave fabrics, analyzing material behavior, and employing a non-orthogonal constitutive model. The comparison between real tests and simulations in this study shows that the simulation results are very close to the real results. The experiments performed, including the mechanical testing of tissue prepregs and the simulation of the deformation behavior during the draping process, led to results that were almost identical to the actual draping behavior. This was true both for the deformation of the prepregs and for the accurate prediction of the bending angle in local areas of the draped structures [15]. In a different study, Efendioglu et al. investigate the applicability of 3D visualization and simulation programs in leather apparel, potentially expanding the use of such technologies in the fashion industry. The Vidya 3D program realistically simulates model and material behaviors using FAST system (Fabric Assurance by Simple Testing) data but requires improvement in accurately depicting the drape parameter [16]. Central to the study of fabric drape is the drape coefficient, a parameter that quantifies the extent of deformation exhibited by a fabric during the draping process. This coefficient serves as a numerical representation of the ratio between the projected surface area of the draped fabric and the corresponding surface area of the undraped fabric [17]. By gauging the drape coefficient, researchers and designers gain insights into how fabrics will cascade and fall when fashioned into garments, providing a critical tool for predicting the aesthetic and functional attributes of the final product. The utilization of 3D simulation software, extends the application of drape coefficient analysis into the digital realm [18]. Virtual draping tests, conducted within the virtual environment of the software, enable designers to visualize and assess the behavior of fabrics as they interact with the human form and other external factors. The software calculates and renders the drape coefficient based on the simulation, thereby allowing designers to experiment with different materials, designs, and scenarios prior to physical production [19].

4 Comparison of Simulation and Real-World Testing

The integration of 3D simulation tools offers a unique opportunity to evaluate the correspondence between virtual representations and real-world outcomes. In the context of fabric properties and drape behavior, this comparison becomes even more pertinent. Virtual simulations, while powerful, are not devoid of discrepancies when juxtaposed with real-world material tests [20, 21]. In the pursuit of accuracy, researchers have undertaken meticulous comparisons between simulation-based material tests and physical investigations. Tensile testing, flexural stiffness testing, and draping tests conducted on real fabric samples provide a benchmark against which simulation results are measured. The visual and qualitative differences between simulated folds and real-world draping are noted, highlighting the nuances that simulations might fail to capture entirely. While simulations can closely replicate topological features and even some aspects of fold depth, the deviation in the number of folds from that measured by the Cusick drape meter accentuates the complexity of fabric behavior. Kamadi et al. investigated the transformation of the traditional paper weighing method into a digital image analysis technique for the precise evaluation of fabric drape. The use of modern software allows the prediction of the Drape Coefficient % (DC%) prior to fabric production, thus saving time and providing more fabric choices [22]. As noted by Davis et al., the hanging drape method offers advantages for evaluating specific physical characteristics, such as the fabric's bending degree. Besides the conventional round sample approach for textile fabric drapability analysis, certain methods involve rectangular-shaped samples [23]. The study by Ghaderi et al. investigated the influence of various physical fabric parameters on the objective and subjective draping of different types of fabrics with different properties. The results showed that the bending length in both directions (lengthwise and crosswise) of the fabrics are the most significant factors for predicting draping coefficients and that there is a good agreement between the model results and the results of the Cusick method [24]. Reinhardt and Schneider develop a systematic approach for the simulation of textile surfaces taking into account material and design parameters. Although the simulation of garments offers advantages such as the reduction of physical prototypes, it is still not fully reliable for an accurate fit analysis and there is a need for a virtual prediction of the material behavior of textiles [25]. The study Spahiu et al. explores the integration of 3D printing onto textile fabrics, aiming to investigate its impact on fabric drapability. By evaluating various 3D printed geometries and their effect on fabric drape coefficients, the research demonstrates that 3D printed patterns can significantly alter the way fabric drapes, with minimal differences of up to 4% compared to traditional methods [26]. Figure 3 demonstrates the creation of a fabric drape histogram. The RGB image is converted to HSV, with the data stored as "double." Within the HSV components, the V-component identifies key regions: support disk, drape, and paper ring.

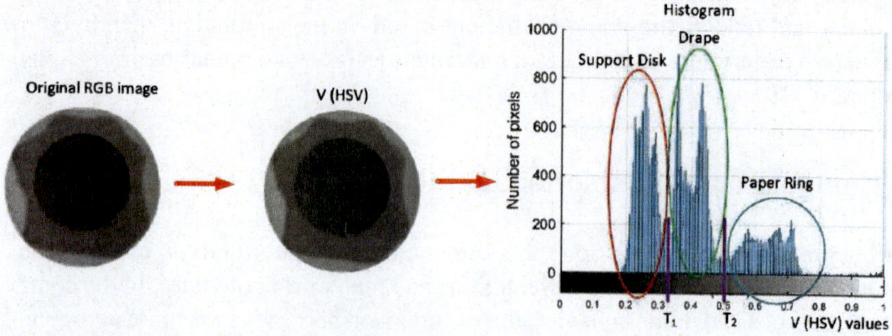

Fig. 3. The process of generating a fabric drape histogram. Initially, the RGB image undergoes conversion into the HSV color space, and the resulting image data are in a "double" format. From [26], originally published under a CC-BY-NC-ND 4.0. License.

These studies show that the results of incorporating 3D simulations are promising but relatively challenging and can be useful for the fashion sector as they facilitate the development of intricate patterns on textiles, customize products and improve the functionality of clothing and technical textiles. This relationship highlights the significant influence of physical and mechanical properties on the accuracy of fabric simulations. These discoveries suggest the prospect of improving simulation accuracy through a more sophisticated understanding and use of material parameters.

5 Conclusions and Future Outlook

To achieve precise dynamic garment simulations, parameters must be extracted from real fabric measurements. This task is intricate due to the complex and anisotropic nature of fabric. Special care is essential during measurements and their application in virtual simulations, particularly when simulating the behavior of physical material parameters in the virtual environment. While disparities between simulation-based material tests and real-world investigations exist, these simulations provide invaluable insights and possibilities for improvement. The trajectory of future research will undoubtedly lead to enhanced simulations that capture the intricate complexities of fabric behavior with even greater precision. As the textile and fashion industries continue to evolve in response to technological advancements, the integration of 3D simulation techniques supported by a comprehensive understanding of material parameters holds the potential to revolutionize design, manufacturing, and decision-making processes. As technology continues to advance and our understanding of material science deepens, the potential for more sophisticated simulations becomes increasingly evident. The capacity to predict garment behavior, aesthetics, and performance prior to physical production holds profound implications for reducing waste, shortening development cycles, and fostering innovative design solutions. In the ever-changing realm of design and innovation, the combination of physical material parameters and 3D simulations showcases human creativity and the pursuit of continuous improvement. As technology rapidly advances, the textile and fashion sectors are on the verge of unleashing new levels of creativity and efficiency, leading to a more sustainable and transformative approach to exploring materials, generating design concepts, and manufacturing garments.

References

1. Acaccia, G.M., Conte, M., Maina, D., Michelini, R.C.: Computer simulation aids for the intelligent manufacture of quality clothing. Comput. Ind. **50**(1), 71–84 (2003). https://doi.org/10.1016/S0166-3615(02)00142-2
2. Volino, P., Cordier, F., Magnenat-Thalmann, N.: From early virtual garment simulation to interactive fashion design. Comput. Aided Des. **37**(6), 593–608 (2005). https://doi.org/10.1016/j.cad.2004.09.003
3. Orlik, J., Krier, M., Neusius, D., Pietsch, K., Sivak, O., Steiner, K.: Recent efforts in modeling and simulation of textiles. Textiles. **1**(2), 322–336 (2021). https://doi.org/10.3390/textiles1020016
4. Long, A.C., Brown, L.P.: 8 - Modelling the geometry of textile reinforcements for composites: TexGen. In Boisse, P. (ed.) Composite Reinforcements for Optimum Performance. Woodhead Publishing Series in Composites Science and Engineering, pp. 239–264. Woodhead Publishing (2011). https://doi.org/10.1533/9780857093714.2.239
5. Lang, T.G., et al.: Simulation-based development of gradient woven fabrics for biomimetic implants to restore tendons and ligaments. Textiles **2**(2), 336–348 (2022). https://doi.org/10.3390/textiles2020019
6. Stephen, C., Behara, S.R., Shivamurthy, B., et al.: Finite element study on the influence of fiber orientation on the high velocity impact behavior of fiber reinforced polymer composites. Int. J. Interact. Des. Manuf. **16**, 459–468 (2022). https://doi.org/10.1007/s12008-021-00808-7
7. Orlik, J., Krier, M., Neusius, D., Pietsch, K., Sivak, O., Steiner, K.: Recent efforts in modeling and simulation of textiles. Textiles **1**(2), 322–336 (2021). https://doi.org/10.3390/textiles1020016
8. Shim, E., Koo, D.S., An, J.S., Yoo, H.J.: Comparison of the silhouette of virtual clothes by fabric characteristics of nylon fabric for the utilization of virtual clothes. ACS Omega **7**(50), 47156–47164 (2022). https://doi.org/10.1021/acsomega.2c06302
9. Liu, Y.M., Jang, H.K.: A study on the functional characteristics of apparel 3D CAD system. AMR **627**, 501–505 (2012)
10. De Silva, R.K.J., Navodhya, P.U., Gill, S.: Analysis of body-to-pattern relationship using traditional pattern drafting techniques: implications for automated digital prototyping. Int. J. Fashion Des. Technol. Educ. (Online first 2023). https://doi.org/10.1080/17543266.2023.2230270

11. Kyosev, Y.K.: Simulation of wound packages, woven, braided and knitted structures. In: Textiles, Simulation in Textile Technology. Woodhead Publishing Series, pp. 266–314. Woodhead Publishing (2012). https://doi.org/10.1533/9780857097088.266
12. Dik, N.Y., Tsang, P.W.K., Chan, A.P., Lo, C.K.Y., Chu, W.C.: A novel approach in predicting virtual garment fitting sizes with psychographic characteristics and 3D body measurements using artificial neural network and visualizing fitted bodies using generative adversarial network. Heliyon **9**(7), e17916 (2023). https://doi.org/10.1016/j.heliyon.2023.e17916
13. Fontana, A., Rizzi, C., Cugini, U.: 3D virtual apparel design for industrial applications. Comput. Aided Des. **37**(6), 609–622 (2005). https://doi.org/10.1016/j.cad.2004.09.004
14. Rudolf, A., Zadravec, M., Stjepanović, Z.: Investigations regarding the effects of simulating parameters during 3D garments' drape simulations. Fibres Text. East. Eur. **24**, 143–150 (2016). https://doi.org/10.5604/12303666.1221749
15. Han, M.-G., Chang, S.-H.: Draping simulation of carbon/epoxy plain weave fabrics with non-orthogonal constitutive model and material behavior analysis of the cured structure. Compos. Part A Appl. Sci. Manuf. **110**, 172–182 (2018). https://doi.org/10.1016/j.compositesa.2018.04.022
16. Efendioglu, N., Mutlu, M., Pamuk, O.: An investigation on usability of 3D visualization and simulation programs in leather apparel. J. Text. Inst. **112**, 1–16 (2021). https://doi.org/10.1080/00405000.2021.1938860
17. Brubacher, K., Tyler, D., Apeagyei, P., Venkatraman, P., Brownridge, A.M.: Evaluation of the accuracy and practicability of predicting compression garment pressure using virtual fit technology. Cloth. Text. Res. J. **41**(2), 107–124 (2023). https://doi.org/10.1177/0887302X21999314
18. Huang, H.Q., Mok, P.Y., Kwok, Y.L., Au, J.S.: Block pattern generation: from parameterizing human bodies to fit feature-aligned and flat-tenable 3D garments. Comput. Ind. **63**(7), 680–691 (2012). https://doi.org/10.1016/j.compind.2012.04.001
19. Mahmood, A., Akram, T., Shenggui, C., Chen, H.: Revolutionizing manufacturing: a review of 4D printing materials, stimuli, and cutting-edge applications. Compos. B Eng. **266**, 110952 (2023). https://doi.org/10.1016/j.compositesb.2023.110952
20. Musoni, P., Melzi, S., Castellani, U.: GIM3D plus: s labeled 3D dataset to design data-driven solutions for dressed humans. Graph. Models **129**, 101187. https://doi.org/10.1016/j.gmod.2023.101187
21. Hilal, B., Turemez, F.: Determining production fabric with three dimensional modeling system (V-Stitcher). Int. J. Sci. Technol. Res. **5**(1), 63–70 (2019). https://www.iiste.org/Journals/index.php/JSTR/article/view/45895
22. Kamadi, N.R., Ghani, S.A., Yahya, M.F., Tulos, N., Yusof, N.A., Yusof, N.J.M.: Framework development of fabric drape evaluation: Paper weighing method transformed to digital image analysis technique. AIP Conf. Proc. **2614**(1), 050067 (2023). https://doi.org/10.1063/5.0153238
23. Davis, A., Bouman, K., Chen, J., Rubinstein, M., Durand, F., Freeman, W.: Visual vibrometry: estimating material properties from small motions in video. IEEE Trans. Pattern Anal. Mach. Intell. **39**(4), 732–745 (2017). https://doi.org/10.1109/TPAMI.2016.2622271
24. Ghaderi, A., Mostahfezian, A., Ahmadi, M.S.: Investigation of the effect of fabric physical parameters on its objective and subjective drape. J. Textile Sci. Technol. (2023) (Online first). https://www.jtst.ir/article_172936.html?lang=en. Last accessed 25 Sept 2023
25. Reinhardt, U., Schneider, T.: System of material and construction parameters for a reliable simulation of textile surfaces in the clothing industry. Eng. Proc. **37**(1), 69 (2023). https://doi.org/10.3390/ECP2023-14632
26. Spahiu, T., Zlatev, Z., Ibrahimaj, E., Ilieva, J., Shehi, E.: Drape of composite structures made of textile and 3D printed geometries. Machines **10**(7), 587 (2022). https://doi.org/10.3390/machines10070587

Comparison of a Computer-Aided Design and Manual Pattern-Making

Maja Jankoska and Ruzica Stevkovska Stojanovska(✉)

Department of Textile Engineering, Faculty of Technology and Metallurgy, University "Ss. Cyril and Methodius" in Skopje, 1000 Skopje, Republic of North Macedonia
{maja,ruzicass}@tmf.ukim.edu.mk

Abstract. Innovation and trends are constantly needed for businesses to grow. For the garment industry Computer-aided design (CAD) is incredibly helpful since it can significantly improve production, reduce costs, and boost profit margins. CAD systems are exploited during the pattern-making process for pattern drafting, pattern grading, and marker creation. The aim of this paper is the compare manual and computer construction of different designs of women's dresses, as well as the difference in construction time and how CAD technology increases productivity and reduces construction time. It was obtained that CAD construction (with AccuMark computer software) is 5 times quicker than manual construction. Also, the conclusion is that as opposed to manual pattern making CAD software gives patterns with a higher level of pattern accuracy since the product lines and curves are considerably more accurate. Additionally, the process of grading sizes and including seam allowances in patterns is quick, precise, and simple to change.

Keywords: CAD software · Comparison · Pattern · Construction · Modeling

1 Introduction

The fashion sector is rapidly changing, bringing opportunities and problems for the garment and clothing industries due to rising demand brought on by an expanding population. The high standards required by the quick advancement of modern apparel manufacture cannot be met by outdated, ineffective traditional, and manual labour [1].

Today's continuously evolving technology means that trends and innovation are always in demand. One of the many products for the apparel industry that can improve garment production is CAD software. Additionally, it has a significant impact on customer satisfaction and encourages customer loyalty, both of which influence a company's reputation [2, 3]. To eliminate routine work in the pattern-making process, CAD/CAM technology is mostly used to design clothing, especially the construction processes. Therefore, the use of 3D scanners and CAD/CAM has made the show possible due to advancements in the field of fit clothing and the current requirement to keep up with the competition in the fashion industry [4]. In the world of fashion, CAM systems are frequently used for automation and efficiency in production. It is also an automation system that is utilized in planning, organizing management, production, and controlling

G. Guxho et al. (Eds.): ITC-ICEE 2023, LNMIE, pp. 461–467, 2024.
https://doi.org/10.1007/978-3-031-48933-4_45

production [5, 6]. Therefore, to survive in the modern world, this software is especially attractive for use in the garment industry for garment design, pattern formulation, and grading activities, but and for marker creations [7, 8].

The use of CAD technology, which has several advantages, enables quick and effective updates to clothing design, including resizing clothes and changing the color or pattern of the fabric. By saving money and time, CAD systems help designers create designs that are more precise and less likely to be mistaken. Due to the automation of repetitive design processes, greater staff productivity, and decreased lead times in the product development process, the industry's adoption of CAD technology has enhanced the efficiency of the design process [9]. CAD/CAM, which is widely used in the textile industry as a medium for designing and a tool for production, is a crucial step in the creative design process but also increases the attractiveness and competitiveness of textile designs and fabric to keep up with the consumer's rapidly shifting taste in fashionable designs [10, 11]. Also, apparel producers must deliver high-quality goods and services with short lead times, low manufacturing costs, and maximum operational efficiencies to remain competitive and grow in the global market, so to accomplish these goals, computer-based technology must be used. The first stage of garment manufacture is design, and computer-aided design has radically transformed the textile and apparel industries [9, 12].

Some studies show that the process times required for CAD systems were shorter than that for manual design in every step that has an important role in the pre-production preparation process [8]. The effectiveness of CAD systems also has been the subject of academic evaluations, thus the efficiency of model complexity when modeling t-shirts was assessed as a comparison of CAD and manual methods [13]. Of course, many other studies analyze the efficiency of different CAD systems and compare costs between manual and computer creation [14–17]. In our previous research, the modeling of clothing is performed by using different methods and computer software as a 2D and 3D CAD system [18–22]. Today, the process of creating patterns is still done manually in many apparel enterprises, even though there is a variety of tools available on the market in this field.

The aim of this study is to compare the efficiency of manual and CAD methods in women‘s dress model complexity when modeling and different the amount of time between manual method and computer methods for creating patterns.

2 Experimental Part

The manual basic construction of a women's dress, modeling, and grading of the patterns was done and later, all of them were created with the use of AccuMark computer software.

The Model 1 is composed of three front and three back parts. The back parts in the lower region, i.e. from the line of the hips, are respectively extended to the length of the cut, the length of which is above the knees. This model has short sleeves (see Fig. 1).

The second dress (Model 2) is like the previous and is composed of three front and three back parts, which in the lower part, i.e., from the line below the hips, are respectively extended to the length of the cut. The length of the dress is also above the knees (see Fig. 1).

Fig. 1. Technical sketch and picture of Model 1 (left) and Model 2 (right)

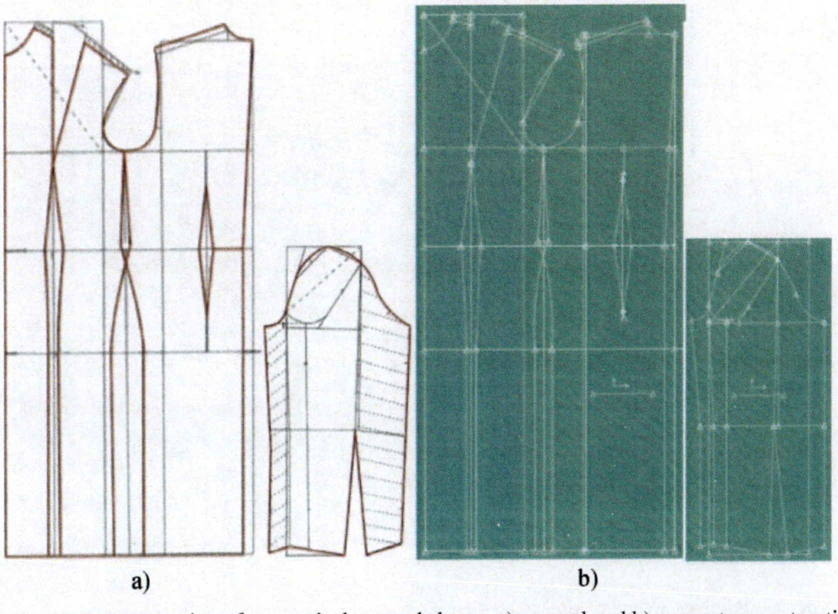

Fig. 2. Basic construction of women's dress and sleeve: a) manual and b) computer construction

Based on the construction measurements and rules for pattern designs of basic constructions of women's dress and sleeve [23–25] are performed basic construction of the women's dress and sleeve, for size 38, regular fit, medium rise. The basic construction of women's dresses and sleeves (see Fig. 2a) and computer construction (see Fig. 2b), were realized with the software AccuMark.

When the construction is finished, the seam allowances are added to all patterns. Therefore, the seam allowances of all patterns are 1cm; only the hem of the front and back of the women's dress and the sleeve is 3 cm. Later, with modeling of the basic constructions, the pattern designs are finalized.

In the meantime, the total time for manual and computer construction was measured.

3 Results and discussion

The design of the dress of Model 1 has three front parts and three back parts. The manual modeling of a woman's dress and sleeves (see Fig. 3a), and modeling by computer (see Fig. 3b) were made.

Mainly, the following modifications on the back part were performed: the opening of the neckline is obtained by determining a point at a distance of 8 cm along the length of the shoulder line, measured to the right of the starting point of the shoulder. This newly obtained point is joined by a curve line with a point obtained 2 cm down, measured from the center point of the neckline. The top point of the sleeve cut is measured 4cm down and this point is connected with a curved line to the top point of the dart.

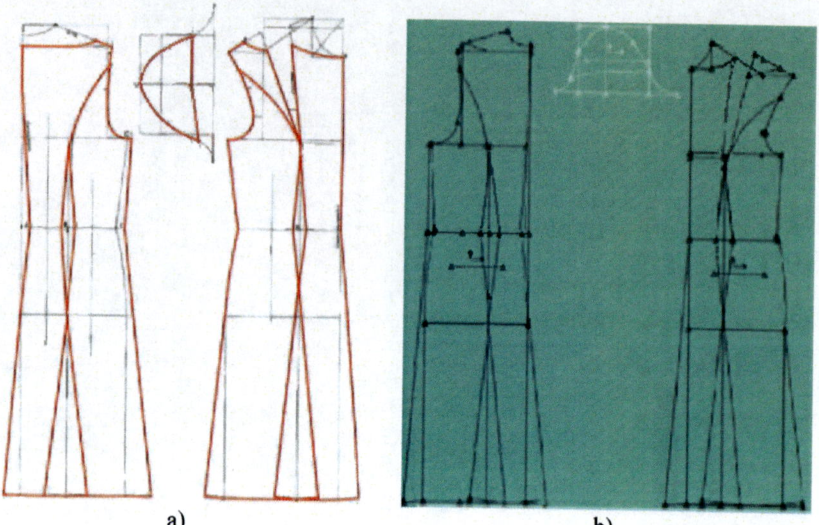

a) b)

Fig. 3. Modeling of women's dress and sleeve - Model 1, a) manual and b) computer modeling

Next, the sleeve line is extended by 1 cm to the right, measured from the lower point of the sleeve curve. This point is connected to the point of the waist, which has been widened due to the narrowing down from the center point to the right by 2 cm. From the center line of the dart at the bottom of the pattern, the dress expands by 5 cm, measured to the left and right, and these points are connected to the bottom point of the dart.

Modeling the front begins by marking a point 5 cm down measured from the top point of the sleeve curve (this point joins the top point of the dart) and 5 cm to the left (this point joins a curved line with a point 5 cm down on the dart line). This modification is transferred to the other line of the dart, where the newly obtained point is connected by a curved line to the central starting point of the neckline. Same as in the back part, indicated above, from the center line of the dart, at the bottom of the pattern, the dress expands by 5 cm, measured to the left and right, and these points are connected to the bottom point of the dart. The upper part of the basic construction of the sleeve pattern is copied. This part is shortened to 5 cm up.

Modeling of the Model 2 (see Fig. 4a - manual and Fig. 4b - computer) was made. The length of this dress is up to the knees. At the back, from the sleeve line, are measured 3 cm from the center point, and this point is marked, and from the same point is marked another, at 4 cm down. In the lower part of the dress, from the line on which the seam is placed, in both directions (both left and right) are measured and marked 12 cm. These points curve at the bottom and connect up to the stitch point (bottom).

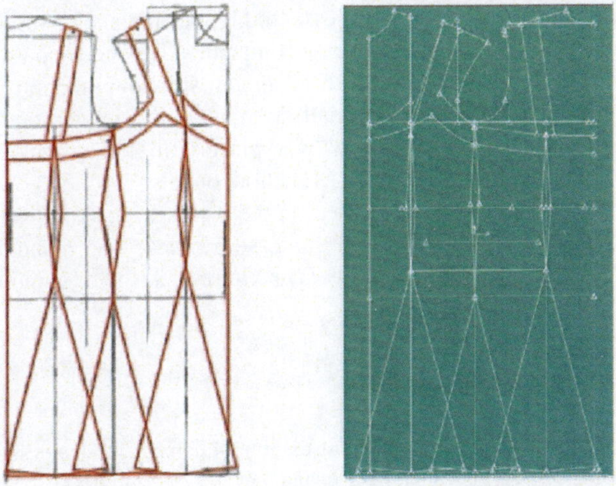

a) b)

Fig. 4. Modeling of women's dress, manual and by computer - Model 2, a) manual and b) computer modeling

In addition, from the lower side point, is measured 10 cm to the right and it is rounded, and in the upper part, it is connected to the side point of the hips. The front part, 1cm is measured from the sleeve line from the center point and marked a point, and another point is marked 4 cm down from the same point. In the lower part of the dress, from the line on which the seam is placed, in both directions (both left and right) are measured and marked points at 12 cm. These points curve at the bottom and connect up to the stitch point (bottom). Then, from the lower side point, 10 cm is measured to the left and it is rounded, and in the upper part, it is connected to the side point of the hips. The width of the strap is 4 cm.

The measured time for creating the patterns modeling for model 1 was 10.5 h (manual) and 2.019 h (by computer) and for model 2 was 9.1 h (manual) and 1.94 h (by computer).

Based on the analysis of the measurements that were monitored and recorded for these two ways of creating pattern designs, it can be stated that the computer way of creation is about 5 times faster compared to the manual creation. Certainly, this depends on the type and complexity of the model that should be created, and if the operator's ability to handle the computer software is greater, the results will be even more satisfactory. The better and more precise visual appearance of the pattern designs was also confirmed.

4 Conclusion

To meet consumer demand for frequent changes in fashion, technology has altered how clothes are created and produced in recent years. Pattern design is the initial stage of garment production, and computer-aided design has completely changed the textile and apparel sectors. Several studies have discovered a strong link between the use of CAD technology and increased sales, high-quality and original designs, employee productivity, cost efficiency, customer responsiveness, production efficiency, short lead times, and the competitiveness of textile manufacturing companies. This study confirmed that CAD systems can help all companies to operate, adopt new solutions, and move forward. The CAD system is more economical, especially in mass industrial garment production as only one worker does all the work at one workplace. Still, the system requires a good knowledge of the program tools and their function, but also a good knowledge of basic manual construction and modeling. Based on the research, the computer construction process takes significantly less time (in this case 5 times) than manual construction, modeling, and grading, so the productivity is maximized. Also, the computer-performed patterns are more accurate and precise.

References

1. Musa, I.A., Agu, M.N.: An assessment of the role of computer-aided design (CAD) in the fashion industry. Int. J. Innov. Sci. Res. Technol. **6**(3), 371–374 (2021)
2. Shazzat, H., Abdus, S., Hasan, A., Jubayer, I., Hazrat, A., Al-Amin, Z. U. B., Mohammad, U., Rezaul, K., Sree, A.A.: An observation toward computer-aided processes in garments production. Comparison and analysis of CAD/CAM software in Bangladesh. Int. J. Textile Sci. **11**(3), 37–46 (2022)
3. Priyanka, S., Neelma, K., Tripti, S.: Comparison of the manual and CAD system for pattern making one time related values for manual and computer aided design. Int. J. Home Sci. **4**(3), 271–273 (2018)
4. Ranjeet, R., Richa, P., Vikrant, K.: Integration of CAD technology for fast and customized production of apparels. Psychol. Educ. **57**(9), 4639–4645 (2020)
5. Yeztova, O.: Computer-aided designing and manufacturing of fashion. In: Goods Innovations in Science: The Challenge of Our Time, 2018. Accent Graphics Communication & Publishing, Ontario, Canada (2018)
6. Kaystha, S.: Use of computer technology in apparel sector. Int. J. Appl. Res. **3**(5), 362–365 (2017)
7. Kılıç, A.Ş.: Comparison of cad and manual system efficiency in pre-production preparation process. Tekstil ve Konfeksiyon **29**(1), 67–77 (2019)
8. Avijeet, D., Mutasim, U., Mahbubur, R., Mazharul, H.: The benefits of using computerized system over manual system in apparel production. GUB J. Sci. Eng. **8**(1), 36–39 (2021)

9. Sarita, C., Pardeep, K., Prashant, J.: Maximizing performance of apparel manufacturing industry through CAD adoption. Int. J. Eng. Bus. Manage. **12**, 1–12 (2020)
10. Kazlacheva, Z.: Fibonacci tilings in fashion design. Ann. Univ. Oradea Fascicle of Textiles-Leatherwork Oradea **17**(1), 77–82 (2016)
11. Indrie, L., Kazlacheva, Z., Ilieva, J., Gherghel, S.: Embroidery-from digital designing to fine art. Industria Textilă **68**(5), 366–369 (2017)
12. Abhishek, D., Avanish, D.: Role of computer and automation in design and manufacturing for mechanical and textile industries: CAD/CAM. Int. J. Innov. Technol. Explor. Eng. (IJITEE) **3**(3), 174–181 (2013)
13. Öndoğan, Z., Erdoğan, Ç.M.: The comparison of the manual and CAD systems for pattern making, grading and marker making processes. Fibres Textiles East. Eur. **14**(1), 62–67 (2006)
14. Özkan, M.: Türkiye'de Hazır Giyim Sektöründe Kullanılan Bilgisayar Destekli Kalıp Hazırlama Sistemlerinden Assyst, Konsancad ve Lectra, Sistemlerinin Karşılaştırılması, Selçuk Üniversitesi Sosyal Bilimler Enstitüsü Yüksek Lisans Tezi, Konya, 93 (2006)
15. Puri, A.: Efficacy of pattern making software in product development. Int. J. Adv. Qual. Manage. **1**(1), 21–39 (2013)
16. Şen Kılıç, A., Yücel, Ö.: Konfeksiyon Üretiminde Konvansiyonel ve Bilgisayar Destekli Kalıp Hazırlama Yöntemlerinin Birim Maliyetler Açısından Karşılaştırılması, Konfeksiyon Teknik 62–66 (2011)
17. Göksel, N., Utkun, E., Öndoğan, Z.: Use of computer-aided pattern preparation method in clothing design. Tekstil ve Konfeksiyon **22**(4), 375–382 (2012)
18. Jankoska, M., Petreska, E.: Computer construction, modeling and fabrication of corset. Tekstilna industrija **67**(4), 11–19 (2019)
19. Jankoska, M., Petreska, E.: Application of CAD system for ladies' dress design with draperies. In: 11 International Scientific-Professional Conference, pp. 89–94, Textile Science and Economy XI, Zrenjanin, Serbia (2020)
20. Jankoska, M.: Design and Modeling of Lady's Dress, Contemporary Trends and Innovations in the Textile Industry. In: 3rd International Scientific Conference, pp. 331–339, CT&ITI Beograd, Serbia, September 17–18 (2020)
21. Jankoska, M.: 3D virtual fitting system on woman's garment. Tekstilna industrija **69**(4), 71–79 (2021)
22. Jankoska, M.: Application CAD methods in 3D clothing design. Tekstilna industrija **68**(4), 31–37 (2020)
23. Cortosheva, S., Jankoska, M.: Construction of women's cloth, 1st edn. Ars Lamina (2021)
24. Winifred, A.: Metric Pattern Cutting for Women's Wear, 6th edn. Wiley (2015)
25. Alison, S.: Dressmaking—The Complete Step-by-Step Guide to Making Your Own Clothes, London, New York, Melbourne, Munich and Delhi (2012)

Contamination Threshold Values for Textile Recycling

Tanita Behrendt[1,2](✉) and Elisabeth Eppinger[1]

[1] School of Culture and Design, Garment Technology, University of Applied Sciences for Technology and Economics Berlin, 12459 Berlin, Germany
{Tanita.Behrendt,Elisabeth.Eppinger}@htw-berlin.de
[2] Faculty V- Mechanical Engineering and Transport Systems, Technische Universität Berlin, 10623 Berlin, Germany

Abstract. Post-consumer textile recycling promises to reduce the amount of textile waste deposed at landfills and incinerated as thermal recycling. It can close loops within the apparel industry and reduce the amount of land and water usage for the production of virgin fibers for new textiles and garments. Despite the advantages of textile recycling, currently only a small percentage of post-consumer textile waste is recycled worldwide. This figure is even less for fiber-to-fiber recycling for new apparel. Multiple recycling technologies are available today, some established, others still at pilot scale, with different requirements about input-materials and acceptable threshold values for contaminants. The following paper examines typical contaminants found in garments and their threshold values for different recycling technologies.

Keywords: Textile recycling · Circular textiles · Sustainable fashion

1 Introduction

Since the development of chemical fibers, the amount of fibers produced worldwide is steeply increasing. In the year 2021, a total of 113 million tons of fibers were harvested or manufactured, estimated to rise up to 149 million tons by the year 2030. Polyester fibers account for about 50% and cotton fibers for about 25% of all fibers produced [1]. As fiber production increases, by default, so does resource extraction for new fibers and the amount of textile waste being generated, making the recycling of used garments more important than ever [2]. To date, only about 33–38% of textiles sold in the European market get collected separately, resulting in about 62–67% of textiles destined to be landfilled or incinerated without prior inspection regarding reusability. This has a number of negative impacts, e.g. the release of greenhouse gasses and the potential release of microplastics or other harmful substances into the environment. [3, 4] The European Parliament even explicates that about 87% of textile waste is being landfilled or incinerated with and without energy recovery within the European Union [5].

With a number of different recycling technologies available and steadily advancing, recycled fibers could soon start to replace virgin fibers in the production of new garments

G. Guxho et al. (Eds.): ITC-ICEE 2023, LNMIE, pp. 468–479, 2024.
https://doi.org/10.1007/978-3-031-48933-4_46

and thus reduce the amount of resources needed for virgin fiber production as well as reducing the negative environmental impacts of both virgin fiber production and disposal of post-consumer textile waste [1, 6]. What sounds theoretically easy proves to be difficult in practice: textile waste is a heterogenous waste stream consisting of many different materials. It includes natural and synthetic fiber-blends, coatings and colorings, as well as haberdashery like buttons or zippers of different materials, such as plastics, metals, wood, and mother-of-pearl. Every added material is a potential contaminant and can disrupt the recycling processes, thus reducing their yield and economic feasibility. This challenge contributes to the issue that less than 1% of the worldwide textile waste was recycled back into fibers in the year 2015 [7]. Even with a recent increase of recycling activities for the textile industry in the year 2021 with about 8.5% of fibers produced from recycled materials, around 7.9% of them were polyester fibers originating from used PET bottles, leaving the amount of fiber-to-fiber recycling at 0.6% [1].

Material-based sorting of post-consumer garments, in addition to clear requirements regarding contamination threshold values, is one of the key factors for establishing fiber-to-fiber recycling. Automating and digitalizing the sorting process could enhance the quality of the textile waste stream, making it suitable as input for various recycling technologies and thus generating a broad variety of high-quality recycling materials suitable for a number of applications in different industries. As every recycling technology allows different fibers or fiber-blends and contaminant threshold values in their processes, a clear overview of those thresholds is necessary for efficient automated sorting, allowing matching of garment and recycling technology. It is also significant for eco-design directives for textiles and environmental labeling of apparel.

To provide an overview of current threshold values for different contaminants, we per-formed a literature analysis of academic, peer-reviewed articles. In the second part of the paper, we explain where fiber-to-fiber recycling is situated in the current lifecycle of a garment and provide an overview of the different fiber recycling technologies. In Sect. 3, we describe our method and Sect. 4 shows the results of different contaminants, their threshold levels and whether they can be removed by different pre-treatments before the recycling process. We conclude with a brief discussion and outlook.

2 Post-consumer Textile Waste Recycling

According to the EU Waste Framework Directive, recycling is better than disposal but less preferred to preparation for reuse [8]. Hence, in theory, recycling should only be done for textiles that cannot be repaired and cleaned for reuse. In practice, garments get sorted for resale but repairing of disposed garments is usually not profitable in European countries.

Figure 1 shows the lifecycle of garments and where fiber-to-fiber recycling is situated. A high percentage of recycled fibers recovered of the current post-consumer garments are either mechanically or chemically recycled. It is clearly shown how fiber-to-fiber recycling can replace virgin fibers in the production of new garments and thus close loops within the textile industry.

Besides the challenges of the fiber-to-fiber textile recycling itself, the collecting and sorting of textile waste streams is a challenge on its own. Used textiles are often collected

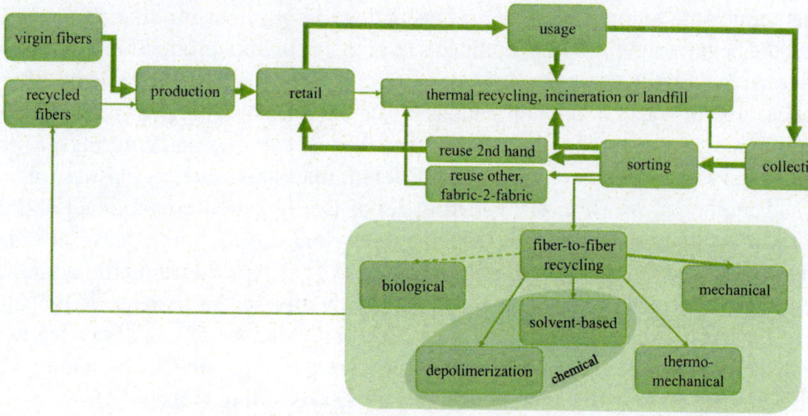

Fig. 1. Fiber-to-fiber recycling as part of the lifecycle of garments

through unsupervised drop-off containers, resulting in consumers frequently getting rid of other waste as well, which can contaminate the textiles, making them unsuitable for reuse or recycling.

After the collection, the textile waste has to be sorted for reusability, such as secondhand usage or usage in other applications like cleaning rags, as defined by the waste hierarchy in the EU Waste Framework Directive 2008/98/EC [8]. This process is mostly done by hand as only trained personal is able to determine the state as well as the fashionability of the garments. During this process, the exact material compositions of the garments are not determined as it is not relevant for the second-hand usage. Trained sorters are able to identify materials usable for e.g., cleaning cloths by sight and touch alone, but for fiber-to-fiber recycling, knowledge of the exact material composition is needed. This information cannot be obtained by checking the care labels of the garments, as this process would take too much time and thus be economically unfeasible, but also because the care instructions fade during laundry and often get removed by the consumers to increase the comfortability of the garments during wear [9]. Aside from that, the EU directive on textile labeling does not require the manufacturers to list the compositions of haberdashery or other possible contaminants, as well as allows 5% in weight of fibers to remain undisclosed, thus giving incomplete and unreliable information needed for high-quality fiber-to-fiber recycling [10]. In addition to that, a recent study has found that the material compositions printed on the care labels are often not accurate or outright false [9].

To reduce the number of textiles going to landfills or incineration and increase the amount of textiles getting recycled, fiber-to-fiber recycling is the next big step toward sustainability and circular economy within the textile industry. There are several recycling technologies available, some established and others still in pilot or lab scale.

The oldest, most common and broadly established textile recycling technology is **mechanical recycling** [11]. During this recycling process, the garments will be cut and disentangled through several carding processes until only the loose fibers remain. Because of the mechanical stress the fibers are exposed to in this process, they will get shortened and lose some of their tensile strength and abrasion resistance, thus reducing their suitability for producing new yarns used for the production of high-quality garments [11, 12]. To reuse the fibers for the production of new garments, virgin fibers are blended in during yarn spinning to balance out the reduced performance of the recycled fibers. Depending on the application, 10–80% of virgin fibers will be blended with the recycled fibers. A large amount of fibers yielded from mechanical recycling are used for e.g., nonwovens and as insulation material. [7, 11–16] Post-consumer garments pre-sorted by color for mechanically recycling have the advantage that bleaching and re-dying can be skipped, thus making the process more environmentally friendly [7, 14].

In terms of amount of possible fibers as input, the second most important recycling technology that is widely applied in the packaging and plastics industry is **thermo-mechanical recycling**. This process is only suitable for thermoplastic materials like polyester and needs pure input streams. Shredded textiles are heated in a screw extruder until melted. The polymer melt is pressed through a perforated plate to produce a strand that is cut into pellets. Those pellets can be remolten and formed into the desired shape for a variety of applications. [17] Due to the low tolerance of contaminants, this recycling process is only applicable for a small percentage of textile waste and requires very thorough sorting of the input material, thus making it economically unfeasible for broad application in the textile industry to date. [15, 18] Post-production waste like cutting waste is best suited as input for this recycling technology as exact material compositions including dyes are still known and no haberdashery has been added yet [14, 19].

Two different recycling technologies are usually summarized under the term **chemical recycling**. As those two technologies follow two different approaches, with one dissolving the materials into their polymers and the other depolymerizing the polymers into monomers, those two technologies should be looked at separately.

Several different processes can be summarized under the term **solvent-based recycling**, also known as polymer recycling. One or several components within the textiles will be dissolved with selective solvents, leaving the other materials intact. The solved components can later be regenerated or disposed of. These recycling technologies are applicable for garments made from a variety of fiber-blends, especially polyester-cotton blends. Cotton can be processed into man-made cellulosic fibers, such as viscose. [14, 15, 20, 21] As in mechanical recycling, virgin cellulosic materials get mixed into the process to improve the performance of the fibers [13].

Similar to solvent-based recycling, the **depolymerization**, also known as monomer recycling or simply chemical recycling, allows the recycling of fabrics made from fiber-blends. During depolymerization processes, the fibers get deconstructed into oligomers or monomers and after cleaning, can be repolymerized and used for the production of either fibers or products for other applications. During the depolymerization step, certain contaminants can get removed, resulting in high-quality monomers. [7, 21, 22] Monomer recycling processes can only be applied to materials made from man-made polymers, like polyester or polyamide, as a repolymerization of natural polymers, like cellulose found in cotton, is not possible. The chemical recycling of cellulosic fibers would thus result in an open-loop recycling, as the cellulose can get depolymerized into e.g., glucose and then be used for other applications like the fermentation into biofuels [15].

Lastly, **biological recycling** technologies are to be named. The input materials will be treated with certain enzymes or microorganisms to deconstruct the input materials into a variety of different materials, e.g. glucose or terephthalic acid [11, 23]. While mostly still in lab-scale, these technologies are promising for ecologically and economically feasible textile recycling [11]. In the case of natural polymers, as with monomer recycling processes, only open-loop recycling is possible. Biological recycling processes often need a pre-treatment step to make the polymers of the fibers accessible for the enzymes or microorganisms and thus enhance the yield. Pre-treatment with ionic liquids, acids and alkaline is possible [13].

To apply the recycling processes with the highest efficiency and economic feasibility for post-consumer textile recycling, high-quality input streams with next to no contaminants would be ideal. As this is not attainable with used garments as input, it is crucial to determine which contaminants, as well as which threshold values are acceptable in the recycling processes for generating high-quality output while simultaneously making the processes economically profitable without harming the environment.

3 Method

To determine the contaminant threshold values for the different textile recycling technologies, a literature analysis has been conducted. The literature should have been published in the last ten years, starting from 2013 to 2023. Basic research literature should have been published in the past 30 years. The literature is preferably written in English, German sources are to be used as well. Literature analysis is conducted with the use of scientific literature databases, such as sciencedirect and wiley online databases. Several key-words have been utilized for finding relevant literature, two of them German. Different combinations of the keywords where used. Utilized keywords are: "recycling", "textile recycling", "contaminant*", "threshold*", "solvent-based", "chemical recycling", "mechanical recycling", "PET", "haberdashery", "plastic*", "Kunststoff*", "Textilrecycling". A total of 13 research papers, five reports and eight books have been selected. Additionally, to complete the list of materials used for the production of haberdashery in Sect. 4, four product catalogs have been used.

4 Results: Contaminants and Pre-treatments to Remove Them

During the production of yarns, fabrics and later garments, a variety of additives, finishing agents and haberdashery is added to produce garments that are functional as well as fashionable. While those additives are necessary to the production processes and most of the haberdashery is necessary to provide wearability, longevity and comfort of the garments, they are contaminants for later recycling processes. Figure 2 depicts typical contaminant-groups added in the various stages of the textile and garment production. The results show a huge variety of contaminants of different materials. Some of them can be washed out or need to be cut out. Others are integrated in the fibers and require chemical treatment.

Common contaminants are zippers, buttons and prints, as well as other haberdashery. Due to the broad range of materials that can be used to produce haberdashery and an equally large range of colorants and dye-stuffs utilized for dying and printing of fabrics, as well as non-visible additives or finishings, a closer examination of those materials

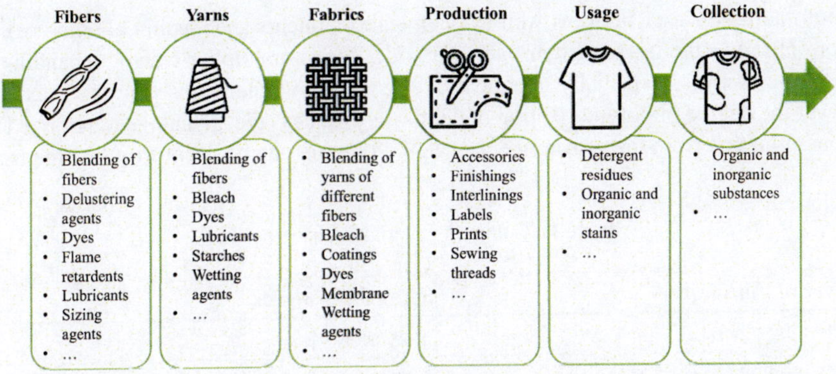

Fig. 2. Possible contaminants added within the different production stages of garments

is necessary. In this section we take a closer look at common materials used for the production of haberdashery.

A zipper alone can be composed of up to nine components. While most of these components can be manufactured from the same material, some might be made from different materials to enhance the durability or to add a fashionable element to the closure. There are several different types of zippers, mainly distinguished by their teeth [24]. The zipper teeth usually consist of different metals or various plastics. The zipper tape is generally made from cotton, polyester, nylon or vinyl, depending on the application [25]. For buttons, they are mainly made from one material, though it is common to have buttons made from two or more materials as well, resulting in a wide variety of material combinations [25]. For multi-layered garments, there is also the issue of added linings, interlinings, paddings and waddings [24]. As it is not required to specify the materials of the used haberdashery on the care labels, the materials are often not disclosed by the manufacturers. Table 1 gives an overview of the most typical haberdashery and the materials they are commonly composed of.

Between the collection and sorting of post-consumer garments and potential fiber-to-fiber recycling, a variety of pre-treatment steps can be conducted. Washing and disinfecting is the most popular pre-treatment step and is used to remove contaminants like organic and inorganic stains from the use-phase of the garments, as well as possible contamination attached to the garments during the collection [16]. To prepare the post-consumer garments for further pre-treatment and recycling steps, size reduction has to be conducted [11, 19]. The garments get either cut into uniformized squares of variable sizes, shredded or ground up to ensure that even contaminants or chemical structures within the fibers themselves can be reached by e.g., solvents [15]. During this step, haberdashery can be removed by e.g. magnets, centrifugal force or air separation [31]. Haberdashery like labels, buttons, zippers, leather patches, but also prints, embroideries, sequins, and beads could also be cut out manually before size reduction [9, 22]. Dyed cotton fabrics can be decolorized with NaOH treatment, though depending on the dyes utilized, the processing times could reach up to 30 h, making the process not feasible for industrial application. Furthermore, dyes on cotton can be removed by hydrolysis, photofading or laser light, as well as chemical treatments such as hot alkaline extraction (HCE) or the usage of hydrogen peroxide, ozone or sulphuric acid, depending on the utilized dyes. Especially chemical dye-removal treatments damage the fibers and lower the degree of polymerization of cellulose. HCE has the advantage to remove PET contamination from the cotton fibers as well. Polyester can be decolorized by the usage

Table 1. Common materials of haberdashery

Type of haberdashery	Common materials
Clothing label	Polyester [9]
Sewing thread	Cotton, elastane, polyamide, polyester, Polytrimethylenterephthalat (PTT), silk [26]
Fastener	
Buttons	
ABS buttons	Acrylonitrile butadiene styrene (ABS) with metal coating [25]
Fabric buttons	Common textile materials, metals and plastics [27]
Metal buttons	Aluminum, brass, nickel, steel [24, 25]
Natural buttons	Coconut, cork, horn, ivory, leather, mother of pearl, tagua nut/vegetable glass, wood [24, 25, 27]
Plastic buttons	Melamine, nylon, polyacrylic, polyamide, polyester, vulcanized fibers [24–27]
Snap buttons (metal and plastic)	Brass (plated, enameled), polyacetal, steel [25]
Zipper	
Coil zipper (also invisible zippers)	Polyester (formerly nylon) [24]
Metallic zippers	Aluminum, brass, nickel [24]
Plastic-molded zippers	Polyacetal resin, other thermoplastics like polyethylene, polyoxymethylene copolymer (Co-POM) [24, 28]
Reverse coil zippers	Like coil zippers with added PVC coating for waterproof finishing and metal pullers [24]
Others	
Elastics	Elastomeric fibers, latex rubber, polyamide, polyester [26]
Hook and eye/bar fasteners	Brass (plated, plastic coated, knit covered), steel [25, 26]
Hook-and-loop tapes (velcro)	Polyamide, polyester, propylene [25, 26]
Toggle	Horn, polyester, wood [27]

(*continued*)

Table 1. (*continued*)

Type of haberdashery	Common materials
Adornments	
Made from textile materials	
Cords, braids, soutache, passementerie	Common textile materials, metallic threads [25, 26]
Embroidery	Common textile materials, ceramic threads, metallic threads, silver threads [26]
Lace	Cotton, elastane, metallic treads, polyacrylic, polyamide, polyester [26]
Patches	Coated cellulosic materials (vegan leather), leather, and textiles [29]
Ribbons, tassel	Cotton, jute, nylon, silk, polyester, polypropylene [24]
Tape, piping	Common textile materials [26]
Made from other materials	
Beads, pearls	Glass, mother of pearl, plastics, wood [30]
Buckles and sliders	ABS, brass (plated), iron (plated), plastics [25, 26]
Eyelets, rivets	Brass (plated, plastic coated) [25]
Sequins	Polyethylene, polyester, polyvinylchloride (PVC) [30]

of certain solvents, ozone and hydrogen peroxide. Good results have been obtained by the utilization of the solvent dimethyl formamide (DMF). [14] An additional sorting of the garments by color can be done before starting any pre-treatment steps and thus potentially omitting the need for decolorization processes as well as re-dying [7, 14]. Additionally, certain selective solvents can be utilized to remove impurities like elastane from the garments [32, 33].

5 Threshold Values and Options to Remove Them in Pre-Treatment

The different recycling technologies allow for different contamination thresholds. In the following section, the recycling technologies and their respective thresholds will be examined.

The **mechanical recycling** has the least restrictions regarding possible textile inputs. As the technology works by disentangling the fibers, garments with a coating cannot be placed into the process. Moreover, fabrics with a high content of elastane fibers as well as filament fibers can be problematic as the fibers may wind around the moving parts of the machinery and thus create disruptions of the process or cannot be properly disentangled because of their length. [12] Atop of those restraints, some other factors regarding the desired output have to be considered. If the fibers are to be re-spun into yarns used for the production of new garments, regulatory requirements like REACH have to be taken into account regarding the chemical composition of the garments [7]. If certain properties or fibers are to be generated, the textile input has to be presorted by fiber-type and/or color.

The **thermo-mechanical recycling** has tight restrictions regarding possible contaminants. Only pure, preferably colorless input can be recycled with this technology [18, 34]. Colorants, additives and other particles can disrupt the process, resulting in a variety of errors, from damaged machinery to reduced fiber properties [15, 23]. This results in only a small quantity of garments being suitable as input. Most of the recycled polyester yarns and fabrics available on the market today are made from PET bottles used as input in thermo-mechanical recycling [1, 35]. As the molecular weight of the polymers can reduce during remelting, especially if water is present and induces hydrolyses processes, the conditions have to be carefully controlled in order to produce output suitable for the desired application of the polymers. Additionally, the quality of the polymers deteriorates during the use-phase as well as during the recycling process, resulting in lower quality polymers. [18, 34, 36] This loss in quality, namely the molecular weight, could be reversed with e.g., solid state post-condensation [37]. As this recycling technology is already applied broadly with PET bottles as input, contaminants usually found in PET bottle waste streams have been identified. The main contaminants found in this waste stream are PVC, PP, PE, PA6, PS, paper and glues. Especially the PVC and paper content should be minimized as far as possible. Additional melt-filtration with 10–30 μm filters should be conducted to remove further contaminants before spinning. [38].

For **solvent-based recycling** processes, the contaminants acceptable depend on the utilized solvents as well as which fibers are to be solved and which are to be retained. The input stream does not have to be of as high purity as for the thermo-mechanical recycling, but depending on the solvents used, certain materials might contaminate the solution or work as inhibitors and thus reduce the yield. [15, 20, 21] As mentioned before, solvent-based recycling technologies are suitable for the recycling of fiber-blends, the utilized solvent thus depends on which fiber type is to be solved and which to be retained as a whole [33]. Typically, the fiber type that is to be utilized for further usage is the one being retained. It is, however, possible to reuse both the solved and retained fibers separately, e.g. in polycotton: retain the cotton fibers and solve the polyester fibers to be repolymerized later on [14, 15, 33]. Especially dye-stuff and additives can disrupt solvent-based recycling processes, as well as metals and other haberdashery [11]. It is possible to remove dyes with selective solvents as a pre-treatment step before solving the fiber components of the input stream [33].

In **monomer recycling** technologies, just like in the formerly named solvent-based processes, the type and amount of acceptable contaminants depends on the type of depolymerization process applied. Typically, the depolymerization is conducted via hydrolysis, glycolysis, methanolysis or alcoholysis. Depending on the utilized process, other fibers and contaminants up to 30% are acceptable [39]. Heavy metals or water contamination of the textile waste can inhibit the depolymerization processes and thus should be prevented [23]. Methanolysis is generally not as sensitive to impurities as e.g. glycolysis and is therefore suitable for input streams with higher impurities [18,

40]. The PVC-content of the input material has to be between 2 and 10 ppm, preferably lower to prevent brittleness of the material and secession of HCl. The threshold for PE is <10 ppm. [17].

As the **biological recycling** processes are still being researched on lab-scale, no contaminant thresholds have been published yet. Biological processes are selectively degrading certain materials in the input stream, thus the recycling of fiber-blends is possible. Due to the fact that in some cases live organisms are utilized for the degradation, it is to be expected that certain chemicals and dye-stuffs can act as inhibitors to the bacterial or enzymatic processes. Some depolymerization products (e.g., oligomers) might also inhibit the activity of certain microbacteria or can be consumed by them, resulting in lower yields [36].

6 Conclusion and Outlook

Contaminants can be differentiated into two main categories: functional and fashionable. The functional contaminants are crucial for either smooth-running production processes and thus higher quality or higher output, or for enhanced comfort and function during the use-phase of the garments. Which of the functional contaminants are necessary or just convenient to increase output, and which materials they should consists of, is debatable. The fashionable contaminants fulfil cultural and sociological requirements, as they help the wearer of the garment to express themself, resulting in a necessity from a very different standpoint. Furthermore, some of the functional haberdashery, such as the design decision to use buttons, zippers or elastics have an overlap of functional and fashionable requirements. Again, the materials and the processes to obtain a certain look vary. Both contaminant categories are non-negotiable in their necessity and thus, they cannot be omitted or their usage forbidden. However, the materials and the amount and level of integration in garments requires an in-depth analysis to advice on eco-design guidelines to enhance the recyclability of garments.

For the development of new approaches and solutions for textile recycling, it is crucial to know which contaminants can be found on garments and what materials they are made of. While we provided some insights into the contaminants and their threshold value for different recycling methods, the academic literature does not provide extensive information to sufficiently advice on eco-design. While all of the presented recycling technologies are subject to further research and development, as well as new pre-treatments are innovated, the insights should be complemented with a patent analysis and an interview study to explicate further threshold values.

Generally speaking, the higher the purity and quality of input streams for recycling technologies, the higher the ecological and economical feasibility. As post-consumer textiles are a highly heterogeneous waste stream with many possible contaminants in a variety of different combinations, especially the presorting of those waste streams is key for a high-quality fiber-to-fiber recycling in the future.

References

1. Textile Exchange (ed.): Preferred Fiber and Materials Market Report (2022)

2. Eurostat, Generation of waste by waste category, hazardousness and NACE Rev. 2 activity. [Online]. Available: https://ec.europa.eu/eurostat/databrowser/view/ENV_WASGEN__custom_6057316/bookmark/table?lang=en&bookmarkId=b07664df-1c64-46d9-aa1a-f7c8761355e7. Last accessed 14 Jul 2023
3. EURATEX (ed.): Facts and Key Figures 2022: Of the European Textile and Clothing Industry (2022)
4. Boiten, V.: Building a Circular Economy for Textiles Supported by Common Rules on Extended Producer Responsibility (EPR) in the EU: Recommendations and Open Questions for the Upcoming Revision of the EU Waste Framework Directive (WFD) (2022)
5. European Parliament, Ed., "The impact of textile production and waste on the environment (infographics)," (Jun. 2023). [Online]. Available: https://www.europarl.europa.eu/news/en/headlines/society/20201208STO93327/the-impact-of-textile-production-and-waste-on-the-environment-infographics. Last accessed 14 Jul 2023
6. Kasavan, S., Yusoff, S., Guan, N.C., Zaman, N.S.K., Fakri, M.F.R.: Global trends of textile waste research from 2005 to 2020 using bibliometric analysis. Environ. Sci. Pollut. Res. Int. **28**(33), 44780–44794 (2021). https://doi.org/10.1007/s11356-021-15303-5
7. Ellen MacArthur Foundation: A New Textiles Economy: Redesigning fashion's future (2017). [Online]. Available: http://www.ellenmacarthurfoundation.org/publications. Last accessed 14 Jul 2023
8. Directive 2008/98/EC of the European Parliament and of the Council of 19 November 2008 on waste and repealing certain Directives. In Official Journal of the European Union, European Parliament and Council of the European Union (eds.) (2008)
9. Wilting, J., van Duijn, H.: Clothing Labels- Accurate or Not. Amsterdam (2019)
10. European Parliament, Council of the European Union.: Regulation (EU) No 1007/2011 of the European Parliament and of the Council of 27 September 2011 on textile fibre names and related labelling and marking of the fibre composition of textile products and repealing Council Directive 73/44/EEC and Directives 96/73/EC and 2008/121/EC of the European Parliament and of the Council Text with EEA relevance," in Official Journal of the European Union, European Parliament and Council of the European Un-ion (eds.) (2011)
11. Ribul, M., Lanot, A., Tommencioni Pisapia, C., Purnell, P., McQueen-Mason, S. J., Baurley, S.: Mechanical, chemical, biological: Moving towards closed-loop bio-based re-cycling in a circular economy of sustainable textiles. J. Clean. Prod., **326** (2021). https://doi.org/10.1016/j.jclepro.2021.129325
12. Dönnebrink, H.: Die Sammlung und Verwertung von Alttextilien: Eine empirische Analyse vor dem Hintergrund des Kreislaufwirtschafts- und Abfallgesetzes. Münster: Forschungsstelle für Textilwirtschaft (1998)
13. Damayanti, D., Wulandari, L. A., Bagaskoro, A., Rianjanu, A., Wu, H.-S.: Possibility routes for textile recycling technology. Polymers **13**(21) (2021). https://doi.org/10.3390/polym13213834
14. Wang, S., Salmon, S.: Progress toward circularity of polyester and cotton textiles. Sustain. Chem. **3**(3), 376–403 (2022). https://doi.org/10.3390/suschem3030024
15. Pensupa, N.: Recycling of end-of-life clothes. In: Sustainable Technologies for Fashion and Textiles, Nayak, R. (ed.), pp. 251–309. Elsevier (2020)
16. Gries, T., Veit, D., Wulfhorst, B.: Textile Fertigungsverfahren. Carl Hanser Verlag GmbH & Co. KG, München (2018)
17. Martens, H., Goldmann, D.: Recyclingtechnik: Springer Fachmedien Wiesbaden (2016)
18. Park, S.H., Kim, S.H.: Poly (ethylene terephthalate) recycling for high value added textiles. Fashion Text., **1**(1) (2014). https://doi.org/10.1186/s40691-014-0001-x
19. Harmsen, P., Scheffer, M., Bos, H.: Textiles for circular fashion: the logic behind recycling options. Sustainability **13** (2021). https://doi.org/10.3390/su13179714

20. Palme, A., Peterson, A., de La Motte, H., Theliander, H., Brelid, H.: Development of an efficient route for combined recycling of PET and cotton from mixed fabrics. Text. Cloth. Sustain., **3**(1) (2017). https://doi.org/10.1186/s40689-017-0026-9
21. Schlummer, M., Fell, T., Mäurer, A., Altnau, G.: Die Rolle der Chemie beim Recycling. Kunststoffe, no. 6 (2020)
22. Roos, S., Sandin, G., Peters, G., Spak, B., Schwarz Bour, L., Perzon, E., Jönsson, C.: White paper on textile recycling. Mölndal (2019)
23. Lee, A., Liew, M.S.: Tertiary recycling of plastics waste: an analysis of feedstock, chemical and biological degradation methods. J. Mater. Cycles Waste Manag. **23**(1), 32–43 (2021). https://doi.org/10.1007/s10163-020-01106-2
24. Chatterjee, K.N., Jhanji, Y., Grover, T., Bansal, N., Bhattacharyya, S.: Selecting garment accessories, trims, and closures. In: Garment Manufacturing Technology, Nayak, R., Padhye, R. (eds.), pp. 129–184. Elsevier (2015)
25. Mitchell, J., McLoughlin, J.: Joining fabrics: fastenings. In: Textiles and Fashion: Materials, Design and Technology, Sinclair, R. (ed.), pp. 413–458. Woodhead Publishing Limited (2015)
26. knauf textil GmbH & Co. KG: Kurzwaren 2023. Esslingen (2023)
27. Union Knopf GmbH: Knöpfe, Boutons, Buttons. No. 04 (unknown)
28. Zhang, Y., Wu, X.Y., Wang, L.L., Ding, X.M.: The industrial water footprint of zippers. Water Sci. Technol.: J. Int. Assoc. Water Pollut. Res. **70**(6), 1025–1031 (2014). https://doi.org/10.2166/wst.2014.323
29. Prym (ed.): Patches and Flickstoffe (2019)
30. knauf textil GmbH & Co. KG: Pailletten- und Strassborten Sortiment (unknown)
31. Barthl, A.: Wissenschaftliche Untersuchungen zur stofflichen Verwertung der textilen Restfraktion durch mechanische Verfahrensschritte und Entwicklung eines geschlossenen Gesamtkreislaufes. Vienna (2010)
32. Muthu, S.S.: Sustainable Innovations in Recycled Textiles. Singapore, Springer Singapore (2018)
33. Sherwood, J.: Closed-loop recycling of polymers using solvents: remaking plastics for a circular economy. Johnson Matthey Technol. Rev. **64**(1), 4–15 (2020). https://doi.org/10.1595/205651319X15574756736831
34. Mwanza, B.G.: Introduction to recycling. In: Siengchin, S., Rajkumar, A. G., Rangappa, S. M., and Parameswaranpillai, J. (eds.) Recent Developments in Plastic Recycling, pp. 1–13. Springer (2021)
35. Sarioğlu, E., Kaynak, H.K.: PET bottle recycling for sustainable textiles. In: Camlibel, N.O. (ed.) Pol-yester—Production, Characterization and Innovative Applications. InTech (2018)
36. Chanda, M.: Chemical aspects of polymer recycling. Adv. Indus. Eng. Polym. Res. **4**(3), 133–150 (2021). https://doi.org/10.1016/j.aiepr.2021.06.002
37. Gantillon, B., Spitz, R., McKenna, T.F.: The solid state postcondensation of PET, 1. Macromol. Mater. Eng. **289**(1), 88–105 (2004). https://doi.org/10.1002/mame.200300289
38. Thiele, U., Gneuss, S.: High-quality materials from PET recycling. Kunststoffe Plast Europe., no. 95 (2005)
39. Greenblue, Ewell, J., Ritch, E., Elsea, A., Dreizen, C., Brister, J.: Chemical Recycling: Making Fiber to Fiber Recycling a Reality for Polyester Textiles (2018)
40. Pudack, C., Stepanski, M., Fässler, P.: PET recycling—contributions of crystallization to sustainability. Chem. Ing. Tec. **92**(4), 452–458 (2020). https://doi.org/10.1002/cite.201900085

The Effect of Aluminum Mordant on the Colour Fastness of Fabrics Made of Cellulose Fibres When Dyeing with Onion Peels

Diana Tuulik(✉) and Merje Beilmann

TTK University of Applied Sciences, Pärnu mnt 62, 10135 Tallinn, Estonia
diana@tktk.ee

Abstract. Natural dyes have become an alternative option for textile dyeing. In the present study, the extract of on-ion (Allium cepa) peels, as food industry waste, was investigated as a natural textile dye. The aim of the study is to compare the colour fastness indicators of cellulosic fabrics dyed with the extract of onion peels, pre-treated with the alum mordant and without pre-treatment. Three different fabrics have been chosen for the study: 100% cotton, 100% linen, and cotton/linen blend. All researched fabrics are dyed in the same conditions. There are two different samples of each fabric. One sample of fabric is dyed, and another is pre-treated with alum mordant before dyeing. Different colour fastness tests were performed on all samples: the colour fastness to rubbing, perspiration, washing and daylight. All tests were performed and evaluated according to ISO standards.

Keywords: Natural textile dyeing · Colour fastness properties · Alum mordant · Textile testing

1 Introduction

The textile sector is one of the most polluting industries in the world, contributing 20% to total industrial pollution of the water resources. In total, 5 billion kg of dyes, pigments and finishing chemicals are currently in use in the textile industry, adding up to more than 8000 different chemicals utilized for garment production. At the end of their life, textiles still contain up to 90% of the chemicals that were initially introduced during manufacturing or finishing. This high level of chemical residues in textiles (only partially washed out during the laundry process) is not only problematic in terms of resource use, but also poses an environmental risk [1].

As people are more conscious of issues related to health and the environment, more and more textile industries are looking for possibilities of using natural dyes and alternative dyeing methods [2, 3].

Sustainable chemistry has become an emerging concept in response to various sustainability challenges associated with the production, distribution, use and discharge of chemicals and products. [1].

G. Guxho et al. (Eds.): ITC-ICEE 2023, LNMIE, pp. 480–490, 2024.
https://doi.org/10.1007/978-3-031-48933-4_47

With increasing awareness for eco-friendly materials from sustainable resources, natural dyes attracted research in traditional and diversified applications to develop effective eco-friendly and cleaner processing technologies [4].

Natural sources such as plants, insects, minerals, agronomic waste and food processing waste could serve as effective sources of natural dyes and are also renewable and biodegradable resources [5]. The purpose of using onion peels was to produce value-added dyed products from kitchen waste as the dye has a very good potential to absorb and adhere to the fabric and has good wash and light fastnesses [6].

The onion can be used to create a warm range of golden colours. As onion peels have natural mordant properties (high in tannin), it will make mordanting optional, adding it will increase colour fastness and longevity of the colours. The dye is inexpensive, easy to collect and quick to extract. [5, 7].

2 Materials and Methods

2.1 Materials

Industrially produced and commercially available cotton, cotton/linen blend and linen fabrics were used for the study. Table 1 gives a more detailed overview of the selection of fabrics.

Table 1. Fabric specification.

Fabric type	Fabric structure	Fabric weight (areal density) (g/m^2)	Remarks
100% Linen	Plain weave	181 ±2	Bleached
50% Linen, 50% Cotton (cotton/linen blend)	Plain weave	128 ±2	Bleached
100% Cotton	Plain weave	142 ±2	Raw fabric, unfinished

The principle of selecting the fabrics was that their density should be relatively similar. All fabrics are industrially produced with a plain weave. Fabrics with different fibre composition and treatment were selected to find out if this could have an effect on colour fastness.

Natural dyes are mostly non-substantive and must be applied on textiles with the help of mordants, usually a metallic salt, having an affinity for both the colouring matter and the fibre [8]. Mordants, except for alum and iron, are considered toxic, and therefore, should be avoided in the preparation of eco-textiles [8]. For this reason, aluminium salt (K2SO4·Al2(SO4)3·24H2O) has been used as a mordant in this study, as well as because the use of aluminium salt does not change the colour during dyeing [9].

Other chemicals known as assistants may be used in addition to dyes and mordants that help in the colouration of the fabric in one way or the other, for example tannic acid and sodium carbonate (washing soda) [8]. Cotton is generally pretreated with tannic acid salt to make it suitable for mordanting [8]. Tannin treatment was omitted in this study because onion peels, as a dyeing agent, already contain a significant amount of tannin [7].

Onion peels were collected as cooking waste in a different household before the test period. Only yellow onion peels were used.

There are many ways to work with plant dyes and many recipes in use according to the experience and preferences of dyers [7]. Many factors are responsible for the dyeing of fabrics, such as fabric/material type, compatibility of the dye and material, pretreatment of material, pH of dye bath, and mordanting [8]. One possible methodology has been applied in this study.

2.2 Methods

Preparation of Fabrics

Natural dyeing requires certain types of pretreatments and/or post treatments. Pretreatment of fabric prior to dyeing mainly involves a combined process generally consisting of preparing the fabric/material for dyeing to increase dyeability, which leads to favourable properties, such as wash fastness [8]. Cotton (as well as other natural cellulosic fibres) is not very porous and will not hold the dye without a more complicated preparation for mordanting; the fibre must be cleaned first [8]. Chemical and softening agents are added to the fibre when it is being manufactured. If these impurities remain in the fabric, they will stop the dye from soaking in fully and produce weak, inconsistent, or blotchy results. In order to scour effectively, the plant fibres require soda ash or soda crystals [7].

To prepare the fabrics for dyeing in this study, they were boiled in 90–95 °C water for 50 min in an aluminium pot in an alkaline solution (adding 1 teaspoon of sodium carbonate) to remove industrial finishing chemicals from the fabrics. In addition to adding substances to a bath for mordanting, the vessel that is used may itself serve as mordant [8]. After boiling, the fabrics were thoroughly rinsed.

Mordanting of Fabrics

All three fabrics were bisected prior to mordant treatment. One part was treated with the mordant and the other part was left untreated. The aim was to find out to what extent the treatment with mordant changes the colour fastness of the fabric.

Mordanting was carried out: 20% alum mordant and 6% sodium carbonate for 100 g of fabric was used as pretreatment. The fabrics were heated in the solution for 20 min at 70–80 °C and the solution was left to stand for 24 h. Control samples without mordant were also prepared for comparison.

Dye Solution

For the dye solution, 50 g of dried, uncrushed onion peel was calculated per 100 g of fabric. The onion peels were placed in a pot, water was poured over them, brought to a boil and boiled for 1 h. It was then left to stand until completely cooled and the onion peels were strained out of the solution.

Dyeing of Fabrics

Dyeing of fabrics with onion peels extract was carried out for 1 h at 70–80 °C. The fabrics were left in the dye solution until cool. The dyed fabrics were thoroughly rinsed with tap water and left to dry hanging in the open air.

Table 2. Fabric samples after dyeing and mordanting.

	Without mordant	With alum mordant
Linen		
Cotton/linen blend		
Cotton		

Table 2 presents two different samples of each fabric, 6 samples in total. One sample of fabric is dyed, and another is pretreated with alum mordant before dyeing. So, all researched fabrics are pretreated and dyed in the same conditions, except for mordanting.

2.3 Colour Fastness Testing

Four different colour fastness tests were performed on all samples. The colour fastness of the dyed samples to conditions like laundering, rubbing and perspiration were determined by standard test methods ISO 105 C06, ISO 105 X12 and ISO 105 E04 [10–12].

Colour fastness to washing was carried out with four different detergents at two different temperatures, 40 °C and 60 °C. Standard detergent (Ece Phosphate Reference Detergent B), laundry soap, Mayer Sensitive washing gel and Ariel Color washing powder were used for washing. The standard detergent was used according to the conditions of the standard for the purpose of comparing the results. The selection of other detergents was based on the detergents used by consumers in their daily lives, to find out the impact of their use on the colour fastness of naturally dyed fabrics. After washing, the fabrics were placed in the drying cabinet to dry.

No standard has been used to determine light resistance. The aim was to determine the colour resistance of the dyed fabrics to daylight according to the actual weather conditions in Estonia, which is why these results are the most subjective in this study. Since the same testing conditions were used for all six fabrics, the results are comparable. Each dyed fabric was divided into nine pieces, and these test pieces were kept on a windowsill in the light for 3 months in such a way that after every ten days one test piece was removed from the light and placed in the dark, so that it would be possible to compare the colour change of the test pieces along the testing timeline during 3 months after every 10 days. In addition, weather conditions (sunny, partly cloudy, cloudy) were documented daily during the three months. Roughly, these conditions were more or less equally distributed during these days.

All tests were performed in the textile testing laboratory of TTK University of Applied Sciences. The method of sampling is done according to Standard EN 12751:2000 [13]. The accuracy of recording the test results is also determined according to the testing standards used.

3 Results and Discussions

3.1 Colour Fastness to Washing (ISO 105 C06)

The colour fastness of the dyed samples to washing has been evaluated in the light cabinet, using an ISO grey scale for assessing the colour changes. The colour change compared to the original fabric was evaluated.

Tables 3 and 4 present the colour fastness evaluation results to washing and according to the grades, the Mayer Sensitive washing gel shows the best results for all fabrics at both temperatures among the detergents. With other detergents, the fabric lost almost all of its colour, and the most when washing with Ariel detergent, which bleached the fabric almost white. As expected, all fabrics lost more colour at 60 °C. Comparing fabrics dyed with and without a mordant, the colour of the fabric dyed with an alum mordant was preserved somewhat better or equally than the fabrics dyed without a mordant. When washing at 60 °C, the numerical indicator of colour fastness of fabrics treated with an alum is higher on cotton/linen blend fabric and cotton fabric compared to fabric without mordant, because the original colour of fabrics treated with a fixer is more intense and the change with the washing result is therefore also more intense. From the results, it can be concluded that the colour fastness of fabrics is better only with mild detergents.

3.2 Colour Fastness to Rubbing (ISO 105 × 12)

Colour fastness to rubbing was tested both dry and wet, as well as in the warp and weft directions. Although it is a single-coloured material with a plain weave, it was considered important to know if the direction of abrasion could affect the colour fastness of the fabric.

Table 5 presents the results of colour fastness to rubbing. Tests showed that dry tests on linen and cotton were better than wet tests. The linen fabric had better colour fastness than dyed fabric without a mordant, and cotton/linen blend and cotton treated fabric with an alum mordant. Contrary to the more common practice of rubbing tests that the results

Table 3. Colour fastness to washing.

Material	Washing temperature °C	Treatment	Standard detergent	Laundry soap	Mayer Sensitive	Ariel color
Linen	40	Without mordant	2	2	4	1/(2)
		With alum mordant	2/3	2/3	4/5	1/(2)
	60	Without mordant	1/(2)	1/(2)	3	1
		With alum mordant	1/(2)	1/(2)	4/5	1
Cotton/ linen blend	40	Without mordant	1	1	4	1
		With alum mordant	2	2	(4)/5	1
	60	Without mordant	1	1	2/3	1
		With alum mordant	1	1	3/4	1
Cotton	40	Without mordant	1/2	2	4/5	1
		With alum mordant	2/3	2	2/3	1
	60	Without mordant	1	1/2	3	1
		With alum mordant	1	1	5	1

of wet tests are worse than dry tests, the evaluation of cotton/linen blend fabric revealed that both without mordant and with mordant treated fabrics give more colour in dry tests and half a shade more in the weft direction. Therefore, there was a slight difference in the warp and weft direction in the colour fastness rubbing tests. The dyed cotton/linen blend fabric without mordant gave off the most colour.

The results of the colour fastness to rubbing tests were very variable and it is complicated to find a correlation in colour fastness depending on the fibre composition and the use of the mordant.

3.3 Colour Fastness to Perspiration (ISO 105 E04)

Table 6 presents colour fastness to perspiration test results. It can be stated that the fabrics treated with alum mordant changed colour less than the fabrics without mordant,

Table 4. Fabric samples after washing.

	Without mordant			With alum mordant		
	Standard detergent	Mayer Sensitive	Ariel color	Standard detergent	Mayer Sensitive	Ariel color
Linen						
Cotton/linen blend						
Cotton						

whereas the linen and cotton/linen blend fabrics became lighter and the cotton fabric samples became darker compared to the original colour in the alkaline condition. The cotton fabric changed colour the least and the cotton/linen blend fabric the most. In an alkaline solution, the colour is more durable for all fabrics than in an acidic solution.

3.4 Colour Fastness to Daylight

Table 7 presents the colour fastness test results to daylight. Colour fastness to daylight is relatively similar for all fabrics, regardless of the fibre composition or whether or not the fabric has been pretreated with a mordant.

Table 8 presents how the colour has almost disappeared on all the test bodies after a three- month test period.

Table 5. Colour fastness to rubbing.

Material	Condition	Direction	Without mordant	With alum mordant
Linen	Dry	Warp	4/5	4
		Weft	4/5	4
	Wet	Warp	3	2/3
		Weft	3/4	2/3
Cotton/linen blend	Dry	Warp	2/3	3/4
		Weft	2	3
	Wet	Warp	(3)/4	4/5
		Weft	(3)/4	4/5
Cotton	Dry	Warp	3/4	4
		Weft	4	4/5
	Wet	Warp	3/4	4
		Weft	3/4	4

Table 6. Colour fastness to perspiration.

Material	pH	Without mordant	With alum mordant
Linen	Alkaline	3	4
	Acid	2/3	3/4
Cotton/linen blend	Alkaline	2/3	4
	Acid	1/2	2
Cotton	Alkaline	4	4/5
	Acid	4	4/5

Table 7. Colour fastness to daylight.

Material	Treatment	Duration		
		10 days	1 month	3 months
Linen	Without mordant	(4)/5	4	1/2
	With alum mordant	(4)/5	4/5	2
Cotton/linen blend	Without mordant	4	2/3	1
	With alum mordant	4	2/3	1
Cotton	Without mordant	4/5	4	1
	With alum mordant	4/5	4	1

Table 8. Fabric samples after daylight test.

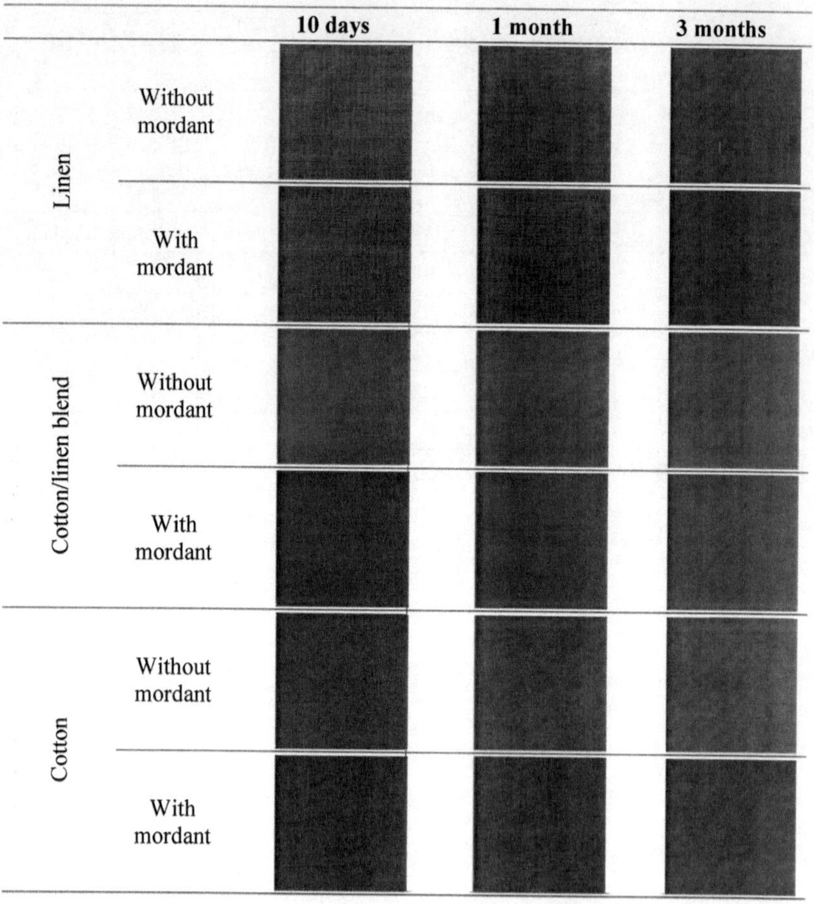

4 Conclusion

Based on the results of the study, it can be concluded that there are no big differences in the colour fastness of fabrics made of cellulose fibres under the influence of various factors (washing, rubbing, sweat and daylight). Although the pretreatment and dyeing conditions were the same for all the fabrics used in the study, no final general conclusions can be made about colour fastness depending on the fibre composition, since colour fastness is also influenced by many other factors due to the structure of the fabric and yarns used to make it, also industrial finishing technology and the chemicals used, which cannot be controlled in this study.

Several earlier studies confirm [5, 6] and it can be confirmed based on the results of this study, that fabrics, incl. Cotton fabrics, dyed with onion peels and pretreated with alum mordant provide a more intense colour during dyeing, and the colour fastness is generally better or at least equivalent to the same fabrics that have not been treated.

The colour fastness to washing results deserve special attention in this study. In this study, it cannot be claimed, as in some previous studies [6], that cotton fabrics treated with a metal mordant and dyed with onion peels have significantly better colour fastness to washing than untreated fabrics. However, it can be strongly influenced by the choice of detergent, about which very detailed information has not been provided in previous studies. It is very important that suitable detergents are chosen when washing naturally-dyed fabrics, for which in the example of this study only mild detergent is suitable. Analogous detergents are produced by many brands around the world, and apparently the development of such detergents is increasingly on the agenda due to the alarming situation in the textile and clothing industry.

There is no standard when dyeing with natural colours. Practitioners dealing with natural dyeing have developed several recipes and methods through trial and error over time to ensure the best possible colour fastness. As natural dyeing becomes more and more relevant and popular as an alternative to synthetic dyes, it is important to continue research and share knowledge and experience in order to increase the colour fastness of natural dyeing.

Consumers' values are certainly changing in relation to the requirements for colour fastness and hopefully it will not be considered a disadvantage in the future if the colour on the textile fades over time but as an opportunity to redefine colours due to the recent fashion as also some researchers have written [14].

References

1. Matthes, A., Beyer, K., Cebulla, H., Arnold, M.G., Schumann, A. (eds.): Sustainable Textile and Fashion Value Chains, 1st ed. Cham, Springer International Publishing (2021). https://doi.org/10.1007/978-3-030-22018-1
2. Samanta, A.K., Agarwal, P.: Application of Natural Dyes on Textiles (2009)
3. Haji, A., Naebe, M.: Cleaner dyeing of textiles using plasma treatment and natural dyes: a review. J. Clean. Prod. **265**. Elsevier Ltd (2020). https://doi.org/10.1016/j.jclepro.2020.121866
4. Shahid, M., et al.: Dyeing, fastness and antimicrobial properties of woolen yarns dyed with gallnut (Quercus infectoria Oliv.) extract. Dyes Pigm. **95**(1), 53–61 (2012). https://doi.org/10.1016/j.dyepig.2012.03.029

5. Amutha, K., Annapoorani, S.G., Sakthivel, P., Sudhapriya, N.: Ecofriendly dyeing of textiles with natural dyes extracted from commercial food processing waste materials. J. Natl. Fibers **19**(15), 10394–10411 (2022). https://doi.org/10.1080/15440478.2021.1993506
6. Vankar, P.S., Shukla, D.: New Trends in Natural Dyes for Textiles. Elsevier (2019). https://doi.org/10.1016/C2017-0-03994-0
7. Davey, K.: Natural Dyeing: Learn How to Create Colour and Dye Textiles Naturally. Hardie Grant Books, UK (2022)
8. Vankar, P.S.: Natural Dyes for Textiles: Sources, Chemistry and Applications. Elsevier (2017). https://doi.org/10.1016/C2015-0-01885-8
9. Amutha, K., Grace Annapoorani, S., Sudhapriya, N.: Dyeing of textiles with natural dyes extracted from Terminalia arjuna and Thespesia populnea fruits. Ind. Crops Prod., **148** (2020). https://doi.org/10.1016/j.indcrop.2020.112303
10. EN ISO 105-C06:2010 Textiles—Tests for colour fastness -- Part C06: Colour fastness to domestic and commercial laundering. Brussels (2010). [Online]. Available: www.evs.ee
11. EN ISO 105-X12:2016 Textiles-Tests for colour fastness-Part X12: Colour fastness to rubbing. Brussels (2016). [Online]. Available: www.evs.ee
12. EN ISO 105-E04:2013 Textiles—Tests for colour fastness—Part E04: Colour fastness to perspiration. Brussels (2013). [Online]. Available: www.evs.ee;
13. EN 12751:2000 Textiles—Sampling of fibres, yarns and fabrics for testing. Brussels (2000). [Online]. Available: https://www.evs.ee
14. Fröse, A., Schmidtke, K., Sukmann, T., Junger, I.J., Ehrmann, A.: Application of natural dyes on diverse textile materials. Optik (Stuttg) **181**, 215–219 (2019). https://doi.org/10.1016/j.ijleo.2018.12.099

The Influence of Bio-Innovative Pretreatment to Cotton Fabric Whiteness

Anita Tarbuk(✉), Tihana Dekanić, Franka Karin, and Irena Šabarić

Faculty of Textile Technology, University of Zagreb, Prilaz baruna Filipovića 28a, 10000 Zagreb, Croatia
anita.tarbuk@ttf.unizg.hr

Abstract. Raw cotton fabric contains added and non-cellulosic impurities, including starch and pigments, which are responsible for its yellow color. To remove the impurities in the raw cotton fabric with maximum efficiency, save energy and minimize the environmental impact, bio-innovative pretreatment and oxi-dative percarbonate bleaching were performed. The fabrics were desized with the enzyme α-amilase and bio-scoured with an enzyme complex containing neutral pectinase, followed by bleaching with an unconventional bleaching agent, percarbonate, at 60 and 90 °C. Since the main purpose of the pretreatment is to prepare the fabric for further processing with as little damage as possible, mechanical damage and hydrophilicity were measured using standard methods. Subsequently, the cotton fabrics were optically brightened in a wide range of concentrations and the spectrophotometrically determined values for whiteness, yellowing and tinting of the fabric were analyzed. The obtained results confirmed that the bio-innovative pretreatment has good efficiency and percarbonate provides good whiteness and absorption to achieve excellent whiteness.

Keywords: Cotton · Desizing · Bio-scouring · Percarbonate bleaching · Optical brightening

1 Introduction

Awareness of sustainable development has led to research into new methods and environmentally friendly substances that are used instead of those that have a negative impact on the environment [1–7]. Enzymatic processes that save energy, reduce greenhouse gas emissions, water consumption, chemical use, and byproduct generation are now being used in wet textile processes, resulting in a more environmentally friendly environment [2–11].

Raw cotton fabric contains added and non-cellulosic impurities. In woven fabrics, starch is added as an impurity, and the scouring process must remove non-cellulosic impurities in the raw cotton fiber, such as pectin, waxes, minerals, natural dyes and others. To avoid the use of various acids, alkalis and oxidants that are not suitable for the environment, amylase enzymes are used that can degrade starch in the desizing process. Scouring process is usually done with 2–3% NaOH at 100 °C for 60 min what

G. Guxho et al. (Eds.): ITC-ICEE 2023, LNMIE, pp. 491–499, 2024.
https://doi.org/10.1007/978-3-031-48933-4_48

is not environmentally friendly. However, it is possible to do enzymatic scouring i.e. bio-scouring process, which has a positive effect on the environment and requires little energy because it can be carried out at lower temperatures and under slightly alkaline conditions. Pectinase is the most important enzyme for bio-scouring because pectin is found in the cuticle of the primary wall of cotton fiber as the outer layer. By removing pectin, other impurities can be removed very easily [7–10]. After the scouring process, the only remaining impurity is natural cotton color, pigments. In addition to hydrogen peroxide, which is usually used at high temperatures to degrade cotton pigments, research is being conducted on oxygen-based chemical bleaching agents that allow effective bleaching at lower temperatures, such as perborate and percarbonate. When is sodium percarbonate in contact with water, it breaks down into sodium carbonate (Na_2CO_3) and hydrogen peroxide (H_2O_2). The released hydrogen peroxide then becomes an active oxidizing agent that can degrade cotton pigments [11–14]. To achieve an even higher degree of whiteness of the cotton fabric, it can be optically brightened. However, one of the most important requirements for achieving high whiteness is the material basic whiteness. In addition, the result of a high whiteness will not be achieved if the concentration of the optical brightening agent (OBA) is not optimal [15, 16].

For all these reasons, in this study, the impurities in the raw cotton fabric were removed by bio-innovative pretreatment and oxidative percarbonate bleaching, with the aim of achieving maximum efficiency, saving energy and minimizing environmental impact.

2 Material and Methods

For this research raw cotton fabric, produced in the company Čateks d.d. Čakovec, 100% cotton, basis weight 140 g/m^2, in plain weave, warp density 29 threads/cm and weft density 26.5 threads/cm was used. The fabrics were desized with the enzyme α-amilase BEISOL T2090 (CHT/Bezema) and bio-scoured with an enzyme complex containing neutral pectinase BEISOL PRO (CHT/Bezema) in the same bath containing 3% owf BEISOL T2090, 3% owf BEISOL PRO, and 1% nonionic surfactant Felosan RG-N (CHT/Bezema) at T = 60 °C, t = 40 min, pH 7. Chemical bleaching with an unconventional bleaching agent, percarbonate (Labud), was performed in bath containing 15 g/l of percarbonate ($Na_2CO_3 \cdot 1.5H_2O_2$), 2 g/l organic stabilizer Contavan ALR (CHT/Bezema), 1 g/l Felosan RG-N. Bleaching was performed at T = 90 °C, t = 30 min, pH 8 and at T = 60 °C, t = 60 min, pH 8 without and with addition of 1.5 g/l tetraacetylethylenediamine (TAED, Labud) as bleaching activator. After treatment, fabrics were rinsed and air-dried.

After pretreatment, the cotton fabrics were optically brightened at a wide range of concentrations to investigate the effect of pretreatment on high whiteness. The optical bleaching agent Uvitex® BHT from Ciba-Geigy AG (Fig. 1) was used in the concentrations FWA 0.1; 0.5; 1; 2; 5 and 20% owf by the exhaust method on the Linitest apparatus (Original Hanau) at a bath ratio of 1:20 for 30 min at 60 °C. After processing, the fabrics were air-dried.

Tensile properties of fabrics were determined on TensoLab Strength Tester (Mesdan, Italy) according to ISO 13934–1:2013 Textiles—*Tensile properties of fabrics—Part*

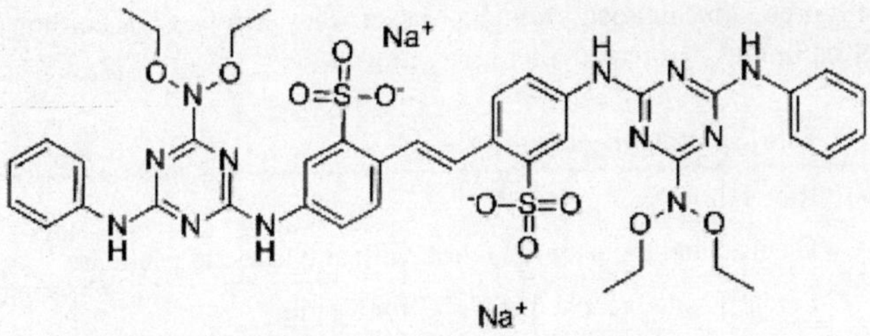

Fig. 1. Uvitex BHT, C.I. Fluorescent Brigthener 113.

1: Determination of maximum force and elongation at maximum force using the strip method. Distance between clamps was 100 mm, bursting speed 100 mm/min and pretension 2 N. From the results mechanical damage according to ISO 4312:1989 *Surface active agents—Evaluation of certain effects of laundering—Methods of analysis and test for unsoiled cotton control cloth* was calculated:

$$U_m = \frac{F_0 - F}{F_0} \cdot 100 \quad [\%] \tag{1}$$

where U_m is mechanical damage (wear) [%], F_0 is breaking force of raw fabric [N], and F is breaking force of treated fabric [N].

Hydrophilicity was determined after pretreatments according to AATCC 79–2018 *Absorbency of Textiles.*

Spectral remission was measured on a remission spectrophotometer Spectraflash SF 300 (Datacolor). The whiteness degree according CIE (W_{CIE}) was calculated from remission in accordance to ISO 105-J02:1997 *Textiles—Tests for colour fastness—Part J02: Instrumental assessment of relative whiteness,* Yellowing Index (YI) in accordance to DIN 6167:1980 *Description of yellowness of near-white or near-colourless materials,* and corresponding Tint Values (TV) and Tint Deviations (TD) with their coloristic meanings according to Ganz–Griesser [17].

3 Results and Discussion

As discussed above, raw cotton fabric contains added and non-cellulosic impurities, including starch and pigments, which are responsible for its yellow color. To remove these impurities with maximum efficiency, energy saving and minimizing the environmental impact, bio-innovative pretreatment and oxidative percarbonate bleaching were performed. Since the main purpose of the pretreatment is to prepare the fabric for further processing with as little damage as possible, mechanical damage and hydrophilicity were determined using standard methods.

The hydrophilicity results expressed as absorbency (Drop test) according to AATCC TM 79–2018 are given in Table 1.

As it can be seen fron the results, the raw fabric is hydrophobic due to the added and natural impurities, and the drop remains on the fabric surface without wetting time. By removing the starch using amylase in desizing and pectinase in bio-scouring, the

Table 1. Labels, pretreatment processes and the absorbency (Drop test) according to AATCC TM 79–2018 of cotton fabrics before and after pretreatments.

Label	Fabric pretreatment	t
R	Raw fabric	>60 min
AP	Desized and bio-scoured fabric with amylase and pectinase	6.8 s
PC_90–30	Percarbonate bleached at 90 °C for 30 min	3.4 s
PC_60–60	Percarbonate bleached at 60 °C for 60 min	3.4 s
PC_TA_60–60	Percarbonate bleached at 60 °C for 60 min with addition of TAED as activator	3.2 s

fabric becomes more hydrophilic and the wetting time improves significantly to 6.7 s. This means that the fabric is wetted but not yet completely hydrophilic. Perca-bonate bleaching removes pigments, but also clean surface, resulting in completely hydrophilic fabric, wetting time <5 s.

After desizing, bio-scouring and chemical bleaching, the mechanical properties of the fabrics were tested. The results of tensile properties, i.e., force (F) and elongation at break (ε), determined on a Tensolab dynamometer according to ISO 13934–1:2013, are presented with mean values and coefficient of variation (CV) in Table 2, and the mechanical damage (U_m) was calculated according to (1).

Table 2. Tensile force (F) and elongation at break (ε), and mechanical damage (U_m) of cotton fabrics before and after pretreatments.

Label	F [N]	CV [N]	ε [%]	CV [%]	*Um* [%]
R	492.2	1.38	11.29	1.45	–
AP	436.4	4.87	25.35	6.29	13.00
PC_90–30	399.9	1.04	24.28	1.48	18.90
PC_60–60	418.3	2.71	23.14	2.99	15.04
PC_TA_60–60	367.5	1.33	18.73	3.32	25.40

Analysis of the mechanical properties of cotton fabrics before and after pretreatment shows that pretreatment reduces the breaking strength and increases the breaking elongation. During pretreatment, cotton is cleaned of added and natural impurities, which reduces its strength, but shrinkage, which occurs in all wet treatment processes, leads to an increase in elasticity, i.e. elongation at break. The reason for this is that cotton cellulose swells during wet treatment and therefore fabric shrinks after drying. Comparing the results of breaking force and breaking elongation (Table 2), the differences between the effects of pretreatment are clearly visible.

The raw fabric has a tensile strength of 492 N. Removal of starch, an additive added to increase warp strength and reduce friction in the weaving process, and removal of pectin in bio-scouring, resulted in the strength loss (mechanical damage) of 13% (436 N) compared to the raw fabric. The results of force and elongation at break indicate an additional 10% reduction in mechanical properties of chemically bleached fabrics. The lower breaking forces compared to raw (S), desized and bio-scoured (AP) cotton fabrics indicate cellulose chain scission by oxidative bleaching. Namely, oxidative bleaching cleaves the double bonds of the colored α and β unsaturated carboxyl compounds (α-diketones), resulting in colorless products that can be easily washed out. The goal of bleaching is to achieve a high degree of whiteness with little damage to the cotton cellulose. However, in oxidative bleaching, the concentration of bleach, temperature, pH of the bath, and time must be controlled, and the amount of heavy metals in the system must be determined, as they can cause catalytic decomposition of the peroxide and thus damage the material [12]. Regardless of a well-stabilized bath and added sequestering agents, the presence of perhydroxyl radicals in the alkaline bath that oxidize cellulose is possible. Due to the presence of –OH groups, formation of oxycellulose can occur at any of these groups. Oxidized cellulose contains electronegative aldehyde and ketone groups, which can lead to cellulose chain scission, resulting in reduced strength of the cotton fabric. The effect of sodium percarbonate at an elevated temperature of 90 °C results in loss of 19% compared to the raw fabric. However, when the temperature is lowered to 60 °C, the tensile strength decreases by 15%, which is only 2% compared to bio-scoured fabric (AP), indicating slightly less damage to the cellulose of the cotton fabric (PC_60–60). When TAED is added for activation at a lower temperature (PC_TA_60–60), the breaking forces are the same as at a higher temperature, indicating a stronger effect of the oxidant.

The cotton fabrics were optically brightened in a wide range of concentrations and from the measured remission on the spectrophotometer, the whiteness degree, yellowing index and tinting of the fabric were analyzed. The results are collected in Tables 3, 4, 5 and 6.

Table 3. Whiteness degree according to CIE (W_{CIE}), yellowing index (*YI*) and corresponding tint deviations (TD) with its coloristic meanings according to Ganz/Griesser of cotton fabrics pretreatments.

Fabric	W_{CIE}	YI	Tint deviation (TD) from the white scale and its coloristic meanings	
R	12.2	27.62	RR	Tinted in red direction
AP	30.4	18.68	R5	Very markedly redder then the white scale
PC_90–30	74.2	4.69		No appreciable deviation in tint from the white scale
PC_60–60	61.0	9.06	R1	Trace redder then the white scale
PC_TA_60–60	61.2	8.76	R1	Trace redder then the white scale

From the results, it can be seen that the raw fabric (S) has the lowest whiteness, containing all added and natural impurities. The raw fabric has a yellowish hue, which is noticeable by tint deviation in red direction compared to the white scale. Removing the starch and all impurities except pigments in desizing and bio-scouring (AP), the whiteness increases from 12.2 to 30.4 and the yellowness decreases from 27.62 to 18.68, but the hue is still redder then the white scale.

As expected, chemical bleaching at 90 °C resulted in the highest whiteness degree of 74.2 and the least yellowing, with no appreciable deviation in tint from the white scale. This result is excellent since it was achieved in a less alkaline medium (pH 8) then if hydrogen peroxide was used. Lowering the temperature to 60 °C resulted in lower whiteness ($W_{CIE} = 61$), regardless of whether the TAED activator was added to the bath or not. Although the achieved whiteness is 15 units lower than at 90 °C, it is still 30 units higher than after bio-scouring.

To determine whether the achieved whiteness was sufficient for higher whiteness, the fabrics were optically brightened with Uvitex BHT in a wide range of concentrations.

From the whiteness, results can be seen that cotton fabric chemically bleached with percarbonate at 90 °C has a slightly yellowish tone although the pigments are destroyed. Optical brightening at a very low concentration of 0.1% owf increases the whiteness considerably to 110.9 and 0.5% owf to 128.6. The highest whiteness is obtained at a concentration of 2% owf but even 1% gives a similar result $W_{CIE} = 138$, so this concentration can be considered optimal. In addition, optically brightened fabric at 1% does not show significant tint deviations from the white standard. Exceeding the concentration of 2% of the optical brightener leads to quenching of the fluorescence, which manifests in lower whiteness, tinting in green direction and stronger yellowing.

The whiteness results of chemically bleached cotton fabric with percarbonate at 60 °C with the addition of TAED as an activator have the same basic whiteness as without it. However, optical brightening achieves similar results as after bleaching at 90 °C. A truly neutral whiteness is achieved at a concentration of 2% owf.

When TAED was used as an activator in chemical bleaching, more impurities were removed, and fabric has highest mechanical damage, so its absorption is better. Therefore, this cotton fabric (PC_TA_60–60) absorbs highest amount of OBA resulting in similar whiteness if 1, 2 or 5% owf was applied. Although the whiteness is somewhat lower than at a concentration of 2%, a concentration of 1% owf is considered optimal because of the lower OBA consumption. As the concentration increases, OBA molecules are layering and fabrics become markedly greener then the white scale. It should be noted that when the highest concentration of 20% is applied, the phenomenon of quenching is most pronounced, indicating the imminent exhaustion of the optical brightener.

4 Conclusions

In this study was researched how to remove the impurities in the raw cotton fabric with maximum efficiency, save energy and minimize the environmental impact. Therefore, bio-innovative pretreatment and oxidative percarbonate bleaching were performed. The fabrics were desized with the enzyme α-amilase and bio-scoured with an enzyme complex containing neutral pectinase, followed by bleaching with an unconventional

Table 4. Whiteness degree according to CIE (W_{CIE}), yellowing index (YI) and corresponding tint deviations (TD) with its coloristic meanings according to Ganz/Griesser of percarbonate bleached cotton fabrics at 90 °C and optical brightening in wide concentration range.

Fabric	W_{CIE}	YI	Tint deviation (TD) from the white scale and its coloristic meanings	
PC_90–30	74.2	4.69		No appreciable deviation in tint from the white scale
PC_90–30 OBA 0.1%	110.9	−9.13	R1	Trace redder then the white scale
PC_90–30 OBA 0.5%	128.6	−18.14		No appreciable deviation in tint from the white scale
PC_90–30 OBA 1%	138.0	−20.39		No appreciable deviation in tint from the white scale
PC_90–30 OBA 2%	139.8	−21.66		No appreciable deviation in tint from the white scale
PC_90-30 OBA 5%	135.3	-19.83	G2	Slightly greener then the white scale
PC_90-30 OBA 20%	114.4	-12.65	G5	very markedly greener then the white scale

Table 5. Whiteness degree according to CIE (W_{CIE}), yellowing index (YI) and corresponding tint deviations (TD) with its coloristic meanings according to Ganz/Griesser of percarbonate bleached cotton fabrics at 60 °C and optical brightening in wide concentration range.

Fabric	W_{CIE}	YI	Tint deviation (TD) from the white scale and its coloristic meanings	
PC_60–60	61.0	9.06	R1	Trace redder then the white scale
PC_60–60 OBA 0.1%	95.5	−3.47	R1	Trace redder then the white scale
PC_60–60 OBA 0.5%	120.3	−13.1	R1	Trace redder then the white scale
PC_60–60 OBA 1%	125.3	−15.34	R1	Trace redder then the white scale
PC_60–60 OBA 2%	132.2	−16.46		No appreciable deviation in tint from the white scale
PC_60-60 OBA 5%	130.7	-18.86	G2	Slightly greener then the white scale
PC_60-60 OBA 20%	106.9	-9.56	G5	Very markedly greener then the white scale

Table 6. Whiteness degree according to CIE (W_{CIE}), yellowing index (YI) and corresponding tint deviations (TD) with its coloristic meanings according to Ganz/Griesser of percarbonate bleached cotton fabrics at 60 °C with TAED addition and optical brightening in wide concentration range.

Fabric	W_{CIE}	YI	Tint deviation (TD) from the white scale and its coloristic meanings	
PC_TA_60–60	61.2	8.76	R1	Trace redder then the white scale
PC_TA_60–60 OBA 0,1%	103.7	−6.37	R1	Trace redder then the white scale
PC_TA_60–60 OBA 0,5%	127.0	−15.68	R1	Trace redder then the white scale
PC_TA_60–60 OBA 1%	130.5	−15.91		No appreciable deviation in tint from the white scale
PC_TA_60–60 OBA 2%	133.4	−18.62	G2	Slightly greener then the white scale
PC_TA_60-60 OBA 5%	132.3	-18.93	G2	Slightly greener then the white scale
PC_TA_60-60 OBA 20%	96.4	-6.11	G5	Very markedly greener then the white scale

bleaching agent, percarbonate, at 90 and 60 °C without and with the addition of TAED as activator. The mechanical damage, hydrophilicity/absorbency and whiteness were analyzed, since the main purpose of the pretreatment is to prepare the fabric for further processing with the least possible damage.

It has been shown ecological enzymatic pretreatments such as desizing and bio-scouring with amylase and pectinase allow the fabric to retain its strength and achieve excellent hydrophilicity. Chemical bleaching at 90 °C slightly reduces strength but increases hydrophilicity and whiteness. Chemical bleaching with percarbonate at lower temperature maintains the strength, while hydrophilicity and whiteness increase, and the whiteness is slightly lower, but still high enough. Optical brightening resulted in high whiteness ($W_{CIE} > 130$).

The obtained results confirmed that the bio-innovative pretreatment has good efficiency and percarbonate provides good whiteness and absorption to achieve excellent whiteness. The process of one-bath desizing and bio-scouring followed by chemical bleaching with percarbonate at reduced temperature can be proposed as an ecologically acceptable method for pretreatment of raw cotton fabric.

Acknowledgement. The work has been supported by Croatian Science Foundation under the project UIP-2017–05-8780 HPROTEX. This research is related in part to Erasmus+ project "Sustainable Design andProcess in Textiles for Higher Education" (GreenTEX, 2021–1-PL01-KA220-HED-000032201). The authors would like to thank the companies Labud, Croatia and CHT/Bezema, Switzerland for donation of chemicals.

References

1. Senthil Kumar, P., Gunasundari, E.: Sustainable wet processing—an alternative source for detoxifying supply chain in textiles. In: Detox Fashion, Textile Science and Clothing Technology, pp. 37–60. Springer (2018)
2. Sutlović, A., Tarbuk, A., Flinčec Grgac, S., Dekanić, T.: Sustainability of wet textile processes. In: Book of Proceedings of the 10th International textile, clothing and design conference—magic world of textiles, pp. 429–434. University of Zagreb Faculty of Textile Technology, Zagreb (2022)
3. Cavaco-Paulo, A., Gübitz, G.M.: Textile Processing with Enzymes. Woodhead Publishing Ltd and CRC Press LLC, Cambridge England (2003)
4. Doshi, R., Shelke, V.: Enzymes in textile industry—an environment-friendly approach. Indian J. Fibre Text. Res. **26**(1–2), 202–205 (2001)
5. Kumar, A., Choudhury, R.: Enzyme applications in textile chemical processing. In: Sustainable Technologies for Fashion and Textiles, pp. 91–115, Woodhead Publishing (2020)
6. Madhu, A., Chakraborty, J.N.: Development in application of enzymes for textile processing. J. Clean. Prod. **145**, 114–133 (2017)
7. Sheikh, J., Bramhecha, I.: Enzymes for green chemical processing of cotton, Chapter 6 In: The Impact and Prospects of Green Chemistry for Textile Technology, pp. 135–160, The Textile Institute Book Series, Woodhead Publishing (2019)
8. Grancarić, A.M., Pušić, T., Lesić-Domšić, B., Plantić, L.J.: The influence of alkaline pectinases treatment to the sewability of cotton knitted fabric (in Croatian). Tekstil **50**(2), 55–62 (2001)
9. Pušić, T., Tarbuk, A., Dekanić, T.: Bio-innovation in cotton scouring—acid and neutral pectinases. Fibres Text. Eastern Europe **23**(1), 98–103 (2015)
10. Grancarić, A.M., Tarbuk, A., Hadžić, S., Simončič, B.: From raw to finished cotton—characterization by interface phenomena. AUTEX Res. J. **23**(2), 184–192 (2023)
11. Vujasinović, E., Tarbuk, A., Pušić, T., Dekanić, T.: Bio-Innovative Pretreatment of Coarse Wool Fibers. Processes 11(1), 103, 1–14 (2023), DOI:https://doi.org/10.3390/pr11010103
12. Pušić, T., Jeličić, J., Nuber, M., Soljačić, I.: Investigation of bleach active compounds in washing bath (in croatian). Tekstil **56**(7), 412–417 (2007)
13. Li, Z., Wang, Y., Wang, Z.: Study on bleaching technology of cotton fabric with sodium percarbonate. In: MATEC Web Conf. International Symposium on Materials Application and Engineering, vol. 67, p. 06048 (SMAE 2016). https://doi.org/10.1051/matecconf/201666770 6048
14. Bahlool, S.O., Saad, H.A.: Comparative performance evaluation of bleaching of egyptian cotton fibers using different bleaching agents. Egypt. J. Chem. **63**(9), 3491–3502 (2020)
15. Soljačić, I.: About optical brigthening agents (in croatian). Tekstil **21**(5), 377–398 (1972)
16. Grancarić, A.M., Tarbuk, A.: Quenching of fluorescence in world of whitenes. In: Book of Papers of 11th AIC 2009, pp. 395–401, Sydney, The Colour Society of Australia (2009)
17. Griesser, R.: Assessment of whiteness and tint of fluorescent substrates with good inter-instrument correlation. Color Res. Appl. **19**(6), 446–460 (1994)

Data-Driven Multi-scale Numerical Homogenization

Georgios E. Stavroulakis(✉) and Eleftheria Bletsogianni

School of Production Engineering and Management, Technical University of Crete, 73100 Chania, Greece

gestavroulakis@tuc.gr

Abstract. Homogenization for complex, nonlinear materials, including composites and textiles, is considered. The effect of the microstructure, including nonlinearity, for various loadings is transferred to the homogenized medium through the Representative Volume Element technique. The classical approach through nested finite element analysis, the so-called FEM2 method, is very expensive. Data-driven techniques have been proposed, where the homogenization is replaced by a surrogate model based on data generated through selected numerical or physical experiments. An artificial neural network surrogate constitutive model leads to a FEANN method. Furthermore data-driven solution methods have been recently proposed, as a development of the LATIN iterative method. A short review of recent contributions is presented, with examples from composites, including woven composites and auxetics. Research topics will be discussed, including the development of Physics Informed Neural Network surrogates.

Keywords: Homogenization · Multiscale analysis · Finite elements · Neural networks · Composites

1 Introduction

Multiscale modeling is based on a simulation technique that describes the behavior at a given length scale based on the physics at a finer scale, which in turned is considered to be better known or easy to be modeled. Structures made of composite materials or materials with microstructure constitute typical candidates for multiscale modeling. In fact, it is impossible to model a whole structure by taking all constituents and their, possibly nonlinear interactions within the model. Therefore a compromise must be considered, where a reduction in precision and an increase in uncertainty that is related with multiscale modeling should be accepted, in order to avoid the complexity of a full-scale model [1]. In engineering, homogenization is a representative example of applied multiscale modeling [2, 3].

In linear problems homogenization is usually based on analytical expressions that give the homogenized properties for a given microstructure or composite material [4, 5]. This approach is powerful, despite certain theoretical difficulties for the calculation of homogenized properties. Numerical homogenization allows for the extension of

G. Guxho et al. (Eds.): ITC-ICEE 2023, LNMIE, pp. 500–508, 2024.
https://doi.org/10.1007/978-3-031-48933-4_49

classical analytical homogenization in order to describe more complicated problems. Homogenized properties are load-dependent or even path-dependent, therefore analytical solutions are difficult or even do not exist. Woven composites and fabrics, with many slender components interacting through unilateral contact and friction and subjected to buckling is a challenging application. An efficient way to store the constitutive relation based on a data-base of responses of a representative volume element is proposed through the construction of a neural network constitutive model surrogate. Subsequently, usage of this surrogate within a nonlinear, upper level finite element model, is straightforward. The proposed method will be outlined for a static analysis problem of a woven composite.

2 Data-driven Numerical Homogenization Replacing the FE2 Multi-scale Analysis

Analytical homogenization replaces a representative volume element (RVE) with several materials (composite) or microstructure with a continuum with homogenized properties [6, 7]. In nonlinear behavior this step must be repeated for each level of stress resp. strain or even load path. Nevertheless this approach has been followed and gives a multi-scale technique which has been called FEM2 [8]. The method is accurate but requires enormous resources, due to the need of solving the detailed model of the RVE for every different combination of loadings that appear during incremental-iterative solution of the homogenized finite element model (Fig. 1).

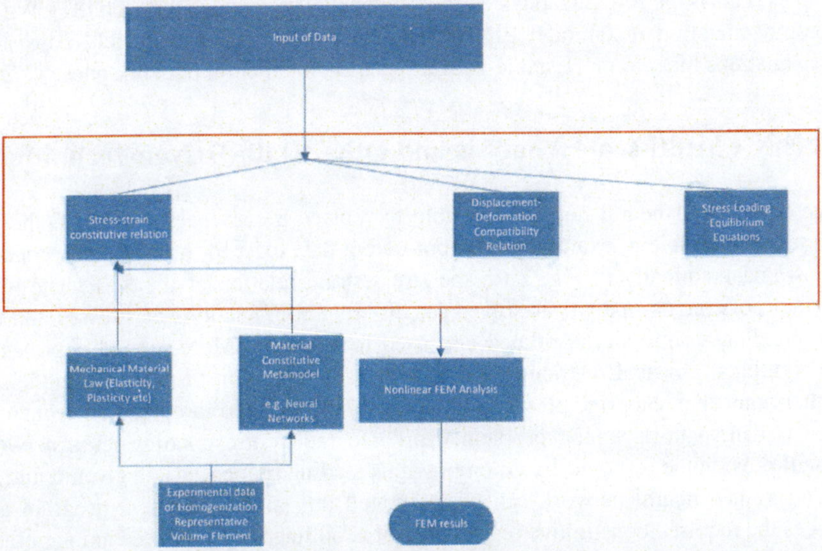

Fig. 1. The concept of FEM2 multi-scale modeling. The material constitutive law is based on experimental or RVE data. It is based either on an established mechanical material law or a metamodel (interpolation), for example a neural network one.

Since the analysis of the RVE is repeated for different points of the homogenized medium and different stress-strain levels, the usage of a surrogate based on selected computations is straight-forward [9]. Classical polynomial surrogates as well as neural network and other ones have been proposed and tested [3, 9–16].

Artificial intelligence methods and especially feedforward neural networks which can be trained by the backpropagation method have been used for the solution of direct and inverse, parameter identification problems in mechanics, by using input-output examples experimentally generated or numerically calculated, [17–20].

A feed-forward neural network has the ability to correlate vector inputs with vector outputs, provided that the parameters involved have been suitably calculated. This step is done with the help of data (examples) within a step called training.

The topology of the neural network (number of layers and nodes) as well as the activation functions (filters) at nodes in correlation with the number of examples define the ability of the network to approximate well an unknown function. Trial and error or optimal design principles can be used towards a satisfactory result [19].

This is a data-based usage of neural networks, which totally avoids the usage of mechanical models for the approximated relation. Classical interpolation or similar methods are also candidates for similar tasks.

In the presented application no specific homogenization model is reguired to apply this method. The surrogate based on representative calculations will provide a 'model free' approximation of the stress-strain relation at RVE level [3, 10, 21].

Furthermore, the differential equations of mechanics can be used, in combination with automatic differentiation of the neural network, in order to train the network by using the governing partial differential equations, without using specific input-output examples. This is the so-called physics-informed neural networks approach [22]. Within the homogenization metamodel PINN could provide thermomechanicaly consistent approximations that are more stable with respect to blind neural network ones.

3 FE-NN Multi-scale Analysis and Other Data-Driven Techniques

Usage of artificial neural networks is able to replace a classical mechanical theory, provided that sufficient 'examples' of input-output data exist for training. The concept is shown schematically in Fig. 2 for the stress-strain relation of the RVE. The RVE for every possible complexity is solved for different combinations of stresses and the corresponding strains are calculated and used for training. More refined approaches pose additional physical restrictions on the corresponding constitutive metamodel, so that thermomechanical principles are not violated [16]. Furthermore, loading sequences can be used in a path-dependent problem, in order to train a corresponding neural network that will provide the response based on previous loading sequences at a given point.

The trained neural network can be integrated into a finite element program and replaces the required constitutive response, as it is outlined in Fig. 1, see among others [1, 3, 4, 23]. This approach is near to the classical usage of constitutive relations and can be integrated with existing commercial finite element packages [24–28]. A MATLAB implementation for two-dimensional problems is available with reference [3].

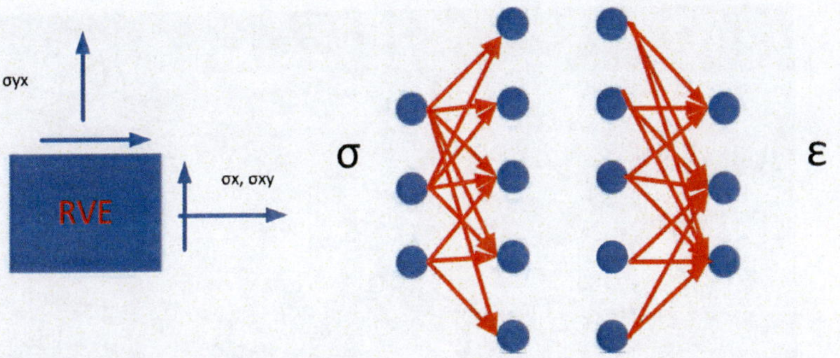

Fig. 2. A figure caption is always placed below the illustration. Short captions are centered, while long ones are justified. The macro button chooses the correct format automatically.

A recent approach, the so-called data-driven technique [29] can be used as well. With reference to Fig. 1, one adopts the compatibility and equilibrium equations of mechanics and tries to solve them with the best approximation of stress-strain couples among the available in a data-base. Again the need of a specific mechanical model is replaced with the existence of data. Approximation is not restricted from any metamodel but can be realized with more general techniques from searching matching data from a big database [30, 31].

4 Analysis of Woven Fabrics at the RVE Level

Wooven fabrics are materials with complicated microstructures involving buckling and contact interactions [23, 33–35]. This example is chosen here in order to outline the proposed multiscale procedure.

A weave fabric unit cell with two yarns per direction has been generated by texgen [36]. The dimensions are: Yarn spacing = 2, Yarn width = 0.8, Fabric thickness = 0.2. An isotropic material with Young Modulus = 220.69 Gpa and Poisson ratio = 0.2 has been considered (Figs. 3 and 4).

A nonlinear finite element analysis has been performed, with unilateral contact and friction effects between the yarns being the source of nonlinearity, with friction coefficient equal to 0.2 (Figs. 5, 6, 7, 8 and 9).

The nonlinear behavior of the RVE is shown in force-displacement plots of Figs. 8 and 9. At each loading level the numerical homogenization is based on the generated data and the result is forwarded to the upper level of the multi-scale procedure. Depending on the analysis, geometric nonlinearities including buckling and slackening of yarns can also be captured with the proposed approach.

5 Conclusions

A numerical homogenization based on RVE data and subsequent multiscale analysis based on FEM and NN surrogate constitutive model has been outlined. An RVE of woven composite has been used here for demonstration. The technique is applicable for any kind of composite material, masonry or metamaterials with given microstructure.

Fig. 3. Model of plane weave fabric unit cell

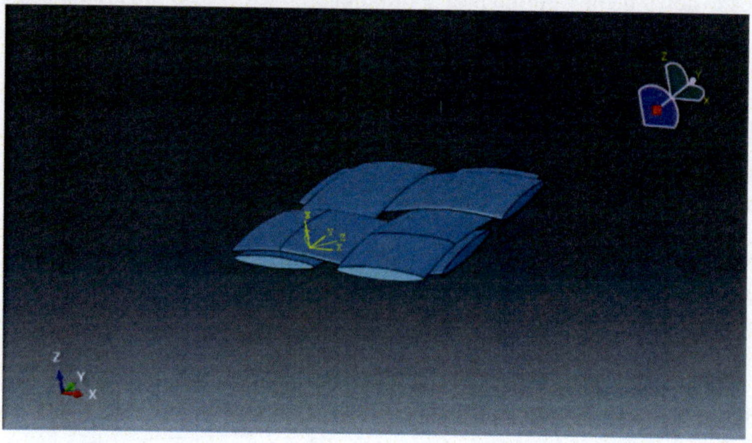

Fig. 4. The plane weave fabric three-dimensional model

Integration within a small displacement homogenized continuum scheme is possible [3, 4]. Nevertheless several practical issues related to local buckling and instabilities must be further investigated.

Extensions of this work in order to take into account path-dependent constitutive relation of the RVE and dynamic effects is possible and constitute the subject of current research investigation. For the solution of this task, tools recently developed within the area of data-driven computational mechanics will be helpful [3, 29, 32]. In a more general setting, the group of equations governing the continuum medium are coupled with iterations so that the best experimental couple of constitutive quantities is used. This technique resembles and emanated from the classical Large Time Increment scheme (LATIN, [2, 37]).

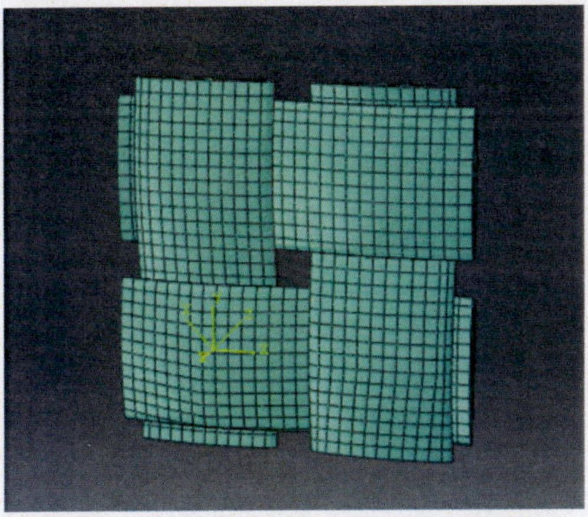

Fig. 5. Finite element discretization of RVE

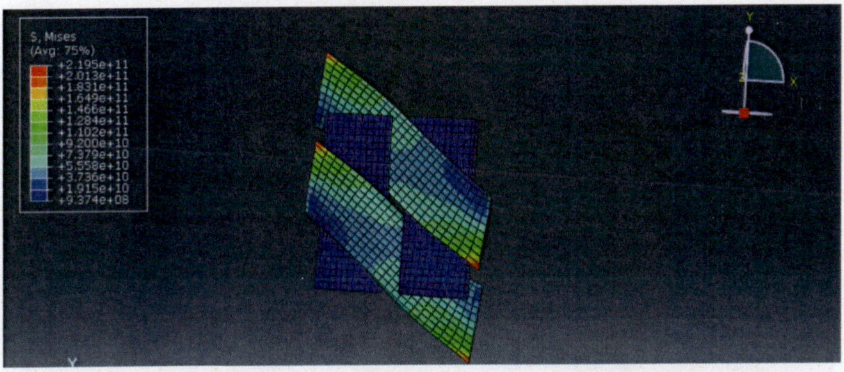

Fig. 6. Representative results of RVE

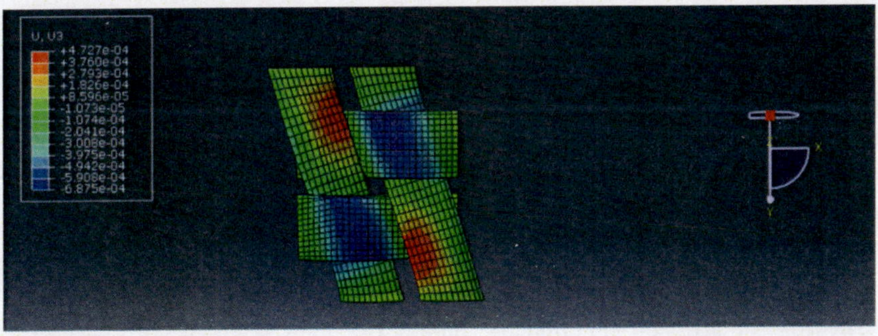

Fig. 7. Out-of-plane component of displacements of the RVE

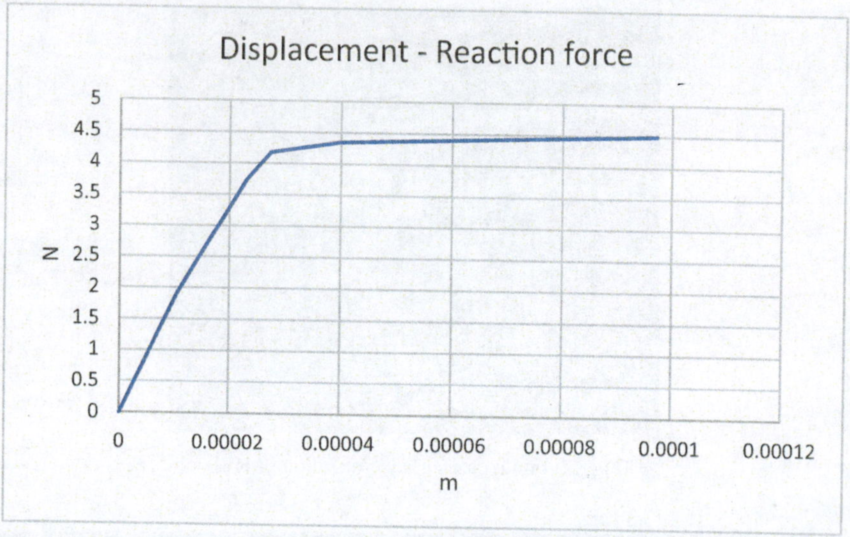

Fig. 8. Nonlinear response of RVE for a loading inclined 33 degrees with respect to horizontal direction

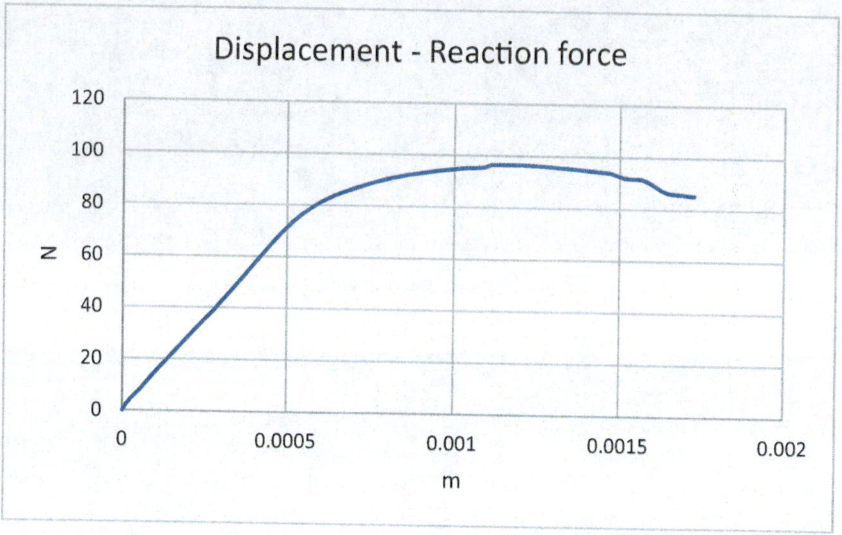

Fig. 9. Nonlinear response of RVE for a loading inclined 45 degrees with respect to horizontal direction

The role of artificial intelligence, interwoven with methods of computational mechanics has been demonstrated to be critical in this work. The advantages in comparison with more classical multiscale approaches, like FEM2, are considerable. Further refinement of the tools could incorporate the investigation of other surrogate models and the refinement of the NN surrogates by including elements of physics and thermomechanics in their creation using the PINN methodology.

References

1. Fish, J., Wagner, G.J., Keten, S.: Mesoscopic and multiscale modelling in materials. Nat. Mater. **20**, 774–786 (2021)
2. Yvonnet, J.: Computational Homogenization of Heterogeneous Materials with Finite Elements. Springer, Berlin (2019)
3. Drosopoulos, G.A., Stavroulakis, G.E.: Non-linear Mechanics for Composite. Heterogeneous Structures. CRC Press, Taylor and Francis (2022)
4. Urbański, A.: The unified, finite element formulation of homogenization of structural members with a periodic microstructure. Cracow University of Technology, (2005)
5. Geers, M.G.D., Kouznetsova, V.G., Brekelmans, W.A.M.: Multi-scale computational homogenization: Trends and challenges. J. Comput. Appl. Math. **234**, 2175–2182 (2010)
6. Michel, J.-C., Moulinec, H., Suquet, P.: Effective properties of composite materials with periodic microstructure: a computational approach. Comput. Methods Appl. Mech. Eng. **172**, 109–143 (1999)
7. Zohdi, T.I., Wriggers, P.: An Introduction to Computational Micromechanics. Springer, Berlin (2008)
8. Tikarrouchine, E., Benaarbia, A., Chatzigeorgiou, G., Meraghni, F.: Non-linear FE2 multiscale simulation of damage, micro and macroscopic strains in polyamide 66-woven composite structures: analysis and experimental validation. Compos. Struct. **255**, 112926 (2021)
9. Drosopoulos, G.A., Giannis, K., Stavroulaki, M.E., Stavroulakis, G.E.: Metamodeling-assisted numerical homogenization for masonry and cracked structures. ASCE J. Eng. Mech. **144**(8), art. no. 04018072, (2018)
10. Drosopoulos, G.A., Stavroulakis, G.E.: Data-driven computational homogenization using neural networks: FE^2-NN application on damaged masonry. ACM J. Comput. Cult. Heritage **14**(1), 1–19 (2020)
11. Yvonnet, J., He, Q.C., Li, P.: Reducing internal variables and improving efficiency in data-driven modelling of anisotropic damage from RVE simulations. Comput. Mech. **72**, 37–55 (2023)
12. Le, B.A., Yvonnet, J., He, Q.-C.: Computational homogenization of nonlinear elastic materials using neural networks. Int. J. Numer. Meth. Eng. **104**(12), 1061–1084 (2015)
13. Urbański, A., Szymon, l., Marcin, D.: Multi-scale modeling of brick masonry using a numerical homogenization technique and an artificial neural network. Arch. Civil Eng. 68(4), 179–197 (2022)
14. Eivazi, H., Tröger, J.-A., Wittek, S., Hartmann, S., Rausch, A.: FE^2 computations with deep neural networks: algorithmic structure, data generation, and implementation (2023). Available at SSRN: https://ssrn.com/abstract=4485434 or https://doi.org/10.2139/ssrn.4485434
15. Fish, J., Yu, Y.: Data-physics driven reduced order homogenization. Int. J. Numer. Methods Engrg. **124**(7), 1620–1645 (2023)
16. As' ad, F., Avery, P., Farhat, C.: A mechanics-informed artificial neural network approach in data-driven constitutive modeling. Int. J. Numer. Methods Engrg. 123(12), 2738–2759 (2022)
17. Mroz, Z., Stavroulakis, G.E. (eds.),: Parameter Identification of Materials and Structures. CISM Lecture Notes vol. 469, Springer, Wien, New York, Lecture notes of CISM Course (2005)

18. Stavroulakis, G.E.: Inverse and crack identification problems in mechanics. Springer, Kluwer Academic (2000)
19. Protopapadakis, E., et al.: A genetically optimized neural classifier applied to numerical pile integrity tests considering concrete piles. Comput. Struct. **162**, 68–79 (2016)
20. Spathopoulos, S.C., Stavroulakis, G.E.: Springback prediction in sheet metal forming, based on finite element analysis and artificial neural network approach. Appl. Mech. **1**, 97–110 (2020)
21. Drosopoulos, G.A., Wriggers, P., Stavroulakis, G.E.: A multi-scale computational method including contact for the analysis of damage in composite materials. Comput. Mater. Sci. **95**, 522–535 (2014)
22. Muradova, A.D., Stavroulakis, G.E.: Physics-informed neural networks for elastic plate problems with bending and Winkler-type contact effects. J. Serbian Soc. Comput. Mech. **15**(2), 45–54 (2021)
23. Lu, X., Giovanis, D.G., Yvonnet, J., Papadopoulos, V., Detrez, F., Bai, J.: A data-driven computational homogenization method based on neural networks for the nonlinear anisotropic electrical response of graphene/polymer nanocomposites. Comput. Mech. **64**(2), 307–321 (2019)
24. Tchalla, A., Belouettar, S., Makradi, A., Zahrouni, H.: An ABAQUS toolbox for multiscale finite element computation. Compos. B Eng. **52**, 323–333 (2013)
25. Wei, H., Wu, C.T, Hu, W., et al.: LS-DYNA machine learning–based multiscale method for nonlinear modeling of short fiber–reinforced composites. J. Eng. Mech., **149**(3) (2023)
26. Su, T.-H., et al.: Multiscale computational solid mechanics: data and machine learning. J. Mech., **38**, 568–585 (2022)
27. Fei, T., Xin, L., Haodong, D., Wenbin, Y.: Learning composite constitutive laws via coupling Abaqus and deep neural network. Compos. Struct. **272**, 114137 (2021)
28. Altair multiscale designer https://altair.com/multiscale-designer
29. Kirchdoerfer, T., Ortiz, M.: Data-driven computational mechanics. Comput. Methods Appl. Mech. Engrg. **304**, 81–101 (2016)
30. Park, H., Cho, M.: Multiscale constitutive model using data–driven yield function. Compos. B Eng. **216**, 108831 (2021)
31. Tang, Z., Guo, L., et al.: A combined machine learning and numerical approach for evaluating the uncertainty of 3D angle-interlock woven composites. Compos. Struct. **294**, 115726 (2022)
32. Huang, W., Xu, R., Yang, J., Huang, Q., Hu, H.: Data-driven multiscale simulation of FRP based on material twins. Compos. Struct. **256**, 113013 (2021)
33. Provatidis, C.G., Vassiliadis, S.G., Anastasiadou, E.A.: Contact mechanics in two-dimensional finite element modelling of fabrics. Int. J. Clothing Sci. Technol., **17**(1), 29–40 (2005)
34. Kallivretaki, A., Vassiliadis, S., Blaga, M., Provatidis, C.: Finite element modelling of the warp knitted structure. Res. J. Text. Appar. **11**(4), 40–47 (2007)
35. Dixit, A., Mali, H.S.: Modeling techniques for predicting the mechanical properties of woven-fabric textile composites: a review. Mech. Compos. Mater. **49**, 1–20 (2013)
36. Brown, L.P., Long, A.C.: Modeling the geometry of textile reinforcements for composites: TexGen. In: Boisse, P. (ed.) Composite Reinforcements for Optimum Performance (Second Edition), pp. 237–265. Woodhead Publishing, 2021 (2021)
37. Stavroulakis, G.E. et al.: Data-driven, data-based and artificial intelligence methods in computational mechanics. In: CESARE'22 3rd Coordinating Engineering for Sustainability and Resillence, May 6th–May 9th, 2022, Irbid, Jordan (2022)

Auxetics, Theory and Textile Applications

Ioannis Ntintakis[1(✉)], Georgios E. Stavroulakis[2], and Stavros Karathanasopoulos[1]

[1] Department of Mechanical Engineering, Hellenic Mediterranean University, Heraklion, Greece
ntintakis@hmu.gr

[2] Computational Mechanics and Optimization Laboratory, Technical University of Crete, Chania, Greece Heidelberg, Tiergartenstr. 17, 69121 Heidelberg, Germany

Abstract. Auxetic materials are characterized by unusual elastic behaviour, which means that when expanded in a specific direction, there are expanded in the perpendicular direction. This unique characteristic is derived from the various shapes of auxetic structures that are often characterized by complex 3D geometries. Aux-etic materials have enchanted mechanical properties such as fracture resistance, high energy absorption, and formability. Due to their advances, auxetic materials may be used in textile design and fabrication. In current manuscript the star shape auxetic microstructure in woven fabrics is studied. The results illustrate the behaviour of the auxetic microstructure in both tensile and compressive loadings. When comparing the results of the auxetic microstructure with those of the solid textile material, it is evident that the applied stresses show a more uniform distribution across the entire surface.

Keywords: Auxetic materials · Star auxetic microstructure · Textile design · Product design · Furniture design

1 Introduction

1.1 Auxetic Materials Background

Natural materials generally thin out in the direction of stretching (positive Poisson's ratio) and increase in the direction of compression. Poisson's ratio is usually a positive number taking values in the interval (0, 0.5). This ratio is defined as the negative fraction of transverse strain $\Delta y/ly$ over the axial strain $\Delta x/lx$ [1]. In contrast, there is a special category of materials called auxetic materials that exhibit the opposite behavior, getting thicker when stretched due to their unique internal structure with a negative Poisson's ratio. The auxetic behavior of these materials arises from the presence of artificial joints within their microstructure, facilitating flexing [2, 3]. This unique characteristic is attributed to their specific internal structure, typically described by star-shaped frames or inclusions, though other models like chiral, perforated, or mechanism-based structures have also been considered (Fig. 2) [4].

G. Guxho et al. (Eds.): ITC-ICEE 2023, LNMIE, pp. 509–517, 2024.
https://doi.org/10.1007/978-3-031-48933-4_50

Auxetic properties are attributed to nonconvex, reentrant microstructures based on star-shaped microstructures [5–7]. In statics the most interesting property of auxetics is their beneficial behavior around contact areas, where the material 'condenses' and supports the place coming in contact [6–11]. In dynamics, the creation of band gaps, where waves of certain frequencies are blocked within an auxetic medium, is the most promising effect, which can be used for vibration suppression and seismic shielding of equipment [6]. Published results demonstrate how woven can be constructed based on reentrant microstructure [12]. Auxetics, including woven materials, can be created by modern additive manufacturing techniques [12–14].

Auxetic materials feature behaviors of indentation resistance which mean that they can resist deformation and penetration when subjected to force or pressure. This unique behavior sets them apart from conventional materials. Indentation resistance offers several benefits in various applications. In protective equipment, auxetic materials distribute forces over a larger area, increasing resistance to impacts and pressures. They are ideal for shock-absorbing applications, as they absorb and dissipate energy effectively, providing better protection from sudden impacts. Auxetic materials also enhance air filter efficiency by withstanding particle impacts without getting easily clogged or deformed. Overall, their indentation resistance makes them promise for diverse engineering and material science applications. Auxetic behaviour can be incorporated within a composite structure, like in honeycomb structures. In conventional honeycombs with hexagonal cells, stretching the network in one direction causes elongation along that axis and closing along the perpendicular axis, which leads to a positive Poisson's ratio. To achieve auxetic behavior, the deformation mechanism (rib hinging) is maintained, but the honeycomb cell geometry is modified to create a reentrant structure. In this new structure, cells feature a different behavior of elongation both along and perpendicular to the applied load direction, exhibiting negative Poisson's ratio. Various tessellations can be constructed from 'arrows' suggesting different arrangements of reentrant cells in tessellated patterns. By manipulating these arrangements, engineers can manipulate various reentrant honeycomb structures with unique mechanical properties, allowing sophisticated characteristics for specific applications.

2 Auxetic Materials in Fabrics

A proposition to use an engineered auxetic material actuated by a reduced number of motors for shape displays. Unlike conventional mechatronic pin-matrix displays, the auxetic material bends in multiple directions, providing a smooth and rigid touch experience. A prototype implementation with nine actuators successfully displays a range of surface curvatures without sharp edges. The study demonstrates the potential of this approach through user experiments, highlighting the advantages of auxetic materials in creating more continuous and realistic simulated surfaces for tactile interaction [15]. This was followed up by a study of auxetic materials that exhibit expansion in the transverse direction when stretched longitudinally, and they have significant implications in biomedical engineering [16]. The review highlights the application of auxetic textile materials in the field of biomedical engineering, considering their relevance for tissue repair and transplant due to their conformance to human body mechanics. Preceding the previous research, an investigation on the auxetic behavior of modified non-woven fabric achieved by laser cutting rotating square units with slits. Two commercial fabrics used in the footwear industry were analyzed. Tensile tests showed lower breaking force in modified samples, thus limiting their application to low tensile loads. Smaller rotating cell sizes exhibited the highest negative Poisson's ratio. Fabrics with equal fiber distribution and orientation in both directions had better auxetic response with smaller out-of-plane rotations [17]. Furthermore, a research focused on the design and production of perforated metamaterials with tailored mechanical properties and specialized characteristics. The metamaterials were created by introducing microstructural cuts in a rubber sheet using direct laser cutting. Experimental tests and Finite Element Analysis were conducted to analyze the mechanical properties and deformation behavior. The study demonstrates the effectiveness of this method for fabricating auxetic metamaterial sheets and achieving a wide range of mechanical properties and load responses through patterned cuts in thin sheets [18]. Following that, they provide a comprehensive review of design methods and advanced manufacturing technologies for auxetic tubular structures. It covers various types of cellular, nonporous, and porous auxetic tubes, as well as macro and micro auxetic tubes. The paper discusses the auxetic behavior, mechanical properties, and potential applications of these structures. It concludes by addressing the challenges and opportunities in auxetic tube research, aiming to inspire future studies [19]. Following up, a proposal of a novel approach to efficiently produce auxetic fabrics using rotating square geometry and lamination technique. The fabrics are manufactured by laser cutting and bonding layers together. Tensile tests revealed that the auxetic behavior depends on the initial modulus difference and square unit parameters. Three-layer laminated fabrics are preferred for avoiding out-of-plane deformation and extending the tensile strain range. This work offers insights for developing efficient methods to produce auxetic fabrics with desirable properties [20].

An exploration of an idea that presents a computational method for interactive 3D design and rationalization of surfaces using auxetic materials. The approach utilizes cuts in inextensible materials to create hexagonal openings that stretch uniformly. Conformal geometry is employed to understand and optimize the auxetic design, dealing with non-trivial topology and non-local dependencies. Design studies and physical prototypes showcase the potential applications of this method [21]. Published research on auxetic materials, known for their unique deformation behavior, expands the opportunities for innovative applications in fashion, design, and architecture. Their flexibility, adaptability, and responsiveness to the environment are yet to be fully explored [22]. Following this, a recent theoretical work at the time hypothesized that the pruning-based design of disordered networks can create materials with unique mechanical properties, including auxetic behavior. However, the experimental realization of these materials has been challenging. This paper introduces a realistic model incorporating angle-bending forces and boundary conditions, enabling precise engineering of mechanical responses through sequential pruning. The model is validated through physical realizations, and optimization algorithms further enhance the auxetic behavior. The findings establish pruned networks as a promising platform for creating mechanical metamaterials with tunable properties [23].

The purpose of this manuscript is to exploit the use of auxetic materials in fabrics. Specifically, it aims to compare the behavior of the fabric with and without auxetic structure under specific loading conditions. The study is conducted using the finite element method. In the comparative study, one of the most common auxetic shapes are going to be used. This shape is known as five-star shape which can be described as "connected stars" as they contain star-shaped units that are interconnected to form two-dimensional periodic structures [24]. Critical conclusions about the use of auxetic shape in the fabric arise from the overall strength, stress developed on the material's surface, and maximum displacement.

3 Design of Auxetic Shapes

The auxetic shapes are defined by several parameters: the height of the cell (h), the length of the cell wall (l), the depth of the cell wall (d), and the angle formed between the horizontal and the leaning cell wall, the angle (θ) is set at -30 degrees (Table 1).

Table 1. Dimensions of auxetic cells

Geometrical parameters	Auxetic star shape	Auxetic arrow shape
Height of the cell wall, h (mm)	8.46	4.00
Cell wall length, l (mm)	4.23	2.30
Depth of the cell wall, d (mm)	0.423	0.23
Cell angle, θ	30	30

In the current manuscript the so-called five-star auxetic structure is considered. The cellular auxetic structure is presented in Fig. 1.

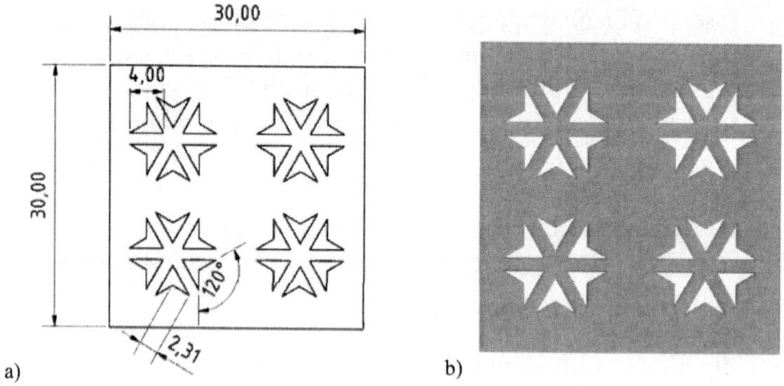

Fig. 1. a The layout dimensions and **b** the 3D CAD model.

4 Results—FE Analysis

To evaluate the behavior of woven textiles, a Finite Element Analysis (FEA) was performed using a model with auxetic microstructure. A FE analysis was executed on a representative volume element of the fabric with surface dimensions of 30 × 30 mm. The cell of auxetic microstructure, as shown in Fig. 1, is repeated throughout the entire size, regardless of the model's overall dimensions. An isotropic material with a Young's Modulus of 220.69 GPa and a Poisson's ratio of 0.2 was considered for the analysis. The auxetic shape can be fabricated using a laser-cutting process or using special knitting techniques. The results of the auxetic microstructure's tensile strength analysis and stress-strain curve are presented in Figs. 2 and 3, respectively. Likewise, the results of the compression FE analysis and the corresponding stress-strain curve are presented in Figs. 4 and 5, respectively.

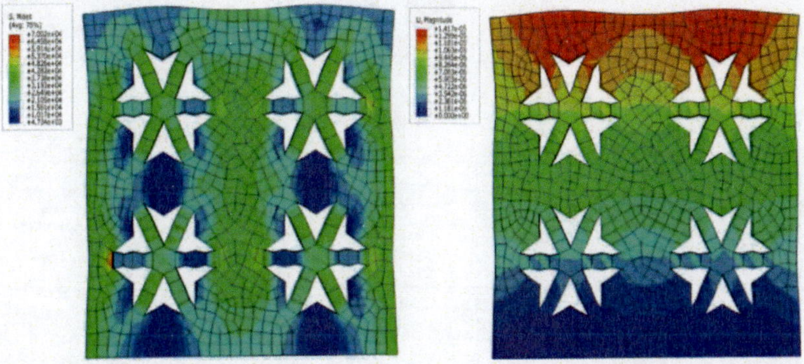

Fig. 2. The Von Mises stress and total displacement of auxetic microstructure (tensile strength analysis)

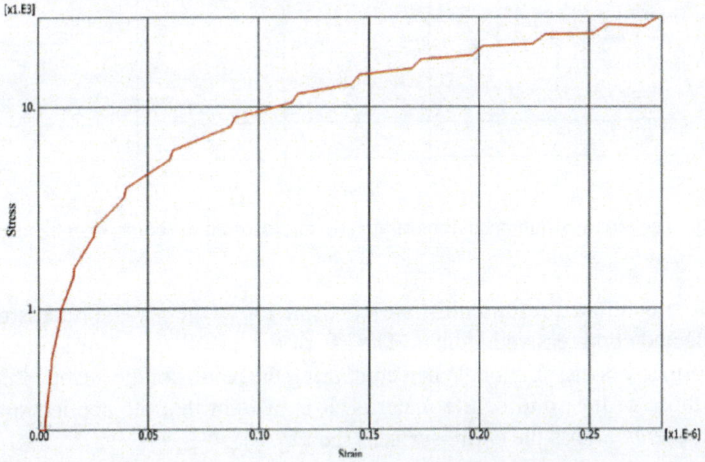

Fig. 3. The stress- strain behaviour of auxetic microstructure (tensile strength analysis)

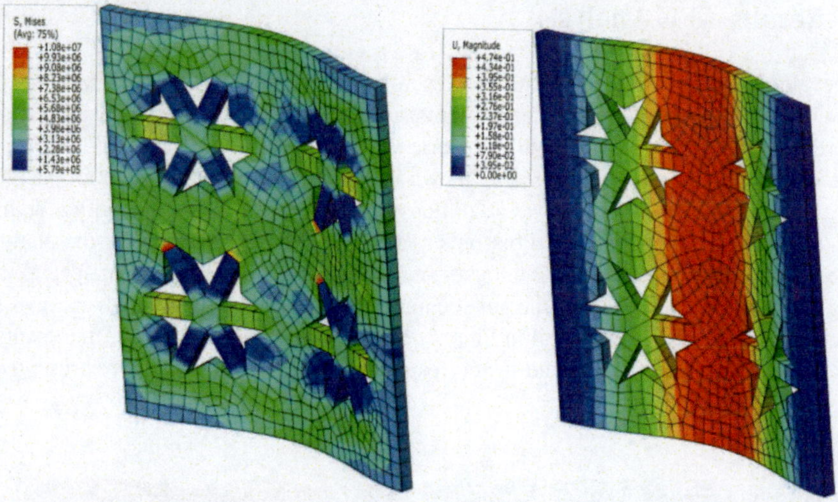

Fig. 4. The Von Mises stress and total displacement of auxetic microstructure (compression analysis)

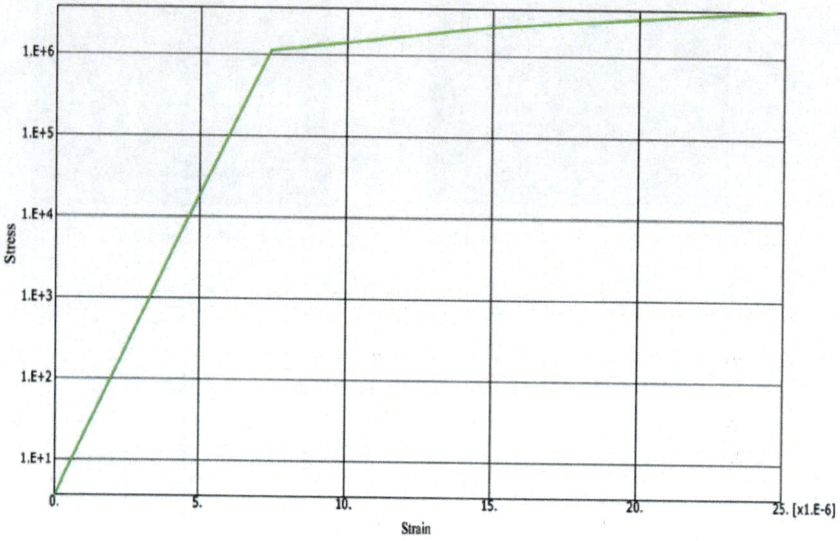

Fig. 5. The stress- strain behaviour of auxetic microstructure (compression analysis)

The aforementioned results illustrate the behavior of the auxetic microstructure in both tensile and compressive loadings. The use of the auxetic microstructure leads to a more uniform stress distribution. When comparing the results of the auxetic microstructure with those of the solid textile material, it is evident that the applied stresses are distributed evenly across the entire surface (see Fig. 6).

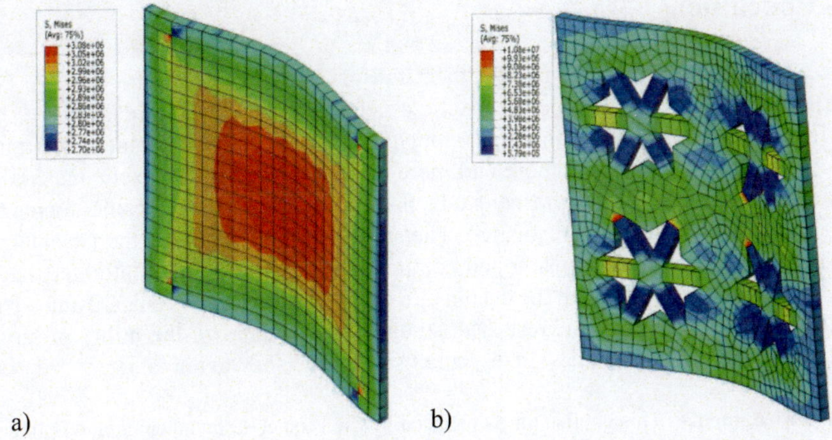

Fig. 6. **a** The (Von Mises) stress behaviour of solid textile model and **b** textile model with auxetic shape.

5 Product Design with Auxetic Structure

The use of an auxetic microstructure in textiles holds vast potential for various applications in product design. In Fig. 7 is presented an example, where fabric with auxetic microstructure is used on the back and on the seat of an armchair. By using the star-shaped auxetic microstructure, tensions are evenly spread throughout the entire seating surface. This attribute presents a noteworthy advantage when utilizing the auxetic structure for designing an array of products. The suggested manufacturing technique for producing the auxetic shape involves laser cutting and special knitting techniques.

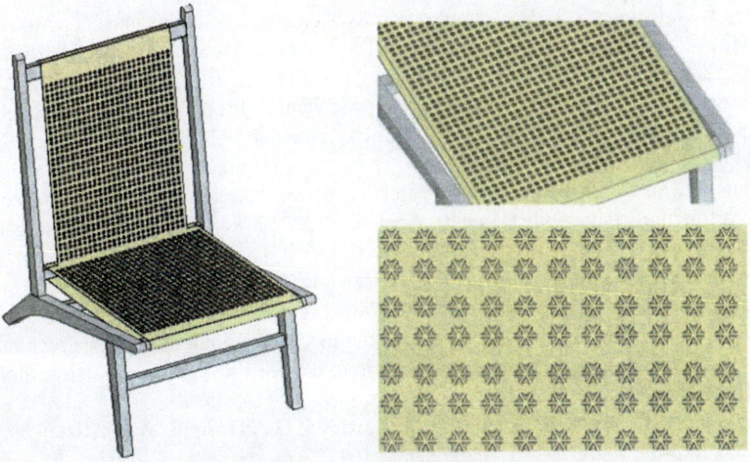

Fig. 7. Utilizing auxetic microstructure in the seat and in the back of an armchair

6 Conclusions

The auxetic materials are known for their unusual behavior due to the negative Poisson's ratio. The main advantages of auxetic structures, among others, are increased shear stiffness, increased plane strain fracture toughness, and increased indentation resistance. These properties are leading more and more researchers to utilize auxetic materials in various applications. In this present study, the use of a star-shaped auxetic microstructure in woven fabric was investigated. The study results highlighted the advantage of auxetic materials through tension and compression studies. Subsequently, the auxetic microstructure was utilized in the design of a product, specifically a chair, using auxetic structure in the seat and backrest. The study results emphasize the utility of auxetic microstructures in product design, especially in products involving fabric.

Acknowledgments. This publication is financed by the Project "Strengthening and optimizing the operation of MODY services and academic and research units of the Hellenic Mediterranean University", funded by the Public Investment Program of the Greek Ministry of Education and Religious Affairs.

References

1. Wan, H., Ohtaki, H., Kotosaka, S., Hu, G.: A study of negative Poisson's ratios in auxetic honeycombs based on a large deflection model. Eur. J. Mech. A. Solids **23**(1), 95–106 (2004). https://doi.org/10.1016/j.euromechsol.2003.10.006
2. Theocaris, P.S., Stavroulakis, G.E., Panagiotopoulos, P.D.: Negative Poisson's ratios in composites with star-shaped inclusions: a numerical homogenization approach. Arch. Appl. Mech. **67**(4), 274–286 (1997). https://doi.org/10.1007/s004190050117
3. Evans, K.E.: Auxetic polymers: a new range of materials. Endeavour **15**(4), 170–174 (1991). https://doi.org/10.1016/0160-9327(91)90123-S
4. Tairidis, G., Ntintakis, I., Drosopoulos, G., Koutsianitis, P., Stavroulakis, G.: Auxetic metamaterials subjected to dynamic loadings. Theor Appl. Mech. (Belgr) **49**(1), 1–14 (2022). https://doi.org/10.2298/TAM211103002T
5. Koustoumpardis, M.C.-G., Stavroulakis, G.: Mechanical Properties of Cookie-Shaped Auxetics using Finite Elements and Soft Robotics Application (2023). https://doi.org/10.24352/UB.OVGU-2023-059
6. Chinis, D., Stavroulakis, G.E.: Band gap analysis for materials with cookie-shaped auxetic microstructures, using finite elements. Appl. Sci. **13**(5), 2774 (2023). https://doi.org/10.3390/app13052774
7. Barchiesi, E., Spagnuolo, M., Placidi, L.: Mechanical metamaterials: a state of the art. Math. Mech. Solids **24**(1), 212–234 (2019). https://doi.org/10.1177/1081286517735695
8. Prawoto, Y.: Seeing auxetic materials from the mechanics point of view: a structural review on the negative Poisson's ratio. Comput. Mater. Sci. **58**, 140–153 (2012). https://doi.org/10.1016/j.commatsci.2012.02.012
9. Montgomery-Liljeroth, E., Schievano, S., Burriesci, G.: 2D. Appl. Mater. Today **30**, 101722 (2023). https://doi.org/10.1016/j.apmt.2022.101722
10. Shukla, S., Behera, B.K.: Auxetic fibrous materials and structures in medical engineering—a review. J. Text. Inst. **114**(7), 1078–1089 (2023). https://doi.org/10.1080/00405000.2022.2116549

11. Daraki, M.-S., Marakakis, K., Stavroulakis, G.E.: Modeling of shunted piezoelectrics and enhancement of vibration suppression through an auxetic interface. Micromachines **14**(2), 289 (2023). https://doi.org/10.3390/mi14020289
12. Yang, L., Harrysson, O., West, H., Cormier, D.: Mechanical properties of 3D re-entrant honeycomb auxetic structures realized via additive manufacturing. Int. J. Solids Struct. **69–70**, 475–490 (2015). https://doi.org/10.1016/j.ijsolstr.2015.05.005
13. Ochoa, O., Cuan-Urquizo, E., Álvarez-Trejo, A., Roman-Flores, A., Silva, R.G.: Additively manufactured auxetic arc-based architected metamaterial: mechanical properties and their directional dependency. Mech. Adv. Mater. Struct., pp. 1–14 (2023). https://doi.org/10.1080/15376494.2023.2231453
14. Li, Y., Yu, W.-R.: Development of a textile structure for multi-directional auxetic deformation. Mater. Des. **223**, 111237 (2022). https://doi.org/10.1016/j.matdes.2022.111237
15. Steed, A., Ofek, E., Sinclair, M., Gonzalez-Franco, M.: A mechatronic shape display based on auxetic materials. Nat. Commun. **12**(1), 4758 (2021). https://doi.org/10.1038/s41467-021-24974-0
16. Shukla, S., Behera, B.K.: Auxetic fibrous materials and structures in medical engineering—a review. J. Text. Inst., pp 1–12 (2022). https://doi.org/10.1080/00405000.2022.2116549
17. Dubrovski, P.D., Novak, N., Borovinšek, M., Vesenjak, M., Ren, Z.: In-plane behavior of auxetic non-woven fabric based on rotating square unit geometry under tensile load. Polymers **11**(6), 1040 (2019). https://doi.org/10.3390/polym11061040
18. Mizzi, L., Salvati, E., Spaggiari, A., Tan, J.-C., Korsunsky, A.M.: Highly stretchable two-dimensional auxetic metamaterial sheets fabricated via direct-laser cutting. Int. J. Mech. Sci. **167**, 105242 (2020). https://doi.org/10.1016/j.ijmecsci.2019.105242
19. Luo, C., Han, C.Z., Zhang, X.Y., Zhang, X.G., Ren, X., Xie, Y.M.: Design, manufacturing and applications of auxetic tubular structures: a review. Thin-Walled Struct. **163**, 107682 (2021). https://doi.org/10.1016/j.tws.2021.107682
20. Yuping, C., Liu, Y., Hong, H.: Deformation behavior of auxetic laminated fabrics with rotating square geometry. Text. Res. J. **92**(23–24), 4652–4665 (2022). https://doi.org/10.1177/00405175221109629
21. Konaković, M., Crane, K., Deng, B., Bouaziz, S., Piker, D., Pauly, M.: Beyond developable: computational design and fabrication with auxetic materials. ACM Trans. Graph., **35**(4), 89:1–89:11 (2016). https://doi.org/10.1145/2897824.2925944
22. Papadopoulou, A., Laucks, J., Tibbits, S.: Auxetic materials in design and architecture. Nat. Rev. Mater. **2**(12), 1–3 (2017). https://doi.org/10.1038/natrevmats.2017.78
23. Reid, D.R. et al.: Auxetic metamaterials from disordered networks. Proc. Natl. Acad. Sci. U.S.A., **115**(7) (2018). https://doi.org/10.1073/pnas.1717442115
24. Mir, M., Ali, M.N., Sami, J., Ansari, U.: Review of mechanics and applications of auxetic structures. Adv. Mater. Sci. Eng. **2014**, 1–17 (2014). https://doi.org/10.1155/2014/753496

Electrospun Nanofiber Mats for Mammalian Cell Growth

Andrea Ehrmann(✉)

Working Group "Research On Materials for Functional Textiles and 3D Printing (MTex3)", Faculty of Engineering Sciences and Mathematics, Bielefeld University of Applied Sciences and Arts, 33619 Bielefeld, Germany
andrea.ehrmann@hsbi.de

Abstract. Electrospinning can be used to prepare nanofiber mats from diverse polymers, polymer blends and embedded nanoparticles. Especially nanofiber mats from biopolymers promote cell growth for wound healing, tissue engineering or other biotechnological applications. These materials, however, are often not stable at high temperatures, which makes sterilization challenging. The paper discusses sterilization methods, a comparison of different biopolymer blends for cell growth, and the influence of vacuum ultraviolet irradiation (VUV) and plasma treatment on cell growth and adhesion on electrospun nanofiber mats. This mini-review thus serves as a base for researchers working on the improvement of nanofiber mats for mammalian cell growth, giving some general ideas as well as interesting special findings.

Keywords: Electrospinning · Poly(acrylonitrile) (PAN) · Biopolymers · Mammalian cells · Sterilization · Surface modification

1 Introduction

Cell cultivation belongs to the standard procedures in biotechnology. Typically, adherent cells are grown in roller bottles or well plates, which are relatively expensive [1], or nowadays using microcarriers or hollow fibers as substrates [2, 3]. Single-use fixed-bed bioreactors, on the other hand, contain macroporous nonwovens or rolled membranes, often from hydrophilized poly (ethylene terephthalate) (PET) microfibers [4].

As a possible alternative, nanofiber mats have been investigated for the adhesion and proliferation of diverse cells [5]. Nanofibrous surfaces can improve and guide cell growth as well as support stem cell differentiation [6]. The increasing possibilities to produce nanofiber mats from different polymers by diverse electrospinning techniques make this approach more and more attractive [7, 8], especially due to the possibility to create membranes with highly oriented fibers to guide cell growth [9].

The paper starts with a discussion of possible sterilization methods for nanofiber mats and sheds a light on the impact of the chosen polymer as well as plasma of VUV treatment on the growth of adherent mammalian cells on nanofibrous membranes.

G. Guxho et al. (Eds.): ITC-ICEE 2023, LNMIE, pp. 518–527, 2024.
https://doi.org/10.1007/978-3-031-48933-4_51

2 Sterilization of Nanofiber Mats

In many cases, sterilization is performed by autoclaving, i.e. by heating up the samples to a temperature of typically 121 °C under a defined pressure [6, 10]. This temperature, however, may be much too high for electrospun nanofiber mats from many biopolymers and other polymers [6].

This is why other authors tested different methods. Machado-Paula et al. sterilized their poly(ε-caprolactone) (PCL) nanofiber mats with UV light for 30 min on each side and did not report any contaminations nor a negative impact on the viability of human dermal fibroblasts (HDF) grown on them [11]. Sefat et al. used γ-irradiation with an external dose range of 25–40 kGy and found a significant reduction of the molecular weight and a significant increase of the polydispersity as well as the average fiber diameter, combined with more brittle samples after γ-irradiation, but no effect on the cell adhesion on the membrane [12]. A simpler method is sterilization by alcohol, as Augustine et al. used it combined with UV treatment [13].

A comparison of different methods was reported, e.g., by Yamaguchi et al. who found no difference in the number of hydroxyl groups in ethanol-sterilized and non-sterilized silicate nanofiber mats, but a reduction of hydroxyl groups by more than half after autoclaving [14].

In a more comprehensive study, Wehlage et al. compared sterilization of poly(acrylonitrile) (PAN)/gelatin nanofiber mats by autoclaving, heat treatment, UV treatment and ozone treatment [6]. On the one hand, autoclaving severely damaged the surface, as visible in Fig. 1a, as compared to the ozone-treated surface which did not change its morphology, as depicted in Fig. 1b. On the other hand, ozone had a significantly negative impact on the growth of Chinese hamster ovary (CHO) cells on these sterilized nanofiber mats, while especially hot air sterilization was found to be supportive for the cell adhesion. Autoclaving, on the other hand, gave the highest sterility, suggesting using this method if possible in spite of the surface modifications.

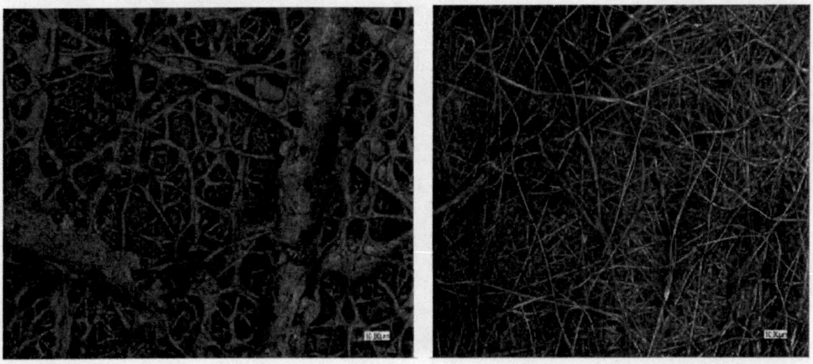

Fig. 1. Comparison of PAN/gelatin nanofiber mats (left panel) after autoclaving and (right panel) after ozone sterilization. From [6], originally published under a CC BY-SA 4.0 license.

Besides the sterilization techniques and the nano-structure, the materials used to prepare the nanofiber mats are highly important for the adhesion and proliferation of mammalian cells.

3 Biopolymeric Nanofiber Mats for Mammalian Cell Growth

In many cases, biopolymers show advantageous properties for the growth of mammalian cells. Several biopolymers, however, are water-soluble, necessitating a subsequent crosslinking step after electrospinning, which can be challenging due to the cell-toxicity of many crosslinking agents [15]. On the other hand, combining biopolymers with water-stable polymers, such as PAN, can be a good strategy to add the advantageous properties of biopolymers without the necessity of crosslinking [6].

Iliou et al. reviewed the possibilities of marine biopolymers for electrospun nanofiber mats with good cell adhesion and proliferation [16]. Amongst the favorable polymers, they mentioned especially alginate, but also carboxymethyl-κ-carrageenan, fucoidan, chitin and chitosan. Generally, they found the broad range of bioactivities shown by marine biopolymers to be positive for their use as scaffolds for cell proliferation and differentiation [16]. The afore-mentioned chitosan and gelatin were also reported by Ranganathan et al. to enhance bone re-generation, making them highly useful for bone tissue engineering [17]. Blends from poly(lactic acid) (PLA) and poly(hydroxyoctanoate) (PHO) with different ratios were investigated by Solarz et al. who mentioned especially the tendency of mouse embryonic fibro-blasts to grow not only on the surface, but also inside the nanofiber mats, as visible in the sample cross-sections depicted in Fig. 2 [18]. This behavior generally makes nanofiber mats preferable to flat substrates as it enables a 2.5D or even 3D cell culture.

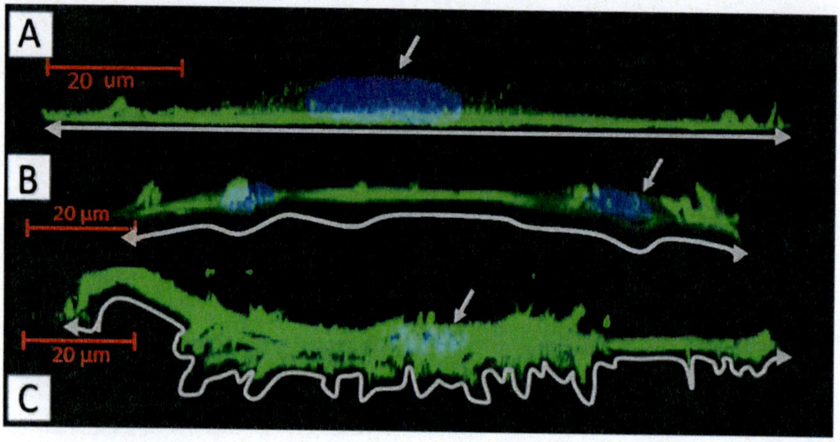

Fig. 2. Side projections of the mouse embryonic fibroblast 3T3 cells, obtained by fluorescent confocal microscopy (actin cytoskeleton-green, nuclei-blue): (A) cell cultured on the glass; (B) cells cultured on the 75% PLA/25% PHO blend; (C) cells cultured on the 40% PLA/60% PHO blend. From [18], originally published under a CC BY 3.0 license.

Bacterial biopolymers were investigated by Esmail et al. who extracted them from the lyophilized biomass of Cupriavidus necator and Pseudomonas chlororaphis, respectively [19]. The so obtained homopolymer P(3HB) and the copolymer P(3HB-co-3HV) as well as mcl-PHA were electrospun in different blends, resulting in different fiber morphologies. Interestingly, most of these electrospun scaffolds showed a higher cell attachment in comparison with a commercial 3D mesh, as shown in Fig. 3; however, the values were significantly below those measured on a common polystyrene well [19].

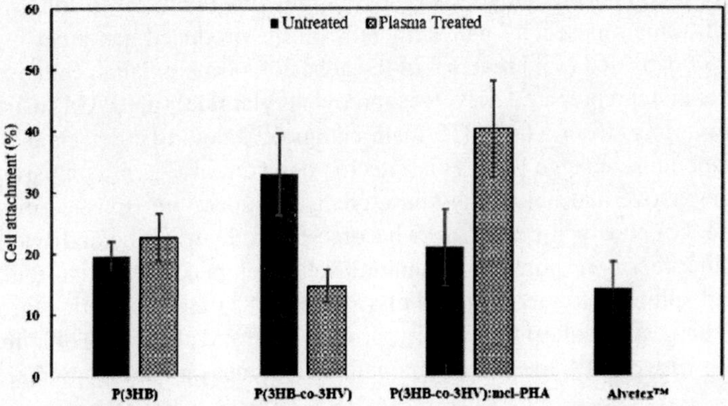

Fig. 3. Comparison of the HDFns adhesion on different types of PHA based scaffolds with and without oxygen plasma treatment, measured by MTT assay. A commercial 3D mesh, Alvetex™, is used as reference. The cell attachment values are the ratio between the absorbance of each sample and the attachment attained in polystyrene wells. From [19], originally published under a CC BY license.

Combining PAN with different biopolymers, Wehlage *et al.* tested the adhesion and proliferation of CHO cells on nanofiber mats [20]. As a negative reference, they used PAN/ZnO on which no cell growth could be observed. Pure PAN used after electrospinning, additional thermal stabilization as well as carbonization enable a small amount of cell growth. PAN blended with chitosan and gelatin showed much higher cell growth, as expected due to the aforementioned literature reports. Blending PAN with maltodextrin and pea protein, on the other hand, did not improve cell adhesion and proliferation as compared to pure PAN, showing that not all biopolymers are generally advantageous to prepare electrospun scaffolds for the adhesion and growth of mammalian cells.

4 Cell Growth and Viability

Depending on the aim of the studies, different cell types are investigated, and different parameters are in the focus of the investigation. Jenab et al., e.g., produced nanofiber mats from PAN blended with kefiran, a polysaccharide from branched glucogalactose, and investigated the attachment and potential morphological changes of PC12 neural stem cells [21]. They found 5% kefiran advantageous for cell growth, while 10% kefiran were assumed to be able to induce cell differentiation.

Ruiter et al., investigated scaffolds for human cornea stromal cell (hCSC) proliferation
and found that blending PLA with poly(di(ethylene glycol) methyl ether methacrylate)-co-poly(di(ethylene-glycol) carboxylate ester ethyl thiol) (PDEGMA/SH) instead with pure PDEGMA resulted in elongated and spread cell morphologies, i.e. clearly better adhesion of hCSC cells [22]. An increase of mammalian cell activity, combined with reduced bacterial growth, was found by Chung et al. when adding selenium nanoparticles to electrospun silk nanofiber mats [23]. Similarly, adding 20–40% zein to PCL spinning solutions resulted in bactericidal properties combined with improved adhesion and proliferation of human mesenchy-mal stem cells on the produced nanofiber mats [24]. A temperature-controlled rapid reversal of the adhesion of mammalian cells was reported by Cicotte et al. who prepared poly(N-isopropyl acrylamide) (pNIPAM) nanofiber mats by electrospinning, from which 3T3 mammalian cells could be detached very fast by a rapid temperature change [25]. This effect is depicted in Fig. 4, comparing different temperatures above and below the phase change temperature (lower critical solution temperature, below which the polymer becomes more hydrophilic and swells) of pNIPAM. Another thermoresponsive cell culture was developed by Aladdad et al. who used blend-electrospinning to incorporate Poly(PEGMA188) into poly(ethylene terephthalate) (PET) and tested adhesion, viability, proliferation and phenotype of human corneal stromal cells as well as some other mammalian cell types on these nanofiber mats [26].

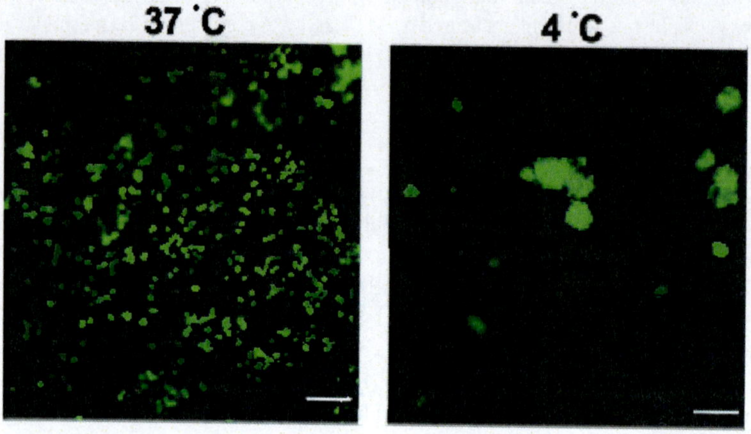

Fig. 4. Fluorescence microscopy image of 3T3 cells tagged with CellTracker. This marker allowed imaging of live cells through the opaque espNIPAM mat. Images of 3T3 cells adhered to electrospun pNIPAM nanofiber mats were obtained above the lower critical solution temperature (LCST) (left) and after the cells have detached below the LCST (right). The scale bar is 100 μm. From [25], copyright 2017 AIP Publishing, reproduced with permission.

Amongst the most important parameters to evaluate a substrate, the cell viability has to be mentioned. Canbolat et al., e.g., compared the viability of 3T3 cells which were integrated into the electrospinning solution with cells on which electrospinning was performed and found that the first died due to dehydration and fiber stretching, while the latter survived [27]. The addition of TiO2 nanoparticles to PLA nanofiber mats resulted in high bactericidal activity of the membrane without toxicity for mammalian cells [28]. Similarly, Sun et al. found very good cell viability for cells grown on electrospun cellulose acetate (CA) nanofiber fabrics containing an N-halamine antimicrobial agent, combined with high antimicrobial efficacy [29]. Electrospun membranes from thermoplastic polyurethane (TPU)/graphene oxide (GO) with 0.5 wt% GO showed high cell viability and proliferation for mouse fibroblast 3T3 and human umbilical vein endothelial cells (HUVECs), while the latter also showed increased viability for an oxygen plasma treatment of the nanofiber mats [30]. Testing aligned poly-(glycerol se-bacate):poly(caprolactone) (PGS:PCL) electrospun nanofiber mats, Masoumi et al. reported high viability of primary human valvular interstitial cells (VICs) on these scaffolds, independent of the PGS:PCL ratio, with a slight decrease during the 10 days of culture, as usual for cells grown on synthetic scaffolds [31].

5 Cell Differentiation

Another important parameter especially in stem cell cultivation is the cell differentiation. Jung et al. embedded gold nanoparticles labeled with cell-adhesive peptides in polymethylglutarimide (PMGI) nanofiber mats and found, besides significantly increased attachment of HeLa cells, increased cardiomyocyte differentiation of human pluripotent stem cells, which they evaluated by observing whether colonies started contracting synchronized [32].

Badami et al., evaluated the influence of the average fiber diameter on the differentiation of osteoblastic cells and found smaller projected areas for cells grown on fibers, as compared to flat surfaces, and a higher aspect ratio for growth on relatively thick fibers with diameters around 2.1 μm, indicating an influence of the substrate morphology not only on cell proliferation, but also on cell morphology [33]. The impact of the fiber diameter was also investigated by Christopherson et al. who evaluated rat neural stem cells grown on nanofiber mats with average diameters between 283 nm and 1452 nm and found especially high oligodendrocyte differentiation on the thinnest fibers, compared to higher cell proliferation and spreading [34], as opposed to the previous paper, indicating that the effect of nanofiber diameter differs depending on the stem cells and the required differentiation (Fig. 5).

Hackett et al., investigated a scaffold for spinal cord injury recovery and found that an electrospun PCL/collagen nanofiber mat could promote cell differentiation in vitro, which they attributed to the similarity to the native extracellular matrix and the porosity of the scaffold which could be used to add growth factors, in this way controlling growth of the desired cell types [35]. Besides pores and fiber diameters, Lins et al. reported a significant impact of the fiber alignment on the differentiation of neural stem cells into neuronal and glial cells, using electrospun poly(vinylidene) fluoride (PVDF) nanofiber mats [36].

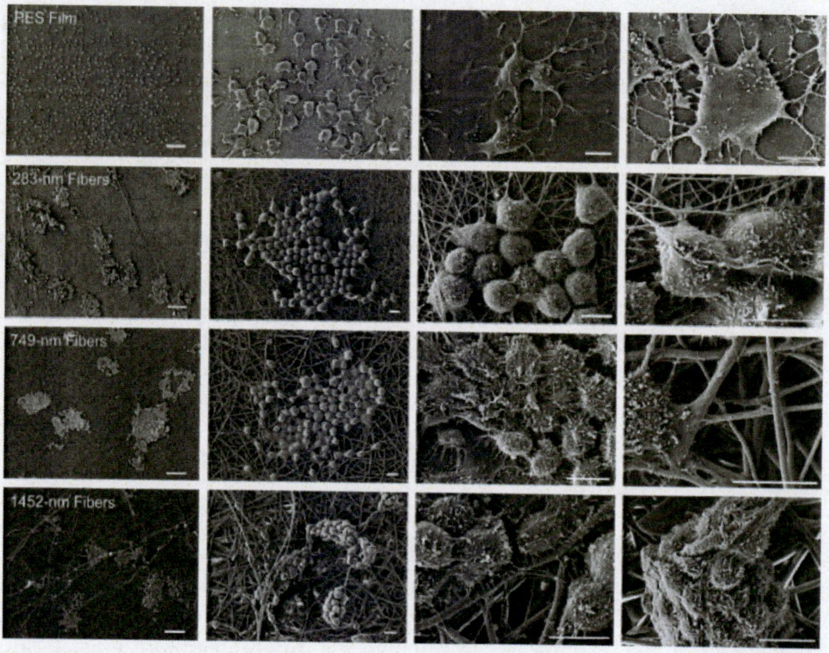

Fig. 5. SEM images of rat neural stem cells cultured in serum free medium and 20 ng/mL FGF-2 serum for 5 days. Scale bars for low magnification images (first column) are 100 μm; all other scale bars are 10 μm. Reprinted from [34], copyright 2009, with permission from Elsevier.

6 VUV and Plasma Treatment

While the chemical and morphological parameters of electrospun nanofiber mats influence cell adhesion and proliferation as well as stem cell differentiation, there are also possibilities to use post-treatments to optimize the desired properties of the electrospun membranes. Atmospheric pressure plasma with Ar and O_2 gas was used by Das et al., to increase the in vitro biocompatibility of poly(vinyl alcohol)/chitosan nanofiber mats, as tested by mouse fibroblast cells [37]. Lopresti et al. used air plasma to increase the long-term stability of a kefiran coating on PLA nanofiber mats, where the coating supported proliferation and metabolic activity of 3T3 cells [38]. A plasma treatment was also used to increase the attachment and proliferation of olfactory ensheathing cells on poly[(l-lactide)-co-(ϵ-caprolactone)] (PLCL) and poly[(l-lactide)-co-glycolide] (PLGA) copolymers nanofiber mats, while reaching high cell viability of more than 95% [39].

VUV treatment combined with applying a mechanical strain was reported to support cell orientation on thin polyester films by formation of parallel microgroves [40], but still needs research regarding the effect on electrospun nanofiber mats [41].

7 Conclusions

This mini-review gave an overview of parameters influencing the growth of mammalian cells on nanofiber mats. Besides discussing possible sterilization methods for nanofiber mats, it described the impact of the polymer material on cell attachment and proliferation for some typical biopolymers and other blends. Possibilities to improve stem cell differentiation were as well described as potential surface improvements by plasma treatment. This paper thus serves as a base for researchers interested in this field of research, giving some main ideas, without being meant to be exhaustive.

Acknowledgment. This research was partly funded by the German Federal Ministry for Economic Affairs and Energy as part of the Central Innovation Program for SMEs (ZIM) via the AiF, based on a resolution of the German Bundestag, grant number KK5129703CR0.

References

1. Chatel, A.: A brief history of adherent cell culture: where we come from and where we should go. BioProcess Int. **17**, 44–49 (2019)
2. Whitford, W.G., Hardy, J.C., Cadwell, J.J.S.: Single-use, continuous processing of primary stem cells. BioProcess Int. **12**, 26–32 (2014)
3. Allan, S.J., De Bank, P.A., Ellis, M.J.: Bioprocess design considerations for cultured meat production with a focus on the expansion bioreactor. Front. Sustain. Food Syst., **3**(44) (2019)
4. Lennaertz, A., Knowles, S., Drugmand, J.C., Castillo, J.: Viral vector production in the integrity iCELLis single-use fixed-bed bioreactor, from bench-scale to industrial scale. BMC Proc. **7**, 59 (2013)
5. Tanzli, E., Ehrmann, A.: Electrospun nanofibrous membranes for tissue engineering and cell growth. Appl. Sci. **11**, 6929 (2021)
6. Wehlage, D., Blattner, H., Sabantina, L., Böttjer, B., Grothe, T., Rattenholl, A., Gudermann, F., Lütkemeyer, D., Ehrmann, A.: Sterilization of PAN/gelatine nanofibrous mats for cell growth. Tekstilec, **62**, 78–88 (2019)
7. Mamun, A.: Review of possible applications of nanofibrous mats for wound dressings. Tekstilec **62**, 89–100 (2019)
8. Gao, S.T., et al.: Stimuli-responsive bio-based polymeric systems and their applications. J. Mater. Chem. B **7**, 709–729 (2019)
9. Bazrafshan, Z., Stylios, G.K.: A novel approach to enhance the spinnability of collagen fibers by graft polymerization. Mater Sci. Eng. **94**, 108–116 (2019)
10. Yamaguchi, T., Sakai, S., Kawakami, K.: Application of silicate electrospun nanofibers for cell culture. J. Sol-Gel Sci. Technol. **48**, 350–355 (2008)
11. M. M. Machado-Paula, M. A. f. Corat, M. Lancellotti, G. Mi, F. R. Marciano, M. L. Vega, A. A. Hidalgo, T. J. Webster, A. O. Lobo, "A comparison between electrospinning and rotary-jet spinning to produce PCL fibers with low bacteria colonization," in Mater. Sci. Eng. C, vol. 111, 2020, no. 110706
12. F. Sefat, R. McKean, P. Deshpande, C. Ramachandran, C. J. Hill, V. S. Sangwan, A. J. Ryan, S. MacNeil, „Production, Sterilisation and Storage of Biodegradable Electrospun PLGA Membranes for Delivery of Limbal Stem Cells to the Cornea," in Proc. Eng., vol. 59, 2013, pp. 101–116

13. R. Augustine, S. R. ur Rehman, Joshy K. S., A. Hasan, "Stromal cell-derived factor loaded co-electrospun hydrophilic/hydrophobic bicomponent membranes for wound protection and healing," in RSC Adv., vol. 11, 2021, pp. 572–583
14. Yamaguchi, T., Sakai, S., Watanabe, R., Tarao, T., Kawakai, K.: Heat treatment of electrospun silicate fiber substrates enhances cellular adhesion and proliferation. J. Biosci. Bioeng. **109**, 304–306 (2010)
15. Ehrmann, A.: Non-toxic crosslinking of electrospun gelatin nanofibers for tissue engineering and biomedicine—a review. Polymers, **13**(1973) (2021)
16. Iliou, K., Kikionis, S., Ioannou, E., Roussis, V.: Marine biopolymers as bioactive functional ingredients of Electrospun Nanofibrous scaffolds for biomedical applications. Mar. Drugs, **20** (314) (2022)
17. Ranganathan, S., Balagangadharan, K., Selvamurugan, N.: Chitosan and gelatin-based electrospun fibers for bone tissue engineering. Int. J. Biol. Macromol. **133**, 354–364 (2019)
18. Solarz, D., Witko, T., Karcz, R., Malagurski, I., Ponjavic, M., Levic, S., Nesic, A., Guzik, M., Savic, S., Nikodinovic-Runic, J.: Biological and physiochemical studies of electrospun polylactid/polyhydroxyoctanoate PLA/P(3HO) scaffolds for tissue engineering applications. RSC Adv. **13**, 24112–24128 (2023)
19. Esmail, A., Pereira, J.R., Zoio, P., Silvestre, S., Menda, U.D., Sevrin, C., Grandfils, C., Fortunato, E., Reis, M.A.M., Henriques, C., Oliva, A., Freitas, F.: Oxygen plasma treated-electrospun polyhydroxyalkanoate scaffolds for hydrophilicity improvement and cell adhesion. Polymers **13** (1056) (2021)
20. Wehlage, D., Blattner, H., Mamun, A., Kutzli, I., Diestelhorst, E., Rattenholl, A., Gudermann, F., Lütkemeyer, D., Ehrmann, A.: Cell growth on electrospun nanofiber mats from polyacrylonitrile (PAN) blends. AIMS Bioeng., **7**, 43–54 (2020)
21. Jenab, A., Roghanian, R., Ghorbani, N., Ghaedi, K., Emtiazi, G.: The efficacy of electrospun PAN/Kefiran nanofiber and kefir in mammalian cell culture: promotion of PC12 cell growth, anti-MCF7 breast cancer cells activities, and cytokine production of PBMC. Int. J. Nanomed., **15**, 717–728 (2020)
22. Ruiter, F.A.A., Sidney, L.E., Kiick, K.L., Segal, J.I., Alexander, C., Rose, F.R.A.J.: The electrospinning of a thermo-responsive polymer with peptide conjugates for phenotype support and extracellular matrix production of therapeutically relevant mammalian cells. Biomater Sci. **8**, 2611–2626 (2020)
23. Chung, S., Ercan, B., Roy, A.K., Webster, T.J.: Addition of selenium nanoparticles to electrospun silk scaffold improves the mammalian cell activity while reducing bacterial growth. Front. Physiol., **7**(297) (2016)
24. Plath, A.M.S., Facchi, S.P., Souza, P.R., Sabino, R.M., Corradini, E., Muniz, E.C., Popat, K.C., Filho, L.C., Kipper, M.J., Martins, A.F.: Zein supports scaffolding capacity toward mammalian cells and bactericidal and antiadhesive properties on poly(ε-caprolactone)/zein electrospun fibers. Mater. Today Chem., **20**(100465) (2021)
25. Cidotte, K.N., Reed, J.A., Nguyen, P.A.H., de Lora, J.A., Hedberg-Dirk, E.L., Canavan, H.H.: Optimization of electrospun poly(N-isopropyl acrylamide) mats for the rapid reversible adhesion of mammalian cells. Biointerphases, **12**(02C417) (2017)
26. Aladdad, A.M., Amer, M.H., Sidney, L., Hopkinson, A., White, L.J., Alexander, C., Rose, F.R.A.J.: A thermoresponsive three-dimensional fibrous cell culture platform for enzyme-free expansion of mammalian cells. Acta Biomaterialia, **95**(427–438) (2019)
27. Canbolat, M.F., Tang, F., Bernacki, S.H., Pourdeyhimi, B., Khan, S.: Mammalian cell viability in electrospun composite nanofiber structures. Macromol Biosci. **11**, 1346–1356 (2011)
28. Toniatto, T.V., et al.: Nanostructured poly (lactic acid) electrospun fiber with high loadings of TiO2 nanoparticles: insights into bactericidal activity and cell viability. Mater Sci. Eng. C **71**, 381–385 (2017)

29. Sun, X.B., et al.: Electrospun composite nanofiber fabrics containing uniformly dispersed antimicrobial agents as an innovative type of polymeric materials with superior antimicrobial efficacy. ACS Appl Mater. Interfaces **2**, 952–956 (2010)
30. Jing, X., Mi, H.-Y., Salick, M.R., Cordie, T.M., Peng, X.-F., Turng, L.-S.: Electrospinning thermoplastic polyurethane/graphene oxide scaffolds for small diameter vascular graft applications. Mater. Sci. Eng. C, **49**, 40–50 (2015)
31. Masoumi, N., et al.: Electrospun PGS: PCL microfibers align human Valvular interstitial cells and provide Tunable scaffold anisotropy. Adv Healthcare Mater. **3**, 929–939 (2014)
32. Jung, D.J., Minami, I., Patel, S., Lee, J.W., B. J., Yuan, Q.H., Li, L., Kobayashi, S., Chen, Y., Lee, K.-B., Nakatsuji, N.: Incorporation of functionalized gold nanoparticles into nanofibers for enhanced attachment and differentiation of mammalian cells. J. Nanobiotechnol., **10** (23) (2012)
33. Badami, A.S., Kreke, M.R., Thompson, M.S., Riffle, J.S., Goldstein, A.S.: Effect of fiber diameter on spreading, proliferation, and differentiation of osteoblastic cells on electrospun poly(lactic acid) substrates. Biomaterials **27**, 596–606 (2006)
34. Christopherson, G.T., Song, H.J., Mao, H.Q.: The influence of fiber diameter of electrospun substrates on neural stem cell differentiation and proliferation. Biomaterials **30**, 556–564 (2009)
35. Hackett, J.M., Dang, T.T., Tsai, E.C., Cao, X.D.: Electrospun biocomposite polycaprolactone/collagen tubes as scaffolds for neural stem cell differentiation. Materials **3**, 3714–3728 (2010)
36. Lins, L.C., Wianny, F., Livi, S., Dehay, C., Duchet-Rumeau, J., Gérard, J.-F.: Effect of polyvinylidene fluoride electrospun fiber orientation on neural stem cell differentiation. J. Biomed. Mater. Res. B Appl. Biomater. **105**, 2376–2393 (2017)
37. Das, P., Ojah, N., Kandimalla, R., Mohan, K., Gogoi, D., Dolui, S.K., Choudhury, A.J.: Surface modification of electrospun PVA/chitosan nanofibers by dielectric barrier discharge plasma at atmospheric pressure and studies of their mechanical properties and biocompatibility. Int. J. Biol. Macromol. **114**, 1026–1032 (2018)
38. Lopresti, F., Campora, S., Tirri, G., Capuana, E., Pavia, F.C., Brucato, V., Ghersi, G., la Carruba, V.: Core-shell PLA/Kef hybrid scaffolds for skin tissue engineering applications prepared by direct kefiran coating on PLA electrospun fibers optimized via air-plasma treatment. Mater. Sci. Eng. C, **127**(112248) (2021)
39. Techaikool, P., Daranarong, D., Kongsuk, J., Boonyawan, D., Haron, N., Harley, W.S., Thomson, K.A., Foster, L.J.R., Punyodom, W.: Effects of plasma treatment on biocompatibility of poly[(L-lactide)-co-(ϵ-caprolactone)] and poly[(L-lactide)-co-glycolide] electrospun nanofibrous membranes. Polym. Int. **66**, 1640–1650 (2017)
40. Barb, R.-A., et al.: VUV treatment combined with mechanical strain of stretchable polymer foils resulting in cell alignment. Appl Surf. Sci. **325**, 105–111 (2015)
41. Chen, L., Yan, C., Zheng, Z.J.: Functional polymer surfaces for controlling cell behaviors. Mater Today **21**, 38–59 (2018)

Assessment of Silk Quality in the Shkodra Region of Albania: A Study on Physico-Mechanical Properties

Ilda Kola, Blerina Kolgjini(✉), and Suela Ademi

Polytechnic University of Tirana, "Mother Tereza" Square 1, 1019 Tirana, Albania
{ikola,bkolgjini}@fim.edu.al

Abstract. This article focuses on a thorough investigation of the mechanical characteristics of silk produced in the Shkodra region, which was formerly well-known for its silkworm farming. A key determinant of silk's quality and suitability for many uses, including textiles and fashion, are its mechanical qualities. Three different samples, each with a different number of filaments, are examined as part of the inquiry. The analysis's findings show that silk has linear density ranging from 0.251 to 0.57 [Tex] and tenacity ranging from 0.223 to 0.280 [N/Tex]; comparable to theoretical values of silk filaments. The study highlights the significance of silk production as a social and economic opportunity as well as a vital aspect of preserving the cultural heritage of the region.

Keywords: Silk fiber · Linear density · Tensile · Elongation

1 Introduction

Silk production in Albania was so common in the past that the country earned the nickname "Tibet of Europe". By the end of the ninth century and the beginning of the tenth, Albania had become an important supply of raw silk, notably for Venice [1–4]. Over the years, albanian silk manufacturers specialized, with Elbasan having 27 silk factories in the 17th century [4] and Shkodra being the primary hub of the province for silk cultivation, particularly during the Ottoman takeover. Between the 16th and 19th centuries, Shkodra became the most significant silk trading hub in the region, shipping to towns such as Cremona, Milan, and Venice in Italy [1–3]. Villages in the Shkodra area, such as Mjeda, Shiroka, Zogaj, GuriiZi, and Zadrima, played a vital role in silk cultivation [3, 4].

Albania is famous for its traditional garments, which are often produced from local materials such as silk, wool, linen, and hemp. These traditional outfits serve as a witness to Albanian textile industry history and are preserved in a variety of ways, including written documents [5, 6], museum items and pictures [7] as presented (see Fig. 1).

Furthermore, the cultural significance of Albanian silk is represented in the presence of traditional silk garments, ornamental components, and scarves in prestigious institutions such as the Boston Museum of Fine Arts and the Metropolitan Museum of Art in

G. Guxho et al. (Eds.): ITC-ICEE 2023, LNMIE, pp. 528–534, 2024.
https://doi.org/10.1007/978-3-031-48933-4_52

a) b) c)

Fig. 1. a) Women in traditional dress made by silk, Picture from Studio Marubi, Shkodra, Albania [7]. b). The riza (pl. Rizat), the traditional silk headscarf of northern Albanian villages. c) Silver embroidered silk blouse, probably early 20th century [5].

New York, USA [3, 4]. These documents serve to highlight the importance of silk as a cultural treasure in Albanian society.

Silk has often been associated with luxury, tradition, and craftsmanship, making it a meaningful and enduring part of cultural heritage. The dynamics of silk farming in the region, however, have evolved dramatically during the last century. Several causes influenced the drop in interest in silk fibers. The growth of synthetic fibers, which had reached an advanced stage in terms of development and manufacturing, created major competition. Furthermore, the centralization of the country's economy and failed attempts to introduce new species of silkworms contributed to the fall [8].

Despite these obstacles, some Zadrima families managed to maintain silk cultivation as a valued practice even throughout the communist era. Today, they continue to cultivate silk and produce a variety of items, keeping on the tradition of Albanian silk craftsmanship.

To revitalize and sustain this rich legacy, efforts are being made to encourage more people in the Shkodra region to engage in silk cultivation. Silk production is not only a viable economic opportunity, but also an important component in conserving the region's cultural heritage.

Considering the importance of silk in the Shkodra region, this study intends to provide a complete assessment of the physical and mechanical characteristics of silk produced by the local producers. Physical and mechanical properties are critical indicators of silk quality. Silk fibers are remarkable materials with unique mechanical qualities such as tensile strength, elasticity, and fineness [9].

Understanding these qualities is critical for determining the potential of Shkodra's silk in a variety of modern applications. This study aims to provide knowledge on the distinct characteristics of Shkodra silk and its potential for revival in both traditional and modern contexts. Understanding the mechanical characteristics of silk can help to preserve and revitalize a valuable cultural practice while also providing social and economic advantages.

2 Materials and Methods

In this study, three samples of Bombyx Mori homemade mulberry silk are considered. The silk is collected from different producers in Shkodra Region (North West of Albania). Each sample is composed by a different number of filaments per yarn, as shown in Table 1.

The samples were prepared to ensure their purity and integrity for testing. This includes cleaning and cutting into uniform specimens. Samples were tested under standard conditions in terms of temperature and humidity.

In order to find the linear density expressed in Tex, 10 repetitions of weighing six meter yarn segments are performed respectively for each sample, and the average value of mass is calculated.

$$Tex_y = \frac{m(gr)}{l(m)} 1000$$

In addition measurements for tensile strength and elongation are performed at the Accredited Physical Mechanical Laboratory, Department of Textile and Fashion in Polytechnic University of Tirana.

Tensile tests are performed according to international standards for tensile tests for multi filaments yarns—EN ISO2062 [10].

For each sample are performed 50 repetitions of tensile strength and elongation testing.

The average of the obtained values and the standard deviation are calculated and presented in Table 2.

3 Results and Discussions

To assess the quality of silk fibers one of the most important parameters is the linear density/fines of each single filament. Therefore the linear density per each filament is calculated by dividing the Linear Density of the yarn to the number of filaments. The results are presented in Table 1.

The linear density per filament of the tested samples varies from 0.224 to 0.57 Tex. Results for Sample 2 and Sample 3 are comparable with existing literature values of 0.1–0.3 Tex [3, 9, 11]. These values refer to the pure silk fibers which mean fibers composed only by fibroin and the sericin is washed out during the extractions process.

Sample 1 show a higher value of linear density, more than double of the other values, and this may be as a result of the presence of sericin in the chemical composition of the fiber.

In this study needs to be highlighted the fact that these fibers are entirely homemade, and they may be subject to extensive tensile applied during the extraction of the fibers from the cocoon or the impurities that may be present in the fiber, such as sericin which needs to be washed out of the clean fibers, in order to have a better quality in terms of physical and mechanical properties of silk filaments [12, 13].

Table 1. The linear density (tex and den) of the silk samples.

Sample	Mass (gram)	Length (meter)	Number of filaments	Linear density of yarn (Tex_y)	Linear density per filament (Tex_f)	Linear of density per filament in ($den_f = Tex_f*9$)
1	0.0411	6	12	6.85	0.57	5.13
2	0.0458	6	34	7.63	0.224	2.02
3	0.01508	6	100	25.1	0.251	2.26

From the results obtained from testing the tensile strength and elongation to be noticed is that the sample presenting the highest tensile strength is the sample 3, at 6.37N as presented in Table 2. This is due to the very large number of filaments composing that sample.

In Table 2, there are presented measurements performed in 1987 for silk yarn produced in the same region but in an organized sericulture of silk, named as sample 4. This sample is used as a reference.

Table 2. Tensile strength and elongation for silk samples tested.

Sample	Tensile strength of yarn (N)	Tensile strength per filament (N)	Tenacity (N/Tex)	Tenacity (g/den)	Elongation (%)
1	1.92 ± 2.6	0.16	0.280	3.17	15.60 ± 3.7
2	1.76 ± 0.3	0.05	0.230	2.61	08.77 ± 2.5
3	6.34 ± 0.3	0.06	0.252	2.85	11.85 ± 0.9
4*	1.44	0.1	0.3	3.42	15.70

*Measurements that are performed in 1987 from V. Pysqyli in her diploma theses. The tested yarns were untwisted and composed by 14 filaments. In addition, in 1987 the silk production was organized and all the processes of cocoon and fiber collection were controlled in so called "Silk Factory in Shkodra"

In order to have a clear assessment of strength for each silk sample studied, the tensile strength is calculated for each filament. Sample 1 has the highest tensile strength per filament at 0.16 N value, comparable to the reference sample, Sample 4. This suggests that the values are attributed mainly to the linear density, with the thickest filaments having the highest strength.

In terms of strength, textile fibers are mainly evaluated by tenacity, which is a specific value of Tensile, considering also the linear density of fibers instead of cross section area, due to the irregular shape of the fibers cross section, and also to the variability of the fibers fines along its length [14, 15].

Therefore, the tenacity is obtained for each filament tested, and still the Sample 4 and 1 have the highest tenacity at 0.3 N/Tex and 0.28N, indicating its overall strength relative to its linear density.

From the graph (see Fig. 2) it is noticed a strong correlation between the linear density and the strength of filaments in both terms of tensile strength per filament and tenacity, with the finest filament showing the lowest values.

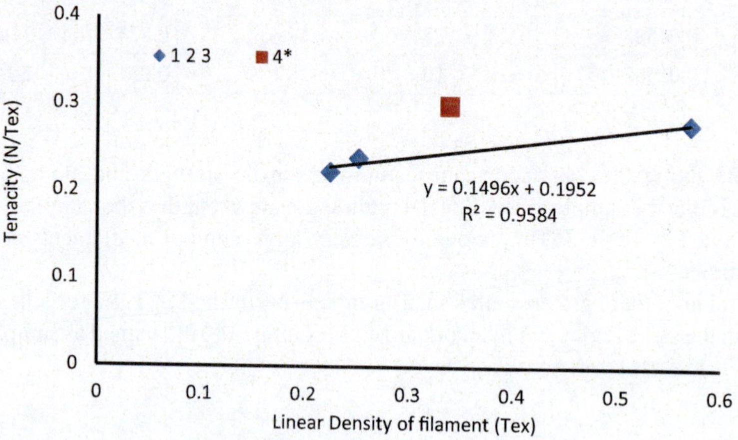

Fig. 2. Correlation between Tenacity of each filament and Linear Density presented for three samples tested in the laboratory and reference sample 4 measured in 1987.

Tensile strength per filament shows a very strong correlation, with the R value at 0.99 (Fig. 3), and this is expected due to the dependence of strength from the fiber thickness. Furthermore, if we compare the values obtained from the samples of silk fibers produced in Shkodra region, with other reference values for tenacity of silk fiber in general, of 0.3–0.4N/Tex, it seems to have comparable values, at 0.3 the highest and 0.223 the lowest, indicating the quality of fibers produced.

In terms of elongation (see Fig. 4), the results obtained show the same trends as for the tensile (see Fig. 3). Sample 1 and Sample 4, have the highest values of elongations, at 15.6% and 15.97%, indicating that the fibers with higher strength tend to have a higher elongation. If compared to theoretical values of elongation, at 15–25% [14, 15], these two fibers are comparable, differently from Samples 2 and Sample 3, which show a slightly lower elongation from the bottom range of elongation (from 8.8 to 11.8%).

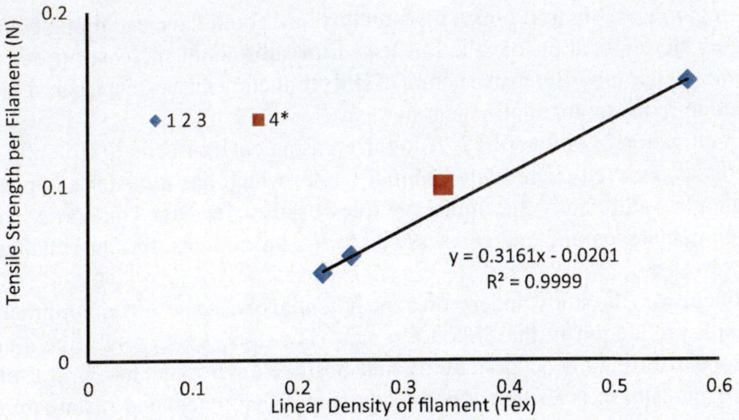

Fig. 3. Correlation between Tensile strength of each filament and Linear Density presented for three samples tested in the laboratory and reference sample 4 measured in 1987.

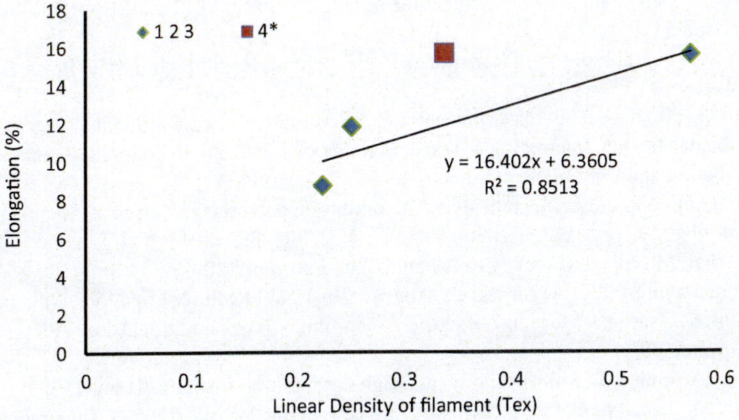

Fig. 4. Correlation between Elongation and Linear Density, presented for three samples tested in the laboratory and reference sample 4 measured in 1987.

4 Conclusions

The current study gathered valuable information regarding the Physical—Mechanical properties of silk fibers domestically produced in Shkodra region. The study's results indicate that the silk fibers have linear density and tensile values that are consistent with existing literature. While there are some differences observed among the samples, these variations still fall within the range of good quality silk. This suggests that the production processes in the region maintain a consistent level of quality.

These results highlight the considerable potential that Shkodra, Albania has in the cultivation of silkworms. Silk is considered to be a luxury natural fiber in terms of comfort, performance and low environmental impact. Having very good overall properties and being completely biodegradable make it an attractive option for various uses.

The region's ability to produce high-quality silk could have economic benefits and revitalizing the cultivation of silk can lead to a substantial increase in employment opportunities. The labor-intensive nature of silk production means that it can significantly contribute to reducing unemployment in the region.

The cultivation of silkworms is recognized as an environmentally friendly activity. It operates as a closed cycle with minimal waste, which has a positive impact on the environment. Additionally, the mulberry trees used as feed for silkworms contribute to environmental cleaning and can serve as fruit-bearing trees, further enhancing their ecological value.

In conclusion, the study underscores the potential economic and environmental benefits of silk production in the Shkodra region. The good quality silk fibers produced there, along with their biodegradability and positive environmental impact, make it a promising industry to invest in for both economic development and sustainability.

References

1. Genti, G.: Mëndafshi. Çfarë duhet të dime për fibrat e luksit. ShBLU, ISBN: 978–99956–43–72–0 Tirana (2011)
2. Mola, L.: The silk Industry of Renaissance Venice. Johns Hopkins University Press, Baltimore (2000)
3. Genti, G., Muja, A.: The past, present and opportunities of cultivation and processing silk in Albania. In: 9th International Textile Conference and 3rd International Conference on Engineering and Entrepreneurship. pp. 80–86. Tirana (2021)
4. Genti, G.: Economic and technical considerations on potential recourses of Albania in textile natural fibers. J. Text. Eng. Fashion Technol. **4**(3), 208–213 (2018)
5. Seneschal, M.: Silk and society in Albania. Am. Entomol. (2005)
6. Kultura popullore 1, 2; Akademia e Shkencave, Instituti i Kulturës Popullore, Albania (1992)
7. Marubi, K.:, National Museum of Pictures—Marubi,. Women traditional dress. Pictures Code: 12X16_RA_29/4
8. Xhoxhi A., Bejtja T., Silkworm growth. High Agricultural Institute. Tirana (1983)
9. Lewin, M.: Handbook of fiber chemistry. 3rd edn. ISBN 0–8247–2565–4. Taylor and Francis Group, LLC. NW (2007)
10. Standard. ISO 2062.: Textiles—Yarns from packages. Specifies methods for the determination of the breaking force and elongation at break of textile yarns taken from packages (2009)
11. Kundu, B., et al.: Silk proteins for biomedical applications: bioengineering perspectives. Progress Polym. Sci. **39**(2), 251–267 (2014)
12. Abbott, A., Ellison, M.: Biologically inspired textiles. ISBN 978–1–84569–508–8 (e-book). Woodhead Publishing Limited and CRC Press LLC. England (2008)
13. Perez-Rigueiro, J., Viney, C., Llorca, J., Elices, M.: Mechanical properties of single-brin silkworm silk. J. Appl. Polym. Sci. **75**(10), 1270–1277 (2000)
14. Morton, W., Hearle, J.W.S.: Physical Properties of Textile Fibers. 4rth edn. ISBN 978–1–84569–442–5. Woodhead Publishing Limited. England (2008)
15. Guxho, G., Shehu, Y., Shehi, E.: Science and Technology of Textile Materials. ISBN: 978–99943–49–99–9. Shtëpia botuese Albin. Tirane (2011)

Characterization of Electrical Conductivity and Electrical Heating of RGO/MWCNT Coated Nonwoven Fabrics

Nergis Demirel Gültekin[1], Özgen Özturan[2,3](✉), and İsmail Usta[1]

[1] Department of Textile Engineering, Marmara University, İstanbul, Türkiye
[2] Institute of Pure and Applied Sciences, Marmara University, İstanbul, Türkiye
ozgen.ozturan@mavi.com
[3] Department of Design, Mavi Clothing Industry and Trade Inc, İstanbul, Türkiye

Abstract. In this study, the light-weight and flexible reduced graphene oxide and multi-walled carbon nanotube coated nonwoven fabrics were obtained by a facile and scalable blade coating technique. The polypropylene and polyester spunbond nonwoven fabric with mass per unit area of 25 g/m^2 were coated with a water based-graphene oxide coating paste. Afterwards, the reduction process with L-ascorbic acid was applied to turn graphene oxide into electrically conductive reduced graphene oxide form. In addition, graphene oxide/multi-walled carbon nanotube nanocomposite coating formulations were also applied to the nonwoven fabrics to investigate the effect of multi-walled carbon nanotubes on the heating property. The obtained reduced graphene oxide-coated nonwoven textiles with and without multi-walled carbon nanotube were characterized by means of FTIR and DSC. The electrical conductivity and heating property of coated nonwoven fabrics were examined. The heating test results revealed that the coating of polypropylene nonwoven fabric with reduced graphene oxide and reduced graphene oxide/multi-walled carbon nanotube was increased the surface temperature by 20.2 K and 38.6 K at 20 V, respectively.

Keywords: Graphene oxide · Blade coating · Spunbond nonwoven fabric · Electro-thermal property

1 Introduction

Carbon-based materials with high electrical conductivity, light-weight, flexibility and large specific surface area have been widely employed as conductive fillers to fabricate functional textiles [1]. Graphene is a low cost, lightweight nanomaterial that make it an attractive candidate for electrothermal heating textiles [2]. Graphene oxide is an oxygenated form of graphene that contains hydroxyl, epoxy, carbonyl and carboxyl groups which converted chemically from graphite [3]. To fabricate graphene based textiles, coating of graphene oxide by the help of oxygenated functional groups onto the textile is a very efficient approach [4]. Carbon nanotubes are one of the most promising nanomaterial in functional textiles, with extraordinary properties such as flexibility, electrical and thermal conductivity, mechanical and chemical durability [5].

G. Guxho et al. (Eds.): ITC-ICEE 2023, LNMIE, pp. 535–545, 2024.
https://doi.org/10.1007/978-3-031-48933-4_53

Textiles are the most ideal substrates for wearable electronics due to their flexibility, light-weight, comfort, stretchability and easy usage [6]. Electrothermal heating textiles have drawn growing attention for the area of functional textiles. Flexible wearable textile heaters with good electrical conductivity meet the market requirement preferably than films [7]. Textile heaters potentially can lead to novel functionalities with high performance [8].

Functional textiles can be prepared by various coating methods, and the blade coating method is one of the suitable technique for the production of heating textiles. Blade coating is an environmentally friendly method and does not requires much time and effort [3]. Nonwoven fabrics can be ideal supporting materials to get lightweight heating textiles. Nonwoven fabrics are widely used in various industrial applications, including technical areas such as medical, military, filtration, sound absorption [9, 10]. Spunbond nonwoven fabrics are used in a wide range of areas due to the environmentally friendly production technique which utilizes controlled heat to bond synthetic fibers together [11]. Spunbond nonwoven fabrics can be obtained with a wide range of product characteristics ranging from very lightweight and flexible structures to heavy and stiff structures [12].

In this study, the light-weight and flexible reduced graphene oxide and multi-walled carbon nanotube coated nonwoven fabrics were obtained by a facile and scalable blade coating technique. The polypropylene and polyester spunbond nonwoven fabric with mass per unit area of 25 g/m^2 were coated with a water based-graphene oxide coating paste with/without multiwalled carbon nanotubes. Afterwards, the reduction process with L-ascorbic acid was applied to turn graphene oxide into electrically conductive reduced graphene oxide form. The heating performance of the coated nonwoven fabrics were evaluated between 5 and 20 V.

2 Experimental

2.1 Materials

Multiwalled carbon nanotubes (MWCNT) were purchased from MK Nano, Canada. External diameters and length of MWCNTs are 10–20 and 10–30 μm, respectively. The purity was higher than 95%. MWCNTs were used without further modification. Graphite flakes and L-ascorbic acid were purchased from Sigma Aldrich. Hydrogen peroxide (H_2O_2, 35%), sulfuric acid (H_2SO_4, 95–98%), phosphoric acid (H_3PO_4), potassium permanganate ($KMnO_4$) and hydrochloric acid (HCl, 37%) were purchased from Merck. All chemicals were of analytical reagent grade and used without further purification. Distilled water was used throughout the experiments. The polypropylene and polyester spunbond nonwoven fabrics with mass per unit area of 25 g/m^2 were used.

2.2 Graphene Oxide Synthesis

The improved Hummer's method was applied to synthesize graphene oxide from flake graphite. Briefly, 3 g graphite flakes was added in to a flask containing a mixture of H_2SO_4/H_3PO_4 with a ratio of 360:40 mL. 18 g $KMnO_4$ was added into the flask slowly. After complete dissolving of potassium permanganate, the reaction was placed to oil bath followed by heating to 50 °C and stirred for 12 h. Afterwards, the reaction was cooled to room temperature and poured onto 400 mL ice. The reaction was terminated by adding 6 mL of 35% H_2O_2. The suspension was purified with 1 M HCl and ethanol (2 ×) using a successive centrifugation process. The centrifugation was continued with washing process until a pH of 4–5 was achieved. The resulting product was dried at 60 °C to obtain solid GO nanosheets.

2.3 Preparation of Coating Pastes

The graphene oxide based coating pastes were prepared with and without MWCNTs. The aqueous graphene oxide coating paste was prepared with a concentration of 2.5 wt.% and the MWCNTs content was 1 wt.%. The determined amount of GO was added to distilled water and mechanically stirred.

2.4 Coating of Nonwoven Fabrics

The polypropylene and polyester spunbond nonwoven fabrics were coated with GO and GO-MWCNT by blade coating technique on a laboratory type coating machine (Ataç GK40 RKL). The distance between the blade and fabric was fixed at 0.2 mm. The samples were dried at 110 °C for 10 min. Each side of nonwoven fabrics were coated one-passage.

2.5 Reduction of GO-Coated Nonwoven Fabrics

The reduction process of GO and GO-MWCNT coated nonwoven fabrics was carried out using L-ascorbic acid as a reducing agent. The coated nonwoven fabric samples were immersed into the reduction bath containing 0.2 M aqueous solution of L-ascorbic acid for about 90 min. The temperature of the reduction process was kept at 95 °C. The resulting fabrics were rinsed with distilled water to remove the remaining reducing agent. The nonwoven fabric samples were dried in an oven at 90 °C. The obtained fabrics after reduction process were coded as RGO and RGO-MWCNT.

2.6 Characterization

The chemical structure of nonwoven fabric samples was recorded using an attenuated total reflection Fourier transform infrared spectrometer (ATR-FTIR, Perkin Elmer Spectrum 100). Differential scanning calorimetry (DSC) curves were obtained using calorimeter (Jade DSC, Perkin Elmer). The nonwoven fabric samples were tested under nitrogen flow at a heating rate of 10 °C/min from 25 to 300 °C. The electrical resistance of the nonwoven fabric samples was measured using a standard four-point probe. The setup consists of a sourcemeter (Keithley 2450 Sourcemeter) and a four point probe station (Everbeing Int'l Corp). The heating performance of coated PES and PP spunbond nonwoven fabrics was measured by applying a DC voltage (Textronix PWS4305 30V, 5A power supply) varying from 0–20 V (in intervals of 5 V) for 60 s through the fabrics prepared as 2.5 × 2 cm sample. The surface temperature of the fabrics was measured using an infrared camera (Testo 868).

3 Results and Discussion

The FTIR has been made to analyze the coating and reduction of GO nanosheets on the fabric surface. Figure 1(a) shows the FTIR analysis of GO, RGO and RGO + CNT coated polyester nonwoven fabric. The peaks at 722 cm^{-1}, 871 cm^{-1}, and 1408 cm^{-1} are attributed to C–H bond in the aromatic group, the peaks at 1713 cm^{-1} and 1239 cm^{-1} were assigned to C = O and C–O stretching vibrations of aromatic ester and ester, respectively [13, 14]. The C–O stretching of glycol was seen at 1017 cm^{-1} and benzene in-plane vibrations was found at 970 cm^{-1}. The peak at 1095 cm^{-1} represents the ester C = O stretching vibration [15]. After the GO coating process, a broad peak between 3600–2800 cm^{-1} shows O–H stretching vibrations of carboxylic acid can be attributed to the water molecules and alcohol groups in GO. The new peak appeared at 1635 cm^{-1} corresponds to the C≡C stretches from unoxidized graphitic domain [16]. After the reduction of GO, the broad peak between 3600 and 2800 cm^{-1} disappeared and the intensity of the sharp peaks decreased.

Figure 1(b) shows the FTIR analysis of GO, RGO and RGO + CNT coated polypropylene nonwoven fabric. The peaks at 2950 and 2868 cm^{-1} correspond to CH_3 asymmetric and symmetric stretching vibrations, respectively. The peaks at 2918 and 2838 cm^{-1} are assigned to CH_2 asymmetric and symmetric stretching vibrations, respectively. The peak at 1455 cm^{-1} is suggested to be due to CH_3 asymmetric deformation vibrations or CH_2 scissor vibrations, whereas the strong peak at 1375 cm^{-1} corresponds to CH_3 symmetric deformation vibrations. The band at 1167 cm^{-1} originates from C–C asymmetric stretching, CH_3 asymmetric rocking and C–H wagging vibrations. The peak at 997 cm^{-1} is attributed to CH_3 symmetric rocking vibrations. The peak at 973 cm^{-1} appears due to CH_3 asymmetric rocking and C–C asymmetric stretching vibrations whereas the peak at 898 cm^{-1} is ascribed to CH_3 asymmetric rocking and C–C asymmetric and symmetric stretching vibrations. The peaks at 841 and 809 cm^{-1} are related to CH_2 rocking vibrations [17]. After the GO coating process, a broad peak between 3600 and 2800 cm^{-1} shows O–H stretching vibrations of carboxylic acid can be attributed to the water molecules and alcohol groups in GO. The intensity of the characteristic peaks of PP decreased and three new peaks appeared at 1732 cm^{-1}, 1625 cm^{-1} and 1051 cm^{-1} which correspond to C = O stretch of carboxyl group, C≡C stretches from unoxidized graphitic domain and C = O stretching vibrations of C–O–C, respectively [16].

Thermal behavior of neat, RGO and RGO + CNT coated PES and PP spunbond nonwoven fabrics was studied by DSC. Figure 2(a) shows the DSC thermograph of the PES spunbond nonwoven fabric. The neat PES nonwoven fabric showed an endothermic peak at 524.15 K with the melting enthalpy of 113.897 J/g, while the PES/RGO sample showed an endothermic peak at 530.15 K with the lowest melting enthalpy of 40.209 J/g due to the presence of reduced graphene oxide nanosheets. In Fig. 2(b), it is shown that the neat PP nonwoven had an endothermic peak at 436.52 K with the melting enthalpy of 106.311 J/g. The melting temperature and the melting enthalpy of the RGO coated PP nonwoven fabric was obtained as 437.81 K and 53.452 J/g respectively. The coating of both PES and PP spunbond nonwoven fabrics with the RGO has decreased

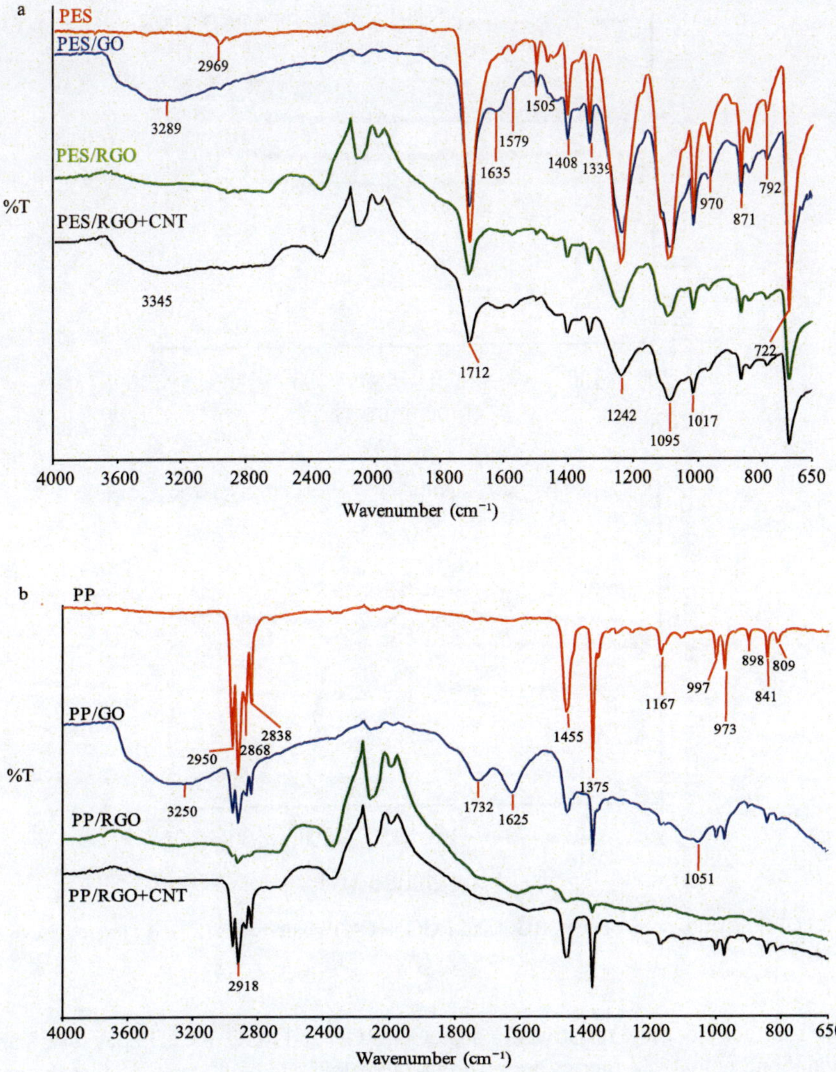

Fig. 1. FTIR spectra of GO, RGO and RGO + CNT coated PES (a) and PP (b) nonwoven fabrics.

the heat release in comparison to the neat PES and PP nonwoven fabrics. Also, it can be understood that the RGO coating has an increasing effect on the melting temperature of the neat PES and PP nonwoven fabrics and providing better thermal stability.

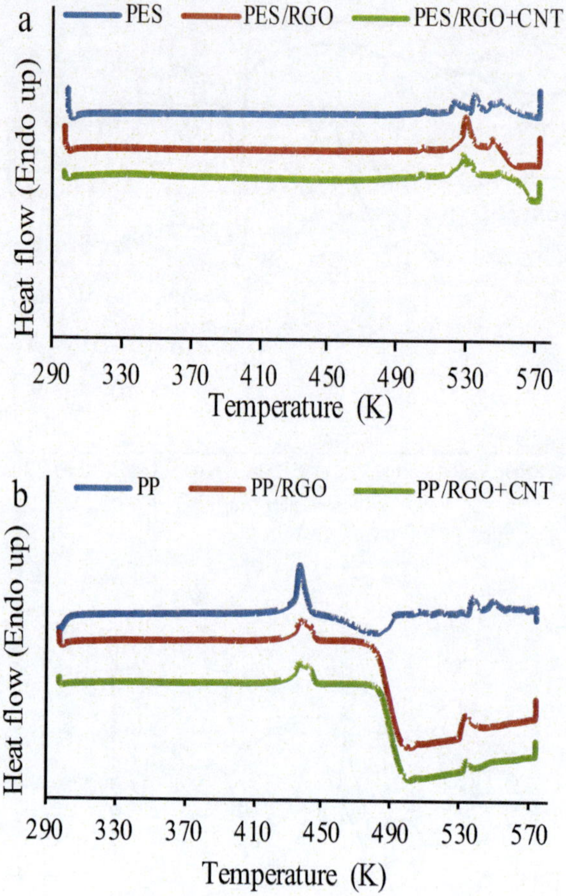

Fig. 2. DSC thermograms of neat, RGO and RGO + CNT coated PES (a) and PP (b) nonwoven fabrics.

The surface electrical resistivity of neat, GO, RGO and RGO + CNT coated PES and PP spunbond nonwoven fabrics are given in Table 1. It can be understood that the neat PES and PP nonwoven fabrics have high surface resistivity values, which means that they do not conduct electricity. Besides, it is also known that GO is an electrical insulator due to the oxygenated functionalities. Therefore, the surface resistivity of PES/GO and PP/GO nonwoven fabrics were obtained in the range of 10^7 Ω/sq. The decrease in surface resistivity after coating process is due to the carbon nanotubes within the GO coating paste. With the reduction process using L-ascorbic acid, the surface resistivity of nonwoven fabrics have shown a remarkable decrease. The surface resistivity of PES/RGO and PES/RGO + CNT was obtained as 573.44 Ω/sq and 560.36 Ω/sq, respectively. The lowest electrical surface resistivity values were obtained at PP/RGO and PP/RGO + CNT nonwoven fabrics as 87.94 Ω/sq and 83.24 Ω/sq, respectively.

Table 1. Surface resistivity of neat, RGO and RGO + CNT coated PES (a) and PP (b) nonwoven fabrics.

Sample	Electrical resistivity (Ω/sq)	
	PES	PP
Neat	1.94E + 09	8.08E + 08
GO	1.03E + 07	1.64E + 07
RGO	573.44	87.94
RGO + CNT	560.36	83.24

The heating performances of RGO and RGO + CNT coated PES and PP spunbond nonwoven fabrics were studied through measuring the surface temperature of the fabric with an applied voltage ranging from 5 to 20 V. The thermal response of the coated nonwoven fabrics is given in Fig. 3. The maximum temperature increase was observed for PP spunbond nonwoven fabric. The RGO + CNT coated PP nonwoven fabric showed the highest surface temperature of 337.65 K at 20 V while the RGO + CNT coated PES nonwoven showed a surface temperature of 303.75 K at 20 V. The heating performance of the coated nonwoven fabrics showed an increasing trend with the increase of the applied voltage from 5 V to 20 V. The difference between the surface temperatures of the coated nonwoven fabrics before and after applying the voltage is given in Fig. 3(b). It can be understood that the highest temperature change was obtained for the PP nonwoven fabric coated with RGO and RGO + CNT. Moreover, the rate of the temperature difference increased gradually with the applied voltage. The maximum surface temperature increase was obtained with the PP/RGO + CNT sample as 38.6 K at 20 V.

The infrared thermograms of PES/RGO + CNT and PP/RGO + CNT samples at applied voltage between 0–20 V are given in Fig. 4. The uniform heat generation was observed throughout the coated nonwoven fabrics after applying 10 V. Also, it is clearly can be seen that the increasing the applied voltage from 5 V to 20 V affects the temperature distribution throughout the nonwoven fabric. The temperature dissipation was observed to happen from the center of the fabric through the edges of the fabric. For the PP/RGO + CNT nonwoven fabric, it can be said that a better distribution of temperature was observed due to the uniform coating of nanoparticles.

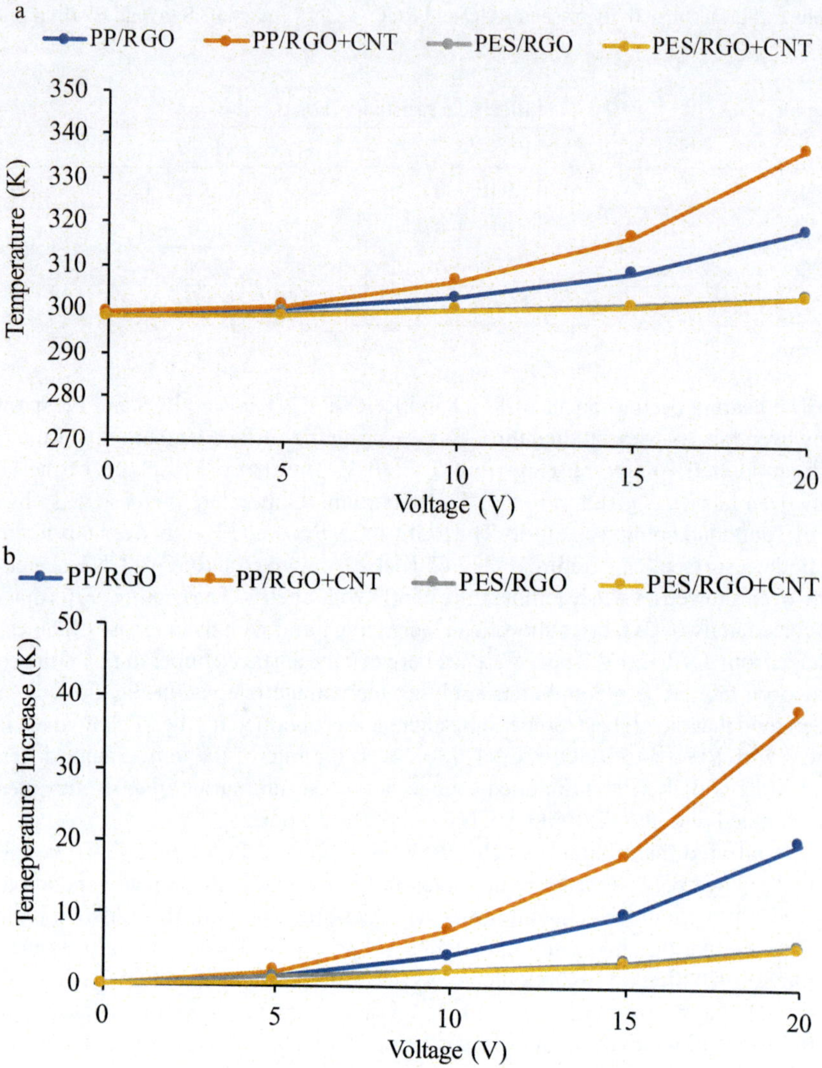

Fig. 3. Thermal response of RGO and RGO + CNT coated PES and PP nonwoven fabrics as a function applied voltage. (a) surface temperature of the samples and (b) temperature increase of the samples.

4 Conclusion

RGO and CNT based electro-thermal heater with high electrical conductivity, flexible and lightweight were prepared through blade coating of GO/CNT paste on the both side of PES and PP spunbond nonwoven fabrics. The blade coating technique has great potential for large scale production of graphene based textiles because it is one of the easy coating process, requires little time and energy, and environmental friendly with the low production of wastes. The RGO and RGO + CNT coated PES and PP spunbond nonwoven fabrics were characterized and electro-thermal heating properties were tested. The characteristics of GO and RGO were confirmed by FTIR. The thermal behavior of RGO and RGO + CNT coated PES and PP nonwoven fabrics was determined by DSC. The electrical resistivity results indicated that the lowest values were obtained with RGO and RGO + CNT coated PP nonwoven fabric as 87.94 Ω/sq and 83.24 Ω/sq, respectively. The electrothermal heating performance of RGO and RGO + CNT coated PES and PP nonwoven fabrics was determined by measuring the surface temperatures using infrared thermal camera under different voltages from 0-20V. The surface temperature of the RGO and RGO/CNT coated PES and PP nonwoven fabrics increased with the increase of applied voltage and the highest steady-state temperatures were obtained with the highest applied voltage in both PES and PP. The PP/RGO + CNT spunbond nonwoven fabric exhibited the highest steady-state temperature (337.65 K) with the highest temperature increase rate (38.6 K).

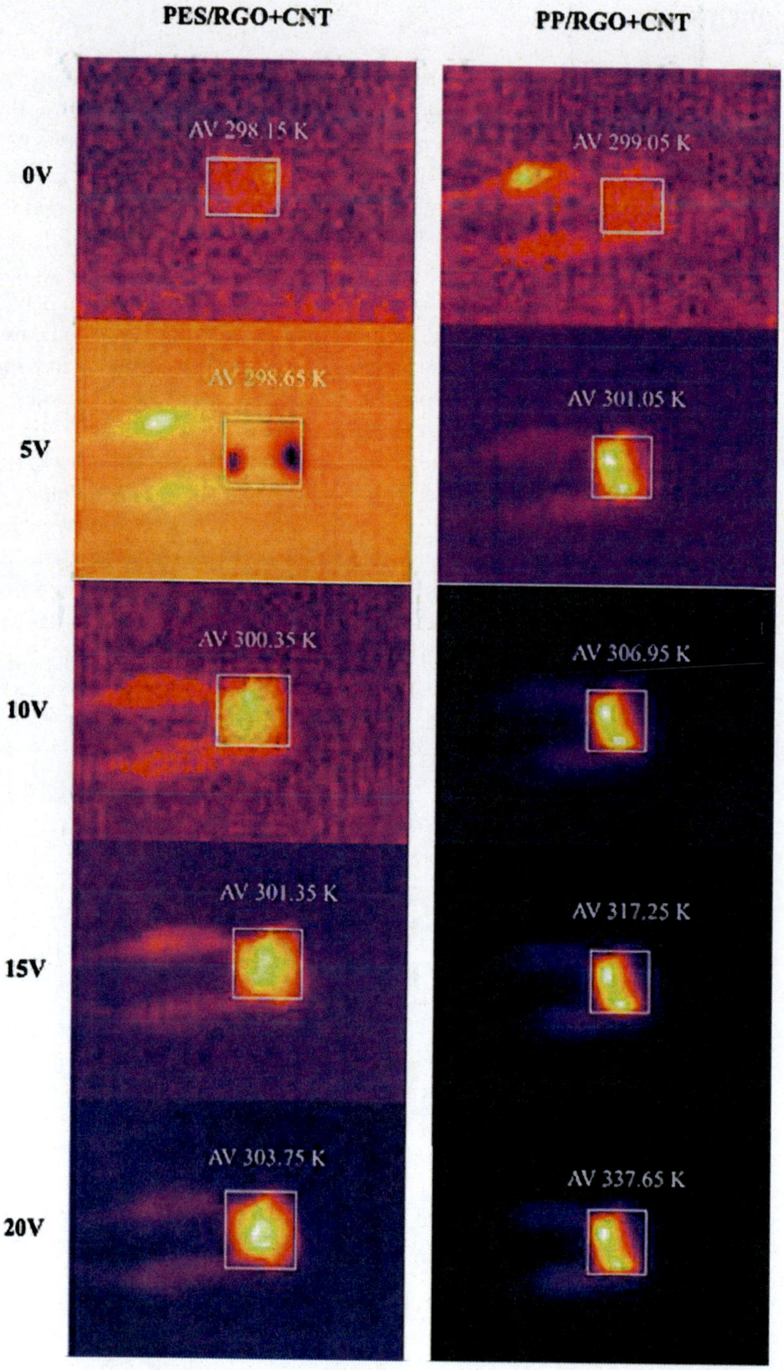

Fig. 4. IR images of RGO + CNT coated PP and PES nonwoven fabrics before and after applying 5V, 10V, 15V, and 20 V for 60 s.

References

1. Tian, M., et al.: Electromagnetic interference shielding cotton fabrics with high electrical conductivity and electrical heating behavior via layer-by-layer self-assembly route. RSC Adv. **7**, 42641–42652 (2017)
2. Lin, S.-Y., et al.: High-performance graphene-based flexible heater for wearable applications. RSC Adv. **7**, 27001–27006 (2017)
3. Gültekin, B.C.: Electrically conductive, hydrophobic, UV protective and lightweight cotton spunlace nonwoven fabric coated with reduced graphene oxide. Turk. J. Chem. **46**, 968–986 (2022)
4. Ji, Y., et al.: Fire-resistant and highly electrically conductive silk fabrics fabricated with reduced graphene oxide via dry-coating. Mater. Des. **133**, 528–535 (2017)
5. Ilanchezhiyan, P., et al.: Highly efficient CNT functionalized cotton fabrics for flexible/wearable heating applications. RSC Adv. **5**, 10697–10702 (2015)
6. Sadi, M.S., et al.: Direct screen printing of single-faced conductive cotton fabrics for strain sensing, electrical heating and color changing. Cellulose **26**, 6179–6188 (2019)
7. Tian, M., et al.: Enhanced electrothermal efficiency of flexible graphene fabric Joule heaters with the aid of graphene oxide. Mater. Lett. **234**, 101–104 (2019)
8. Neella, N., et al.: Low cost, disposable and wearable body warmer using RGO sheets coated on cloth substrate as heating element. In: Proceedings of the 12th IEEE International Conference on Nano/Micro Engineered and Molecular Systems 2017, pp. IEEE (2017)
9. Stolyarov, O., Ershov, S.: Characterization of change in polypropylene spunbond nonwoven fabric fiber orientation during deformation based on image analysis and Fourier transforms. J. Strain Anal. Eng. Des. **52**, 457–466 (2017)
10. Venkataraman, D., et al.: Advancement of nonwoven fabrics in personal protective equipment. Materials **16**, 3964 (2023)
11. Altay, P., et al.: Comparison of conventional and ultrasonic method for dyeing of spunbond polyester nonwoven fabric. Ultrason. Sonochem. **42**, 768–775 (2018)
12. Midha, V.K., Dakuri, A.: Spun bonding technology and fabric properties: a review. J. Text. Eng. Fash. Technol **1**, 1–9 (2017)
13. Shao, F., et al.: Fabrication of polyaniline/graphene/polyester textile electrode materials for flexible supercapacitors with high capacitance and cycling stability. Chem.—Asian J. **11**, 1906–1912 (2016)
14. Rathinamoorthy, R., Balasaraswathi, R., S.: Effect of surface modification of polyester fabric on microfiber shedding from household laundry. Int. J. Cloth. Sci. Technol. ahead-of-print (2022)
15. Ouadil, B., et al.: Surface modification of knit polyester fabric for mechanical, electrical and UV protection properties by coating with graphene oxide, graphene and graphene/silver nanocomposites. Appl. Surf. Sci. **414**, 292–302 (2017)
16. Emiru, T.F., Ayele, D.W.: Controlled synthesis, characterization and reduction of graphene oxide: a convenient method for large scale production. Egyptian J. Basic Appl. Sci. **4**, 74–79 (2019)
17. Markovic, D., et al.: Impregnation of corona modified polypropylene non-woven material with thymol in supercritical carbon dioxide for antimicrobial application. J. Supercritical Fluids **101**, 215–221 (2015)

Polypyrrole Treated Leather and Triboelectricity

Aristeidis Repoulias[1](✉), Majlinda Hylli[2], Ilda Kazani[2], Sotiria F. Galata[1], Sindi Cela[2], and Savvas Vassiliadis[1]

[1] Department of Electrical and Electronics Engineering, University of West Attica, P. Ralli and Thivon 250, 12241 Athens, Greece
arepoulias@uniwa.gr

[2] Textile and Fashion Department, Polytechnic University of Tirana, Mother Tereza Square, No. 1, 1001 Tirana, Albania

Abstract. Leather has been used for various applications for thousands of years and it remains popular today for the making of apparel, footwear, furniture and accessories, thanks to its durability, flexibility and good insulation. But beyond its traditional properties, leather can also accept conducting polymers to enrich its electrical properties. In this study, leather samples were treated with polypyrrole to increase their electrical conductivity. Their triboelectric potentials were studied using a measuring triboelectric generator which combines sliding and vertical contact motions. The results were characterized by low voltages in the range of 53 mV to 115 mV. Thus it was seen that the hereby applied in situ polymerization of pyrrole on leather did not provide significant triboelectric outcomes. The polypyrrole treated leather samples which underwent the double in situ polymerization provided slighter lower voltages. This study constitutes a trigger for further research on the area of the triboelectric potentials of leather.

Keywords: Leather · Polypyrrole · Conductivity · Triboelectricity · Triboelectric generator

1 Introduction

During the last years, research has risen in the direction of studying the potential of using textile materials to exploit the phenomenon of triboelectricity. In the area of clothing and wearables, this might offer the option of providing sustainable electric energy to low-consumption devices or self-powered sensors.

Triboelectricity is a natural phenomenon which describes the generation of electric charges between two surfaces upon their contact and separation [1]. Triboelectric generators (TEGs) are structures of various sizes and designs which are built to reproduce and exploit the aforementioned phenomenon based on four different modes of motion [2]. The intensity of the phenomenon depends significantly on the materials which have been selected to use on the TEG. A classification of the materials according to their tendencies to acquire electric charges has been presented in various tables called triboelectric series

G. Guxho et al. (Eds.): ITC-ICEE 2023, LNMIE, pp. 546–551, 2024.
https://doi.org/10.1007/978-3-031-48933-4_54

[3, 4]. Among the various natural materials which are used to build TEGs, not much attention has been paid to leather.

Leather is a natural material that most commonly comes from cows after leather tanning or chemical treatments. Thanks to its resistance to wear and tear, aesthetic appearance, and nice touch-feeling it is used widely in personal protection gear, clothing accessories, decoration and furniture, footwear etc. Moreover, its process is supported by mature production technology and wide technical knowledge.

Lately leather has gained researchers' interest for its potential to enhance its electrical properties [5, 6], opening up possibilities for new smart leather products and functionalities [7]. Since it is an insulator, to make it conductive various processes must be applied like printing, polymer coating technology, sewing or knitting of conductive yarns etc., using conductive polymers like polypyrrole (PPy), polyaniline (PANI), poly(3,4-ethylene-dioxythiophene) polystyrene sulfonate (PEDOT:PSS) or carbon black [8]. Polypyrrole (PPy) is an intrinsically conducting polymer which comes from the in situ polymerization of pyrrole. Using the polymer coating technology, PPy is used to enrich smart textiles [9, 10] or leather [11] with electrical conductivity.

Taking these into account, in this study leather samples were treated with PPy to increase their conductivity, by applying single and double in situ polymerization of pyrrole, and they were afterwards brought in contact with Polytetrafluoroethylene (PTFE) to produce triboelectric charges and measure them.

2 Materials

Four different cow leather samples were used (samples 1, 2, 3 and 4) of thicknesses 1.3 mm, 1.4 mm, 1.5 mm and 1.5 mm respectively. The leather samples were cut into specimens of identical size (5 cm × 5 cm). Then a single in situ polymerization treatment was applied on samples 1 and 2, by dipping them in aqueous solutions containing pyrrole monomer and an acid, AQSA (Anthraquinone-2-sulfonic acid sodium salt monohydrate), and they were left for an hour rotating with 10 rpm. After that, Ferric Chloride (FeCl3) was added to the mixture so that pyrrole would polymerize. The polymerization was carried out for two hours at 5 °C and 10 rpm. Finally, the leather samples were rinsed with distilled water and then they were left to dry properly. On the other hand, a double in situ polymerization treatment was applied on samples 3 and 4. Samples 3 and 4 were treated with the same reactants and followed the same procedure as the single in situ polymerization, but applied two times. Optical images of their front (grain) and back (flesh) sides are presented in Figs. 1 and 2. Although samples were black both front and back, a whitish colour was captured in the optical images of the front sides (grain) due to the light reflection on the glossy leather surface. The in situ polymerization treatments were applied at the Polytechnic University of Tirana.

A PTFE-coated glass fabric (5 cm × 5 cm) was bought from Fiberflon and used as the second triboelectric material. This was made of woven glass fibers which have been coated with PTFE. The PTFE material was selected to be used as the second triboelectric material of the TEG as it has a very strong tendency to obtain a negative charge.

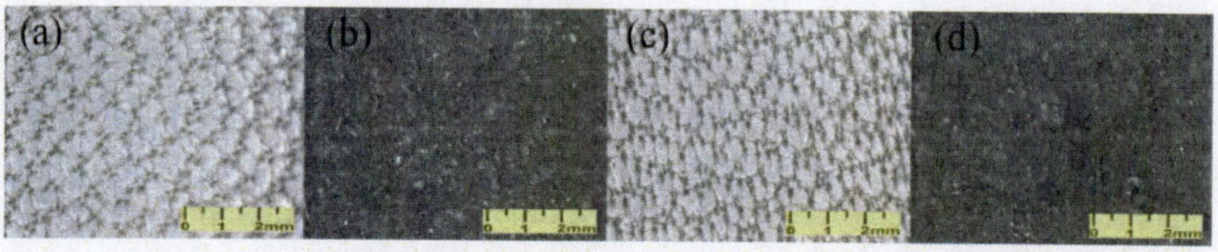

Fig. 1. Optical images of the front (grain) and back (flesh) side of the samples 1–2.

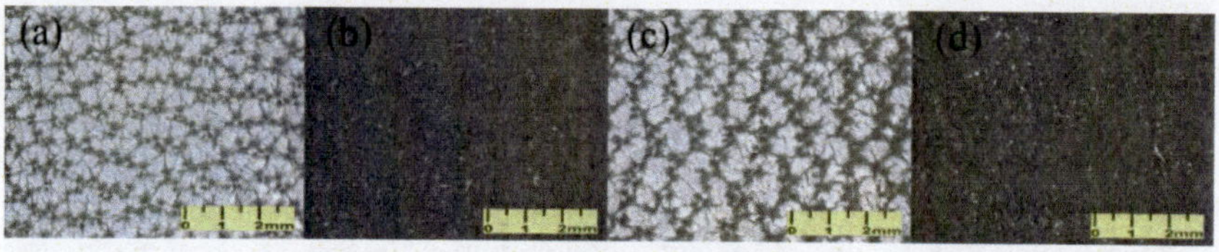

Fig. 2. Optical images of the front (grain) and back (flesh) side of the samples 3–4.

3 Method

The measuring TEG device which has been designed and developed at the University of West Attica, consisted of an upper and a lower specimen holder which include an upper and lower electrode respectively made of copper sheet (5cm × 5 cm) [12]. The design of the measuring TEG was such that the upper holder could move and execute a quasi-elliptical orbit over the lower holder (Fig. 3). On the upper specimen holder, a leather sample was positioned, while on the lower specimen holder, the PTFE-coated glass fabric was positioned. The appropriate position of the upper specimen holder over the lower was set so that they could contact each other under a load of 20 grf. The procedure of contact, sliding and separation was repeated 750 times with a 1 Hz frequency.

After the PPy treated leather samples were prepared, they were paired successively one by one with the PTFE sample. Consequently, each time the upper specimen holder was executing a quasi-elliptical orbit, the two positioned samples were coming in contact, sliding and separation. The result of the above-mentioned motion was the creation of triboelectricity.

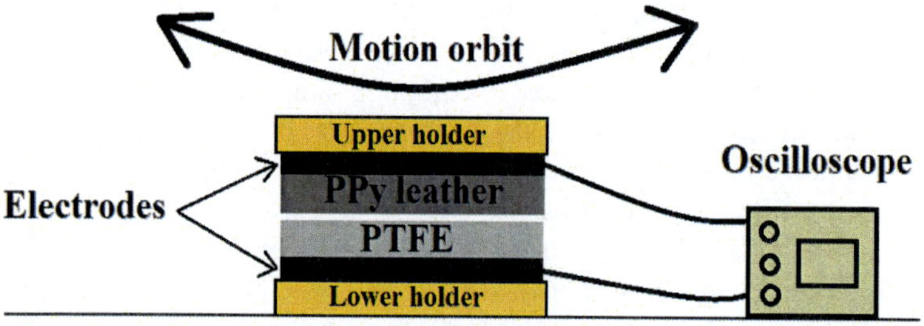

Fig. 3. Sketch of the measuring TEG device with the upper and the lower specimen holder which include an upper and lower electrode. The upper holder executes a quasi-elliptical orbit.

To measure the electric outcome, an oscilloscope was used. An oscilloscope consists of an electronic measuring unit and two probes, which must be connected to a circuit to measure the coming voltage. Hereby, each of the two probes was connected to each of the two electrodes of the two specimen holders. Consequently, at the moment that the two samples were in contact, rubbed and separated, triboelectricity was created and the coming V_{oc} (open circuit voltage) was measured on the oscilloscope. To record the measurements, a computer was connected to the oscilloscope. The conditions during the measurement were 23 ± 1 °C temperature and 40 ± 2% relative humidity.

4 Results and Discussion

The examination of PPy treated leather samples 1, 2, 3 and 4 with the PTFE-coated glass fabric provided voltages in the range of 83–105 mV, 89–115 mV, 53–66 mV and 51–59 mV respectively (Fig. 4). These outcoming voltages are considered very low, however not negligible considering the small contact area surface and the low applied load.

It is concluded that in the hereby applied experiments, PPy acted as a barrier in the triboelectric phenomenon. This contradicts a possible initial expectation that PPy would help in the transportation of the triboelectric charge into the body of leather and provide high outputs. A possible explanation for this is that PPy acted as a conductor which short-circuited the two surfaces upon each contact of the participating samples, thus reducing the intensity of the triboelectric phenomenon.

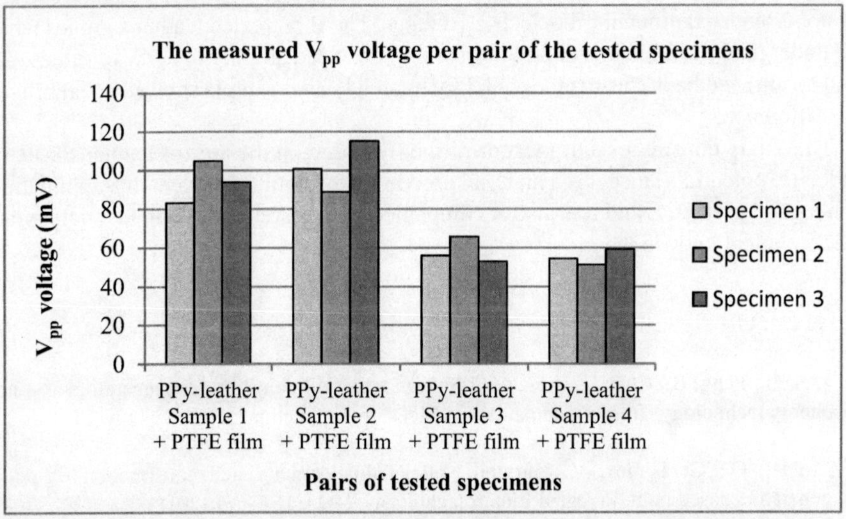

Fig. 4. The measured peak to peak (V_{pp}) voltage per pair of the tested specimens after 750 repeats of contact, sliding and separation (weight load 20 grf, 23 ± 1 °C, RH 40 ± 2%).

A comparison of the results between the samples coming from the single and double in situ polymerization treatment showed that the double in situ polymerization treatment provided slightly lower voltage values. However, these values were low and this difference was not significant.

Additionally, a comparison was made between the front and back sides of a PPy treated leather sample however no significant difference was detected in the measured Vpp.

5 Conclusions

Over the last years, research has been applied to the study of triboelectric properties of natural materials (e.g., cellulosic) in the direction of using them in TEGs and self-powered sensors. Such a natural material is leather.

Hereby an attempt is made to approach the triboelectric potentials of cow leather after it was treated with PPy, a material used to enrich the electrical conductivity of smart textiles.

More specifically, four cow leather samples were treated with a single and a double in situ PPy polymerization to examine the triboelectric results which would come upon their contact, rubbing and separation with another PTFE-based material. A measuring TEG was used to bring in contact, slide and separate the PTFE-based sample with each of the PPy treated leather samples. This motion was repeated 750 times and the V_{pp} was measured with a connected oscilloscope.

The results were characterized by low voltages in the range of 53–115 mV. Thus it was seen that the hereby applied in situ polymerization of pyrrole on leather surfaces did not provide significant triboelectric voltages. The PPy treated leather samples which had undergone a double in situ polymerization, provided slightly lower V_{pp} values. The front (grain) and back (flesh) sides of a PPy treated leather sample showed no significant V_{pp} difference.

This study constitutes a trigger for further research on the area of leather for its triboelectric potentials since it is a material used in many clothing or wearable applications with a good aesthetic, and it is also accompanied by mature production technology and wide technical knowledge.

References

1. Zhu, G., Peng, B., Chen, J., Jing, Q., Lin Wang, Z.: Triboelectric nanogenerators as a new energy technology: from fundamentals, devices, to applications. Nano Energy **14**, 126–138 (2015)
2. Molnar, O., Gerasimov, V., Kurytnik, I.P.: Triboelectricity and construction of power generators based on it. Przegląd Elektrotechniczny **94**(1), 167–171 (2018)
3. StaticWorx GroundSafe ESD Flooring Homepage: What is the Triboelectric Series, https://staticworx.com/esd_glossary/triboelectric-series. Last accessed 20 Sept 2023
4. Hatta, F.F., Mohammad Haniff, M.A.S., Mohamed, M.A.: A review on applications of graphene in triboelectric nanogenerators. Intl. J. Energy Res. **46**(2), 544–576 (2022)
5. Kazani, I., Hylli, M., Berberi, P.: Electrical resistivity of conductive leather and influence of air temperature and humidity. TEKSTILEC **64**, 298–304 (2021)

6. Hylli, M., Shabani, A., Kazani, I., Drushku, S., Guxho, G.: The color fastness properties of conductive leather improved by the use of mordants. IOP Conf. Ser: Mater. Sci. Eng. **827**(1), 012036 (2020)
7. Mitilineos, S., et al.: A wearable NFC antenna sewn on leather substrate for immersive IOT applications. Text. Leather Rev. **5**, 70–84 (2022)
8. Hylli, M., Shabani, A., Kazani, I., Guxho, G.: Investigating the influence on the electro-conductive properties of leather as a result of natural aging. Text. Leather Rev. **6**, 401–416 (2023)
9. Shabani, A., Hylli, M., Kazani, I.: Investigating properties of electrically conductive textiles: a review. TEKSTILEC **65**, 194–217 (2022)
10. Datta, M., Chaudhuri, A., Mitra, M., Nath, D., Das, D.: Development of biodegradable conductive cotton yarns by in-situ polymerisation of pyrrole. J. Textile Inst. **110**(1), 10–15 (2019)
11. Hylli, M., Shabani, A., Kazani, I., Beqiraj, E., Drushku, S., Guxho, G. Application of double In Situ Polymerization for changing the leather properties. In: Proceedings on 8th International Textile Conference. Polytechnic University of Tirana, Tirana, pp. 42–47 (2018)
12. Repoulias, A., et al.: The effect of ambient humidity on the performance of a wearable textile triboelectric generator. Energy Tech. **11**(7), 202201433 (2023)

Author Index

G. Guxho et al. (Eds.): ITC-ICEE 2023, LNMIE, pp. 553–554, 2024.
https://doi.org/10.1007/978-3-031-48933-4

Author Index

G. Guxho et al. (Eds.): ITC-ICEE 2023, LNMIE, pp. 553–554, 2024.
https://doi.org/10.1007/978-3-031-48933-4